AF862501

Safety of Thermal Water Reactors

This Conference was organised by the Commission of the European Communities, Directorate General Science, Research and Development Directorate of Nuclear Research and Development Brussels, Belgium

Programme Committee

Members:
S. Finzi, CEC/BXL, General Chairman
H.M. ApSimon, IMP. COL/London
R. Deruaz, CEA/CEN-G
A.R. Edwards, UKAEA/SRD-C
J.G. KRETZSCHMAR, Janssen Pharmaceutica
M.L. Riethmuller, VKI/R.S.G.
J.P. Schon, EC-CENTR/Lyon
G.L. Shires, UKAEA/AEE-W
E. Skupinski, CEC/BXL
B. Tolley, CEC/BXL
J. Vilain, CEC/BXL
K. Wolfert, GRS/München
J.P. Zeeuwen, TNO/P.M.L.-Rijswijk

Local Organisation: L. de Vestibule, CEC/BXL

Scientific Secretary: E. Skupinski, CEC/BXL

Safety of Thermal Water Reactors

Proceedings of a Seminar on the Results of the European Communities' Indirect Action Research Programme on Safety of Thermal Water Reactors, held in Brussels, 1-3 October 1984

Edited by
E. Skupinski, B. Tolley and J. Vilain
Commission of the European Communities, Directorate-General Science, Research and Development, Brussels, Belgium

Graham & Trotman

Published in 1985 by
Graham & Trotman Limited
Sterling House
66 Wilton Road
London SW1V 1DE
United Kingdom

Graham & Trotman Inc.
13 Park Avenue
Gaithersburg
MD20877
USA

for the Commission of the European Communities,
Directorate-General Information Market and
Innovation,
Luxembourg

EUR 9903
© ECSC, EEC, EAEC, Brussels and Luxembourg, 1985
Softcover reprint of the hardcover 1st edition 1985

ISBN-13: 978-94-010-8701-8 e-ISBN-13: 978-94-009-4972-0

DOI: 10.1007/978-94-009-4972-0

Legal Notice
Neither the Commission of the European Communities, its contractors nor any person acting on their behalf; make any warranty or representation, express or implied, with respect of the accuracy, completeness, or usefulness of the information contained in this document, or that the use of any information, apparatus, method of process disclosed in this document may not infringe privately owned rights; or assume any liability with respect to the use of, or for damages resulting from the use of any information, apparatus, method or process disclosed in this document.

All rights reserved. No part of this publication may be reproduced, stored in a retrieval system, or transmitted in any form or by any means, electronic, mechanical, photocopying, recording or otherwise, without prior permission of the publishers.

CONTENTS

Key:
C. No (I) SR (N) - Contract No (I) on safety research with country (N)
IPA 1 - Invited paper in Area A No 1
IPB 1 - (Analogous)
IPC 1 - (Analogous)

Introduction

The European Community's Indirect Action Research Programme on the Safety of Thermal Water Reactors had as main objectives to execute useful fundamental research, complementary and confirmatory to on-going work in national programmes, and to improve collaboration and exchange of information between laboratories in the Member States.

The Seminar was aimed to report on work performed during the last five years and to identify useful further research areas with a tentative assessment of the state of the art for future work in certain issues of LWR-safety.

The results obtained in 33 research projects executed in different national laboratories of the European Community were presented, evaluated and discussed, together with a number of invited papers on topics related to the research programme.

Topics covered mainly within 3 distinct research areas or sub-programmes:

- Research Area A: The loss of coolant accident (LOCA) and the functioning and performance of the emergency core cooling system (ECCS).
Fundamental work on thermalhydraulics and heat transfer during refill and reflood of an uncovered core after a LOCA.

- Research Area B: The protection of nuclear power plants against external gas cloud explosions.
Study of the impact on plant structure and systems of external explosions of dense combustible gas clouds due to accidental releases of hydrocarbons in the vicinity of the plant.

- Research Area C: The release and distribution of radioactive fission products in the atmosphere following a reactor accident.
Study of the advection and diffusion in the atmosphere and of the possible ground deposition of active fission products released from a damaged reactor core during the course of a severe accident.

The seminar was organized in 3 groups of parallel technical sessions, respectively for the 3 distinct research areas and introduced and concluded by plenary sessions. A panel discussion which completed the meeting involved important scientists and managers in nuclear safety research and attempted to define the state of the art of some main issues of LWR safety research.

The proceedings include all presented papers with related discussions, as well as a report on the concluding panel.

ABBREVIATIONS:

BATT.I./Frankfurt	Battelle Institut E.V., Frankfurt
Brit. Meteorolog. Off./Bracknell	British Meteorological Office/-Bracknell
CEA/CEN	Commissariat à l'Energie Atomique/ Centre d'Etudes de l'Energie Nucléaire
Cad	Cadarache
Far	Fontenay-aux-Roses
G	Grenoble
Sac	Saclay
CEA/CESTA	Commissariat à l'Energie Atomique/ Centre d'Etudes Scientifiques d'Aquitaine
CEC	Commission of the European Communities,
BXL	DG XII Brussels
JRC	DG XII/Joint Research Centre, Ispra
LUX	DG V Luxembourg
CEGB	Central Electricity Generating Board
BNL	Berkeley Nuclear Laboratories
CERL	Central Electricity Research, Leatherhead
CEN/SCK-Mol	Centre d'Etude de l'Energie Nucléaire/Studiecentrum voor Kernenergie, Mol
ECN/Petten	Stichting Energieonderzoek Centrum Nederland, Petten
EC.CENTR./Lyon	Ecole Centrale de Lyon, Ecully
EDF	Electricité de France, Paris
DER	Direction des Etudes & Recherches
SEPTEN	Service d'Etudes & Projets Thermiques & Nucléaires
E.P.F./Lausanne	Ecole Polytechnique Fédérale de Lausanne
ENEA/Casaccia	Comitato Nazionale per la Richerca e per lo Sviluppo dell'Energie Nucleare e delle Energie Alternative, Casaccia
ENSMA/Poitiers	Ecole Nationale Supérieure de Mécanique & Aérotechnique, Poitiers
Finn.Meteorolog.I./Helsinki	Finnish Meteorological Institute, Helsinki Helsinki
FRAUNH.I./Pfinztal	Fraunhofer Institut für Treib-und Explosivstoffe, Pfintzal
GRS/München	Gesellschaft für Reaktorsicherheit mbH, München
HSE/Sheffield	Health & Safety Executive, Sheffield
IMP.COL./London	Imperial College of Science & Technology, London
KWU/Erlangen	Kraftwerk Union Aktiengesellschaft, Erlangen
Norw.I.Air.Res./Lillestrøm	Norwegian Institute for Air Research, Lillestrøm

Risø N.L./Roskilde	Risø National Laboratory, Roskilde
RWTH/Aachen	Rheinisch - Westfälische Technische Hochschule, Aachen
Swed.Meteorolog.Hydr.I/ Norrköpping	Swedish Meteorological and Hydrological Institute Norrköpping
T.N.O./P.M.L./Rijswijk	Organisatie voor Toegepast Natuurwetenschappelijk Onderzoek - Prins Maurits Laboratorium, Rijswijk
U/Hannover	Universität, Hannover
TU/München	Technische Universität, München
UKAEA	United Kingdom Atomic Energy Authority
AEE-W	Atomic Energy Establishment, Winfrith
AERE-H	Atomic Energy Establishment, Harwell
SRD-C	Safety and Reliability Directorate, Culcheth
U/Bourges	Université de Bourges
U/Clermont	Université de Clermont Ferrand
U/Liège	Université de Liège
U/Uppsala	University of Uppsala
V.K.I./Rhode-St-Genèse	Institut Von Karman de Dynamique des Fluides, Rhode-St-Genèse

Opening Plenary Session

Chairman: S. Finzi (CEC/BXL)

Session secretary: E. Skupinski (CEC/BXL)

Welcome address by Prof. P. Fasella, presented by S. Finzi

Ladies and Gentlemen,

Professor Fasella, Director General for Science, Research and Development and the Joint Research Centre at the CEC was not able to open this seminar due to an unforeseen other commitment. He asked me to deliver to you the following message.

"On behalf of the Commission of the European Communities, welcome to the seminar on the results of the recently concluded programme of contractual research in Safety of Thermal Water Reactors. The role of nuclear energy in our society is, and will remain, a very important one. A strengthened understanding of the various factors which are taken into account in ensuring the safe operation of nuclear reactor installations will play an important part in improving the effectiveness with which nuclear power is provided. Because this improvement is recognised as important and is sought by utilities and authorities in most of our Member States and because the research work required is of a high order it makes obvious sense for scientists in the field of the various Member States to cooperate. This was the reason for, and the great merit of, this research programme.

I believe that the programme has been a success, both in the domain of acquisition of knowledge and in the promotion of dialogue and understanding between scientists in different laboratories and institutions. But that success is not yet complete and ready to be judged. This final seminar is an essential step on the road to a full programme achievement. It provides an occasion for the results of the different research actions to be presented to a wide audience of workers in the field and for the significance of those results to be assessed. I regard this event as important. I have been very pleased by the wide cross section of participants which the seminar has attracted; I am confident that you will play a full part in the proceedings; I thank you for your participation and I hope that you will find the event both useful and enjoyable".

THE CEC RESEARCH ACTION PROGRAMME ON THE DEVELOPMENT OF NUCLEAR FISSION ENERGY

S. FINZI
Commission of the European Communities

Summary

For its "Framework Programme for a Scientific and Technical Strategy for Europe 1984-1987" the Commission has defined the major goals for a European Scientific and Technical Strategy (1). One of the means to reduce the energy dependence of the Community, which is an important objective, is to favour the development of nuclear fission energy. As electricity production by nuclear reactors has reached industrial maturity, Community activities are directed mainly to safety aspects, in order to ensure the protection of workers and the general public against hazards linked to operations in the nuclear fuel cycle.

After a short overview on the proposed Framework Programme a description of the main features of the five sub-programmes on the development of nuclear fission energy is given below; these programmes are:

- Reactor safety;
- Nuclear fuels and actinides research;
- Management of radioactive waste;
- Safeguarding and management of fissile materials;
- Decommissioning of nuclear installations.

The research and development work is carried out either by the Commission's Joint Research Centre (direct action) or by organizations and companies of the Member countries, with the Community's financial support (shared cost action).

1. Framework Programme for a Scientific and Technical Strategy for Europe (1984-1987)

In June 1983 the Council of Ministers of the Community approved, after several months of elaboration, a Commission proposal for the development of a strategy of R&D. It is aimed to contribute to reach the objectives of socio-economical nature which are listed in Table 1, i.e.: the promotion of agricultural and industrial competitiveness; the improvement of the management of raw materials and energy resources - including the development of nuclear fission energy - as well as the improvement of the living and working conditions; the reinforcement of the aid to developing countries and the increase of the effectiveness of S/T potential.

Weight was given to the relative importance of Community R&D actions to be developed to reach the different objectives, taking into account specific criteria to determine the need of Community intervention in R&D with regard to the whole European R&D.

The importance appears in the amounts foreseen for the envisaged programme, covering 4 years. Nearly 4 billion units of account are involved (1983 values).

The correct implementation of the R&D framewok programme requires a substantial increase of the Community budget on scientific and technical activities. In spite of the fact that the Council of Ministers recognized the need to increase the part of the budget attributed to R&D, the present difficult situation in the Community budget has created a blockage in the process of the R&D framework programmes' implementation.

In the global picture, the part connected with energy research has - compared with the past - decreased in importance: it is now less than 50%. For the energy part, nuclear fission research represents about 30%. This means that the research activities of the Community for the development of nuclear fission energy are about 15% of the total R&D.

2. Research Action Programme on Nuclear Fission Energy

The Research Action Programme (RAP) Nuclear Fission Energy (2) is implemented through programmes, legally approved by the Council of Ministers, on the 5 areas described in Table 2. Not all the programmes are approved at the same time: in fact, the 4 programmes of the JRC, i.e. reactor safety, actinides, waste management and safeguards, are approved. The Shared-cost Action programmes are, for the time being, not yet approved, with the exception of the programme on decommissioning of nuclear installations approved in December 1983. The proposals for the future programmes on reactor safety (continuation of the one presented at this seminar) and on waste management (continuation of a programme which will be terminated by the end of 1984) are presently not yet approved by the Council of Ministers.

The two largest research areas are reactor safety and management of radioactive waste. This confirms that the main focus of programmmes in the field of R&D on fission energy is in line with the Commission's concern for public interest.

The following is a survey of the main activities of the RAP, showing where the accent is put in the different areas and referring to some achievements of the recent past which constitute starting points for the future.

3. Reactor Safety

Overall objectives of this programme (3) are:
- Accident prevention;
- Accident analyses, control and mitigation of consequences.

Concerning accident prevention, it is obvious that an increase in operational reliability intensifies the safety and economy of Nuclear Power Plants.

In the LWR field, many plants are in operation. The use of the large operational experience in order to improve reliability and early prevention of possible accidents is the logical way to go. For LMFBR, the study of mechanical behaviour of structures and components is essential to prevent accidents.

Concerning accident analyses, the exploration of the accident scenario connected to low probability/severe consequence events is aimed to quantify the existing safety margins and to enrich the technical data in order to ensure more efficient design and operation.

The future programme of the CEC on Reactor Safety has been explained in detail in the report of Mr. Holtbecker. So, the following is limited to some highlights.

The programme is a logical extension of previous activities, but the following points should be outlined:

- Already well-established projects on LWR safety like Reliability and Risk Evaluation, LOBI, are strengthened and/or reorientated in order to respond better to requirements which have been revealed by operational experience and licensing procedures.

- In the field of severe fuel damage and fission product source terms, the Commission proposes to gain access via Shared Cost Actions to experimental data from relevant programmes under way in Member Countries and elsewhere and to stimulate international collaboration on the analyses of such phenomena.

- Research relevant to LMFBR accident modelling such as EAC (European Accident Code) development, sodium thermohydraulics and PAHR (Post Accident Heat Removal) will be continued at JRC in order to respond to the specific request of the Member States.

- Through the Shared Cost Actions the acquisition at Community level of data on fuel motion and melting phenomena, for LMFBR, will increase the common understanding on accident analyses.

The main topics concerning accident prevention are:

- Reliability and risk assessment;
- Integrity of LWR components and systems;
- Integrity of LMFBR components and systems.

JRC launched 4 data banks in the E.R.D.S. (European Reliability Data System), making use of a large quantity of operational data. It is aimed to have the 4 data banks fully operational in 1987. This system constitutes the most relevant piece in the framework of a strategy for the development of P.R.A. methodologies, with benchmark exercises, including common cause/common mode failure, human errors, and data collections on human factors.

On the very important topic of integrity of components and systems, both in LWR and in LMFBR, the activities of JRC will be supported by shared cost actions.

The programme for Inspection of Steel Components (PISC II) sponsored jointly by CEC and OECD, and with JRC Ispra as the operating agent, is aimed at the assessment of the performance of NDT techniques.

For LMFBRs some studies are intended for material properties, structural analysis and fracture mechanisms.

The main topics concerning accident analyses, control and mitigation of accident consequences are:

- The study of abnormal behaviour of cooling systems and of improbable severe accidents, both for LWR and Fast Breeders, and
- The study of structures under dynamic loading as far as the Fast Breeder Reactor is concerned.

Both separate effects and integral tests are foreseen in these fields, including the possible participation in in-pile tests (LWR and LMFBR) executed in Member States' laboratories.

Concerning experiments on the abnormal behaviour of the LWR primary system, the LOBI facility simulates with a reduced scale of 1/700 a 4 loop primary cooling system of a 1300 MWe PWR reference plant.

It is aimed to study loss of coolant accidents and other special transients. The test loop has been modified recently into the LOBI/MOD 2 configuration more appropriate for the study of Small Break LOCA scenarios. The final objective of this research is the assessment of best-estimate codes, describing the behaviour of the primary circuit under abnormal conditions.

4. Nuclear fuels and actinides research

Research activity on fuel and actinides is carried out at the Karlsruhe establishment of the JRC, the Transuranium Institute (T.U.I.). The main objectives of this activity are:

- The improvement of nuclear fuels for LMFBR;
- The safety of the fuel cycle, both for LWRs and LMFBRs.

The actions developed are:

. Operational limits of nuclear fuels
This contains the evaluation of the impact of operating and extreme working conditions on the mechanical stability of advanced fuel for fast reactors and the determination of thermophysical properties of selected reactor materials at very high temperature, as needed for the core accident analyses.

. Transient behaviour of oxide fuels and fission products release under severe fuel damage conditions

. Actinide cycle safety
This project aims at answering questions arising from the impact of transuranium elements on the safety analysis of possible mechanisms of accidental release of aerosols formed during fuel fabrication and investigation of key problems for reprocessing of Pu-fuels.

. Actinides researches
This means study of the physics and chemistry of actinides.

A major task at the Transuranium Institute is the determination of the thermophysical properties at high and very high temperatures of selected nuclear materials, which are important for severe core accident analyses. An outstanding achievement is the extremely difficult determination of the thermal conductivity of UO_2 in solid and liquid state under well defined and pure conditions.

5. Management of radioactive waste

The main objective of R&D activities of the Community in this field is to help to create a comprehensive waste management system, looking at the economical, safety and environmental problems.

A twelve-year Community Action Plan for Radioactive Waste Management, approved by the Council of Ministers in 1980 and lasting until 1992, ensures the necessary concertation on the policy followed within the Community as well as the orientation and the continuity of the R&D programmes (4).

The specific objectives of the R&D, implemented at the JRC, and within shared cost actions are:

- treatment, conditioning and handling of wastes:
  . to improve the existing methods of management taking advantage of the technological progress;
  . to develop new treatment and conditioning processes for high level wastes and long lived wastes (+ bearing wastes);
  . to develop processes with similar purpose for gaseous wastes;
  . to work out methods of characterization, criteria of acceptance and quality control for wastes to be stored;

- final storage of high level wastes and long lived wastes:
  . to demonstrate the feasibility of final storage in continental geological formations;
  . to evaluate the feasibility of final storage involving burial under ocean sediments;
  . to evaluate the safety of the different options of final storage;

- management of radioactive waste as a whole:
  . the optimization of radioactive waste management with respect to safety and cost;
  . the integration of waste management into a global strategy for the fuel cycle downstream of the reactor;
  . the study of the plurinational dimensions of waste management.

The specific future actions which are partially decided (as far as the JRC is concerned) and partially to be decided by the Council of Ministers (as concerns the Shared Cost Action), are the following:

- Characterization and quality of conditioned waste;
- Demonstration of treatment and conditioning processes in view of an optimized management of wastes. Pilot scale plants are foreseen;
- Conception of final storage of waste in geological formations. Underground laboratories will be installed, namely the salt mine at Asse, an experimental clay vault at Mol, a site in France for executing tests in real conditions. The aim is the demonstration of the practical feasibility of the final storage of highly active waste in different geological formations;
- Long term migration of radioactivity in continental geological storage. This project is based on the continuation of the MIRAGE project (Migration of Radionuclides in the Geosphere), started in

1983;
- Safety evaluation of waste final storage, following different concepts. This action will be based mainly on the PAGIS project (Performance Assessment of Geological Isolation Systems);
- Optimisation of radioactive waste management, involving a global strategy for the fuel cycle downstream of the reactor.

6. Safeguards and Fissile Material Management (JRC Direct Action only)

Objectives:
The establishment and the upholding of a credible safeguards system for nuclear materials in circulation in Europe is among the fundamental elements which condition the acceptability and the development of nuclear energy and is one of the means to respond to the concern of the proliferation risk of nuclear weapons.

Considering also the importance of the economic impact of a safeguards system, it is necessary to establish a European strategy in this field for the 1990's and this involves an important Community action and a concertation with the national efforts.

Proposed actions cover 3 projects:

- Methods and instrumentation for nuclear material assay and containment and surveillance

This project includes the development of the methods and procedures, as well as the evaluation of their performances, and the establishment of validation and training laboratories.

- Safeguards data processing, transmission and evaluation

Two sub-projects are included in this project:
. Nuclear material accountancy data management;
. Automatic processing, transmission and evaluation of inspection data, produced by nuclear material assay and containment and surveillance instrumentation.

- Integration of safeguard activities

Four sub-projects are considered here:
. Study of safeguard practices;
. Development of a conversational simulator for the analysis and assessment of safeguard practices;
. Application of EXPERT system techniques to safeguards studies;
. Analysis of safeguards procedures through the fuel cycle.

This sub-programme on safeguards represents also a substantial technical support to the implementation of safeguards by the Commission services, following the duties of the Euratom Treaty, in connection with the IAEA.

7. Decommissioning of Nuclear Installations (Shared Cost Action only)

Existing experience in the decommissioning of nuclear installations is limited. The aim of Community action in this field is the joint development of a system of management of nuclear installations finally shut down and of the radioactive wastes produced in their dismantling. This system, at its various stages, will ensure the safety and protection of man and the environment against the potential hazards involved in the decommissioning of these installations.

Thus, the following specific objectives are pursued within the just completed and agreed new programmes:

- To establish efficient surface decontamination techniques producing little secondary waste;
- To establish appropriate techniques for the dismantling of thick-walled steel components and reinforced concrete structures;
- To establish techniques for conditioning the special radioactive waste materials arising from the dismantling of nuclear installations;
- To establish more reliable predictions on the consequences of decommissioning as to radiation exposures, costs and waste repository requirements: this involves optimization of overall decommissioning schemes;
- To establish certain cost-effective improvements for the design of future installations, which facilitates their decommissioning:
- To provide guiding principles for decommissioning.

The new 5-year programme, just started at the beginning of 1984, corresponds to 3 actions, different in scope and nature (5):

- The first one includes the study on long-term integrity of the buildings and structures, problems of decontamination and dismantling techniques, treatment and transport of waste material. It includes also the estimation of quantities of radioactive waste arising from decommissioning of nuclear installations in Europe and finally the influence of plant design features on decommissioning.

- The second action covers the identification of relevant guiding principles on design and operation of nuclear installations, with a view to simplifying their subsequent decommissioning and includes the identification of a basis for a Community decommissioning policy.

- The third action includes the testing of new techniques within the framework of large-scale decommissioning operations undertaken in Member states.

The results of the decommissioning programme obtained in the now completed first programme have been presented at the "Conference on the Decommissioning of Nuclear Power Plants", on 21-24 May 1984 at Luxembourg, organized by the CEC.

8. Conclusions

The foregoing indicates the large variety and amount of work and interest covered by the CEC in the nuclear field.

From the "Development" point of view, CEC activities address the complete fuel cycle and its satisfactory closure, in line with the Community policy on nuclear energy.

The research action programme on the development of nuclear fission energy, executed by the JRC since 1973 and by shared cost action since 1979 have demonstrated that such Community actions add two essential elements to the research executed at national level. The first element, of a technical nature, pertains to the nature of the acquired common understanding, which benefits from the cooperation which is stimulated by Community action. The second element, of a more political nature, has to do with the acceptance of nuclear energy: public opinion tends to accept more easily results which are presented as a result of cooperation and critical evaluation performed by experts of the Community within the frame of Research and Development effort coordinated by the Commission.

9. References

(1) A scientific and technical strategy for Europe, Framework Programme (1984-1987), XII/126/84.

(2) COM(83)300 final. Research Action Programme on the development of nuclear fission energy (1984-1987).

(3) COM(83)299 final. Proposal for a Council decision adopting a research programme on reactor safety (1984-1987).

(4) Orlowski, Cricchio, Girardi. The programme of the CEC and the objectives of the Community in the field of radioactive waste. IAEA-CN-43/59, International Conference on Radioactive Waste Management, Seattle, Wa., USA, May 16-20, 1983.

(5) Council (1984) Council decision of January 31, 1984, adopting a research programme concerning the decommissioning of nuclear installations. O.J. No L36 of February 8, 1984, p.23.

Table 1

Framework Programme for European Community Scientific and Technical Strategy (1984-87)

	Million ECU
'romoting Agricultural Competitiveness, including Fish	
* Sub-goal Agriculture	115
* Sub-goal Fish	15
'romoting Industrial Competitiveness	
* Removal and Reduction of Hindrances	30
* Conventional Industries	350
* New Technologies	680
mproving the Management of Raw Materials	80
mproving the Management of Energy Resources	
* Development of Nuclear Energy - Fission	540
* Controlled Thermonuclear Fusion	480
* Development of Renewable Energy Sources	310
* Rational Use of Energy	520
Reinforcing Aid to the Developing Countries	150
Improving Living and Working Conditions	
* Improvement of Safety and Protection of Health	110
* Protection of the Environment	160
Improving the Effectiveness of Scientific and Technical Potential	5% of the total amount of appropriations
Horizontal Action (including the Utilization of R&D)	110
TOTAL	3,750

Table 2

RAP Nuclear Fission Energy
(1984-1987)

	MECU (1982 PRICES)
Direct Action on Reactor Safety (JRC)	192,2
Programme of Shared-cost Action on Reactor Safety	*
Implementation of the Council Resolution of 22 July 1975 on Technological Problems of Nuclear Safety	3,3
Direct Action on Nuclear Fuels and Actinides Research	65,7
Direct Action on the Management of Radioactive Waste	48,9
Programme of Shared-cost Action on the Management and Disposal of Radioactive Waste (1985-89)	*
Direct Action on the Safeguarding and Management of Fissile Materials and Support for the Safeguards Directorate	45,4
Programme of Shared-cost Action on the Decommissioning of Nuclear Plants (1984-88)	11,0
TOTAL:	470,0

* These future programmes are not yet approved by the Council of Ministers.

THE CEC INDIRECT ACTION RESEARCH PROGRAMME, SAFETY OF THERMAL WATER REACTORS (1979-1983)

AIMS, CONTENTS AND IMPLEMENTATION

J. Vilain
Commission of the European Communities
Brussels

Community research can be performed in several ways. One is direct action, 100% funded and done at one of the J.R.C. establishments. Next is indirect action, called now shared-cost. It is done under contracts in the Member States, taking advantage of available competences and facilities, with partial E.C. funding (up to 50%). Concerted actions are conceptually similar, but on a voluntary basis, no E.C. funding of the projects. Finally associations, between Euratom and large national programmes, may cover one development, fusion for instance (or in the past fast breeders or high temperature reactors).

This programme was an indirect action one. It was elaborated early on with the help of the Working Group No 2 on safety research. As it was designed at the point to be complementary to the corresponding national and J.R.C. programmes, on topics where it was felt that not quite enough was done, this explains the choice of its three topics, albeit largely unrelated. Those are, in our alphabetical familiar order:

A. the rewetting and reflooding of fuel subassemblies, damaged or not, in a post-LOCA, recovery phase,
B. the impact on nuclear plant structures of an external hydrocarbon cloud explosion, more specifically the dispersion, explosion and wave-structure interaction aspects,
C. the advection, diffusion and deposition of active products following a major reactor accident, beyond D.B.A. In practice partial effects and some regional - and mesoscale modelling.

The corresponding 33 projects are listed in the seminar's programme, no need to reiterate at length. Your prime choice will be then between parallel sessions A, B or C, but note that overview presentations of all three will also be given Wednesday morning.

Speaking money now, the budget of this programme was 6.3 million ECU for five years, all told, and some 5.5 million covered our contributions to the projects. About half went to Area A, understandably with expensive out-of-pile simulations, and a quarter each to B and C, unfortunately too little to support regional-scale field trials on passive dispersion. It was indeed a tight budget, but indirect actions have a multiplicative advantage: the total amount spent on the projects by all parties was probably between 15 and 20 MECU. On top, in some instances, only running costs, not capital ones, were charged.

For indirect action, the origination of projects rests on a competitive call for proposals. Selection is a rather laborious iterative process, again with assistance and expertise of national delegates, in ad-hoc groups, and equally laborious contracts placement follows. In our case this took place early and once, as most projects were of quite long duration, 3 to 4 years, but diverses extensions for complementary work were also formulated in 1983.

Once work started, contractors and national delegates were gathered in three specialized working groups, one for each area, meeting at regular intervals, say 2 or 3 times a year for 2 or 3 days. Most of each meeting was devoted to the projects, reviewing and discussing progress over the last six months, also based on extensive six monthly reports (which are a contractual requirement). For the remainder topical or invited presentations, on-site visits of projects and laboratories were regularly arranged. One may note that meetings were indeed held, in turn, at contractors laboratories and centers, not in Brussels. The parallel sessions, next, will illustrate the working method, which was certainly quite effective to stimulate performance, create a degree of team spirit and ensure a steady exchange of information and ideas. Above and beyond that, much supplementary effort was also devoted to circulating all manners of relevant information to participants and to slowly enlarge the groups.

Progress of the whole programme was also monitored, at regular intervals, by a higher-level experts group, the Advisory Committee for the Management of the Programme. This group, reporting to the Council and the Commission, also gave advice on the safety research part of the J.R.C. programme, direct action. The last such meeting will immediately follow this seminar.

In most respects then, this programme was a steady and collective exercise, with much goodwill extent. The method, working by way of groups, developed somewhat empirically and achieved much in-depth and thoroughly documented work. However, it is also characteristic of a certain "size", with a limited number of projects and participants in each area, a critical optimum. It would certainly be quite difficult or physically impossible to transpose it to a larger programme, with many more projects or participants.

Projects are now all terminated or just terminating. Note that, apart from the proceedings of this meeting, all final reports will appear in extenso as EUR reports, available from the Office for official publications of the European Communities.

Parallel Sessions Research Area A

The loss of coolant accident (LOCA) and the functioning and performance of the emergency core cooling system (ECCS)

Session A1

Chairman: K. Wolfert (GRS/München)

Session secretary: E. Skupinski (CEC/BXL)

Important thermohydraulic aspects during refilling and reflooding of an uncovered LWR core

E.F. Hicken

Gesellschaft für Reaktorsicherheit / FRG

The topic of this presentation is an overview over the reflood process during a loss-of-coolant accident and - to be more specific - mainly during a LOCA resulting from the double ended break of the cold leg of the main coolant pipe of a PWR.

In 1969 a test in the first SEMISCALE test facility was performed with the result that the injected water did not cool the core adequately. This test, the "Interim Acceptance Criteria" in 1971 and the following hearing resulted in large efforts to study physical phenomena and system behavior during LOCA's. The large break was chosen as a design basis accident (DBA), because it results in the highest loads on the reactor internals, the fastest depletion of water inventory and the fastest core heat up. The cold leg side was selected due to core mass flow reversal effects and the higher core temperatures when compared with a hot leg side break.

My presentation will cover

- an overview over different PWR ECC-systems
- physical phenomena during refill and reflood
- selected aspects of related system behavior
- an overview about the important integral or large test facilities and
- computer code simulation of physical phenomena and system behavior.

This presentation also covers only PWR's of commercial type with the exception of CANDU-Reactors because research and knowledge exchange concentrates on these PWR's.

Different PWR ECC-Systems

In fig. 1 an overview is given about the main ECC-systems. In case of large cold leg LOCA the different systems react in the following manner:

Due to the fast water depletion prepressurized accumulators inject water into the pressure vessel to cool the heated core.

Pumps take over long term water injection and core cooling.

In fig. 1 the four main designs are shown. The most common design is the socalled "cold-leg-injection-system". Several accumulators inject water into the cold legs to refill the lower plenum and later reflood the core. The injection starts at a pressure between 20 and 50 bars (design dependent). It is evident that at this pressure there is still mass outflow of the break resulting in an upward two phase flow in the downcomer. This upflow causes some of the injected water to flow directly to the break (bypass flow).

Another important effect is the socalled "steam binding": During the reflood process water droplets are entrained into the steam generators. Most or all of the droplets will be evaporated due to heat addition from metal surfaces and the steam generator secondary side. The increased volume flow results in some pressure difference between the inlet and outlet of the steam generators which also effects the driving head between downcomer and core. Details about this effects and others will be described later.

To avoid the "steam binding" effect B*W incorporated vent valves in their design. At very low pressure differences those valves open to allow core steam flow into the downcomer.

A quite different system is the "upper-head-injection" (UHI) system. At a relative high pressure (about 70 bars) water is injected via guide tubes into the region above the core to precool the core already during the blowdown phase.

The ECCS of a german vendor is based on the injection of cold water into the hot legs as well as into the cold legs. The injection pressure is about 26 bars. As the hot leg side injection starts during blowdown with downward core mass flow parts or the whole core will be precooled. In addition, steam produced in the core region will be condensed by the injected water.

Worldwide guidelines related to ECC-effectiveness concentrate mostly on limitations of core temperature and cladding oxidation. The upper limits, however, do allow for some rod failures with release of volatile fission products. The objectives of ECCS-effectiveness

- maintain a coolable geometry
- limit fission product release to the containment

have resulted mainly in core temperature studies. I like to point out that the release of fission products to the containment is more important. For radiological dose and release calculations it is important if the soluble fission products are released into the containment atmosphere or are within the liquid phase. Fission products in the liquid phase results in much less doses which can be achieved by condensation of steam on subcooled water. Assessing the capabilities of the ECC-systems, shown in fig. 1, the combined injection into the hot and cold legs has by far the highest capability to condense steam from the reactor core within the primary system. The different steam flows are shown in fig. 2 to fig. 5.

Relevant phyiscal phenomena during refill and reflood

The systematic approach starts at the injection locations (fig. 6). At this location it is of some importance, if the injection line enters the main coolant pipe at an angle of 90° degree or smaller. Both designs are possible and can be found in NPP's. In the latter case, some momentum is available for easier injection into the downcomer. It should be pointed out that an oszillatory behavior of a slug in the main coolant pipe has been seen experimentally. At the injection location subcooling is highest and condensation of steam takes place. The condensation rate may influence slug formation and its frequency.

Between the injection location and the reactor vessel further condensation occurs with steam flowing countercurrent to the ECC water.

The water enters the downcomer jet-like and impinges the core barrel. Heavy condensation occurs resulting in relatively high upward steam flow. This upward steam flow is directing some water to the leak. In addition, an oszillatory behavior of water downfall have been experienced. In general, two-dimensional and partly three-dimensional flow pattern exist depending also on the injection locations, break location, subcooling and mass flows. Up to about 50 percent of the injected water can be bypassed during the first part of the injection phase. Therefore, this behavior cannot be neglected.

The water will be further heated up on its way into the lower plenum. Experiments have shown that only in small downcomers the heat addition from the walls will generate some additional steam resulting in a delayed refill of the lower plenum by some seconds (Hot-wall-effect).

It could be shown, analytically and experimentally, that also in the case of a large break LOCA residual water in the lower plenum exists. The licensing assumption with an empty lower plenum is therefore conservative.

The injected water, sometimes still subcooled, will - depending on the design of the lower plenum and its internals - mix with the residual water, but could also form some stratification with subcooled water.

At high steam flows the water level increase may be stopped and all injected water bypassed to the break.

If the water level enters the core region, fig. 7, stored internal energy in the fuel rods and the decay heat results in a heat up of the water to saturation and evaporation. The degree of water level increase (between several cm/s and more than 10 cm/s) influences quenching, the axial temperature profile in the rods and the entrainment behavior of droplets. Because of the axial power profile these phenomena vary with the level in the core.

Entrained droplets continue to evaporate on their way upwards. The heat transfer mechanisms are

- heat exchange from the wall to contacting droplets
- heat transfer from the wall to the steam
- heat transfer from the steam to the droplets
- radiation heat transfer from the wall to steam and droplets.

It is evident that the steam must be superheated; the degree of superheat depends mainly on the heat transfer rate, the number and surface area of the droplets and the heat transfer between wall/steam and steam/droplets, which is influenced by the velocity difference of steam and droplets. Radiation heat transfer usually does not play an important role. Superheat may be as high as 500 degree K.

Spacers within the bundle influence reflood behavior by

- changing the droplet distribution to smaller sizes
- increasing the velocity difference between droplets and steam
- act sometimes as the beginning location for new quench fronts

A core is still coolable when the flow area in the bundle is reduced. It is the common understanding that with reductions up to about 80 % coolability can be maintained. Research has been done to study heat transfer in the vicinity of blockages.

Entrained droplets usually hit the upper end box. If the steam velocity is not too high, a water film will flow down the fuel rods and quench the rods.

At high steam velocities water droplets will be entrained into the upper plenum. Three possibilities exist (fig. 8 and fig. 10)

- a water pool will be formed
- entrained droplets imping the upper plenum structures. Some droplets will be entrained into the hot legs. The rest will form a falling liquid layer on the structures
- both possibilities are mixed.

One of the most important phenomena is the counter current flow behavior near the interface core/upper plenum. Several experiments with single holes, multiple holes or original end boxes have been performed. Up to now it is clear that no extrapolation from one hole to multiple holes can be done with good confidence. Obviously an instability exists where in few holes downward water flow while through the other holes steam and droplets flow upwards. With real end boxes it could be shown that the bundle and the spacers below the end box influence the overall behavior.

The pool behavior in the upper plenum has not jet been studied experimentally. Related tests are planned in the UPTF-test facility in 1986.

Selected aspects of related system behavior

System behavior is influenced by oszillatory effects. The most known effect is a core/downcomer oszillatory behavior which can be traced also in the loops, e.g. by slug formation in the cold legs. Frequency and amplitude are not yet very well understood. However, core heat transfer is usually increased.

In case of hot leg injection the water distribution in the upper plenum was studied in a full scale air/water test facility. It was found, as shown in fig. 9, that the injected water impinges the supporting columns in the upper plenum and changes its flow direction several times. As a result the horizontal component of its momentum decreases rapidly and the water fall down is concentrated to a small area near the injection point. This leads to a 3-D water distribution in the upper plenum.

Special loop oszillation has been calculated for combined injection ECCS. A slug will be formed in the hot and cold leg. Condensation of steam within the volume between these slugs causes both slugs to move in the direction of the steam generator. When the slug of the hot leg enters the steam generator, heat addition from the secondary side evaporates some water. The pressure increase finally results in the slugs to be injected into the upper plenum and downcomer. This effect will be studied in the UPTF-test facility. Calculations with the TRAC-code show that the different loops behave nearly independent from each other.

Calculations with the 3-d-TRAC-code have shown that the pressurizer mass outflow precools parts of the core. Then, at the beginning of reflood always an asymmetric core temperature profile would exist. Codes with parallel channels (including those in the upper plenum) or with 3-d-capability would be necessary.

Overview about important integral or large test facilities

Physical phenomena and system behavior have been studied in separate and integral test facilities. Three-dimensional effects create sometimes scaling problems. Therefore, three countries (US, Japan, Germany) decided a few years ago to build a test facility of nearly full scale.

An overview about different integral facilities is given in figs. 11 and 12. With the exception of LOBI, BETHSY und ROSA IV, refill and reflood behavior is studied in all other facilities.

The next fig. 13 gives an overview about the number of tests performed and planned. It can be seen that still a lot of experiments will be performed and must by analysed.

In the following figures (14 to 22) some facilities are shown.

Computer Code Simulation of Physical Phenomena and System Behavior

Because only a small part of experimental results can be directly used for the assessment of ECCS effectiveness computer code simulation of physical phenomena and system behavior is mandatory.

In fig. 23 the requirements for the codes are listed with respect to location and phenomena. Within the RPV two-fluid-codes are required to be able to simulate the phenomena.For loop behavior less than six equations would be sufficient.

The advanced codes TRAC, RELAP5/Mod2, FLUT and CATHARE meet most of the requirements, however, with different degree of verification. Only the TRAC code has a 3-dimensional capability. The general question is if a parallel channel model is adaquate to model 3-d-effects. My assessment would be that in most cases the parallel-channel capability would be sufficient to model 3-d effects. This assessment ist based on the knowledge that 3-dimensional two phase flow incorporates a lot of uncertainties. Experiments within the 2D/3D-projekt should show this capability of 1-d-codes.

In fig. 24 a proposal for an assessment matrix for PWR system codes is shown. This matrix should enable to compare different codes and to increase confidence in computer code calculations.

Assessment

- Main and relevant physical phenomena during refill and reflood are known. 3-dimensional effects will be studied in the 2D/3D-project during the next few years. After these studies it can be assessed that the thermalhydraulic phenomena during refill and reflood should be well understood.

- System behavior depends on the specific ECCS design. With known physical phenomena system behavior can be assessed.

- Advanced computer codes meet most of the requirements, however, with different degree of verification. With existing and planned activities (development and verification) reasonable accuracy will be achieved.

- The study of the interaction between thermohydraulics and fission product release and transport within the primary system should be further supported.

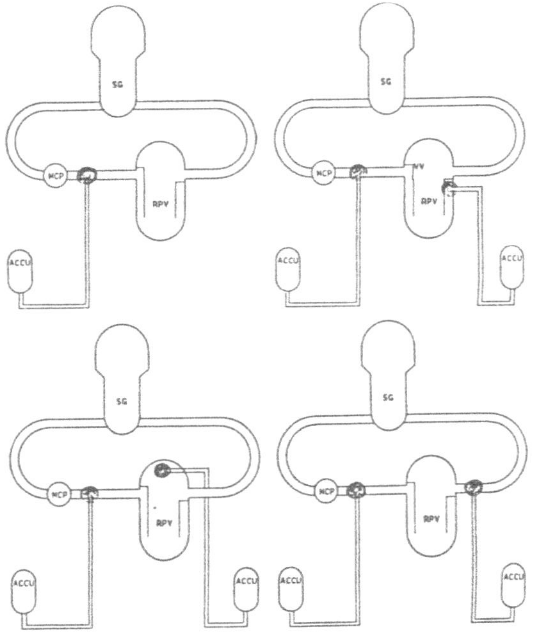

FIG. 1 **Overview about Different PWR ECC-systems**

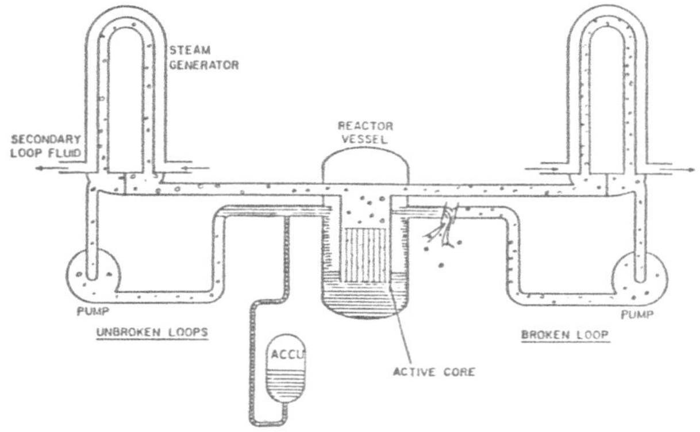

FIG. 2 **COLD LEG INJECTION**

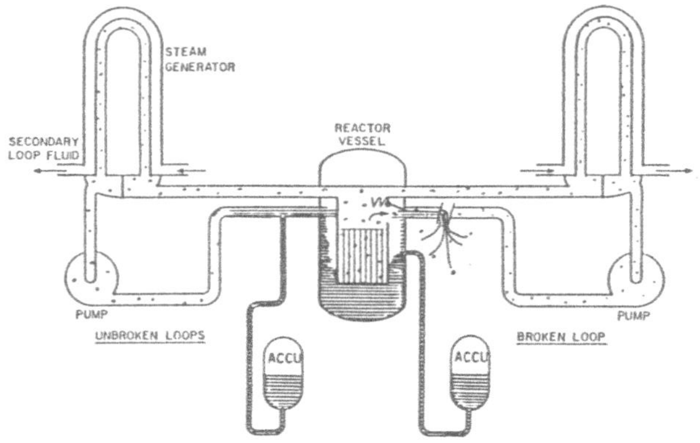

FIG. 3 **COLD LEG INJECTION associated with Vent Valve**

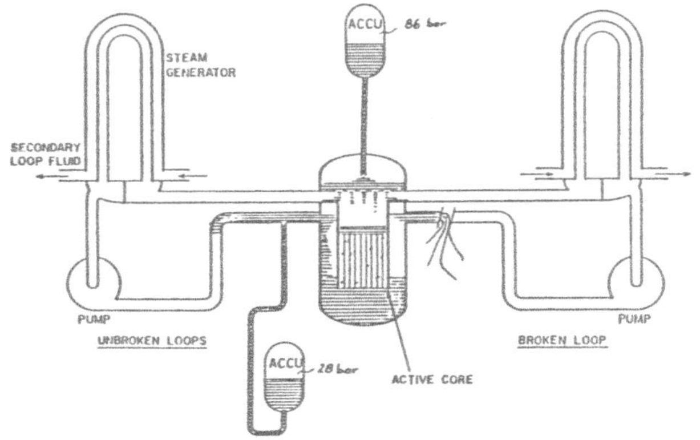

FIG. 4 **UPPER HEAD INJECTION**

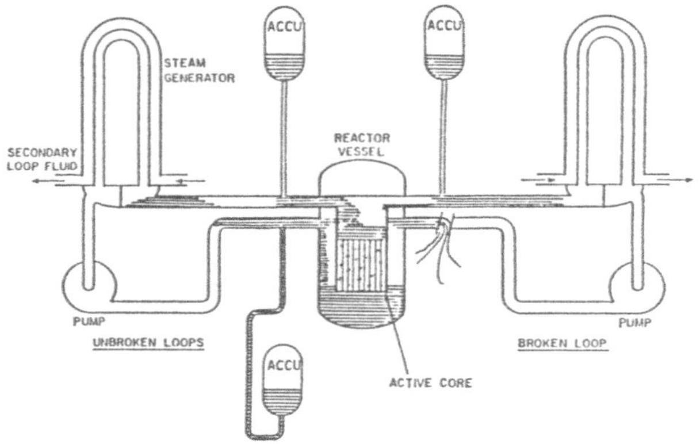

FIG. 5 **COMBINED INJECTION**

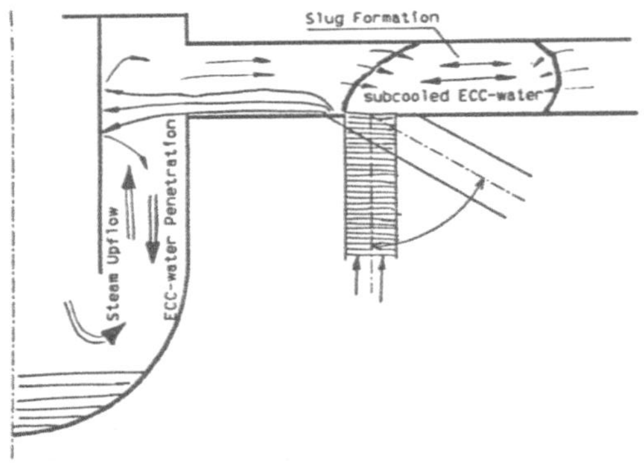

FIG. 6 **Phenomena in Cold Leg and Downcomer**

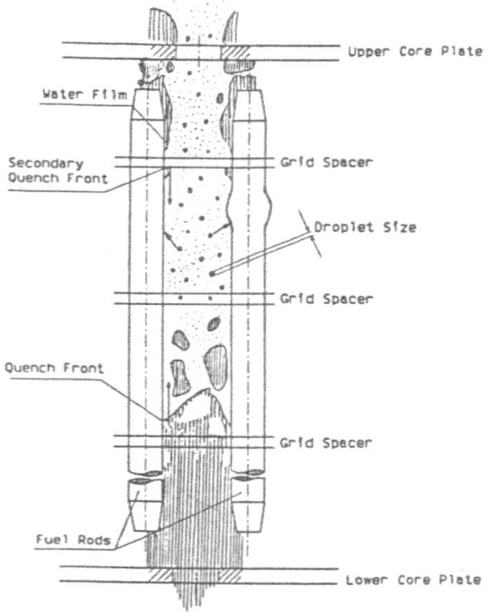

FIG. 7 **Phenomena in Core Region**

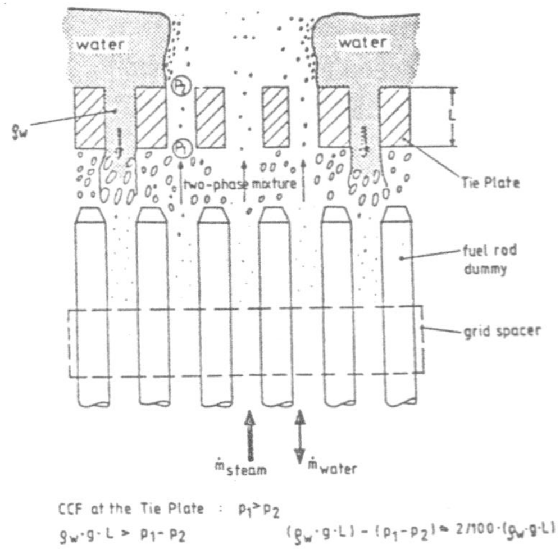

FIG. 8 Flow Phenomena at the Tie Plate in the Flow Regime Simultaneous Steam Upflow and Water Downflow (CCF)

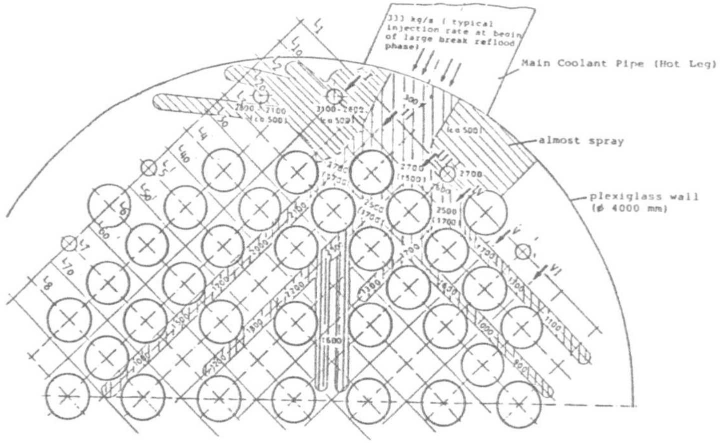

FIG. 9 Heterogeneous ECC-water-distribution in upper plenum
(1 of 4 hot leg injections)
Full scale air-water test facility (KWU) at atmospheric conditions

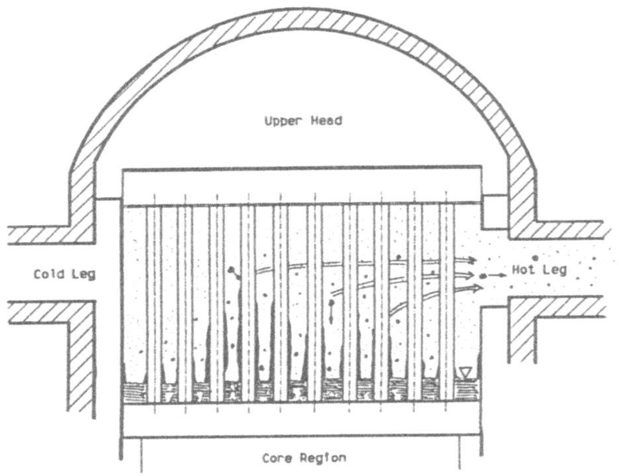

FIG. 10 **Phenomena In Upper Plenum**

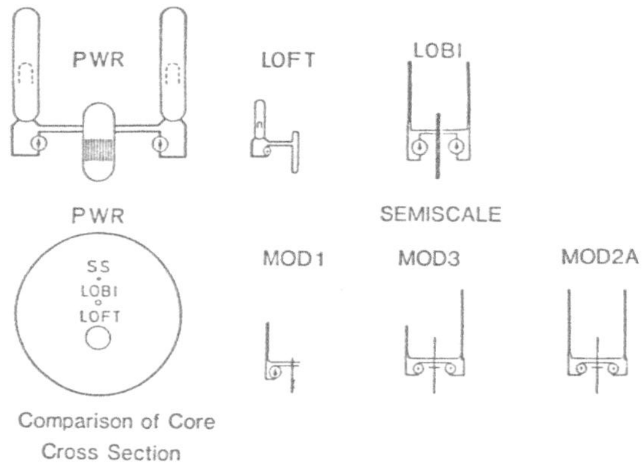

FIG. 11 COMPARISON OF PWR WITH SOME INTEGRAL TEST FACILITIES

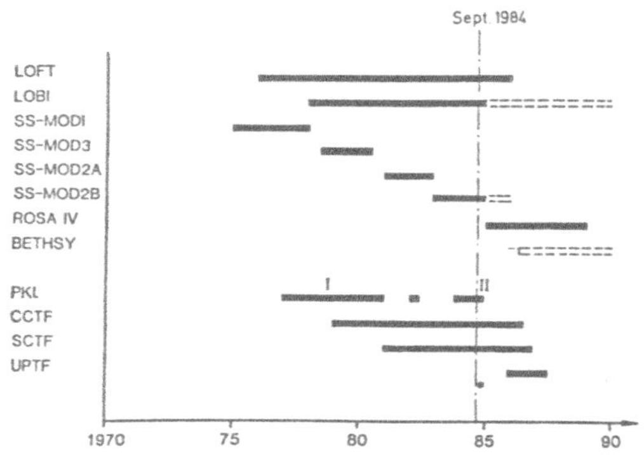

FIG. 12 TIME SCHEDULE OF INTEGRAL TESTS

Large Break LOCA - Refill and Reflood

Tests performed (planned) .

LOFT	11	-
SS-MODI	48	-
SS-MOD3	12	-
PKL	31	(7)
CCTF	74	(6)
SCTF	30	(40)
UPTF	-	(30)
BETHSY	-	(FEW)

FIG. 13

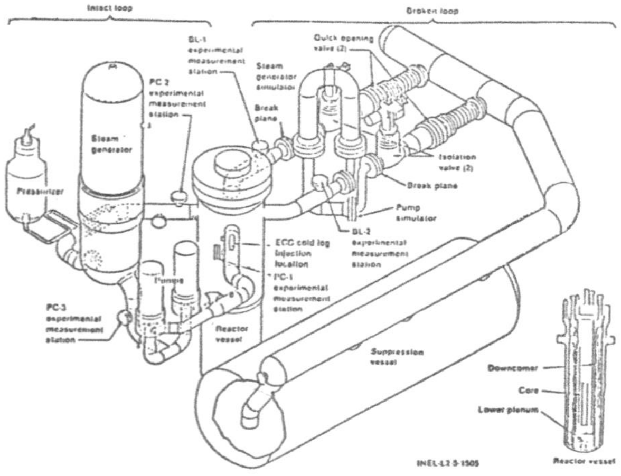

FIG. 14

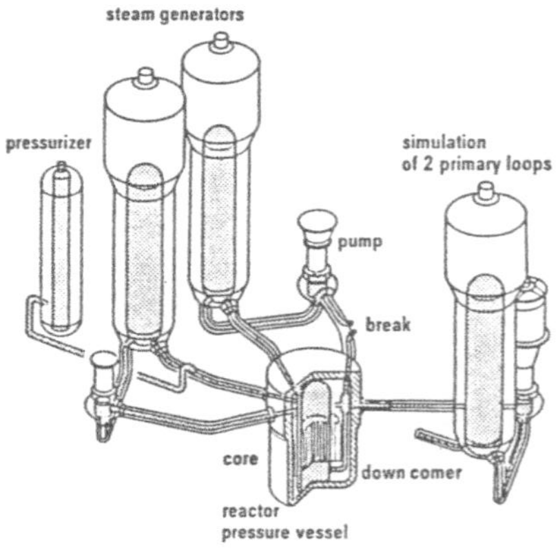

PKL Simulation of 4-Loop PWR

PKL-Systems Effects Tests:

- interplay between components
- code verification

	PWR	PKL
number of rods	45548	340
volume scale	134	1
elevation scale	1	1
primary loops	4	2+1 double
ECC-injection	8	4+2 double

FIG. 15

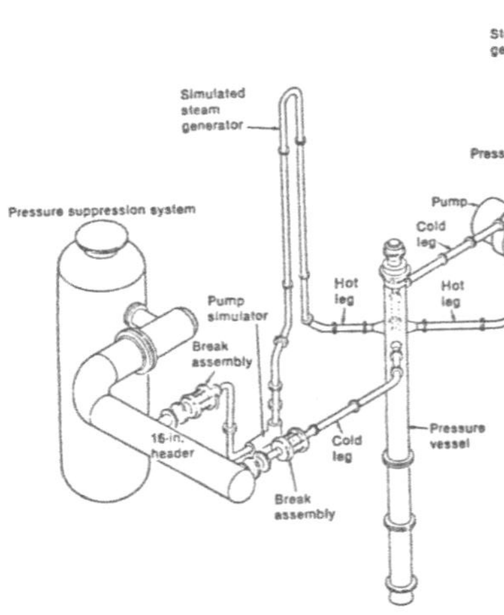

Semiscale Mod-1 configuration

FIG. 16

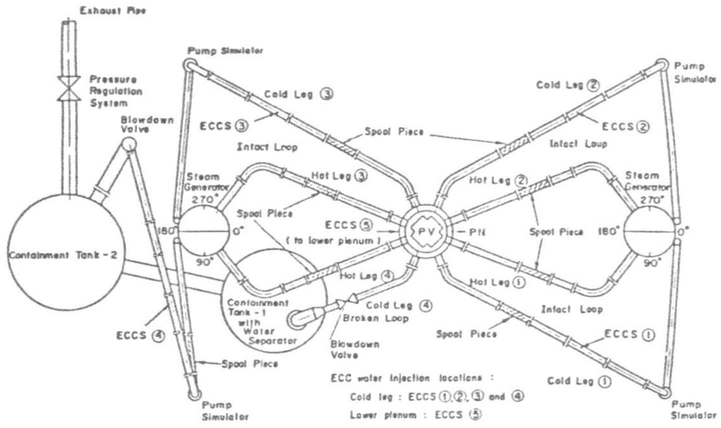

FIG. 17 **CCTF Primary Loop Pipings**

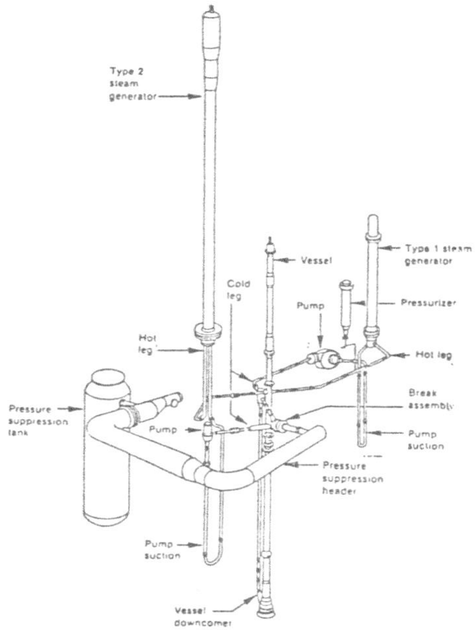

FIG. 18 **Semiscale Mod-3 system configuration**

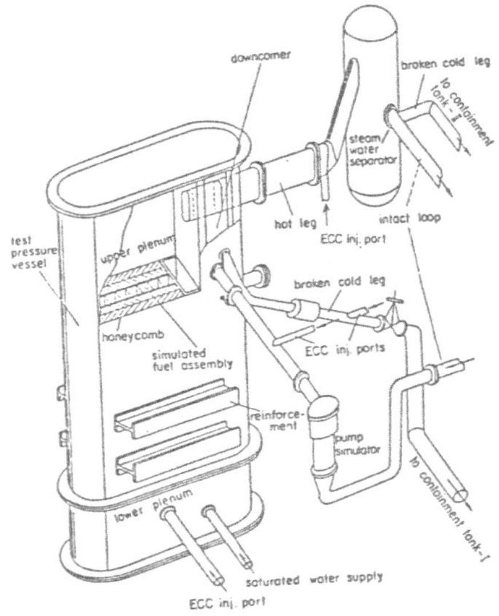

FIG. 19 **View of the SCTF Facility**

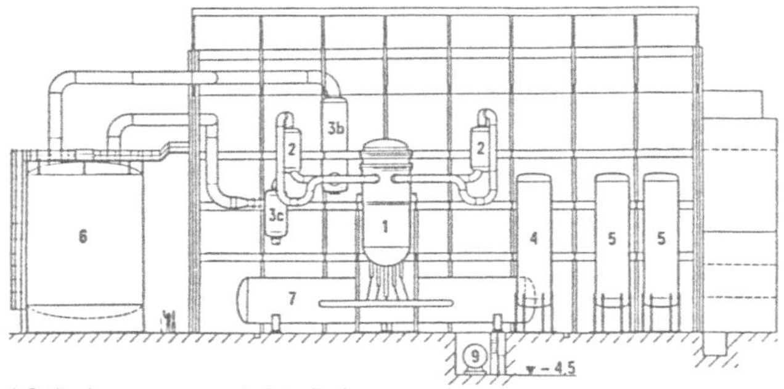

1 Test Vessel
2 Steam Generator Simulator, (Intakt Loop)
3b Water Separator (Broken Loop, Cold Leg)
3c Drainage Vessel for Hot Leg Break
4 Hot Water Storage Tank
5 Accumulators
6 Containment Simulator
7 Steam Storage Tank
9 Drain Storage Tank

FIG. 20 **UPTF Layout Front View**

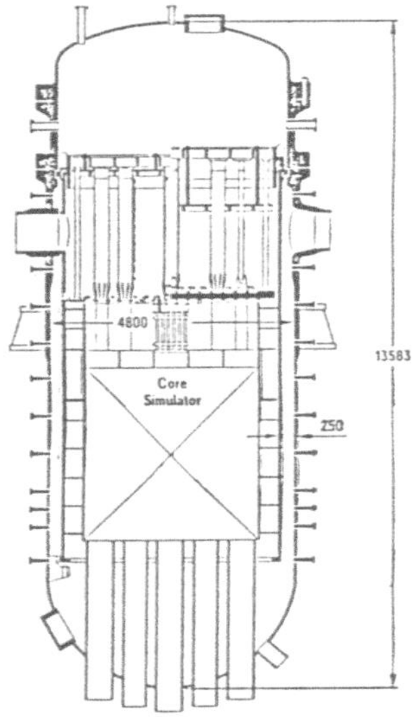

FIG. 21 UPTF – Test Vessel

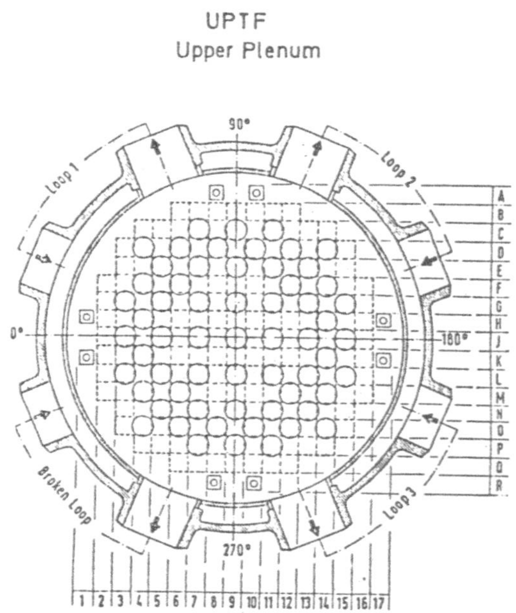

FIG. 22 CORE MIXER INJECTION ZONES

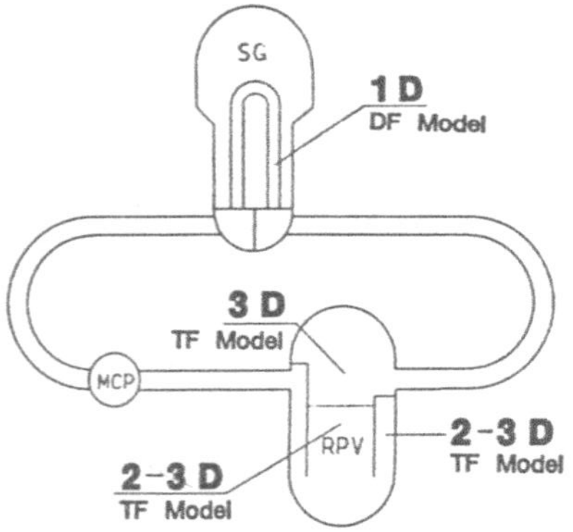

FIG. 23 Requested Dimensionality in Code Simulation

		Transients	Small/ Intermediate Break LOCA	Large Break LOCA
Separate Effect Tests	no. of facilities	–	6	10
	no. of tests	–	13	20
System Behavior Tests	facilities, (no. of tests)	LOFT (3) ROSA-IV (open) LOBI-II (open) BETHSY (open) SPES (open)	LOFT (7) ROSA-IV (open) PKL (4) LOBI-II (2) SEMISCALE (2) BETHSY (open) SPES (open)	LOFT (4) CCTF (3) PKL (3) LOBI-I (4) SEMISCALE (2) (BETHSY open) (SPES open)

1408K PROPOSAL FOR AN ASSESSMENT MATRIX FOR PWR SYSTEM CODES

FIG. 24

Discussion

Question: M.K. Denham - UKAEA/AEE-W

How important is it to study the effect of system - induced hydraulic oscillations on the reflood process?

Answer: E. Hicken - GRS

In system tests, e.g. in LOFT, oscillatory behaviour has been identified. Usually this behaviour enhances core heat transfer by increasing entrainment. Therefore, calculations without an oscillatory behaviour are conservative. An oscillatory behaviour during reflood exists especially for combined ECCS-injection. This phenomena will be studied in the UPTF test facility.

Question: T.C. deBoer - ECN Petten

1) Which computer codes can be used for fission product transport and deposition calculations for postulated DBA and beyond DBA accidents? Which is the specific role of the "LOFT fission product release experiments" for code assessment?

2) What safety issues have been solved by integral system experiments and in which way the different facilities can be complementarily used for code assessment?

Answer: E. Hicken - GRS

1) For the LOFT fission product experiments the TRAP-MELT-code will be used for predictions related to fission product behaviour within the primary system. However, it is generally agreed that this code has to be improved.

The TRAP-code is focussing mainly on behaviour beyond DBA. For the behaviour within the DBA case the same code will be used. However, the fission product bahviour during this sequence is mainly influenced by the thermohydraulic behaviour. Therefore, the commonly used codes will be needed for predictions and post calculations.

2) Integral tests as well as separate effect tests are necessary for code assessment. Integral tests are required to combine the different phenomena, which are not independant from each other. Within the CSNI an effort is underway to set up an international agreed validation matrix. The tests within this matrix should be used for the validation of the different codes.

Different facilities are necessary to assess scaling and to increase the confidence in code calculations.

Question: R. Pochard - CEA

You have presented advantages and disadvantages of several ECCS systems in particular related to fission product release. There is also another aspect which is the pressurized thermal shock. Could you make some comments on that?

Answer: E. Hicken - GRS

The guidelines for ECCs effectiveness are focussing on core temperatures, oxidation of cladding material and to a lesser extent on fuel rod ruptures.

In reality we do want to know the release of fission products into the containment. In case of a leakage in the containment integrity this would threaten the public. In the FRG the number of fuel rod ruptures is limited to 10 percent of the total number of rods.

The release of fission products into the containment atmosphere is influenced by the location of ECCs injection, for low pressure as well as for high pressure sequences. In the FRG also the high pressure ECCs are injecting into the upper plenum and thus condensing some steam. This injection location minimizes, in addition, the pressurized thermal shock. Due to the hot leg side injection of the accumulators with the KWU-PWR design, the cold leg side accumulators are isolated after a specified time to avoid a pressurized thermal shock load on the RPV in case of a small break.

Two tests in the LOFT facility will study the fission product behaviour within the primary system.

Question: R. Deruaz - CEA/CEN-G

Separate effect tests have been performed at Grenoble both on a 6x6 and the 7x51 full length heater rod bundles, regarding core inlet flow rate oscillations during reflooding. No effect on cooling efficiency was found, even in the case of oscillations including reverse flow. Two questions are however remaining:

- how to get an idea of the amplitude of oscillations which could be got on a PWR?

- can we really extrapolate the above mentioned result to fuel rods? Our calculations show that, in this case, the quench front might oscillate and then influence heat transfer in the dry region (directly or/and through possible fallback from the upper plenum to the core). Some tests are planned with respect to this problem.

Answer: E. Hicken - GRS

As already stated, an oscillatory behaviour enhances entrainment in the core. The droplets will decrease steam superheat and therefore enhance core heat transfer. The amplitude is obviously dependant on system behaviour and can only be calculated for large PWRs. This behaviour is, however, a phenomena to tbe studied in the UPTF and CCTF facilities. These tests should enable us to assess large PWR behaviour.

STUDIES ON THE EFFECTS OF BLOCKAGE UPON LWR EMERGENCY CORE COOLING SYSTEMS

S.A. FAIRBAIRN and B.D.G. PIGGOTT
Central Electricity Generating Board
Berkeley Nuclear Laboratories
Berkeley, Gloucestershire, GL13 9PB.
United Kingdom

Summary

The aim of this work is to investigate heat transfer effects of localised coplanar ballooning in an electrically heated rod bundle under conditions of single and two phase cooling. Preformed blockage shapes are used to decouple the problems of clad deformation and heat transfer. Data is derived from tests in which the various thermohydraulic effects are investigated separately in a series of tests of increasing complexity proceeding from single to two phase conditions. Drop size and velocity data are presented, for several axial positions in the bundle, which are derived from high speed cine and T.V. video recordings. It is shown that single phase flow diversion can be up to 85% for a 90% local flow blockage. In two phase conditions the majority of the water is observed to pass through the blockage. For the severe blockage there is evidence of significant coalescence and entrainment in the blockage region. The reflood tests demonstrate enhanced coolability of the blockages such that quenching at the blockage elevations occurs earlier than in an unblocked bundle provided that the blockage shapes have a representative low thermal capacity.

1. INTRODUCTION

Ballooning of the zircaloy cladding of PWR fuel pins could occur during a conservatively calculated large break Loss of Coolant accident (LOCA). A localised group of pins could balloon in a coplanar fashion to form a partial flow blockage around which emergency coolant flows could be diverted, this might lead to higher temperatures in the blockage vicinity than would occur for undistorted fuel. This paper describes experimental data obtained in a 44 rod bundle with a localised coplanar blockage under conditions relevant to the reflood phase of a LOCA. The balloons are simulated by preformed sleeves attached to the electrically heated rods and are unpressurised to avoid thermohydraulic-clad deformation interactions. The bundle is housed in a transparent silica shroud to enable photography of water drops. The aim of the investigation is to provide a data base for modelling dispersed two phase flows in the vicinity of a partial blockage. A sample of data from four blockage shapes and an unblocked bundle are given. Each bundle is subjected to a series of tests which aim to measure the various thermohydraulic effects separately as far as possible. (i) Isothermal air flow tests to quantify diversion around the blockage and compare with flow diversion models e.g. [1] (ii) Steam heat transfer tests, important since convection to steam is the dominant factor in dispersed flow heat transfer, these also enable quantification of rig specific features, such as thermal entry and grid effects that are essential to enable attention to be focussed on the drop behaviour where the major uncertainties lie. (iii) Steam and drop tests to investigate heat transfer in a well defined two phase flow. (iv) Conventional reflood tests. An extensive investigation of droplet dynamics was carried out during the steam and drop tests using T.V. video and H.S. cine techniques with back lighting. All experiments were carried out at atmospheric pressure and are fully reported in [2].

2. DESCRIPTION OF APPARATUS

The test section (figures 1 and 2) consisted of 44 electrically heated rods 12.2 mm diameter, 1 m heated length each with a maximum uniform power of 2 kw, the rod cladding was Inconel 600 with a mean thickness of 0.9 mm. Some rods had 6 mineral insulated 0.5 mm diameter thermocouples spot welded at their junctions to the inner surface of the cladding. The rods were arranged on a square pitch (P/d = 1.33) and enclosed in a cylindrical transparent silica shroud. In order to preserve the correct ratio of rod to flow-space in the periphery of the bundle 8 unheated dummy rods were installed. A grid was included at bundle mid-height in order to keep the rods in a regular array. This grid was simply a Nichrome wire (0.72 mm diameter) mesh which caused a 32.4% subchannel blockage. Air, steam or water flows could be introduced into a steel lower plenum chamber and passed over an unheated calming length of 28 equivalent diameters before the start of the heated section. For the steam and drop tests water could be injected, close to the start of the heated length, through capillary tubes located at the centre of each subchannel. Eight 0.5 mm diameter mineral insulated thermocouples were used to measure subchannel steam temperatures at various elevations in the bundle and were supplemented by three pulse-heated thermocouple steam probes. Sixteen blockage sleeves were fitted to the central 4 x 4 group of rods, see figure 1. Four different types of sleeve were used, see figures 3 and 4. The first two were machined and gave a 90% maximum subchannel blockage, and the final two were hydraulically formed at UKAEA Harwell having thin walls to give representative fuel pin balloon thermal capacities and produced maximum subchannel blockages of 90% and 61%. Sleeve temperatures were measured with 0.5 mm diameter mineral insulated thermocouples. For Blockage 1 they were set in grooves on the outer surface of the sleeve and for Blockage 2 they were spot welded to the inner surface. For the thin-walled sleeve they were brazed into the sleeve wall so that they were flush with the outer surface, this work was carried out at UKAEA Winfrith.

3. AIR FLOW TESTS

These tests were carried out with the upper plenum chamber removed and air from the laboratory compressed air supply was passed into the lower plenum. A Pitot-static tube of 2 mm o/d was introduced into the subchannels between the rods to measure the axial variation in air velocities and static pressures. Measurements were made on the undistorted bundle (no blockage sleeves) in the Reynolds number range 2000-9000. The spacer grid loss coefficient and rod skin friction factors were thus deduced without the complicating effects of flow diversion around a blockage. For the blocked bundles detailed axial Pitot traverses were carried out in the central (blocked) subchannel and one peripheral (unblocked) subchannel. The results from such a traverse for Blockage 3 is shown in figure 5. The velocities quoted are centre-of-subchannel values from which a subchannel mean was deduced and after corrections for the effect of introducing the

Pitot tube a mass flow distribution was obtained. The steep rise in velocity along the length of the blockage implies a change from turbulent to laminar flow. Figure 6 illustrates the change in radial velocity profile due to the blockage. The effect of varying Reynolds number was investigated for Blockage 1 which showed that the flow diversion pattern is the same within ±5% over the range 2900-8600. For Blockages 2-4 a single Reynolds number of about 8000 was therefore used. The single phase flow diversion was 42% for Blockage 4 and 85% for Blockage 3. These results are in good agreement with a flow distribution model [1] fitted to Blockage 1 and other data [2].

4. STEAM HEAT TRANSFER

These tests were performed using transition flow type Reynolds numbers (2000-6000) as such flows cover the majority of steam convection conditions found in reactor simulations e.g. FLECHT-SEASET [3]. The tests were at steady state with an initial temperature, usually 400°C, set at a control elevation 18 equivalent diameters from the start of the heated length. The resulting axial temperature profile for Blockage 3 is shown in figure 7 together with calculated steam temperatures which are in reasonable agreement with measured values. This agreement supports the use of the mass flow distribution inferred from the air tests. For Blockage 1 the rod temperature goes down in the region of the blockage indicating an improvement in heat transfer. For Blockage 3 a corresponding heat transfer enhancement at entry is present but the small thermal capacity of the steam leads to a rapid rise in throat temperature. There is a similar but smaller rise for the Blockage 4. Figure 8 compares heat transfer coefficients for Blockage 3 with those for the undistorted bundle. For the undistorted bundle the Nusselt number was found to vary as $Re^{0.45}$. The fully developed Nusselt numbers in the blockage throat tend to a value of about 4.2 which is similar to values given by laminar flow correlations for a circular tube. It is somewhat lower than the value of 7.8 for fully developed laminar flow in a rod bundle. The results from a range of blockages show that heat transfer to steam is good so that a short blockage improves cooling despite flow diversion. An axially extended blockage, however, can lead to temperature penalties due primarily to the limited thermal capacity of the steam.

6. STEAM AND DROP TESTS

Some drop dynamics results for Blockage 4 are illustrated in figures 9-12 obtained from tests in which water was injected into the centre subchannel only in order to simplify the field of view. Rod temperatures were similar to those in figure 15 and the inlet steam and water superficial velocities were 12 m/s and 5 mm/s. A Sauter mean diameter (SMD) and momentum averaged velocity was calculated for each data set. Capillary water jet break up produced drops of uniform velocity (2.7 m/s) in the size range 0.93 ± 0.22 mm diameter with a few small satellite drops which were soon absorbed into larger drops. Below the grid at the 290 mm elevation the drops have accelerated to 3.45 m/s and coalescence has increased the SMD to 1.49 mm. The maximum drop size was about 2 mm and when coalescence produced larger drops (rarely) they rapidly broke up. Drop impact at the spacer grid produced a large increase in the number of small drops. Most drops penetrated through the blockage and only a limited number of the smallest drops (<0.5 mm) has significant transverse velocities. On passage through the blockage only the largest drops appear to break up with a corresponding fall in the SMD. It is interesting to note that for Blockage 3 (x 4 smaller flow area) many slow moving small drops were observed above the blockage. Once again there was no significant diversion of water or dramatic change in the drop size distribution in passage through the blockage. We believe that there are significant agglomeration events in or above the throat of Blockage 3 together with entrainment of small drops which are not observed for Blockage 4.

Figures 13-16 compare some representative results for Blockages 3 and 4, with water injection into all subchannels, at two steam superficial velocities (time point A = 5.7 m/s and time point B = 12.6 m/s) and an equivalent reflood rate of 9.8 mm/s. It is interesting to note that at the lower steam flux conditions better cooling is observed for the 90% blockage whereas at the higher steam flux it is the 61% blockage which is better cooled. This and other data suggests that the good cooling observed in the 90% blockage is the result of an enhanced liquid fraction which is reduced when the steam flow increases. We expect these data to provide a critical test for models of dispersed flow cooling, our own model is described in [4].

6. REFLOOD

Forced flow, constant power reflood tests were carried out after setting the bundle to a steady temperature at the control elevation. To simulate decay heat power levels in some tests we added 1 kw to all rods prior to initiating reflood. The results illustrated in figures 17-20 for Blockages 2-4 are for an initial temperature of $800^{o}C$ and a reflood rate (water close to saturation temperature) of 20 mm s^{-1}. Quench times for rod and sleeve thermocouples relative to the quench time for the lowest thermocouple in the bundle (from the same rod in all builds) are shown versus elevation from the start of the heated length. For Blockage 2 the sleeves quench much later than the peripheral or by-pass rods which are in good agreement with undistorted bundle data. FEBA Series 3 results [5] show a similar delay in sleeve quenching. For Blockage 3 the sleeves quench (from the top downwards) much earlier than the by-pass due to their much lower thermal capacity. When 1 kw is added to each of the Blockage 3 rods just prior to reflood the sleeves again quench downwards but at time much closer to the by-pass rods and some central rods above the blockages quench later than the by-pass. In the THETIS tests [6] also with low thermal capacity sleeves quenching of the blockage occurs upwards from the bottom of the sleeves at about the same time as the by-pass. Finally for Blockage 4 the sleeves quench very early compared with the by-pass and above the region of the blockage all rods quench earlier than the undistorted bundle. Similar early rewetting is observed in FEBA Series 4 tests.

7. CONCLUSIONS

Air flow tests have been used to obtain estimates of blockage mass flow diversion which lead to calculated steam flow temperatures in reasonable agreement with measurements. This flow diversion has been modelled successfully using a simple model which is now being extended in particular to cover 2 phase conditions. Single phase steam flow tests have shown that at 90% blockage a temperature penalty results. From steam and drop heat transfer tests we observe that blockage cooling depends on steam velocity and can fall for a rise in steam velocity, an effect which we associate with a corresponding drop in throat liquid fraction. Analysis of these tests is continuing. For representatives steam velocities (15 m/s) we do not find that blockages cause much nett drop break up and that only a few small drops are diverted around a blockage. Under reflood conditions all blockages tested have quenched well before the arrival of the main bundle quench front, except for Blockage 2 which had an unrepresentative large thermal capacity.

REFERENCES

[1] ARDRON, K.H. and FAIRBAIRN, S.A. Nuc. Eng and Des. 72, 391-409 (1982)
[2] FAIRBAIRN, S.A. and PIGGOTT, B.D.G. CEGB Report TPRD/B/0458/N84, TPRD/B/0510/N84, TPRD/B/0511/N84 (1984)
[3] LEE, N. et al. Westinghouse Corp.Report WCAP-9891 (1981)
[4] CLARE, A.J. and FAIRBAIRN, S.A. Paper IPA2, presented at this Seminar
[5] IHLE, P. and RUST, K. KfK Reports 3657/8/9 (1984)
[6] PEARSON, K.G. et al. UKAEA Report AEEW-R (1983)

ACKNOWLEDGEMENTS

This paper is published by permission of the Central Electricity Generating Board and the work was partly supported by the Commission of the European Communities under the auspices of its cost sharing programme for LWR safety research.

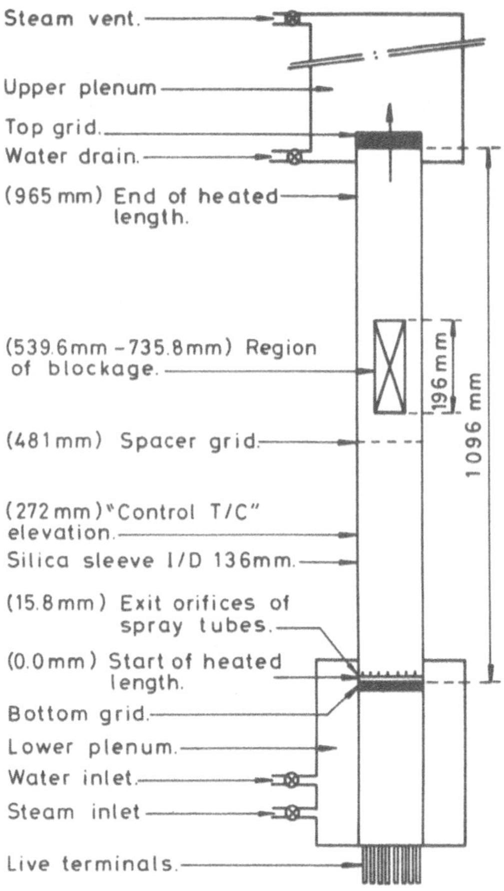

FIG.1. Test Section Details.

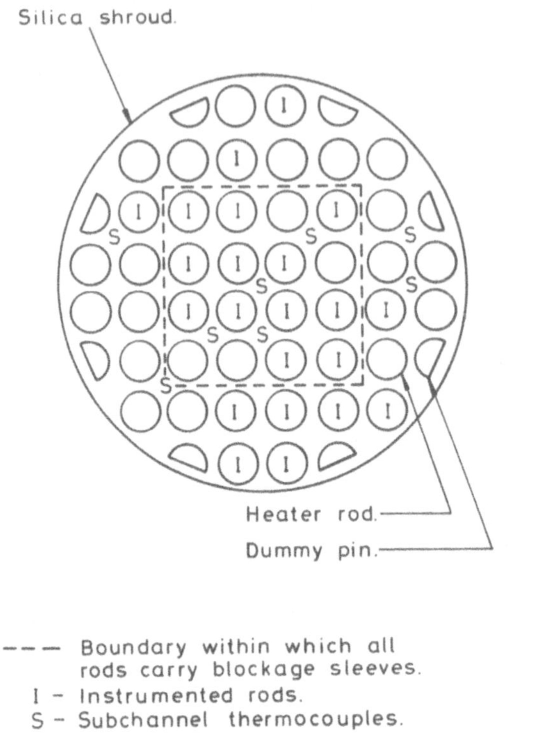

FIG. 2. Cross Section of Test Bundle.

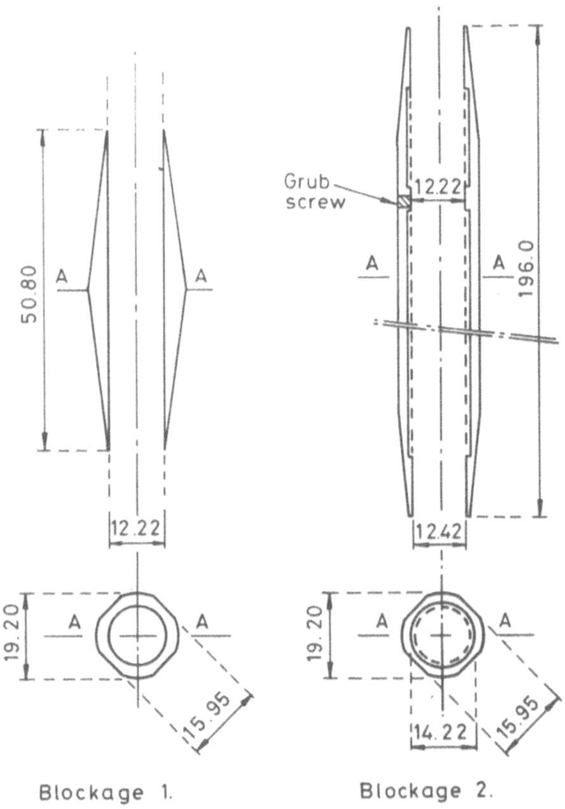

FIG. 3. Sleeve Details.

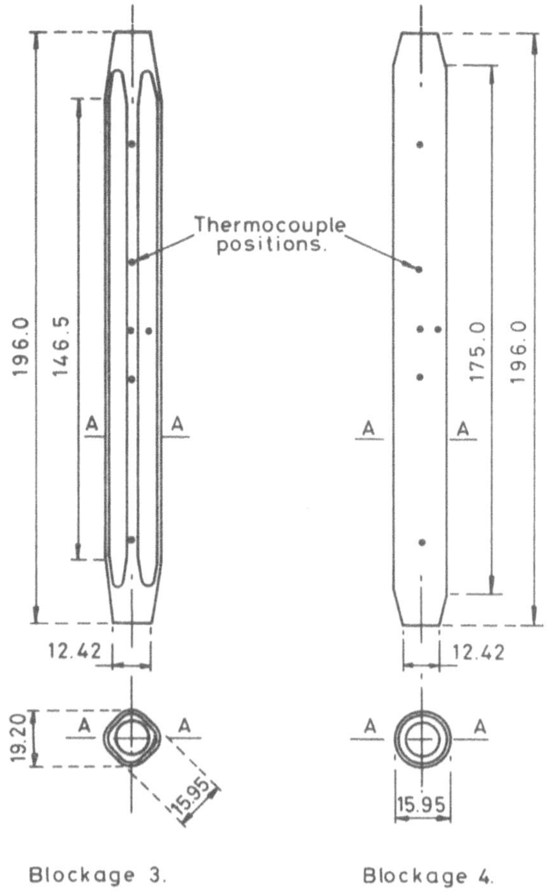

FIG. 4. Sleeve Details.

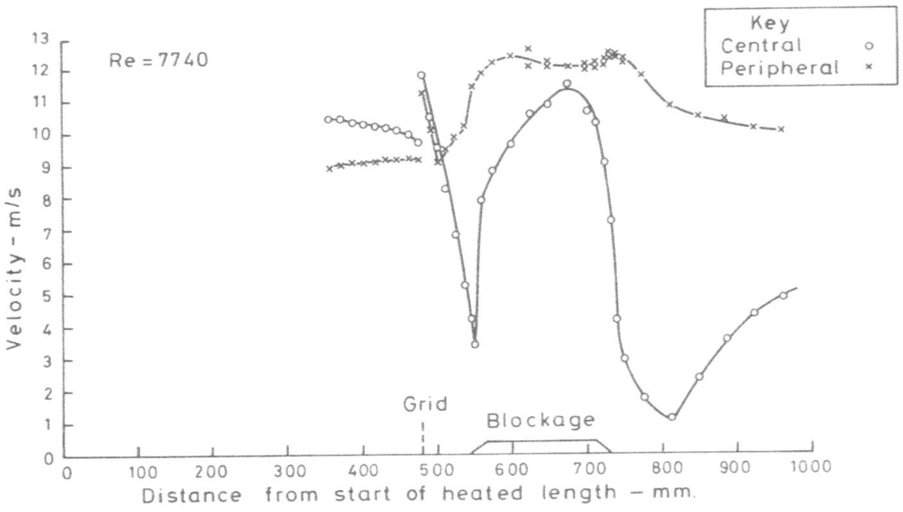

FIG.5. Axial Velocity Profiles for Blockage 3.

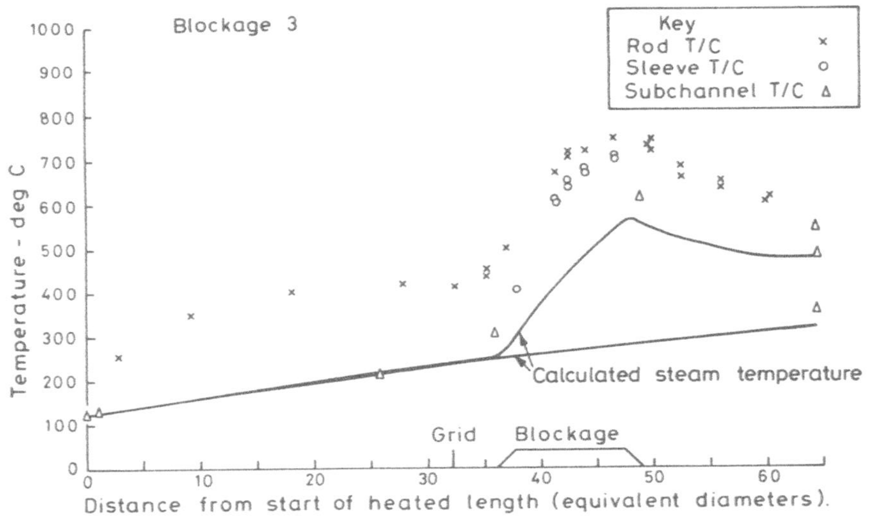

FIG.7. Axial Temperature Profile for Steam Test.

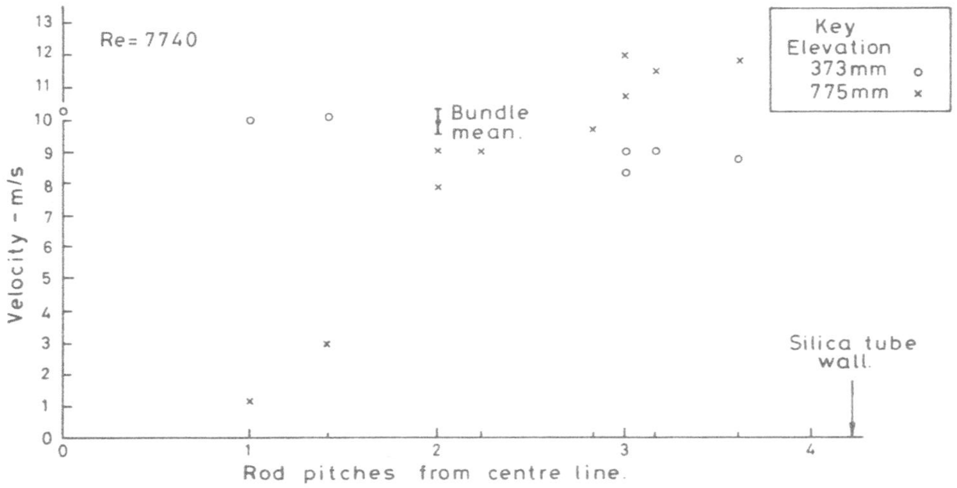

FIG.6. Radial Velocity Profiles for Blockage 3.

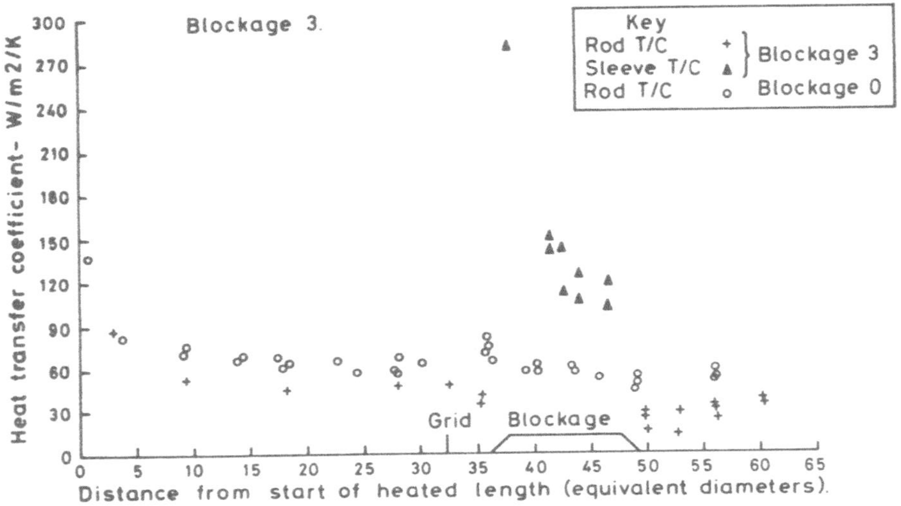

FIG.8. Axial Heat Transfer Pattern Steam Tests.

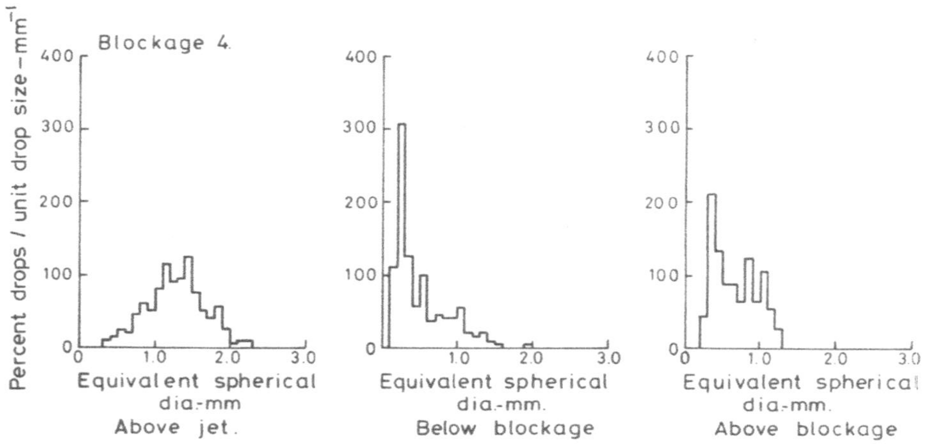

FIG.9. Drop Sizes from H.S. Movie of an 800°C Steam and Drop Test.

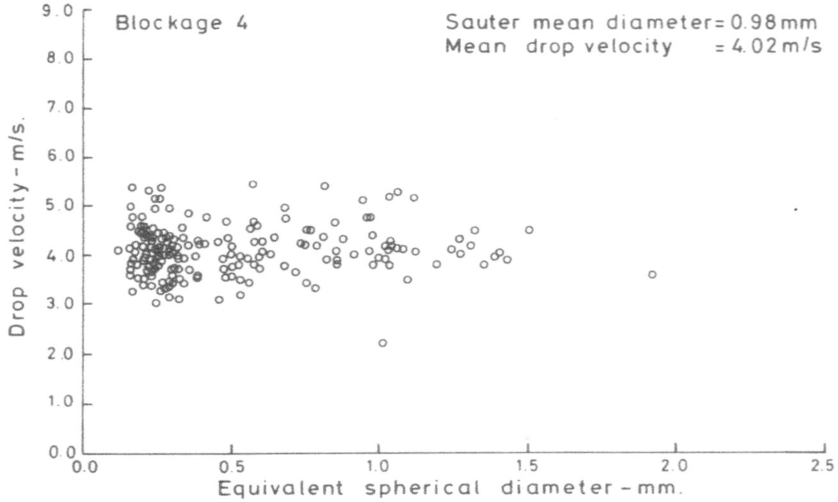

FIG.11. Drop Size and Velocity from H.S. Movie (Below Blockage: 525 mm from Start of Heated Length.)

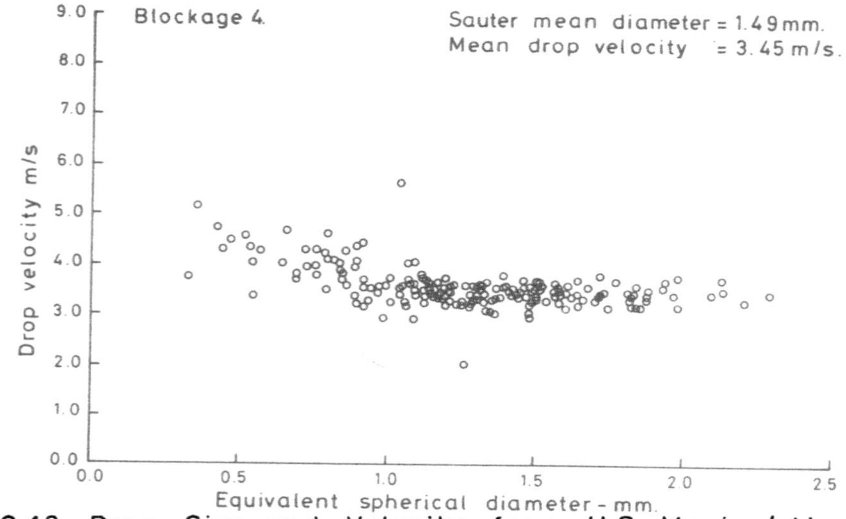

FIG.10. Drop Size and Velocity from H.S. Movie (Above Jet: 290 mm Above Start of Heated Length.)

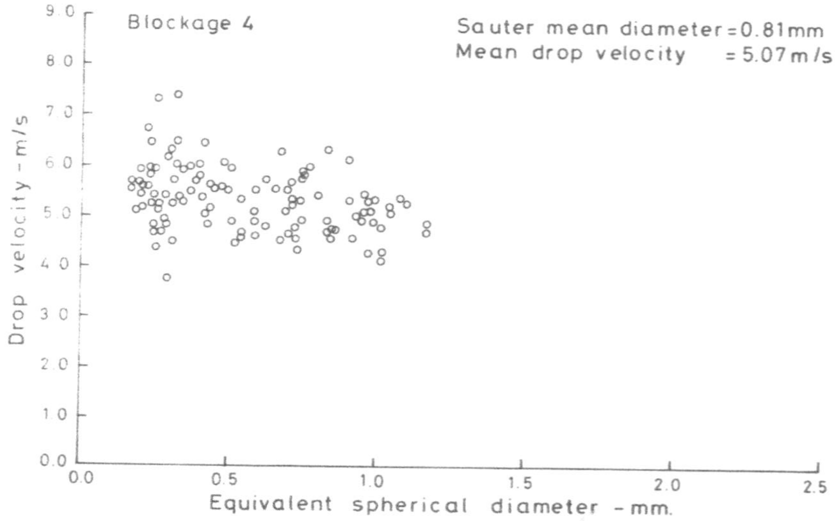

FIG.12. Drop Size and Velocity from H.S. Movie (Above Blockage: 750 mm from Start of Heated Length.)

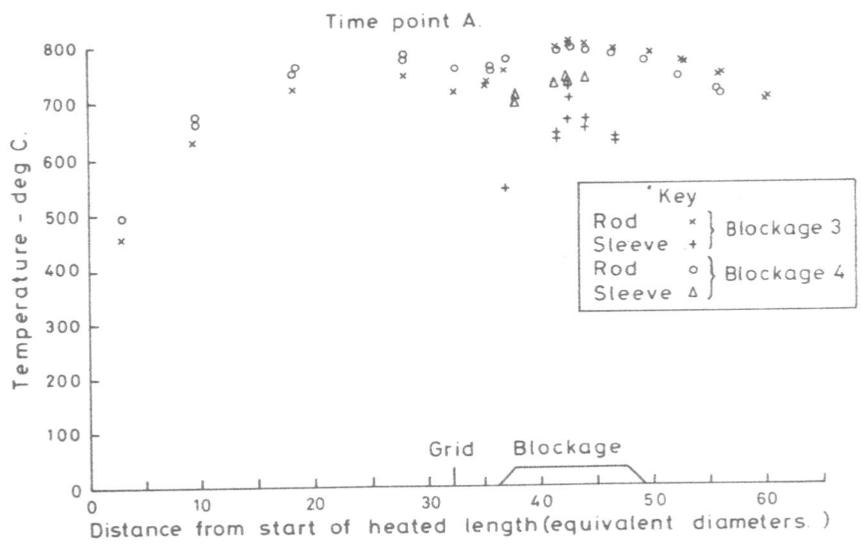

FIG. 13. Axial Temperature Profile for Steam and Drop Test.

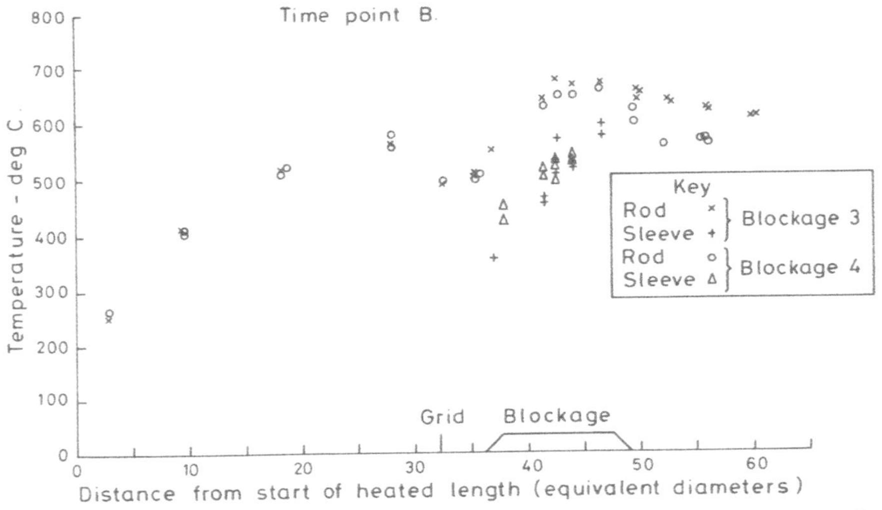

FIG.15. Axial Temperature Profile for Steam and Drop Test.

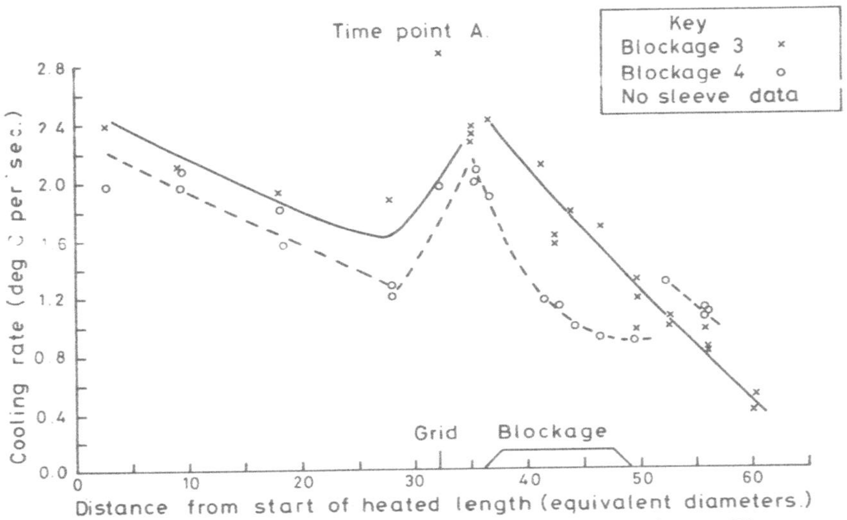

FIG.14. Axial Variation in Cooling Rates for Steam and Drop Test.

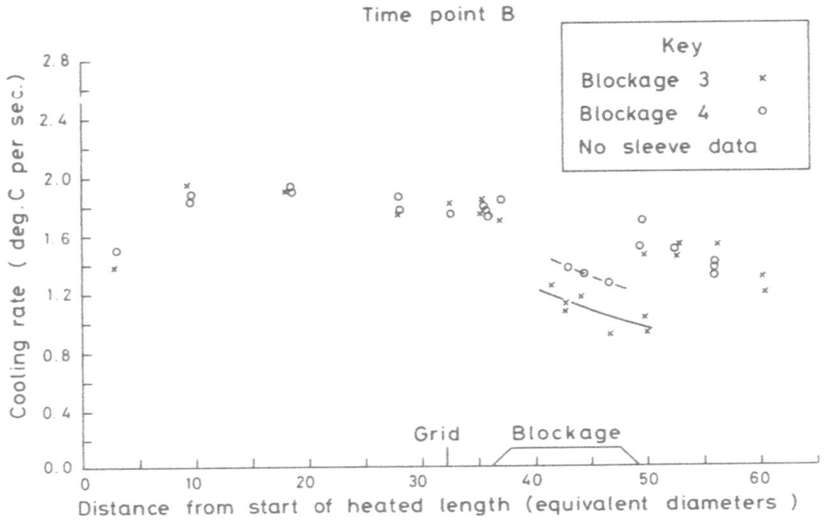

FIG.16. Axial Variation in Cooling Rates for Steam and Drop Test

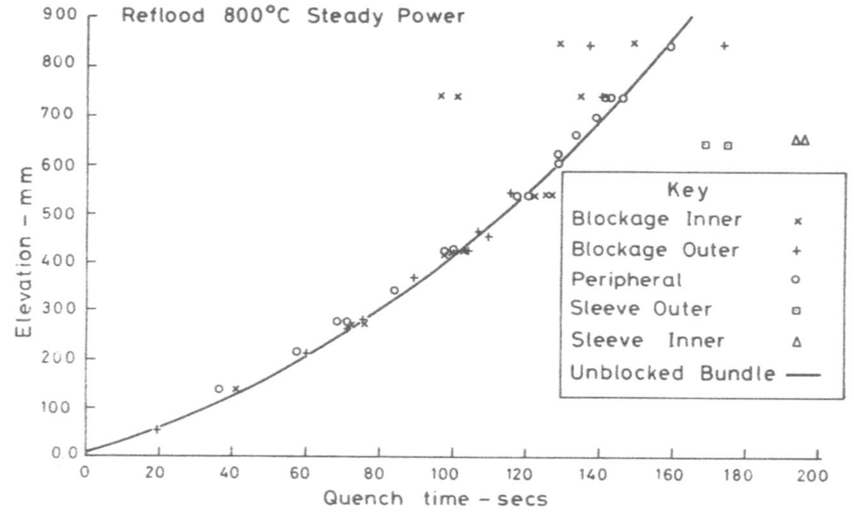

FIG.17 Quench Time Versus Elevation for Blockage 2.

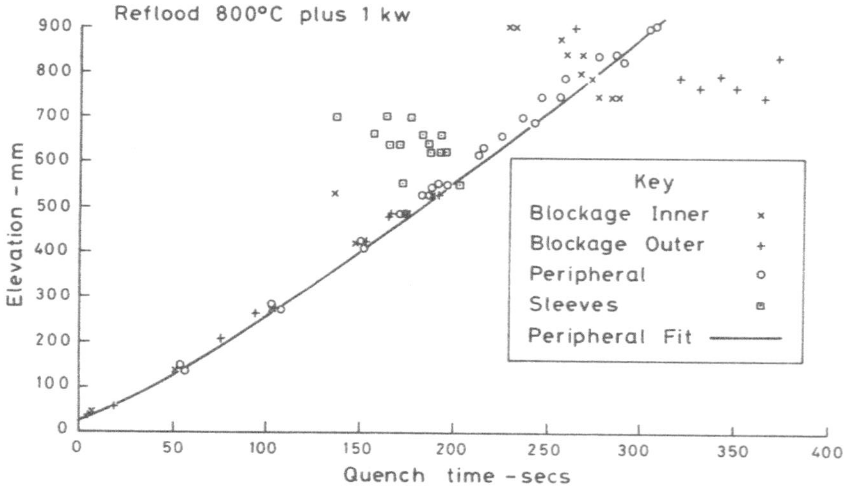

FIG.19. Bundle Quench Times Versus Elevation for Blockage 3.

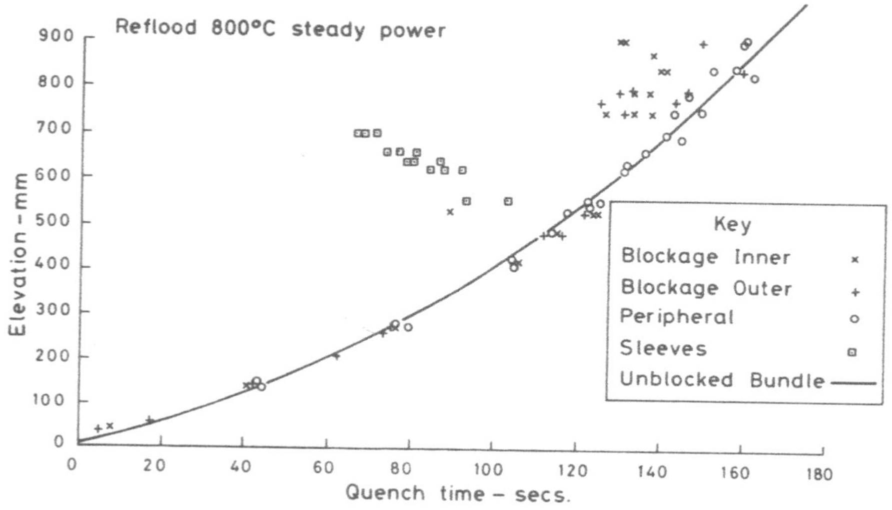

FIG.18. Quench Times Versus Elevation for Blockage 3.

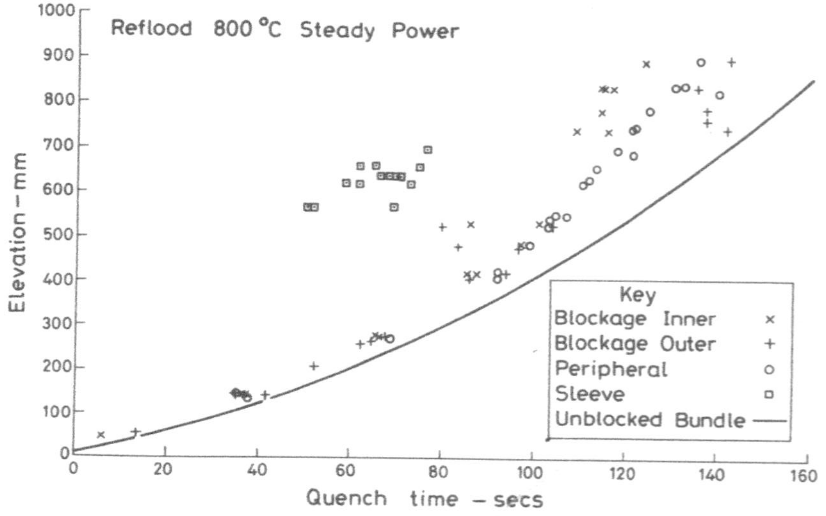

FIG. 20. Quench Time Versus Elevation for Blockage 4.

iscussion

uestion: ·R. Deruaz - CEA/CEN-G

mong others an interesting result of your work is that no gross liquid iversion takes place around the blocked area. Did you succeed in ιeasuring and/or calculating also steam by-pass, and if yes, does this team by-pass depend upon the ratio of the flow area blockage?

ιnswer: BDG Piggott - CEGB/BNL

le could not measure steam flows within the bundle, but we could calculate ιy-pass steam flow for single phase conditions by using the air flow test :esults. We would expect by-pass steam flow to increase as blockage :raction is increased, but for our bundle the effects will not be ιensitive due to the relative large by-pass flow area compared with the ιlockage flow area.

Question: D. Mewes, U/Hannover

?lease, can you give us some information about the method you used to neasure drop-sizes.

Answer: BDG Piggott, CEGB/BNL

Backlighting was used to produce an accurate drop image in silhouette on the film. The maximum drop chard length and the maximum length at right angles were measured from projected images and used to calculate an equivalent, spherical diameter. It is to be noted that only small drops were spherical.

Question: D. Mewes, U/Hannover

Do you have some indications about the mechanism of droplet growth near the blockage and its relation to heat transfer?

Answer: BDG Piggott - CEGB/BNL

Small drops were accelerated preferentially and were absorbed into slower moving large drops. The level of drop interaction depends on test conditions, but large drops or slugs formed and broke up due to aerodynamic forces. It was not possible in these experiments to relate heat transfer to drop size, as a model for drop cooling is needed. This will be presented in the next paper. Blockage cooling did, however, depend on the liquid fraction in the vicinity of the blockage.

Question: E. Skupinski, CEC/BXL

Is the problem of cooling of blockages settled now?

Answer: BDG Piggott, CEGB/BNL

These experiments show good coolability of a severe (90%) blockage if the balloon low thermal capacity is correctly modelled. The experiments were aimed at providing data on drop dynamics, etc. and were not intended to be prototypic. We must look at other data such as that from the Thetis facility to be presented later in this Seminar to answer the question.

Parallel Sessions Research Area A

The loss of coolant accident (LOCA) and the functioning and performance of the emergency core cooling system (ECCS)

Session A2

Chairman: R. Deruaz (CEA/CEN-G)

Session secretary: E. Skupinski (CEC/BXL)

DROPLET DYNAMICS AND HEAT TRANSFER IN DISPERSED TWO PHASE FLOW

A.J. CLARE and S.A. FAIRBAIRN
Central Electricity Generating Board
Berkeley Nuclear Laboratories
Berkeley,
Gloucestershire,
UK.

Summary

A mechanistic model of heat transfer in a dispersed steam and droplet flow has been compared with experimental data from a multi-rod facility designed to allow detailed measurements of droplet size and velocity distribution. For the conditions analysed the droplet velocity was reasonably well described by a single drop equation of motion, but agreement would be improved by incorporation in the model of observed droplet coalescence and breakup processes. Nevertheless, the steam desuperheating effect of the drops was well predicted by use of the Lee-Ryley steam-drop heat transfer correlation. It was found that enhancement of steam convective cooling by droplet-induced turbulence can be significant and modelling refinements in this area are needed.

1. INTRODUCTION

Heat transfer in dispersed flow film boiling conditions is important in the cooling of a PWR core during the reflood stage of a postulated LOCA. In recent years some advanced reflood codes have incorporated mechanistic non-equilibrium models of dispersed flow in which the dynamics and heat transfer for each phase are considered separately. The success of such methods depends to a large extent on a good calculation of droplet size and velocity along the flow path, since these determine the interfacial area and heat transfer coefficient and hence the evaporation rate. Consequently it is important to compare predictions with experimental observations of droplet behaviour which are sufficiently detailed to indicate whether droplet sizes and velocities change due to interactions between drops or with obstacles such as grids in the flow path. Additionally, both wall heat flux and vapour temperature measurements are needed so that the evaporation rate model can be properly assessed.

This paper describes a mechanistic model of dispersed flow film boiling developed to analyse experiments performed in a multipin cluster facility at Berkeley Nuclear Laboratories. The facility, the range of experiments carried out in it and a sample of the data obtained from it are reported in a separate paper (1) to this seminar. The analysis has so far concentrated on tests in a cluster incorporating a region of 90% flow blockage, although since the model represents only undistorted channels conditions in the blocked region are not calculated.

The emphasis has been to assess the adequacy of a relatively simple droplet model based on a single average drop size and neglecting drop size changes arising from breakup or coalescence processes. Measurements of drop size and velocity distributions at several elevations in the cluster enable the droplet dynamics modelling to be checked. The droplet contribution to heat transfer at the pins is usually considerably smaller than the steam heat transfer contribution so the latter needs to be calculated accurately. For this reason data from preliminary steam heat transfer tests has been used instead of standard correlations so that rig specific thermal entry and grid effects are taken into account. It has been recognised, however, that the presence of droplets may enhance the level of turbulence in the steam flow so that single-phase data may underestimate the convective wall heat transfer in dispersed flow. The paper discusses possible ways of modelling this effect and comments on its significance in these tests.

2. DESCRIPTION OF THE MODEL

2.1 Basic Assumptions

In its present stage of development the model calculates the cooling rate of a single unballooned pin given the inlet fluid conditions and the axial temperature profiles of the rod and its neighbours. The time rates of change of all parameters in the tests analysed are sufficiently low to permit quasi-steady state equations to be employed. The spatial variations are treated by a marching method which proceeds stepwise from inlet to outlet.

The control region of the model, shown in figure 1, is centred on the rod whose cooling rate is to be calculated and includes the fluid of the surrounding four subchannels and the facing elements of the neighbouring pins.

All the droplets are assumed to have the same size and velocity. Their initial velocity is derived from the inlet liquid mass flow rate and the injector internal diameter (d_j). The initial drop size (d_o) is calculated from the wavelength of maximum instability of a capillary jet given by Chandrasekhar (2) as

$$\lambda = \frac{\pi d_j}{0.697},$$

$$\text{giving} \quad d_o = \left(\frac{3}{2} d_j^{\,2} \lambda\right)^{1/3} = 1.891\, d_j. \qquad (1)$$

This agrees well with measurements of the drops produced by the injector. Changes in droplet size resulting from coalescence or breakup processes are not modelled so that only small drop size changes are calculated as a result of evaporation.

2.2 Basic Equations

The basic conservation equations solved are:

(i) mass conservation,

$$W_L + W_V = \text{constant}; \qquad (2)$$

(ii) momentum conservation for each phase,

$$\frac{1}{A}\frac{d}{dz}(W_L u_L) = -(1-\alpha)\frac{dP}{dz} + F_D - F_{GL} - (1-\alpha)\rho_L g \qquad (3)$$

and

$$\frac{1}{A}\ \frac{d}{dz}\ (W_V u_V) = -\ \alpha\ \frac{dP}{dz} - F_D - F_W - F_{GV} - \alpha\rho_V g, \tag{4}$$

where W_L, W_V are the subchannel mass flow rates of the liquid and vapour phases and where the F_i (described below) are forces per unit volume of the channel accounting for interphase drag (F_D), wall friction (F_W) and momentum losses at the grid (F_{GL}, F_{GV}) where they occur;

(iii) energy conservation,

$$\frac{d}{dz}\left[W_L(h_L + \frac{1}{2}\ u_L^2) + W_V\ (h_V + \frac{1}{2}\ u_V^2)\right] = Q_f, \tag{5}$$

where Q_f is the total heat transferred to the fluid per unit length of channel. An equation for the evaporation rate,

$$\frac{dW_V}{dz} = Q_L/h_{fg} \tag{6}$$

where Q_L is the heat transferred to the (saturated) droplets per unit length of channel, completes the set of basic equations. At each space step solutions to these equations are obtained by an implicit numerical scheme.

2.3 The Momentum Terms

Interphase Drag

The drag force on a single droplet in the vapour stream is

$$f_D = C_D\ \frac{\pi d^2}{4}\ \frac{1}{2}\ \rho_V(u_V - u_L)\ \left|u_V - u_L\right|, \tag{7}$$

and since there are

$$N_D = \frac{6(1-\alpha)}{\pi d^3} \tag{8}$$

droplets per unit volume, the drag force per unit volume is

$$F_D = N_d f_D = C_D\ \frac{3}{4}\ \frac{1-\alpha}{d}\ \rho_V(u_V - u_L)\ \left|u_V - u_L\right|. \tag{9}$$

The drag coefficient C_D is computed as a function of the droplet Reynolds number

$$Re_d = \frac{\rho_V d}{\mu_V}\ \left|u_V - u_L\right| \tag{10}$$

according to empirical results for smooth spheres.

Grid Model

The grid at 481 mm elevation is a simple wire mesh of small axial extent; it introduces a form loss into the vapour momentum equation,

$$F_{GV} = K_G \frac{1}{2} \rho_V u_V^2 \qquad (11)$$

where the coefficient $K_G = 0.53$ is obtained from the grid pressure drop measured in air flow tests. The momentum loss of the liquid phase is given by

$$F_{GL} = f_{imp} \cdot x_{mom} \cdot W_L u_L, \qquad (12)$$

where f_{imp} is the probability of a drop impacting on the grid wire, calculated as a function of grid geometry and incident drop size, and x_{mom} is the mean fractional momentum loss of a droplet as a result of impact. It is assumed that for a given impacting drop the fraction of the incident axial momentum which is normal to the surface of the grid wire at each point is destroyed. The momentum loss is averaged over all displacements, relative to the grid, of impacting drops, and the mean fractional momentum loss of a drop is found to be

$$x_{mom} = \frac{2\, d_g}{3(d_g + d)}, \qquad (13)$$

where d_g is the diameter of the grid wire. The grid is assumed to wet and to desuperheat the steam by evaporation of the liquid film. A heat transfer coefficient appropriate to cylinders in cross flow is employed.

2.4 Heat Transfer

Wall/Steam Convection

Since the main objective of analysing the steam and droplet tests was to assess the modelling of the droplet contribution to the overall heat transfer it was necessary to minimize errors in calculating the steam cooling contribution. Therefore, the functional dependence of the steam convective heat transfer coefficient used in the model was that derived from measurements in steam cooling tests in the unblocked cluster. The variations of Nusselt number both with axial location in the cluster and with Reynolds number are reported in detail in (3). Use of these results in the model effectively bypasses the difficulties of predicting the facility - dependent entry and grid effects, and minimises the uncertainties associated with transition region Reynolds number dependence.

Droplet-induced Enhancement

The presence of drops in the steam flow provides a source of turbulence additional to that generated by wall shear, and this will enhance the steam convective heat transfer as deduced from steam only experiments. The calculation of this enhancement is problematic, and since the steam convective heat transfer component is usually dominant this uncertainty could become very significant if the droplet number density were high.

If the steam flow is fully turbulent the convective heat transfer is determined within a thin boundary layer at the wall. In this case, since the boundary layer will be considerably smaller than the mean drop size, the boundary layer will be mainly affected by drop-induced turbulence in the free stream, rather than by submergence of the drops in the boundary layer. Blair (4) has investigated experimentally the effect of free stream turbulence intensity on convective heat transfer and has expressed the results in the following modification of Reynolds analogy:

$$St = \frac{f_w}{2} (1.18+1.3\ T), \tag{14}$$

where T (= $\sqrt{u'^2}/u$) is the turbulence intensity of the free stream. Similarly Simonich and Bradshaw (5) report an enhancement of the skin friction coefficient with increasing turbulence intensity in the form

$$f_w = f_{w_o} (1+a\ T) \tag{15}$$

where a is length scale dependent and has a value a ≃ 2 if the length scale is comparable with the boundary layer thickness.

The prediction of turbulence intensities in two-phase dispersed flows has been treated by Theofanous and Sullivan (6) who propose that the familiar relationship between turbulence velocity, shear velocity and shear stress

$$u' \simeq u_w^* = \sqrt{\frac{\tau_w}{\rho}}$$

remains valid if τ_w is replaced by the total shear stress on the continuous phase. Theofanous and Sullivan introduced a turbulence energy distribution parameter δ, and found that a value δ=2 resulted in predicted turbulence intensities in good agreement with measured values in both dispersed bubble and dispersed particle flow experiments. Their analysis for mechical equilibrium flows has been adapted for the conditions of interest here, where droplets may not be travelling at their terminal velocity, to give

$$T^2 = \frac{f_w}{2} (\alpha+\delta(1-\alpha)) + \delta \frac{F_D\, d_h}{4\, \rho_V u_V^2} . \tag{16}$$

Equations (14), (15) and (16) are used, together with equation (9), in the standard form of the model to calculate an enhanced steam convective heat transfer coefficient. For single phase conditions the values of f_w required for agreement with the measured heat transfer coefficient are in excellent agreement with values derived from air flow tests; this gives some confidence in the validity of the approach. It is anticipated that at lower flows, when conditions may not be fully turbulent, and the boundary layer may occupy a significant fraction of the subchannel area, this model will increasingly underestimate the enhancement since the direct interaction of droplets with the boundary layer will no longer be neglible.

A simple alternative model which can be expected generally to overestimate the heat transfer enhancement in the transition region treats the interphase drag as a direct augmentation of the wall shear stress. In this case an effective friction faction is obtained from the total shear stress on the gas phase:

$$f = \frac{d_h (F_w+F_D)}{2\, \rho_V u_V^2} . \tag{17}$$

The effect of the drops is treated as an effective increase in wall roughness with a "rough wall" friction factor given by equation (17).

Kays (7) has suggested that the reported heat transfer effect of rough walls is well represented by the relationship

$$\frac{h}{h_{smooth}} \quad \alpha \quad \sqrt{\frac{f}{f_{smooth}}} \ , \tag{18}$$

rather than the normal Reynolds analogy. Equations (17) and (18) provide a simple, "rough wall" model of the droplet enhancement of steam convective cooling.

Steam/Droplet Convective Heat Transfer

The convective heat transfer between the phases, per unit volume, is given by

$$q_{VLC} = h_{VLC} \ \frac{6(1-\alpha)}{d} \ (T_V - T_{sat}). \tag{19}$$

A number of correlations of steam/droplet heat transfer coefficient have been reported, of which two have been used in the model to examine the sensitivity to this parameter. These are the Lee-Ryley (8) correlation

$$Nu_{VLC} = 2 + 0.74 \ Re_d^{\ 0.5} \ Pr_V^{\ 0.33} \tag{20}$$

evaluated using free stream properties, and the Yuen-Chen (9) correlation

$$Nu_{VLC}(1+B) = 2 + 0.6 \ Re_d^{\ 0.5} \ Pr_V^{\ 0.33}, \tag{21}$$

where

$$B = (h_v - h_g)/h_{fg},$$

and where properties are evaluated at film conditions with the exception of the density in Re_d which is the free steam density. The Lee-Ryley correlation is used in the standard form of the model.

Radiation

In the radiation model the central rod of the control volume (Fig. 1) is considered to be surrounded by a cylindrical surface having an area equal to the included elements of the neighbouring rods and a mean temperature corresponding to their averaged radiative power. Radiative absorption by the steam is described in terms of a mean beam length (taken to be 0.85 of a subchannel hydraulic diameter) and an attenuation coefficient which treats the steam as an optically thin grey medium. The liquid component is similarly represented as an optically thin medium with an attenuation coefficient given by

$$a_L = \alpha_{eff} \cdot \frac{\pi d^2}{4} \cdot N_d \tag{22}$$

where N_d is given by equation (8) and the effective absorptance of the drops ($\alpha_{eff} = 0.9$) is taken from Harpole (10). Axial radiative heat transfer is ignored, as is axial conduction.

2.5 Flow Diversion

In the axial region of the blockage the flow in the peripheral unblocked part of the rod bundle is enhanced by diversion of flow from the central blocked region. In the present model this is represented by a step increase in the peripheral steam flow at the elevation where the blockage begins, and a step decrease at the end of the blockage. The magnitude of the diversion is calculated on the basis of the flow areas in the blocked and peripheral regions, and is applied to the steam component only. This approximation is a good one for the peripheral region since the blockage flow area is a small proportion of the total, but the analysis of blockage throat conditions requires a proper calculation of the flow diversion. Such a calculation is currently under-development.

3. COMPARISON OF MODEL WITH EXPERIMENT

3.1 Droplet Dynamics

Photographic studies have been carried out in tests with water injection into only the central subchannel in order to simplify analysis of the highspeed cine films. The data shown in figure 2 are taken from detailed drop size and velocity distributions, reported in ref. (11), for tests at a steam flow of $\sim$6.5 kg m^{-2} s^{-1} and single injector flow of 0.675 gm s^{-1}.

Droplet size and velocity calculated by the model for the region upstream of the blockage in the central subchannel are also shown in figure 2. It can be seen that the calculated reduction in drop size due to evaporation is neglible, but the actual changes in drop size due to coalescence upstream of the grid and breakup at the grid are considerable. Acceleration of the droplets in the model is somewhat more rapid than observed in the tests and this is attributed to the absence of droplet coalescence in the model. The errors of small drop size and large drop velocity partially compensate in the calculation of convective heat transfer to the drops, but the drop size effect is the more powerful.

It is also apparent from figure 2 that the calculated droplet momentum loss at the grid is large compared with the small loss indicated by the measurements. In spite of these discrepancies the calculated droplet velocities are within ± 15% of measured values and this is encouraging in view of the simplifications of single drop size and no interactions made in the model.

3.2 Heat Transfer

Pin cooling rates and steam temperatures calculated by the standard form of the model are compared with experimental data in figures 3 and 4 for a test with clad temperature $\sim$800°C, and steam and droplet flows at inlet equivalent to a reflood rate of $\sim$10 mm s^{-1}. Additional calculations are presented in figure 5 to illustrate the sensitivity to droplet heat transfer modelling, and in figure 6 to show the sensitivity to turbulence enhancement modelling.

It can be seen that the standard model prediction of rod cooling rate and steam temperature is generally good, particularly for the centre rod calculations. For these cases, where uncertainties in steam convective cooling and radiation cooling are small, the good agreement between calculated and measured steam temperatures implies that droplet evaporation is well predicted. For the peripheral rods in the region upstream of the blockage, the calculated cooling rates at both times are high by up to 25%. It is apparent from the 'no drops' result in figure 5 that the error is mainly in the steam cooling and/or radiation heat loss calculations. The source of the error is not yet certain, but the radiative heat losses for peripheral rods are considerable in the test considered here and are possibly not well modelled. Bundle steam velocity non-uniformities are also neglected.

It is clear from figure 5 that the droplet contribution to rod cooling is small at low elevations but becomes very considerable in the upper half of the bundle. In part this arises from the decreasing importance of bundle entry effects on steam convective cooling, but it also reflects the progressive increases in steam temperature and phase velocity difference which lead to a corresponding increase in droplet evaporation. The comparison in figure 5 of different droplet heat transfer models shows differences of rod cooling rate comparable with the experimental scatter, though the outlet steam temperature is better predicted by the Lee-Ryley correlation used in the standard form of the model.

Figure 6 shows that the effect of drop-induced enhancement of convective rod cooling is small but not negligible. For the droplet flow of this experiment (1.35 gm s^{-1} per subchannel) the enhancement of the convective component lies in the range 10%-20% depending on the model used. Discrimination between the models considered here would require data from experiments with considerably greater drop induced turbulence.

The droplet and steam flows in the experiment described above are typical of PWR reflood simulation experiments and the model results presented here indicate that for these conditions the modelling uncertainties associated with droplet related heat transfer are not large in the unblocked part of the bundle. However, it is expected that in the blocked region droplet concentrations will rise by up to a factor 10 since only the smallest droplets will be diverted into peripheral subchannels. In these conditions the modelling uncertainties both of droplet heat transfer and droplet induced turbulence enhancement will be very significant and prediction of rod cooling in the blockage throat is expected to be considerably more difficult as a result.

4. CONCLUSIONS

Comparison of predictions of a detailed droplet dispersed flow heat transfer model with measured rod cooling rates and droplet behaviour in a rod bundle has led to the following conclusions:

(i) A droplet equation of motion representing non-interacting drops of uniform size has calculated drop velocities correct to within 15% of measured values. Future improvements should include a refinement of the model of momentum losses at grids, and representation of observed droplet coalescence and breakup processes.

(ii) Rod cooling rates were well modelled in the centre of the cluster and steam temperatures were well predicted. This indicates that the Lee-Ryley correlation used for steam-drop heat transfer provides an adequate prediction of the desuperheating effect of the drops. Residual errors in predicted cooling rates for peripheral rods are probably accounted for by neglect of bundle radial steam velocity profiles and inadequate modelling of radiative losses.

(iii) Enhancement of steam convective cooling by droplet-induced turbulence was found to be small but significant at the low droplet flux of these tests. Currently proposed models differ by a factor 2 in predicting the effect, but discrimination between them requires experimental data for higher liquid fraction and /or higher steam flows.

ACKNOWLEDGMENT

This paper is published by permission of the Central Electricity Generating Board.

REFERENCES

(1) FAIRBAIRN, S.A. and PIGGOTT, B.D.G., (1984), Paper to this Seminar, Session A1.
(2) CHANDRASEKHAR, S., (1961), Hydrodynamic and Hydromagnetic Stability, Oxford University Press.
(3) FAIRBAIRN, S.A. and PIGGOTT, B.D.G., (1984), Flow and Heat Transfer in PWR Rod Bundles in the Presence of Blockage due to Clad Ballooning; Experimental Data Report - Part 1. CEGB Report TPRD/B/0458/N84.
(4) BLAIR, M.F., (1983), "Influence of free-stream turbulence on turbulent boundary layer heat transfer and mean profile development, part II - Analysis of Results", Trans ASME J Heat Transfer, vol. 105, pp 41-47.
(5) SIMONICH, J.C. and BRADSHAW, P., (1978), "Effect of free stream turbulence on heat transfer through a turbulent boundary layer", ASME Journal of Heat Transfer, vol 100, No. 4, pp 671-677.
(6) THEOFANOUS, T.G. and SULLIVAN, J., (1982), Turbulence in two-phase dispersed flows, J. Fluid Mech., vol. 116, pp 343-362.
(7) KAYS, W.M., (1966), Convective Heat and Mass Transfer, McGraw Hill.
(8) LEE, K. and RYLEY, D.J., (1968), The evaporation of water droplets in superheated steam, Trans, ASME J. of Heat Transfer, Nov. 1968, pp 445-451.
(9) YUEN, M.C. and CHEN, L.W., (1978), Heat Transfer Measurements of Evaporating Liquid Droplets, Int. J. Heat Mass Transfer, Vol 21, pp 537-542.
(10) HARPOLE, G.M., (1980), Radiative absorption by evaporating droplets, Int. J. Heat and Mass Transfer, Vol 23, No. 1, pp 17-26.
(11) FAIRBAIRN, S.A. and PIGGOTT, B.D.G., (1984), Flow and Heat Transfer in PWR Rod Bundles in the Presence of Blockage due to clad Ballooning; Experimental Data Report - Part 2. CEGB Report TPRD/B/0511/N84, in preparation.

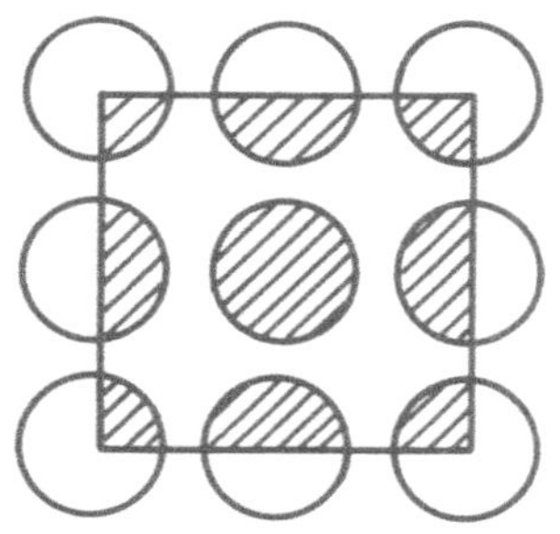

FIG.1. Control Region used in the Model.

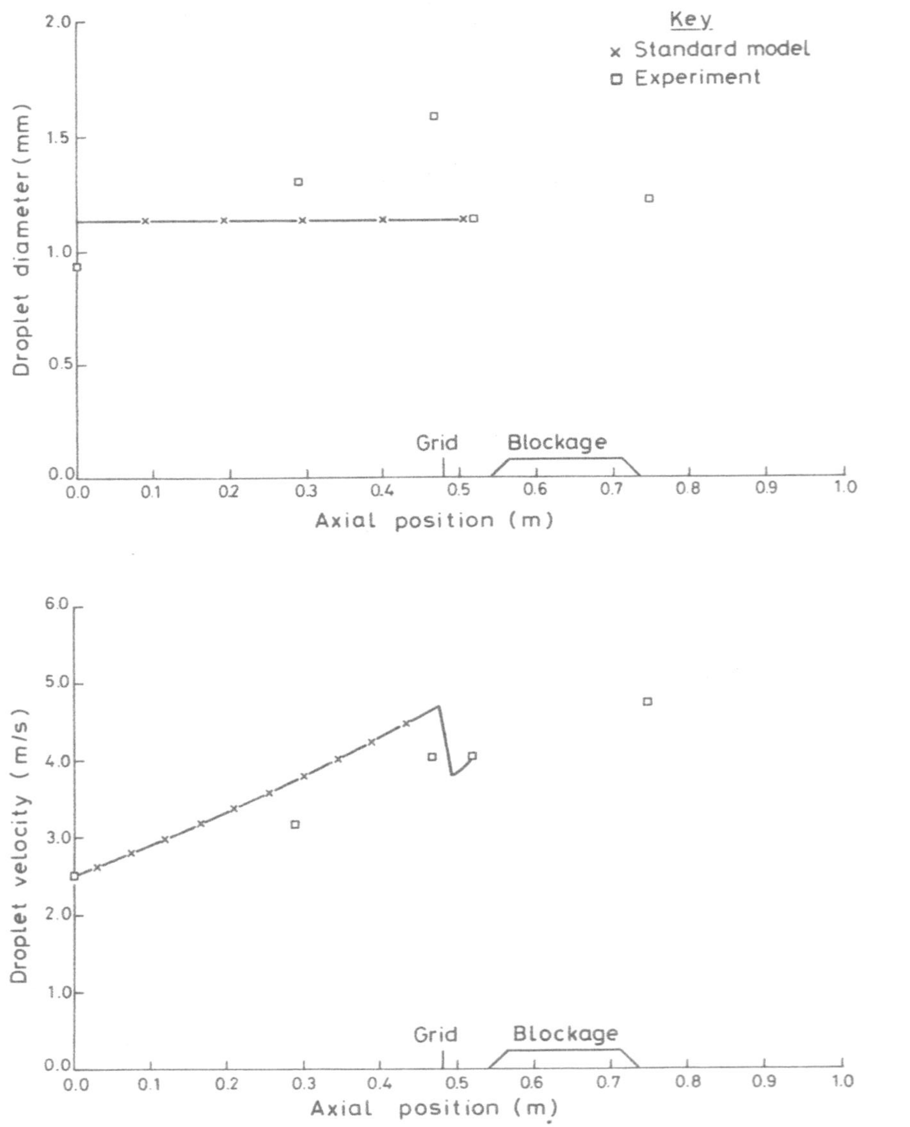

FIG. 2. Comparison of Measured and Calculated Mean Droplet Velocity and Size.

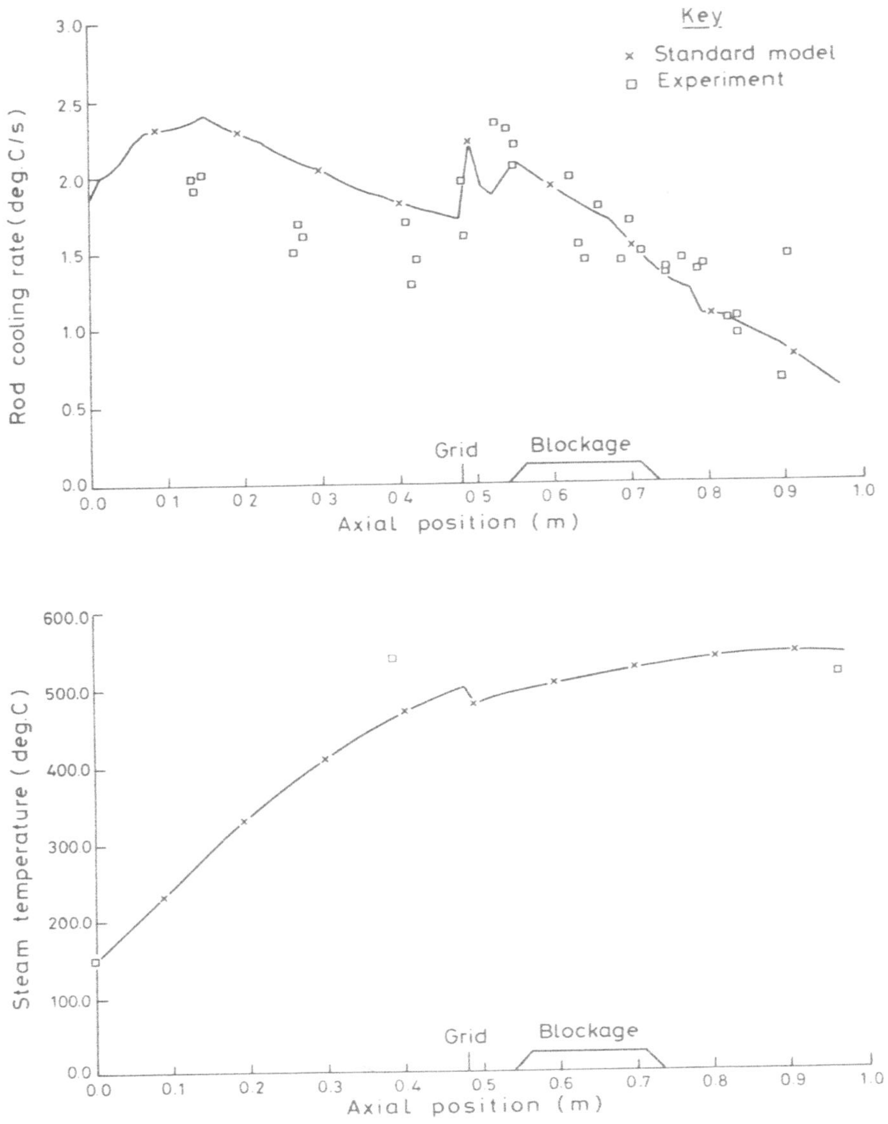

FIG. 3. Model vs. Experiment, Peripheral Rod.

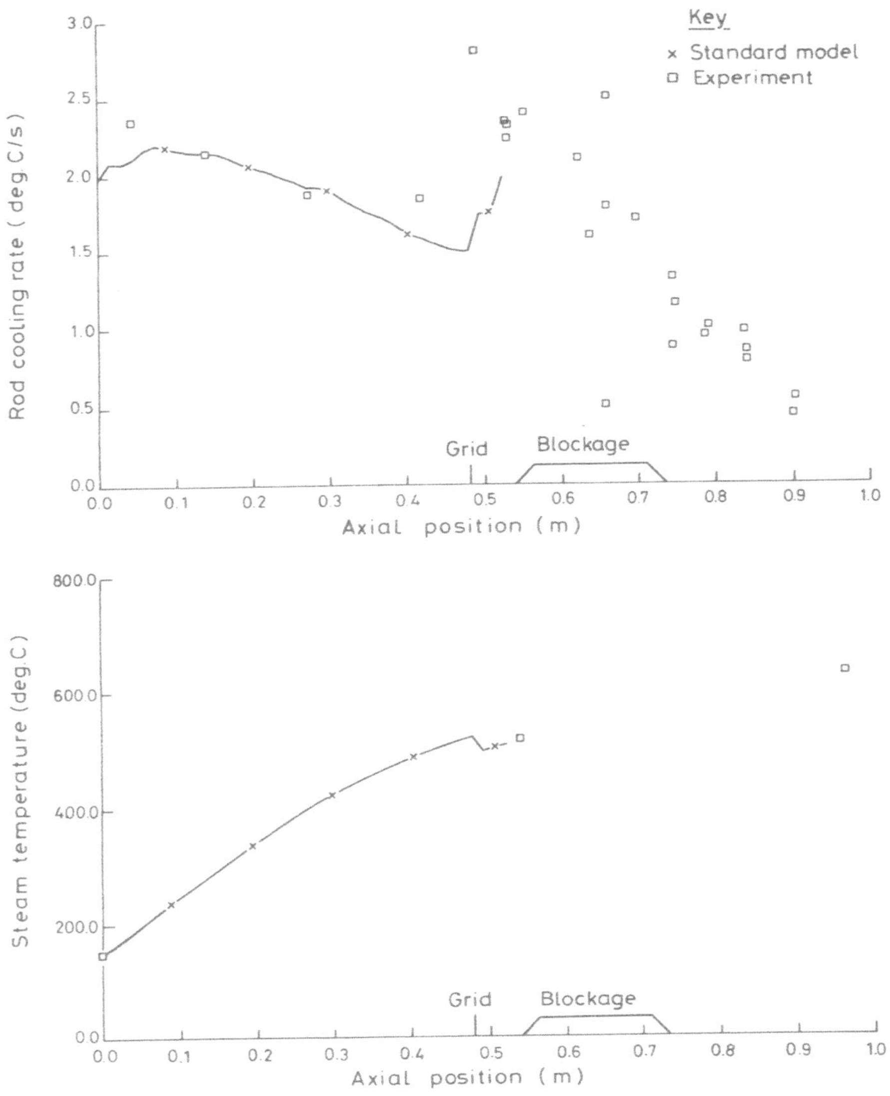

FIG.4. Model vs. Experiment, Central Rod.

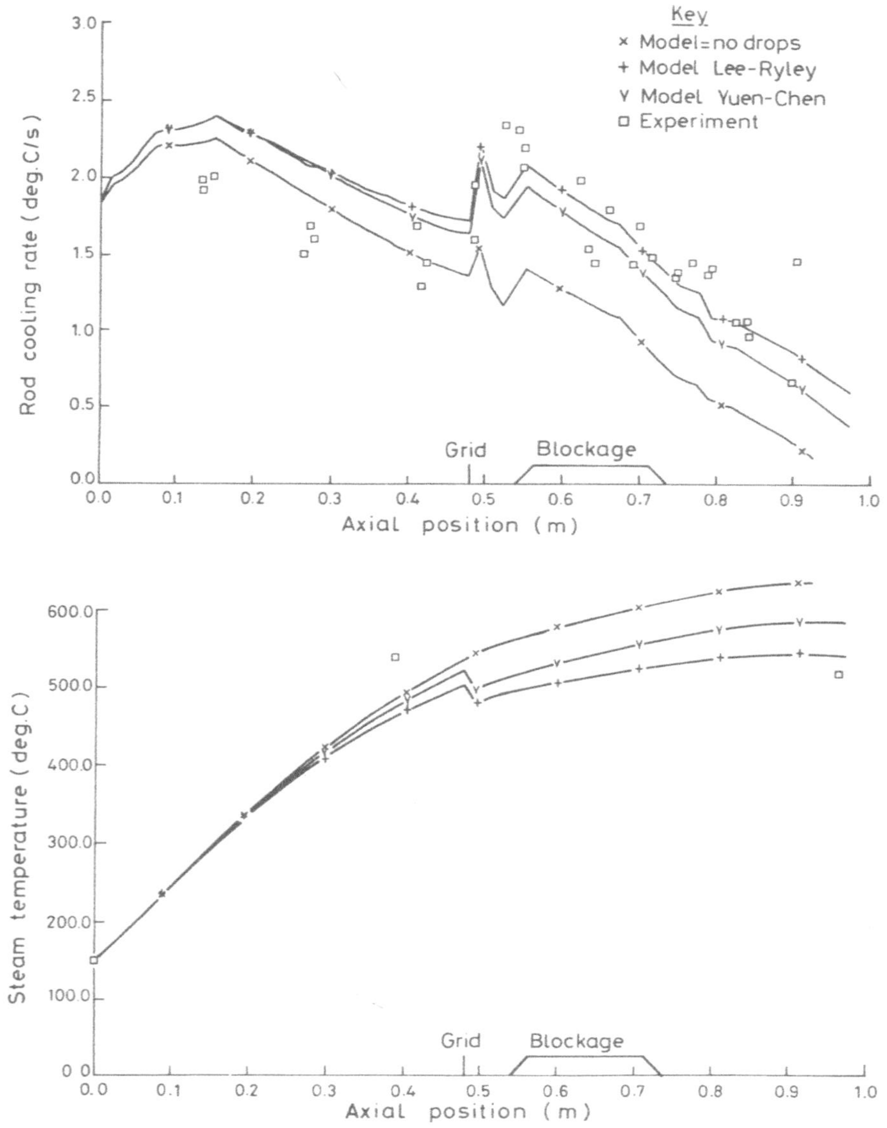

FIG.5. Sensitivity to Droplet Heat Transfer Modelling.

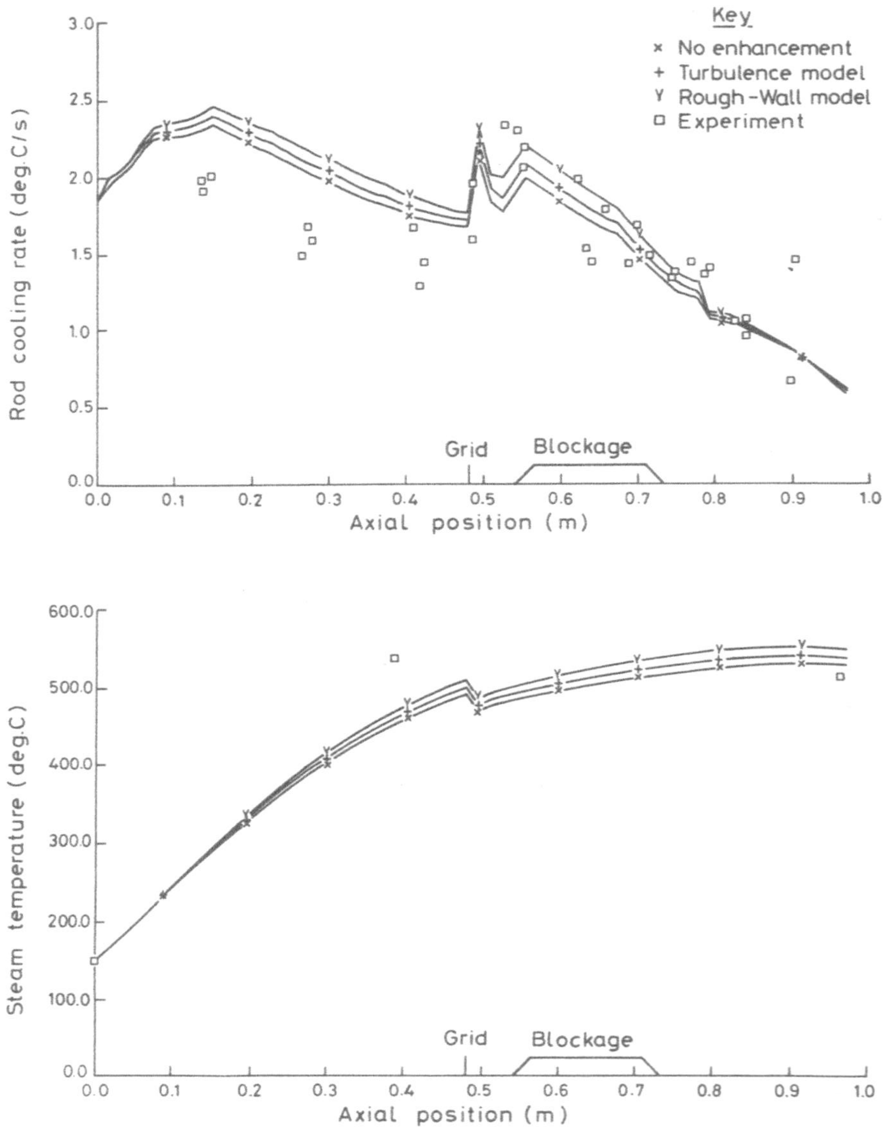

FIG.6. Sensitivity to Drop-Induced Turbulence Modelling.

Discussion

Question: K. Wolfert - GRS

You have presented a model, which describes the fluiddynamic and heat transfer phenomena within one channel in a very detailed manner, e.g. the channel was represented by more than 40 mesh cells in the axial direction. Would you please explain your idea, in which way your model or your findings can be transposed to models used in large system codes with a relatively coarse noding?

Answer: S.A. Fairbairn - CEGB/BNL

The general conclusions outlined in our paper are of direct relevance to the reactor codes. For modelling balloons we need a finer noding at least in the ballooning zone itself than is used in some reactor calculations. We could investigate the result of coarsening the axial mesh in our calculations but we have not done so yet.

Question: R. Deruaz, CEA/CEN-G

On one hand there is experimental evidence for strong droplet interactions, on the other hand your model gives very good results despite the fact it does not take into account this aspect. Does this mean that droplet interactions have no effect on heat transfer?

Answer: S.A. Fairbairn - CEGB/BNL

We are interested in developing a drop coalescence model to represent our data more accurately. In particular, at high levels of drop coalescence we will get a change of flow regime which we would like to model. In our present model the underprediction of drop size leads to use of higher drop/steam convective heat transfer coefficients. There is, however, a compensating effect in that smaller drops are accelerated more rapidly so that we overpredict, by 15%, the drop velocities and so underrestimate the relative velocities between the phases. The effects of errors in calculating drop sizes and velocities, therefore, compensate one another which is one reason why good predictions can be made even though drop interactions are ignored.

Question: D. Mewes - U/Hannover

1) In one of your slides you explained that drop velocity was independent of the drop diameter. What was the velocity relative to the continuous gas-phase or the absolute velocity?

2) How did you recalculate the relative velocity?

Answer: S.A. Fairbairn

1) It was the absolute velocity.

2) The drop velocities shown are absolute velocities. We can produce relative velocities in 2 ways:
(i) Using single phase flow diversion data together with steam temperature measurements from two phase tests.
(ii) Using a calculational model for the two phase flow. Our model does not yet include blockage effects, so the relative velocities are available through route (i) only at present.

AN EXPERIMENTAL INVESTIGATION OF THE EFFECT OF CLAD BALLOONING ON THE EFFECTIVENESS OF PWR EMERGENCY CORE COOLING

C A COOPER, K G PEARSON and D JOWITT
UKAEA, AEE Winfrith, Dorchester, Dorset, UK

Summary

In Pressurised Water Reactors one of the safety issues is the problem of clad ballooning. This is the phenomenon where, following a loss-of-coolant accident, the cladding of some of the fuel elements could swell to partly block the coolant flow passages. The emergency cooling might then be impaired. To assist in the understanding of the influence of such blockages on emergency cooling, experiments have been performed on a full length, electrically heated representation of a fuel element containing a severe partial blockage. The experiments investigated single-phase cooling, forced reflooding, gravity reflooding and level swell phenomena. This paper summarises the findings of the experimental programme.

1. INTRODUCTION

The fuel rods in a Pressurised Water Reactor comprise Zircaloy cans filled with uranium dioxide fuel pellets. The cans are sealed and pressurised with helium to partly offset the high external reactor pressure and to minimise the effect of dilution by gaseous fission products, which would reduce the thermal conductivity of the filling gas.

In the unlikely event of a fracture occurring in one of the larger items of pipework the reactor would depressurise rapidly and the normal coolant would be lost. Cooling would be restored by emergency systems which reflood the core from below, but by the time these take effect the dried out core may be already very hot. Overheated Zircaloy fuel cans can be very ductile and the loss of the external reactor pressure could cause them to swell outwards. This phenomenon is known as clad ballooning. On the basis of extreme assumptions neighbouring cans could swell sufficiently to interact with one another and cause substantial blockages of the coolant flow passages. Heat transfer during the reflooding process is crucial in limiting the temperature which the fuel reaches before it is quenched by a rewetting front, which rises through the core as the reflooding proceeds. It is therefore important to quantify the extent of any deterioration in reflood heat transfer caused by these blockages.

Reflood heat transfer in clusters of electrically heated fuel rod simulators, with various degrees of blockage, has been investigated in the FLECHT [1], FEBA [2] and THETIS [3] experiments. A second series of THETIS experiments has now been completed using a geometry identical to the first except that the severity of the blockage was reduced from 90% to 80%. The comprehensive programme involved four types of experiment. The results are presented in six reports dealing with the

THETIS rig [4], single phase experiments [5], forced reflooding experiments [6], gravity reflooding experiments [7], level swell experiments [8] and an overview [9].

This paper briefly summarises the results of the programme. First the THETIS rig and the experimental cluster are described. Then each of the four types of experiment is reviewed in turn. Finally the main conclusions from the experiments are summarised.

2. THE THETIS RIG AND 80% BLOCKED CLUSTER ASSEMBLY

The rig, test section, instrumentation and data processing are fully reported in Reference 4. The rig consists of a vertical pressure vessel which houses the test section, a downcomer simulation for gravity reflood experiments and a weir vessel for maintaining a steady level in boil-up level swell experiments.

The test section consists of a 7 x 7 square array of electrically heated fuel rod simulators in a square shroud tube (fig. 1). The rods have a heated length of 3.6 m and an outside diameter of 12.2 mm. The axial power profile approximates to a chopped cosine with a peak form factor of 1.25. A 4 x 4 group of rods is fitted with coplanar thin-walled balloons to form the blockage. These balloons are based on clad swellings formed under realistic reflood conditions in the REBEKA rig [10], but the degree of blockage is exaggerated. They consist of a 200 mm long parallel section, where the subchannel blockage is 80%, with a 200 mm entry taper and a 50 mm exit taper (fig. 2). They therefore occupy almost a complete grid interval, and are located between grids at approximately mid-height as illustrated in fig. 3. This figure also shows the grid positions and gives an indication of the detailed axial coverage provided by the rod thermocouples. The grids are of a cellular box structure, are not equipped with mixing vanes and do not directly simulate any proprietary design.

3. SINGLE PHASE COOLING EXPERIMENTS

These were approximately steady-state experiments using air as the coolant. They were performed over a range of inlet Reynolds numbers from 1000 to 9000 and provided useful data on the level of heat transfer within the cluster and the influence of spacer grids.

Remote from the grids heat transfer could be correlated by the relationship between Nusselt number and Reynolds number shown in fig. 4. At Reynolds numbers above 3500 Nusselt numbers were in good agreement with Dittus-Boelter [11], although that correlation was developed for round tubes. At lower Reynolds numbers the dependence on Reynolds number was weaker, but the Nusselt number never became independent of Reynolds number as one would expect in laminar flow (eg Kim [12]).

At higher Reynolds numbers the turbulence induced by the grids produces a clear enhancement in heat transfer. This begins just upstream of the grid, reaches a maximum a little way downstream of the grid and appears to die away before the next grid is reached, as shown in fig. 5. At the higher Reynolds numbers the THETIS enhancements are greater than the correlations of Yao, Hochreiter and Leech [13] or Hassan and Rehme [14] predict for the degree of blockage caused by these grids.

4. FORCED REFLOOD EXPERIMENTS

The forced reflood experiments were performed by first heating the test section to the desired initial temperature and allowing time for the shroud tube to be heated by radiation from the rod cluster. Then water was pumped directly into the bottom of the cluster at a controlled rate to reflood the cluster from below. The experiments covered a range of pressures (1.3 to 4 bar), reflood rates (1 to 6 cm/s), inlet temperature (50 to 90°C), initial peak cluster temperature (550 to 650°C) and cluster power (75 to 200 kW).

The cooling of the cluster which results from the introduction of reflood water is illustrated in fig. 6. This shows the axial variation of temperature along the unballooned rods at different times into the reflood. A rewetting front is seen to move upwards through the cluster, identified by the abrupt fall in temperature to approximately saturation temperature. For some distance above the rewetting front cooling is good and temperatures fall steadily. Higher in the cluster cooling is worse. Temperatures rise at first but then begin to fall as the rewetting front approaches. Superimposed on this behaviour there is exceptionally good cooling at and just downstream of the spacer grids.

Fig. 7 shows the progress of the rewetting front on the unballooned rods. It travels smoothly upwards with its speed initially equal to the reflood rate, which was 2 cm/s. Then it slows progressively until about mid-height in the cluster, where it begins to accelerate again. It travels quickly through grids, particularly in the upper half of the cluster, and then slows along the interval to the next grid. Slightly premature rewetting occurs in the top grid interval.

The spacer grids appear to improve heat transfer in two main ways. They introduce turbulence into the flow which improves the heat transfer between the rods and the steam, rather as they do to the air in the single phase flow experiments. Also, when wet, they cool the steam passing through them so that the temperature difference between the rods and the steam downstream of the grids is increased. An example of the cooling of steam by a wet grid is shown in fig. 8. This shows steam temperatures measured just before and just after passing through a grid and the temperature of the grid itself. While the grid is dry there is very little cooling, but after the grid rewets the steam is cooled by about 100°C in passing through the wet grid.

The effect on rod temperature of the improved cooling caused by the grid is shown in fig. 9. There is negligible cooling both below and above the grid until about 15 s, when liquid starts to be entrained in the coolant flow. Then cooling improves as the evaporation of some of this liquid increases the steam flow. The downstream location is better cooled because of the turbulence induced by the grids. The grid rewets at 140 s which causes a further improvement in downstream cooling resulting from the cooling of the steam as it passes through the wet grid. In spite of this better cooling the downstream location rewets later, but from a lower temperature.

In spite of the severity of the blockage, heat transfer in the blocked region was found to be good early in the reflood, but it deteriorated later in the reflood. Fig. 10 compares the balloon surface temperature in the hottest part of the blockage (the top of the central blocked subchannel) with the unballooned rod temperature in the hottest part of the bypass (which was at the same elevation in the cluster). The reflood rate is 2 cm/s and the pressure 2 bar. The bypass temperature peaks relatively early and then decreases steadily until rewetting occurs. The blockage temperature is initially lower than the bypass temperature, but later it climbs above it, exhibiting a late peak which is, in this case, higher than the earlier peak in the bypass temperature. The temperature remains high until shortly before rewetting occurs. Rewetting of the blockage is slightly premature under these conditions. At higher reflood rates or lower pressures it is more premature. The blockage never rewets later than the bypass at the same elevation.

Fig. 11 shows how the balloon surface temperature varies along the central blockage subchannel. Cooling becomes much worse towards the top of the subchannel. The axial temperature gradient develops only slowly at first, but then at an increasing rate.

While the behaviour exhibited in figs. 10 and 11 is typical of all the forced reflood experiments it should be noted that the late peak in blockage temperature only exceeded the earlier peak in bypass temperature when the reflood rate was less than about 4 cm/s. This temperature penalty became larger as the reflood rate was further reduced or the cluster power increased.

If results with the 80% blockage are compared with those from the THETIS 90% blockage reported in Reference 3 it is found that, over a range of conditions most relevant to safety studies, the 80% blockage was no better cooled. However, it was better cooled than the 90% blockage at high reflood rates, due to premature rewetting of part of the blockage, and at a low reflood rate of about 1 cm/s. An explanation for this surprising behaviour is given in Reference 6.

5. GRAVITY REFLOOD EXPERIMENTS

The gravity reflood experiments were performed to investigate some of the steam binding and inlet flow oscillation effects which are predicted to occur during reflooding of the reactor system. These system effects were crudely modelled using a tubular downcomer, with scaled flow area and correct height, and a simple restrictive orifice at the exit from the test section, through which only the steam generated in the test section had to pass. The experiments were performed in the same way as the forced reflood experiments except that the downcomer shared the total inlet flow with the test section. An arrangement of valves allowed the experiment to start with the downcomer empty, partly full or completely full.

The experimental programme explored the influence of pressure (2 to 4 bar), inlet flow rate (0.15 to 0.6 kg/s), inlet water temperature (50 to 90°C), downcomer initial level (full and empty) and restrictive orifice size (24 to 34 mm).

Fig. 12 summarises the behaviour of the system in an experiment with the downcomer initially full. The total inlet flow rate is 0.16 kg/s, the inlet temperature is 90°C and the pressure is 2 bar. Initially, some of the downcomer contents are dumped into the test section. This causes an abrupt initial rise of about 0.4 m in the test section collapsed liquid level and immediately rewets the bottom 0.2 m of the cluster. Thereafter the test section level oscillates violently. These oscillations have an initial period of about 5 s and diminish in period and amplitude over about 90 s. Then they cease rather abruptly although the signal remains irregular. As the test section level builds up the downcomer level also rises to furnish the necessary pressure difference to overcome the resistance of steam flow through the orifice. At about

180 s steam production begins to decrease, reducing the pressure drop through the orifice. This allows the downcomer level to fall although the test section level continues to rise. Throughout the transient the rewetting front moves upwards through the cluster. Its progress is slower around mid-height and faster at and downstream of grids in the upper part of the cluster. Slightly premature rewetting occurs at the upper grid. Apart from the instantaneous rewetting of the bottom 0.2 m of the cluster, this behaviour is the same as one observes in forced reflood experiments.

The changes in system behaviour produced by changes in test parameters were straightforward. Increasing the pressure reduced the restrictive effect of the orifice allowing more water to enter the test section. So the cluster enjoyed the benefit of both increased pressure and increased flow, giving faster rewetting, more steam production and hence better cooling. Increasing the total inlet flow tended to keep the downcomer fuller, so maintaining a higher driving head between the downcomer and the test section. Thus more flow entered the test section with consequent improvements in rewetting and cooling. However, the benefit was only felt up to the time when the downcomer overflowed, any additional water being wasted. Reducing the water inlet temperature caused less steam to be formed in the early part of the transient, reducing the pressure drop through the orifice. Later the lower average voidage in the cluster allowed the test section collapsed level to rise higher, requiring the downcomer level to rise higher also. But since the total inflow was not enough to cause the downcomer to overflow there was no detrimental effect on the inflow to the test section. Rewetting was faster because heat transfer at the rewetting front improves as the subcooling at the rewetting front increases. Starting with the downcomer initially empty caused much of the inlet flow to go into filling the downcomer to develop enough driving head to force water into the test section. The low inlet flow to the test section during this period had a very detrimental effect on rewetting and heat transfer. The absence of an initial surge of water into the test section also meant that the spacer grids did not rewet instantly as they did with the downcomer initially full. This further degraded the cooling of the cluster. Increasing the size of the restrictive orifice reduced its pressure drop and allowed water to flow more readily into the test section from the downcomer. Conversely, reducing the size of the orifice increased the pressure drop and caused the water level to rise higher in the downcomer. If it rose high enough to overflow, the inflow to the test section was reduced with a detrimental effect on rewetting and heat transfer.

Oscillations occurred in the test section collapsed level in the early stages of all the gravity reflood experiments. With the downcomer initially full these started immediately and ceased rather abruptly. With the downcomer initially empty there was a delay before they started, but they still ceased abruptly. They were accompanied by oscillations, with reversals, in the test section inlet flow.

Temperatures within the cluster, on both the balloons and the unballooned rods, showed the same trends as in the forced reflood experiments. However, there was a strong influence of the variation of reflood rate with time produced by the system effects. Also the instant rewetting of the grids with the downcomer initially full was very beneficial. Nevertheless, it appears that the same phenomena were important in each type of reflood.

Fig. 13 compares the balloon surface temperature at the top of the blockage with the bypass rod temperature at the same elevation. The test conditions are the same as for the forced reflood experiment in fig. 10, except that the inlet flow is now shared between the test section and the downcomer. Because of the insurge of water from the full downcomer, and the instant rewetting of the grids, the bypass is better cooled than in the forced reflood experiment and rewetting is faster. The blockage temperature is initially lower than the bypass but it exhibits a late peak which is higher than the maximum temperature recorded in the bypass. However, the peak blockage temperature is lower than in the forced reflood experiment of fig. 10.

Fig. 14 shows the variation of balloon surface temperature along the central blockage subchannel. It may be compared with fig. 11 which shows the results from the corresponding forced reflood experiment. The initial fall in temperature following the start of liquid entrainment occurs immediately and temperatures are generally lower. At about 120 s improved cooling is observed about half way along the blockage throat. This leads to premature rewetting of the top part of the blockage. References 6 and 7 suggest that this is due to the accumulation of a plug of water in the blockage. The effect is also observed in forced reflood and so is not unique to gravity reflood.

The oscillations in the test section inlet flow and collapsed liquid level, though spectacular, did not appear to have much influence on the temperatures in either the blockage or the bypass regions or on the rewetting of the cluster. This was established by performing forced reflood 'mimics' of some of the gravity reflood experiments. These had approximately the same variation of reflood rate with time but the oscillations were absent. The differences between the forced reflood mimics and the gravity reflood originals were small and could easily be due to slight differences in test conditions.

6. LEVEL SWELL EXPERIMENTS

The level swell experiments were designed to investigate the voidage profile in the cluster under conditions representing partial core uncovery in the reactor. Two types of experiment were performed, each covering a range of pressures from 2 to 40 bar. In the 'boil-up' experiments the mass in the cluster was kept constant by connecting the bottom of the test section to another vessel which contained an overflowing weir. The height of this weir was varied to cover a range of cluster levels from 30% to 70% of the heated length. In the 'boil-down' experiments the bottom of the test section was connected to the downcomer (as in gravity reflood), the two vessels were filled and the contents were then boiled away without further addition of water.

Voidage profiles were determined by differential pressure measurement along the cluster. They were compared with various correlations [15], [16], [17], [18] and [19] for voidage as a function of superficial steam velocity. The correlation of Gardner showed the best overall agreement with the data, followed by that of Cunningham and Yeh. Jowitt's correlation was best up to 10 bar but overpredicted the void fraction at higher pressures. The correlations of Wilson and of Zuber and Findlay seriously overpredicted the void fraction at all pressures. Figs. 15 and 16 compare the various correlations with data from boil-down experiments at 2 and 40 bar.

It should be noted that these correlations are essentially one-dimensional, but the flow diversion caused by the blockage introduced two-dimensional effects which could not be predicted simply by accounting for the overall change in flow area. Upstream of the blockage the measured voidage was lower than the trend of the correlations would suggest and downstream it was higher.

7. SUMMARY

1 A series of single-phase cooling, forced reflood, gravity reflood and level swell experiments has been performed on a full length, electrically heated 7 x 7 rod fuel cluster containing a blockage simulating very severe clad ballooning.

2 The blockage was formed by fitting thin-walled Inconel balloons of length 455 mm to a 4 x 4 group of rods. These reduced the subchannel flow area by 80% over a length of 200 mm.

3 The single phase cooling experiments provided data on the level of heat transfer within the cluster and the enhancement produced by turbulence created by the spacer grids.

4 The forced reflood experiments have led to a better understanding of the rewetting processes during bottom reflooding, the very important influence of spacer grids in two-phase flow and the complicated heat transfer processes within the blockage. The insight obtained has been used to develop a mechanistic model which gives quantitative support to the explanations which have been put forward.

5 The gravity reflood experiments used a very simplified representation of the reactor system to investigate steam binding and inlet flow oscillation effects. They indicate that essentially the same phenomena are important in both forced reflood and gravity reflood. The variation in inlet flow produced by the variations in system parameters was the dominant influence. The inlet flow oscillations which occurred in gravity reflood appeared not to influence overall rewetting or heat removal performance.

6 The level swell experiments investigated the relationship between void fraction and superficial steam velocity up to 40 bar and compared the data with various correlations. The correlation of Gardner was found to be most satisfactory.

REFERENCES

[1] LOFTUS, M. J. et al. NUREG/CR-2444, EPRI NP-2014, WCAP-9992, Vol 1.

[2] IHLE, P., RUST, K. KfK 3657.

[3] PEARSON, K. G., COOPER, C. A., JOWITT, D., KINNEIR, J. H. AEEW - R 1591.

[4] JOWITT, D., COOPER, C. A., PEARSON, K. G. AEEW - R 1763.

[5] COOPER, C. A., PEARSON, K. G., JOWITT, D. AEEW - R 1764.

[6] COOPER, C. A., PEARSON, K. G., JOWITT, D. AEEW - R 1765.

[7] PEARSON, K. G., COOPER, C. A., JOWITT, D. AEEW - R 1766.

[8] JOWITT, D., COOPER, C. A., PEARSON, K. G. AEEW - R 1767.

[9] PEARSON, K. G., COOPER, C. A., JOWITT, D. AEEW - R 1768.

[10] ERBACHER, F. J., NEITZEL, H. J., WIEHR, K. 5th International Conference on Zirconium in the Nuclear Industry, 4-7 August 1980, Boston USA.

[11] DITTUS, F. W., BOELTER, L. M. K. University of California, Berkeley, USA. Publication Eng 2, pp 443-461 (1930).
[12] KIM, J. H. Fluid Flow and Heat Transfer over Rod Bundles, ASME 1979, pp 155-161.
[13] YAO, S. C., HOCHREITER, L. E., LEECH, J. J. J Heat Transfer, Vol 104, pp 76-81 (1982).
[14] HASSAN, M. A., REHME, K. Nucl Technology, Vol 52, pp 401-414 (1981).
[15] CUNNINGHAM, S. P., YEH, H. C. Trans ANS, Vol 17, pp 363-370 (1973).
[16] GARDNER, G. C. Int J Multiphase Flow, Vol 6, pp 399-410 (1980).
[17] JOWITT, D. AEEW - R 1488
[18] WILSON, S. F. et al. Allis-Chalmers Atomic Energy Report No ACNP-65002 (1965).
[19] ZUBER, N., FINDLAY, J. A. J. Heat Transfer, Vol 87, pp 453-466 (1965).

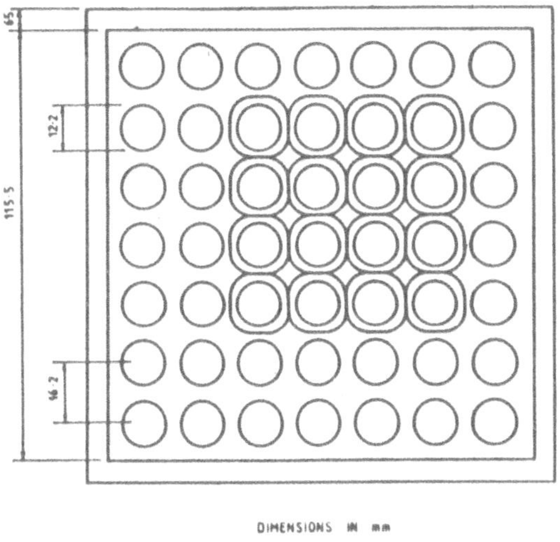

G.1. THETIS 7 x 7 REFLOOD CLUSTER WITH 4 x 4 BLOCKAGE

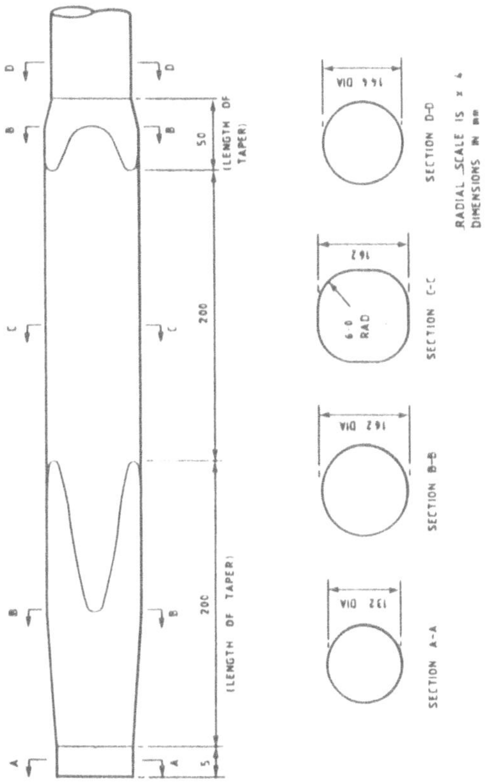

FIG. 2. BALLOON GEOMETRY

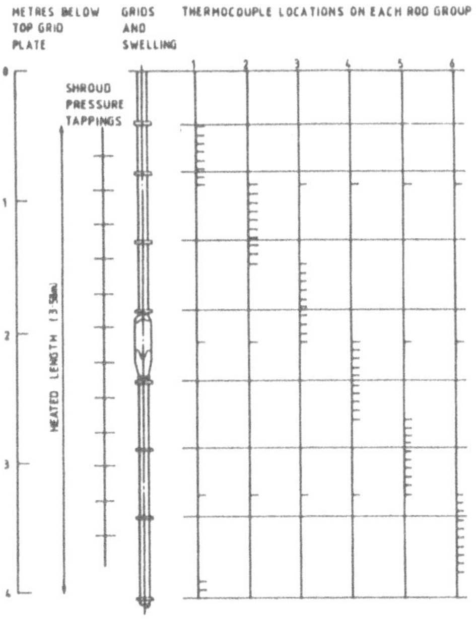

FIG. 3. RELATIVE LOCATIONS OF THERMOCOUPLES TO GRIDS, BALLOONS AND HEATED LENGTH

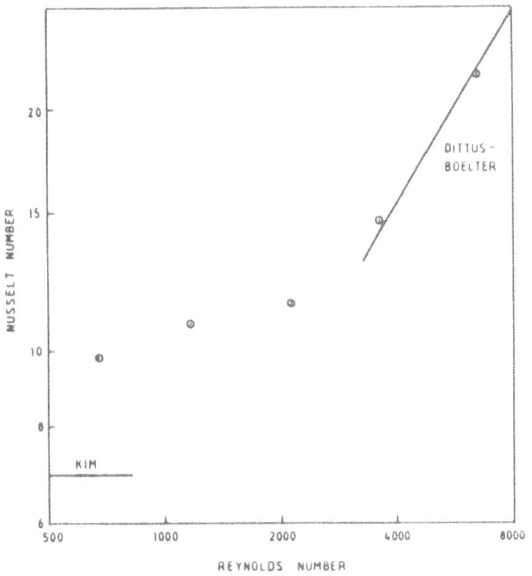

FIG. 4. SINGLE PHASE COOLING REMOTE FROM GRIDS

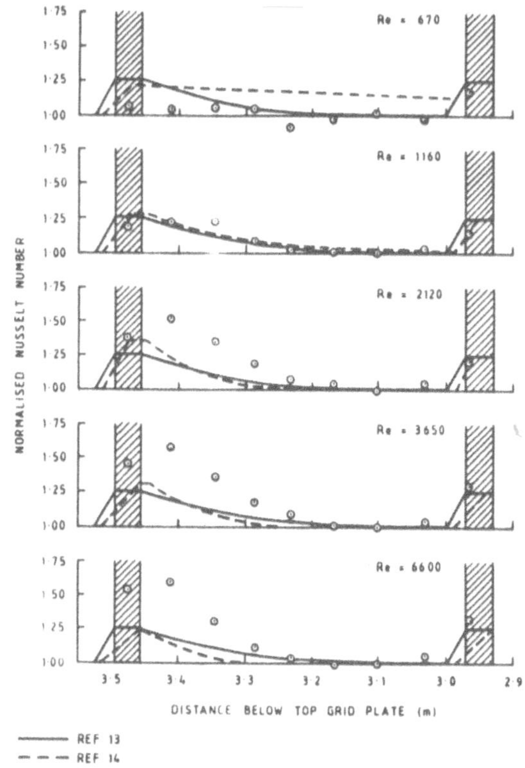

FIG. 5. SINGLE PHASE GRID ENHANCEMENT

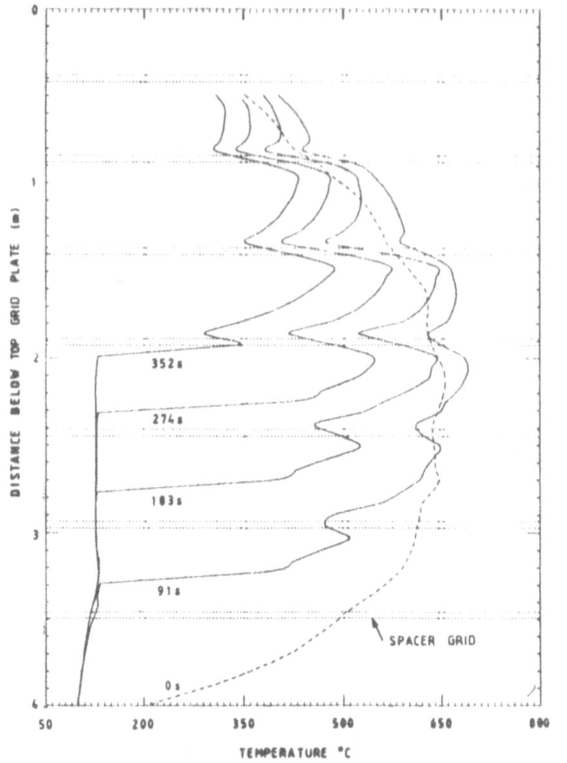

FIG. 6. CLUSTER TEMPERATURES DURING REFLOODING

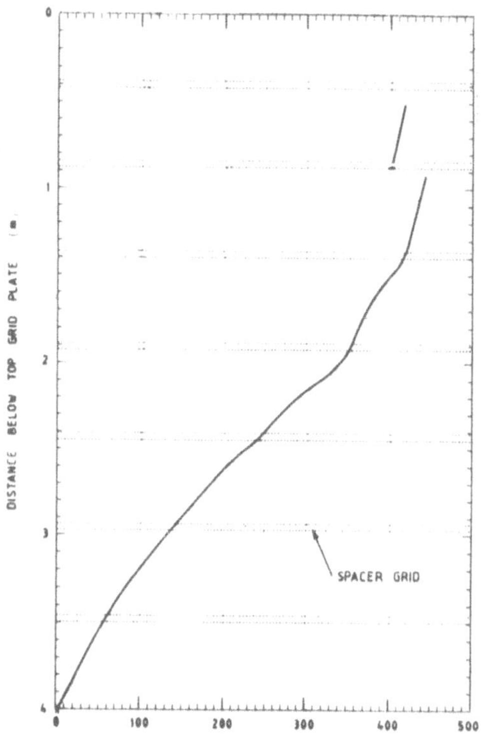

FIG. 7. PROGRESS OF REWETTING FRONT

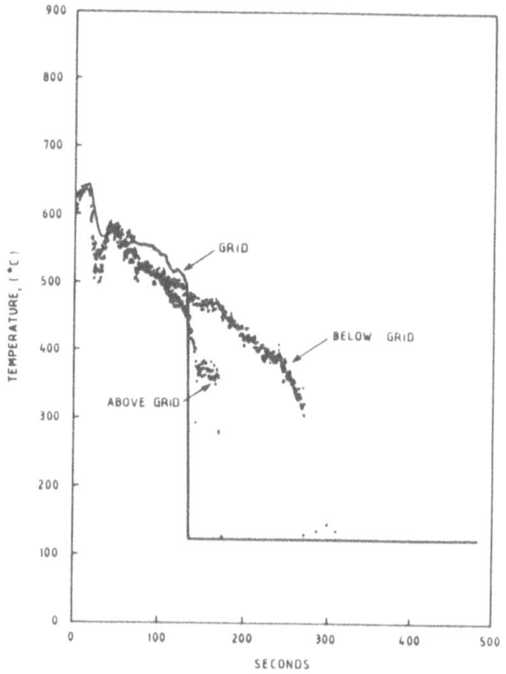

FIG. 8. STEAM DESUPERHEATING BY WET GRID

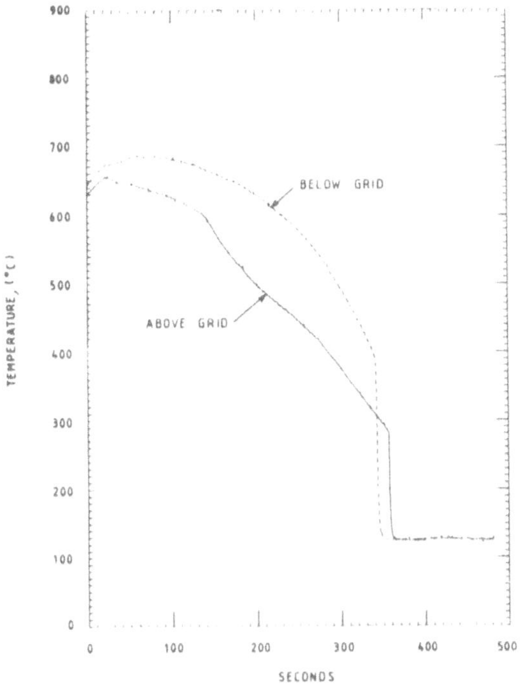

IG. 9. TEMPERATURE HISTORIES ABOVE AND BELOW GRID

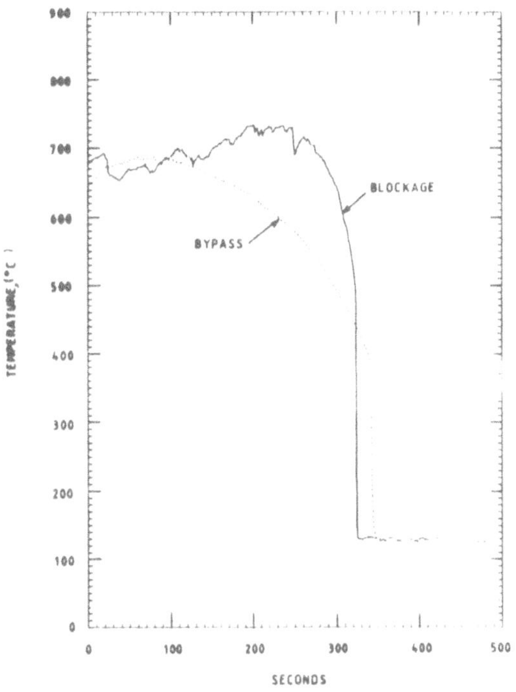

FIG. 10. MAXIMUM BLOCKAGE AND BYPASS TEMPERATURES (FORCED REFLOOD RUN T2R020)

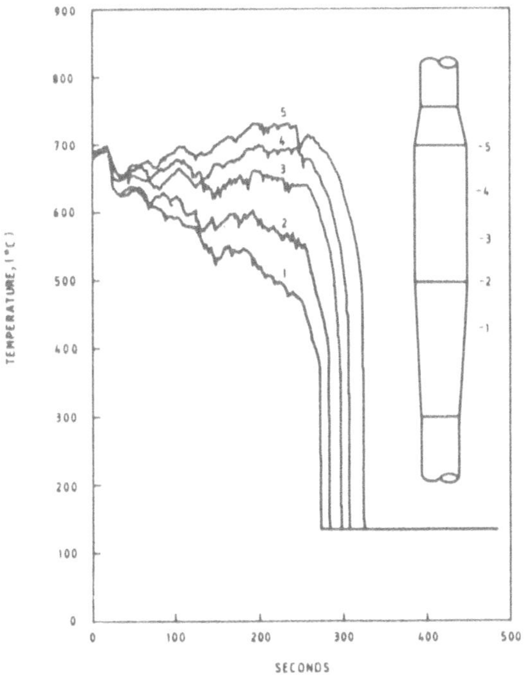

.11. TEMPERATURES ALONG BLOCKAGE SUBCHANNEL (FORCED REFLOOD RUN T2R020)

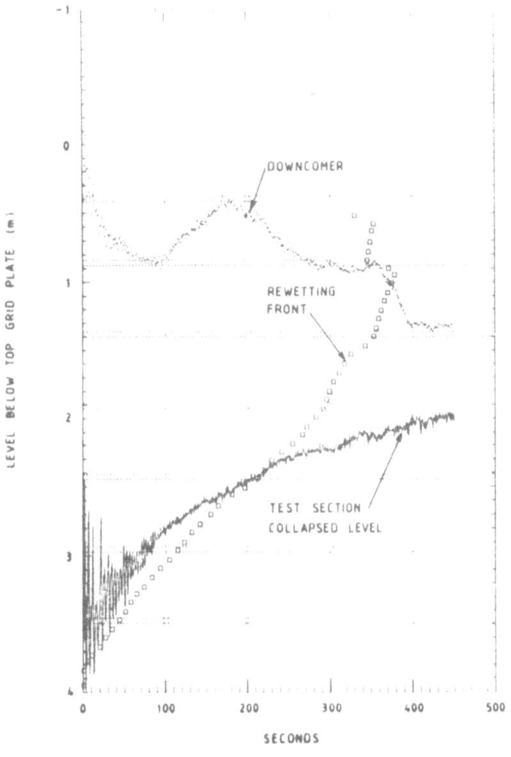

FIG. 12. LEVELS DURING GRAVITY REFLOODING

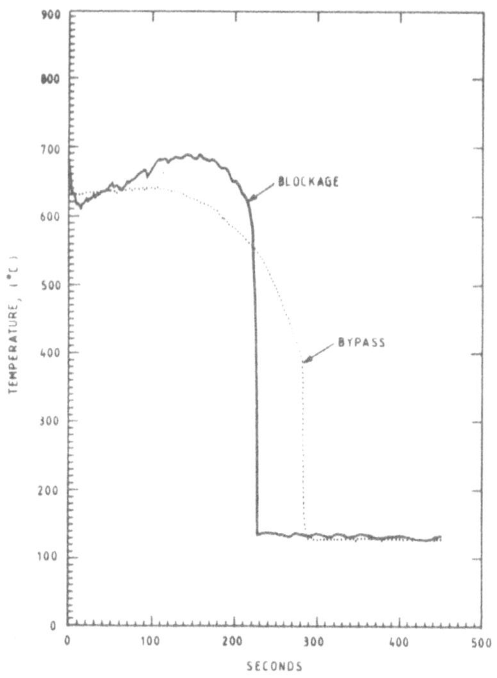

FIG. 13. MAXIMUM BLOCKAGE AND BYPASS TEMPERATURES (GRAVITY REFLOOD RUN T2G068)

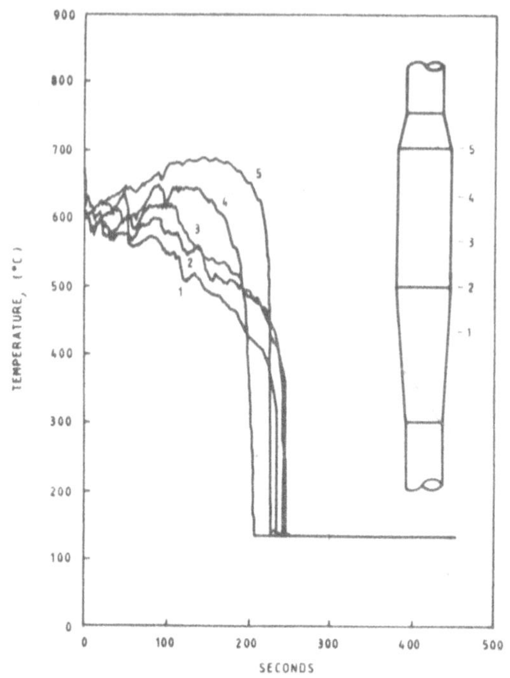

FIG. 14. TEMPERATURES ALONG BLOCKED SUBCHANNEL (GRAVITY REFLOOD RUN T2G068)

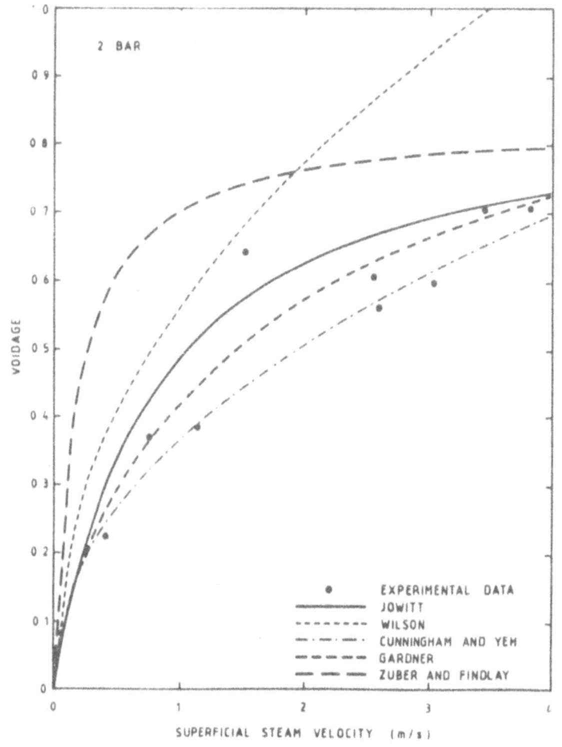

FIG. 15. LEVEL SWELL DATA AT 2 BAR

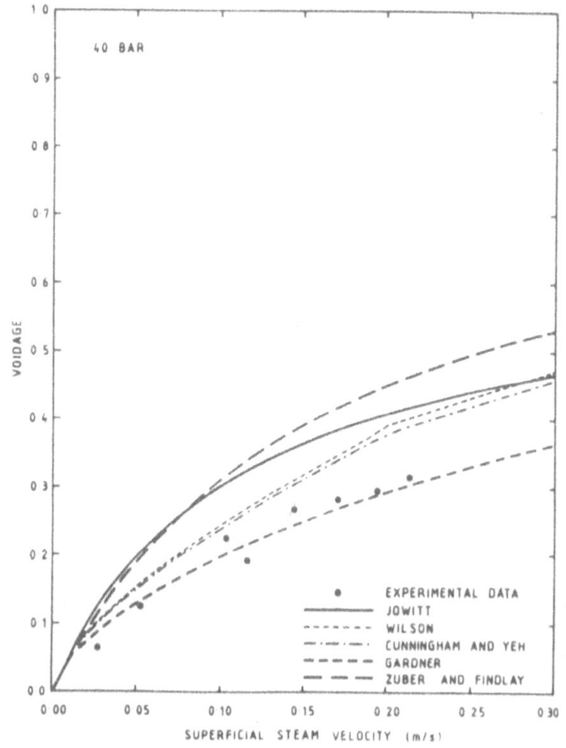

FIG. 16. LEVEL SWELL DATA AT 40 BAR

)iscussion

}uestion: C. Prato - NIRA

1) You showed a couple of pictures with temperature profiles in 5 different sections of a blocked channel. Are there available similar profiles without blockage?

2) How do you think the experimental results on blockage are affected by the type of experiment (geometry of the channel, length, fuel simulation)?

Answer: C.A. Cooper - UKAEA/AEEN

1) No. The series of experiments described in my presentation was the second series of experiments in the THETIS rig containing a long severe blockage but we have not performed experiments with an unblocked cluster. However, we do have figures, corresponding to those shown for the blockage, showing temperatures at 5 elevations at blockage height on undeformed rods in the bypass region around the blockage, but it must be noted that the cooling of these bypass rods would have been better than for rods in an unblocked bundle due to flow bypass effects.

2) I would expect the results of blockage heat transfer experiments to be significantly affected by the length and severity of the blockage, the type of fuel rod simulators used and, to a lesser extent the amount of flow bypass in the blocked cluster. The THETIS experiments were performed with long, severe blockages and using fuel rod simulators with no gas gap simulation (except in the blockage itself where the use of thin-walled balloons allowed the large gas gap of ballooned fuel to be well simulated). Reduction in the length of the blockage would cause more flow to pass through the blockage giving improved cooling within the blockage relative to the bypass. The comparison of the results of the THETIS 90% (200 mm throat) and FEBA 90% (65 mm throat) blockage experiment supports this conclusion. Reduction of the degree of severity of the blockage produces more complex behaviour. Comparison of results from the THETIS 90% and 80% blockage experiments shows that under some conditions the 90% blockage was better cooled. As I explained in my presentation we believe that this surprising result has arisen because of the behaviour of the liquid in the long, severe THETIS blockages where significant deceleration of the liquid is calculated to occur. In experiments with short blockages of lesser severity significant flow diversion around the blockage will not occur and the blockage will be better cooled than the bypass due to enhancement of turbulence and flow acceleration caused by the blockage. This is in marked contrast to the THETIS experiments where the presence of the long, severe blockages caused significant diversion of vapour (and liquid) to occur. This large flow diversion in turn meant that the highest temperatures in the cluster were recorded in the blockage at low reflood rates. The amount of bypass will have a relatively small effect on blockage cooling since it will affect the flow passing through the blockage to some extent. The design of fuel rod simulators can very significantly affect the results of forced reflood experiments in general, as has been demonstrated by the KfK SEFLEX experiments. In these experiments REBEKA rods, which are Zircaloy clad and have a simulation of the reactor fuel rod gas gap, rewetted significantly faster than FEBA rods which are clad in nichrome and have no gas gap simulation. Although the SEFLEX experiments were performed on unblocked bundles the marked

difference in rewetting and precursory cooling ahead of the rewetting front would strongly affect blockage cooling in blocked clusters. Thus we see that many factors of experimental design can strongly influence the results of blockage heat transfer experiments and this emphasizes the danger of attempting to apply results of such experiments directly to reactor safety analyses.

OUT OF PILE BALLOONING AND BURSTING TESTS ON ZIRCALOY CLADDING RODS WITH DIRECT AND INDIRECT HEATING

R. VESPRINI, L. LOMBARDI
TERM-ISP-TECNOLOGIE DEI COMPONENTI, ENEA-CENTRO STUDI CASACCIA - ROMA,ITALY
P. JONES, F. CASADEI
JRC - ISPRA, ITALY

BACKGROUND

The ballooning of Zircaloy clad LWR fuel rods during a hypothetical loss of coolant accident has been investigated for some years. The concern has been that emergency cooling of the fuel might be inhibited if the coolant channels become appreciably blocked. In a PWR adjacent rods will touch at about 30% strain of the cladding and the sub-channels would be completely blocked if all the rods strained to about 70% at the same axial level.Although such a situation is actually impossible, ballooning of this magnitude in many rods might cause a problem.

Laboratory experiments have shown that ballooning Zircaloy can achieve strains between less than 10% and more than 150% depending upon the temperature, the heating rate, the heating method and the environment. It has thus been important to determine which conditions were relevant to a "real" reactor accident and ascertain whether or not a problem exists.

Currently the "Design Basis Accident", the large-break LOCA (LB-LOCA) appears not to be a problem. All experiments performed realistically simulating this event using bundles of internally heated fuel rods or simulators have produced blockages which have been shown to be coolable. However it is not possible to test every permutation of conditions and computer codes have been written to enable the gaps to be filled. These contain various degrees of mixing of physical models and empiricism and have to be validated against the experiments performed.

The experimental facility described here was originally designed to investigate the effect of heating method and heat transfer conditions on ballooning. It enables ballooning with direct and indirect heating to be compared both with and without a heated shroud and presents a good opportunity to test the modelling in the RODSWELL[1] code. This code was specifically written in an attempt to quantify the interaction between heat transfer and clad strain which is now believed to be the major factor determining the failure strain in alpha-phase Zircaloy.

For the wider range of conditions which can potentially exist in a small-break (SB) LOCA, ballooning in the mixed alpha-beta and beta phase temperature regimes is also of interest. Here existing experimental evidence is more scarce and physical modelling more difficult. Again testing of the code was necessary. The main concern in the mixed phase temperature range (1100-1250 ^{0}K) is whether a high ballooning strain is achievable under "real" accident conditions - most experimental data show a trough in failure strain here but superplasticity has also been demonstrated. At even higher temperatures the interest is in the ductile/brittle transition caused by oxidation and the possible total inhibition of failure caused by oxide strengthening.

ANALYSIS AND DISCUSSION

Group I Tests - Alpha Phase Zircaloy

As explained earlier the objectives of the first group of bursting tests was to test the ballooning/heat transfer modelling of the RODSWELL[1] code. This model assumes that for alpha-phase Zircaloy the local strain at the point of rupture of a ballooned tube is very high and is a constant. The measured total strain depends on the distribution of strain around the circumference which for an internally heated tube depends primarily on the temperature distribution.

The pellet stack within the fuel-rod or simulator is assumed to be eccentric just touching the clad at one side. For an internally heated rod this side of the clad will be the hotter. It is further assumed that when straining starts this side of the clad remains in contact with the pellets because of the so-called "hot-side-straight" effect. This effect arises because of the anisotropic creep properties of alpha-phase Zircaloy which causes axial shortening of the clad relatively more on the more highly strained side with resultant bowing towards the heater rod. A circumferential temperature gradient is thus established which increases as the balloon develops by cooling of the opposite side of the clad, localising the strain and limiting its final measured value.

In a recent paper [2] RODSWELL calculations were reported which showed that the main parameters determining the magnitude of the temperature gradients and the failure strains are the surface heat transfer coefficient, the coolant temperature, the clad heating rate and the time scale of the ballooning process. In particular it was predicted that the failure strain should correlate with the nominal surface heat flux and failure time under approximately isothermal straining or with the nominal gap flux (surface flux plus thermal inertia term for clad heating rate) in a temperature ramp situation.

Rig Heat-Transfer Characterisation

Originally it was intended to model the thermohydraulics of the rig in detail so that accurate input data could be used for RODSWELL (heat transfer coefficient and coolant temperature histories and radiant heat exchange betwen rod and shroud). Two attempts were made at this. In the first a code was written which included models for convective (laminar and turbulent flow) and radiant heat transfer between rod, shroud and outer vessel. Measured flow and power levels from the pre-operational steady state calibration tests were input and emissivities adjusted to try to match the observed temperatures. In general this code overpredicted the temperatures and underpredicted the axial temperature gradients.

The second method tried used the measured flow, power and temperatures (at three axial levels) to determine mean heat transfer coefficients and coolant temperatures for the two axial zones thus defined. A large number of permutations of rod, shroud and vessel emissivities were then tested until the "best fit" was found. Three possibilities were allowed - the heat transfer coefficient could be constant in both spans, it could be constant for clad and shroud but vary between spans or it could be different for clad and shroud but constant between spans. More success was found with this approach in that several of the pre-operational and burst tests gave good "best fits". However the emissivity values found were not consistent and the heat transfer coefficients were rather higher than would be expected (50-250 W/m^2 K compared with 30-100 W/m^2 K predicted from the first code).

The inconsistencies in the results are not surprising when the data is studied. Figs.1a,1b,1c show the measured temperature distributions and Fig. 2 the measured power levels for the various types of test. A large scatter is evident. The reasons for this are thought to be a) that the tests did not represent true steady state conditions because of the large thermal capacity of the rig itself, b) the inlet steam flow was not always superheated as was intended and c) that the flow regime was in many cases marginal between laminar and turbulent flow when minor differences in geometry (e.g. thermocouple attachments) could have a large effect.

Because of the above difficulties this approach was abandoned. The real object of the exercise was of course to check the RODSWELL predictions not necessarily every detail of the calculations. Since the nominal surface heat flux can be obtained directly from the measured rod power comparisons were made on this basis. This was thought to be admissible because the earlier calculations [(2)] had shown coolant temperature variation to be a second order effect. However the partitioning of the flux between convective and radiative heat transfer might be expected to have an effect and this will be explored in a subsequent analysis.

It is noteworthy that the range of rod power required to achieve the desired clad temperature for the different flow and shroud heating configurations corresponds very well to the range of gap heat fluxes of interest in the ballooning phase of "real" accidents. Thus the typical 5% decay heat for a LB-LOCA corresponds to 600-800 W for the 0.4 m long specimen tested here and is well simulated by the high flow/heated shroud test. The low flow/heated shroud test is equivalent to about 1% decay heat typical of a SB-LOCA steady state or the LB-LOCA during adiabatic heating (when 80% of the power is used in raising the fuel pellet temperature). Finally the high flow/unheated shroud configuration represents the peak flux achieved during the temperature turn around during the reflood phase of the LB-LOCA.

Results versus Predictions

Fig. 3 shows the measured failure strains as a function of total heat flux derived from the measured rod power in the indirectly heated clad bursting tests. The times to fail in the nominally isothermal tests and the nominal temperature ramp rates for the ramp tests are indicated in parentheses by each point. Predicted curves summarizing RODSWELL predictions are shown for comparison. The trend of the measurements is seen to be in good agreement with the shape of the predicted curves but the observed strains are in general higher than predicted and also the expected reduction of strain with increased failure time is not found. There are two probable explanations for the under-prediction of the failure strains. Firstly the "hot-side-straight" effect assumed in the calculations is associated with a fully eccentric heater and should predict the maximum temperature gradients and minimum failure strains possible. Secondly the presence of the heated shroud in some tests will reduce the circumferential temperature gradients caused by ballooning allowing a higher total strain at rupture. It can be seen in Fig.3 that it is these results which are those most underpredicted. RODSWELL is being modified to model this phenomenon so that the effect can be quantified. The reason for the lack of confirmation of the predicted variation with time is not clear. Stochastic variations of the tube wall thickness and the heat transfer parameters are still being studied. However the discrepancy is not large compared with the heat flux effect itself.

Fig.4 plots the failure strain against the maximum circumferential temperature gradients measured in the tests and shows the RODSWELL model for comparison. Although there is a considerable scatter again a general agreement can be seen. The probable explanation for the discrepancies is the fact that the temperature measurements are not in general at the actual failure position.

Also shown in Figs.3 and 4 are the results for the directly heated specimens. These have failure strains significantly higher than the indirectly heated samples confirming findings from other sources (e.g. Hindle (3), and Chung and Kassner (4)). The reason is that the temperature gradients caused by ballooning now tend to reduce any initial non-uniformity in creep rate because of the more rapid cooling of the larger surface area on the more highly strained side of the cladding. Although this fundamental mechanism is clear the actual simulation by the model has turned out to be much more complex than for the indirect heating situation. Whereas it has been shown that with indirect heating the pellet/clad eccentricity dominates in determining the temperature non-uniformities, the same does not apply with direct heating. Here the pellets merely act as a damper on the clad temperature changes during ballooning because of their thermal capacity although they can also cause cold spots on the cladding during the initial heating to the test temperature. Thus other sources of non-uniform creep must be studied. These include clad wall thickness and material property variations as well as non-uniform heat transfer characteristics. Analysis of the many possibilities is still in progress aimed at finding a consistent and physically reasonable set of assumptions which will fit the current data as well as the other published data. Fig. 5a and 5b illustrate predictions for two sets of cases based on the rod powers measured in the current tests and showing the predicted effect of rod power and time assuming firstly a non-uniform cladding with a uniform heat transfer and a uniform tube with non-uniform heat transfer. The latter is seen to give by far the better agreement with the measurements in both the actual strain level and in the trend with power and time. This suggests that at least in this rig non-uniformity in heat transfer conditions dominates over the tube variability. The agreement in strain level with other published data suggests this is generally true also.

Group II and III tests

The aims of the Group II and III tests were to provide well characterized data in the temperature regions where the material properties are complex and improvements in the modelling were thought to be necessary, and to confirm that burst strains are generally low under relevant conditions. The anisotropy of Zircaloy creep decreases rapidly once the pure alpha phase temperature is exceeded thus removing the main source of the difference between direct and indirect heating. Circumferential temperature gradients are even more important in limiting failure strains in the mixed phase region so to demonstrate the maximum ballooning the majority of tests were done with low power direct heating. Both inert and oxidising atmospheres were used to help development of the oxidation/creep interaction model.

Fig.6 shows the times to fail for the various test temperatures as a function of the internal pressure. Some RODSWELL predicted curves are shown for comparison but the newly installed oxidation/creep interaction model has not yet been operated successfully in the intermediate temperature range. The points to note in Fig.6 are: only a small effect of oxidation on strength below 1273K, reasonable agreement of the predictions particularly at the lower temperatures and shorter times, and a predicted and observed cut off of the failure curves (at about 10 and 8 bars pressure) due to oxidation strengthening. Some improvement of the model can be made here but the overall behaviour seems reasonable.

Fig. 7 shows the failure strains as functions of temperature in both inert and oxidising atmospheres. Fig. 8 shows the same data with summaries of the direct heating test data of Chung and Kassner [4] for ramp tests at 5K/s, Hindle [3] for isothermal tests and Hoffman and Raff [5] for furnace heating isothermal tests. The various data sets are consistent in showing a rapid drop in failure strain above 800°C for short time isothermal tests or ramp tests with very low values in oxidising conditions. High failure strains are however seen in inert atmosphere tests at long times to fail. This is illustrated further in Fig. 9 which show the failure strains as a function of time for these tests and for some early tests of Jones et al. [6] and give RODSWELL predictions for the inert atmosphere mixed-phase temperature region for comparison. The predictions shown assume tubing of +/- 2% wall thickness variation both in and out of phase with a +/- 20% heat flux variation. The trend of the measurements with time is predicted quite well suggesting that the creep modelling, which includes creep of each individual phase as well as superplastic grain-boundary sliding accounting for the increased ductility at low strain rates is reasonably adequate. However the strains are generally overpredicted. One reason for this is thought to be the assumption in the model of a constant strain-rate/stress sensitivity (n = 5) in the creep equation for alpha-phase Zircaloy. This parameter determines the degree of localisation of strain for a given temperature or wall thickness non-uniformity and has in fact been shown to vary with temperature for example by Garde et al,[7] Donaldson [8] and Raff [9]. Fig.8 includes their 'n' values versus temperature and the corresponding RODSWELL calculated failure strains for isothermal furnace heating conditions. The use of a temperature dependent "n" in the model is seen to account for some of the observed variation of strain with temperature.

An alternative explanation for low total failure strains considered was that the local failure strain, which again is assumed constant in the model (at 300%) in fact is low in this temperature region. However a study of metallographic sections at the burst position revealed very high local strains in all these tests - Fig. 10 is an example and calculations showed the total strain to be very insensitive to this parameter, so this possibility was excluded.

As regards calculations in aid of the high temperature tests in steam unfortunately the code did not perform well because of numerical difficulties. This part of the code had been installed only recently and had not been tested extensively before. Although, when satisfactory operation was achieved the results were qualitatively encouraging in predicting a failure cut-off and a ductile/brittle transition it is clear that further modelling improvements are needed. The critical point is thought to be the number of active cracks formed in the oxide layers during ballooning because these, rather than temperature variations are dominant in localising the strain. Fig.11 shows examples of failed tubes with just one and several cracks and Fig.12 plots the number of active cracks against the failure strain.

Axial Distribution of Strain

In addition to the ballooning failure strain itself a further point of interest is the axial extent of the strained portion of the cladding because this affects the probability that balloons in neighbouring rods can interact to produce a blockage. Hindle[3] has demonstrated that axially extended balloons can be formed in directly heated tubes over a wide range of temperatures and rod pressures provided the temperature is quasi-steady-state and the time to fail is greater than 30-50 seconds. The stabilising mechanism is again thought to be the better cooling of the more highly strained portion of the tube which reduces the strain rate giving extra time for strain elsewhere. The effect is predicted to increase with surface heat flux and time as Fig. 13 illustrates. This shows for example how the strain in a portion of a tube with original wall thickness 1% more than at the failure position (or which was 1.6 K lower in temperature) would increase with time for heat transfer conditions equivalent to the three cases considered here. Considerable stablisation of strain is expected under the two higher heat transfer conditions with negligible effect with the lowest heat flux. Fig. 14 shows the actual measured strain distributions for the three situations. The stabilisation pattern is seen to follow the expected trend with heat flux but once again the predicted effect of time is not apparent.

Axial distributions of strain for the group II tests with lower heat flux are shown in Fig. 15. Here the low stability observed for alpha-phase Zircaloy is again seen at short times to failure but lower strain rates give much more extended balloons. This can be attributed to the superplastic strain mechanism which dominates at low strain rates and temperatures near to 900 ^{o}C and has a very low stress and temperature sensitivity. This is confirmed by the 950 ^{o}C tests in which the maximum strain at low strain rates is found in bulges away from the peak temperature position indicating a minimum strength at about 900 ^{o}C. This strength minimum was originally found in tests on Zircaloy 2[6] on which the current creep model was based but was not confirmed by later work on Zircaloy 4[5]. The point is however of only academic interest because although capable of generating very stable balloons the strains are low in a realistic steam environment.

To sum up, the conditions favouring high strain and high axial extension of balloons appear to be mutually exclusive.

SUMMARY AND CONCLUSIONS

The experimental programme of ballooning tests described was designed to demonstrate the variability of failure strain with temperature, heating method and heat transfer from the cladding surface. The intention was to provide a consistent data set (same Zircaloy material, same experimental rig) which could be used to test the current understanding of the factors determining clad strain, test the modelling capability of the RODSWELL ballooning code, clarify areas where further modelling is required and provide suitable data to assist such developments. The heat transfer conditions were chosen to cover the range expected in "real" accidents to enable prediction of those conditions leading to large balloons to be checked.

The following observations and conclusions can be made:

The rig design and choice of experimental parameters achieved very well the aim of covering the range of conditions of interest in both small and large break LOCA's.

In spite of difficulties with the repeatability of the heat-transfer characterisation tests which precluded detailed modelling of individual tests meaningful comparisons of results and calculation were possible.

The test results at about 800° C confirmed the expected difference in clad failure strain between direct and indirect heating. They also confirmed current ideas and modelling predictions on the effect of clad heat flux on circumferential temperature gradients and on the failure strain and its distribution in indirectly heated cladding. They did not confirm the expected influence of failure time in reducing failure strains in indirectly heated rods and increasing them in directly heated ones. This has been partially explained by the effect of other stochastic variables and the investigation is continuing.The measurements confirmed that the maximum failure strains obtained in a steam atmosphere are to be found in the high alpha-phase temperature regime. The existence of superplasticity was confirmed at low strain rates in the mixed phase temperature range in inert atmosphere but not in steam at 900° and higher. Modelling of creep in inert atmosphere in the mixed phases was found to be generally adequate but the model for the interaction with oxidation needs further development.

The use of a variable strain-rate/stress sensitivity in the creep equation for alpha-phase Zircaloy would improve the predictions at temperatures below 750° C and in the mixed phase region.

In the high temperature (beta-phase) region the results confirmed that high strains are not to be expected in steam and that oxidation can prevent failure completely. The model, although showing qualitative agreement needs improvement, particularly in order to predict the number of cracks formed in the oxide layers. The tests provide useful data 'for model development in this field.

The work reinforces the prediction that balloon strains sufficient to be a potential problem should only be expected at temperatures near to 800° C during the quasi-adiabatic heat up phase of a large-break LOCA or during the quasi-steady-state phase late in a small-break LOCA. Such balloons are expected to be localised axially, reducing the potential for coolant channel blockage.

1. Casadei F., Laval H., Donea J., Jones P.M., Colombo A. RODSWELL: A Computer Code for the Thermomechanical Analysis of Fuel Rods under LOCA Conditions. Commission of the European Communities Report EUR 8920 EN (1984).

2. Jones P.M., Casadei F, Laval H. Modelling of Azimuthal Effects arising from Interactions between Clad Deformation and Heat Transfer under LB-LOCA Conditions. J. Nuclear Engineering and Design Vol. 79, No. 3, pp 267-276, (June 1984).

3. Hindle E.D., Mann C.A., Reynolds A.E. The Deformation of Zircaloy PWR Cladding Tubes with Low Internal Pressure under mainly Convective Cooling by Steam. CSNI Specialist Meeting, Espoo, Finland, September 1980.

4. Chung H.M., Kassner T.F. Deformation Characteristics of Zircaloy Cladding in Vacuum and Steam under Transient-Heating Conditions: Summary Report. Argonne National Laboratory Report ANL-77-31.

5. Hofmann P, Raff S. Verformungsverhalten von Zircaloy-4-Huellrohren unter Schutzgas im Temperaturbereich zwischen 600 and 1200 ^{o}C. Kernforschungszentrum Karlsruhe. KfK 3168,(July 1981).

6. Jones P.M., Gittus J.H., Hindle E.D. CANSWELL: A Computer Model of Clad Behaviour during a Loss-of-Coolant Accident. CSNI Specialist Meeting. Spatind, Norway, September 1976.

7. Garde A.M., Chung H.M., Kassner T.F. Uniaxial Tensile Properties of Zircaloy containing Oxygen, Summary Report ANL-77-3-.

8. Donaldson A.T. The Effect of Temperature on the Stress Sensitivity of Creep in Alpha-Phase Zircaloy. CEGB Report RD/B/N4701 (December 1979).

9. Raff S. Dissertation fuer Doktor-Ingenieur, Universitaet Karlsruhe 1982.

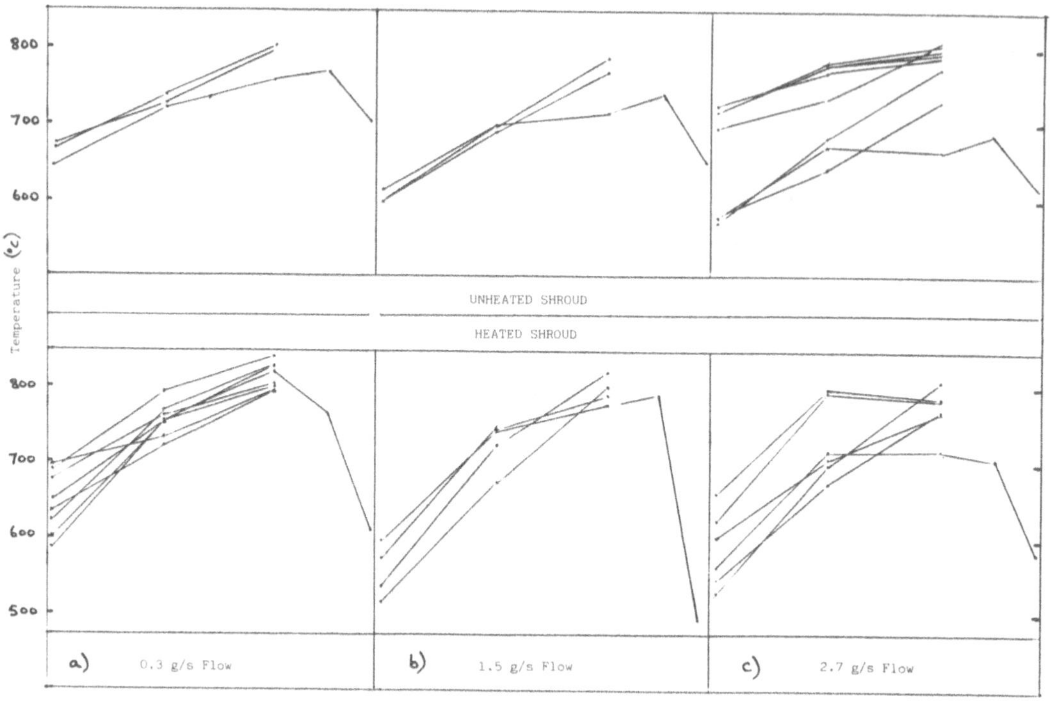

FIG. 1 AXIAL TEMPERATURE DISTRIBUTIONS IN PRE-OPERATIONAL TESTS AND ISOTHERMAL BURST TESTS (800°C NOMINAL)

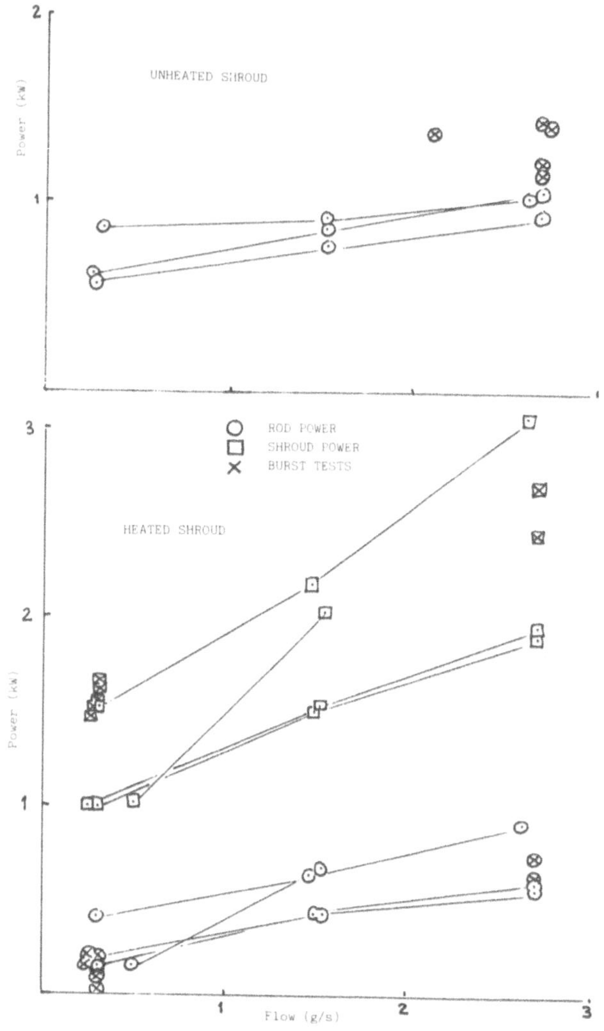

FIG. 2 POWER VERSUS STEAM FLOW FOR PRE-OPERATIONAL AND BURST TESTS AT 800°C NOMINAL TEMPERATURE

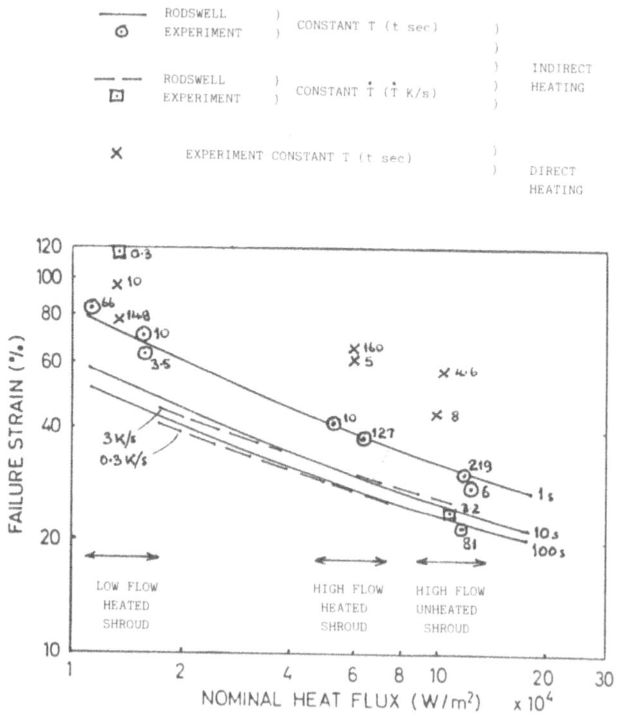

FIG. 3 COMPARISON OF OBSERVED AND CALCULATED FAILURE STRAINS FOR INDIRECTLY HEATED CLADDING

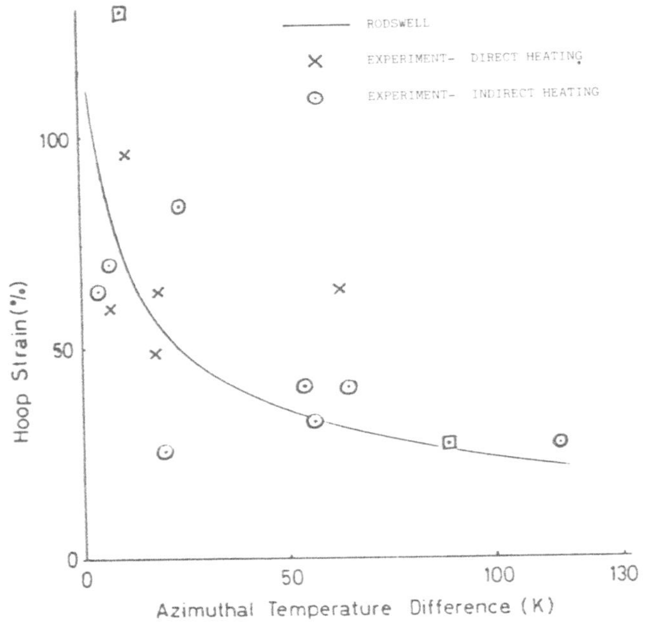

FIG. 4 COMPARISON OF OBSERVED AND MEASURED EFFECT OF TEMPERATURE DIFFERENCE FOR DIRECT AND INDIRECT HEATING TESTS

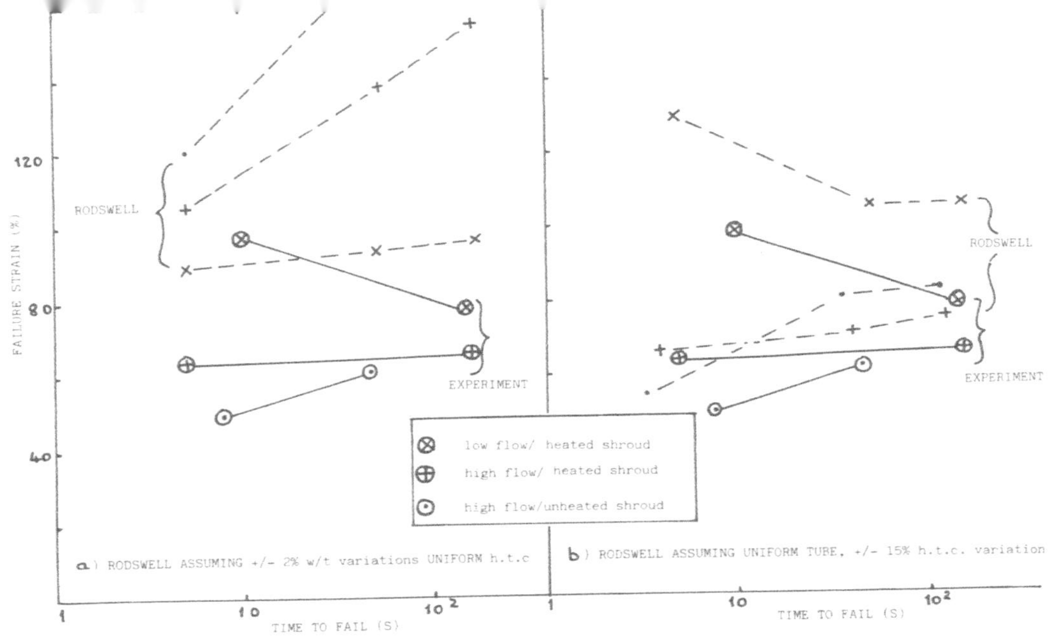

FIG. 5 COMPARISON OF OBSERVED AND CALCUATED FAILURE STRAIN VERSUS TIME FOR DIRECT HEATING TESTS

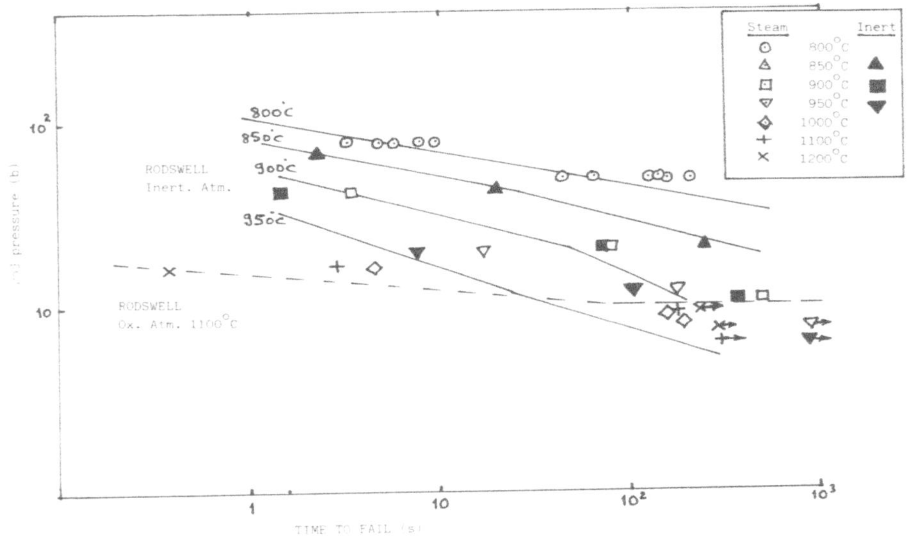

FIG. 6 COMPARISON OF CALCUATED AND OBSERVED TIMES TO FAIL AS FUNCTIONS OF ROD PRESSURE AND TEMPERATURE

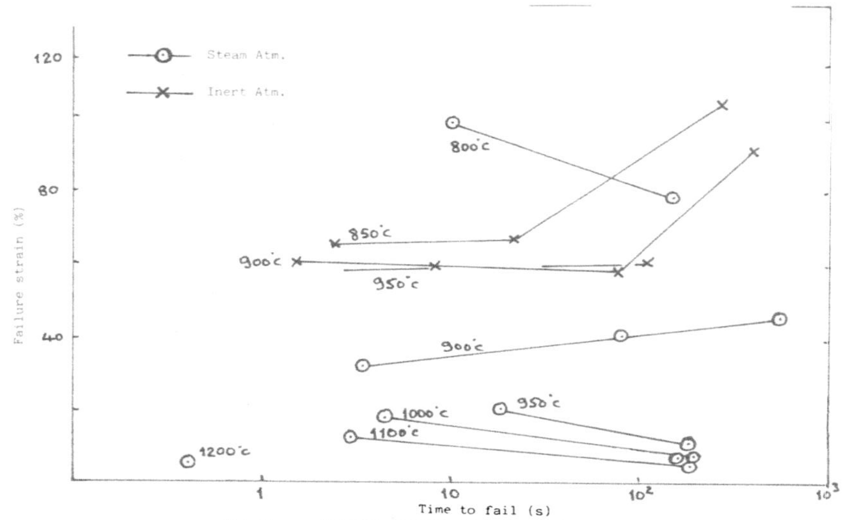

FIG. 7 OBSERVED FAILURE STRAIN AS FUNCTIONS OF TIME AND TEMPERATURE FOR DIRECT HEATING TESTS.

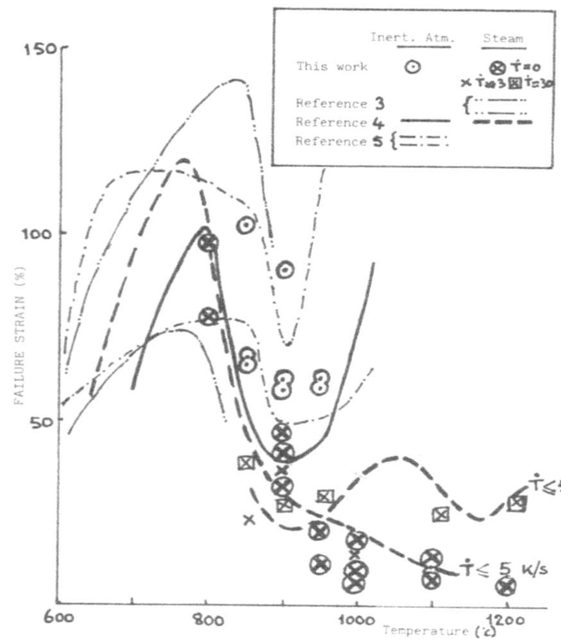

FIG. 8a SUMMARY OF FAILURE STRAIN VERSUS TEMPERATURE DATA

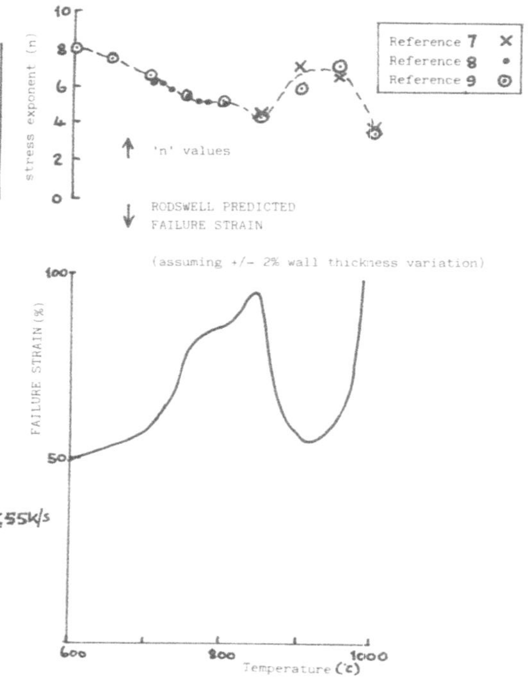

FIG. 8b VARIATION OF 'n' WITH TEMPERATURE AND CALCULATED EFFECT ON FAILURE STRAIN

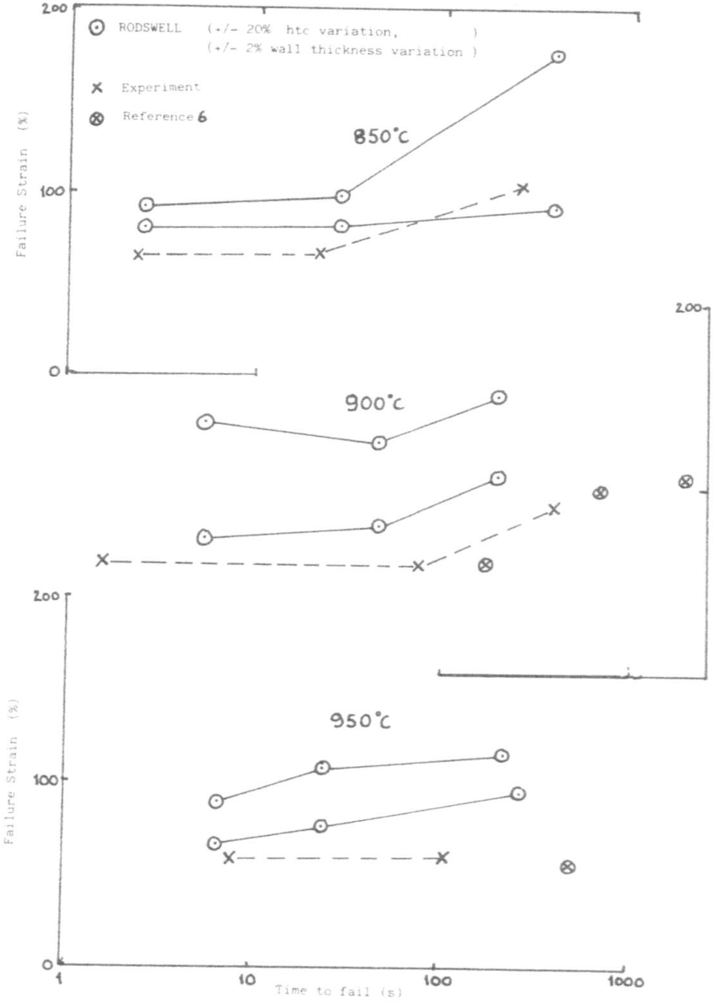

FIG. 9 COMPARISON OF PREDICTED AND MEASURED FAILURE STRAINS FOR DIRECTLY HEATED INERT ATMOSPHERE TESTS IN MIXED-PHASE REGION

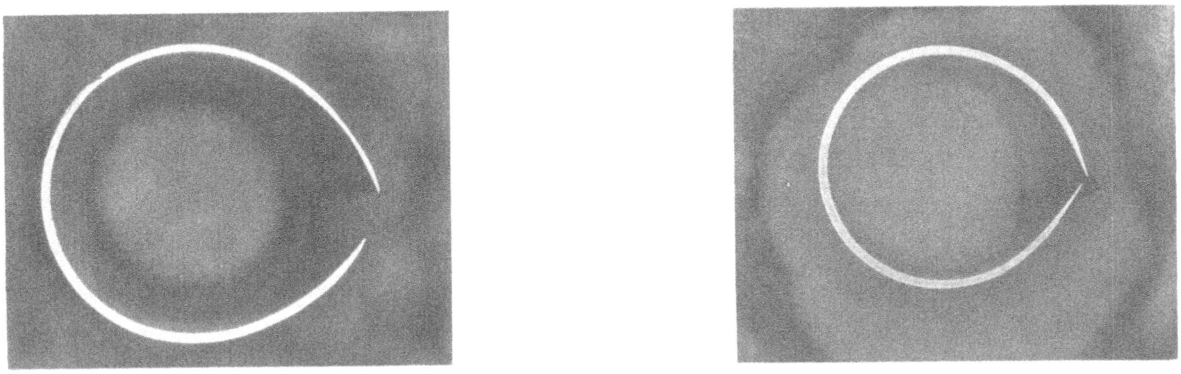

900°c 74s total strain = 58%   890°c 4K/s total strain = 38%

FIG. 10 WALL THICKNESS VARIATION OF FAILED TUBES IN MIXED PHASE REGION

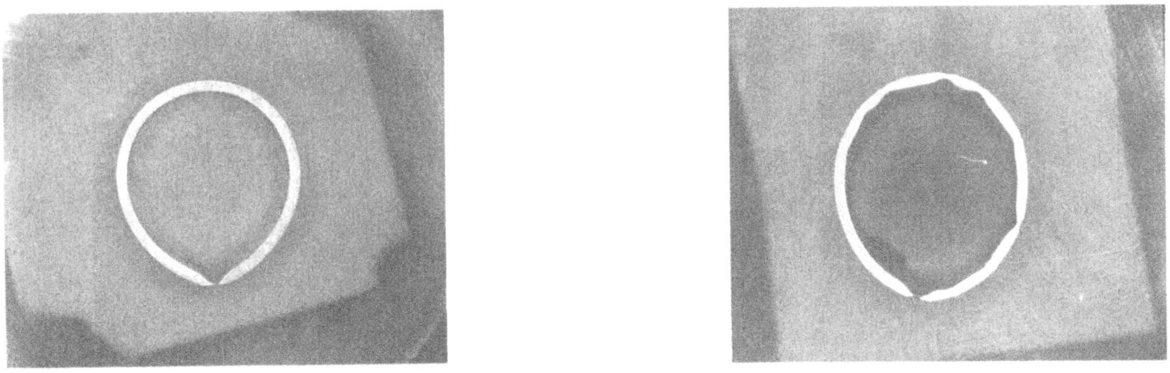

1000°c, 158 s total strain = 7.9%   1000°c, 4.5 s total strain = 18%

FIG. 11 LOCALISATION OF STRAIN BY OXIDE CRACKING

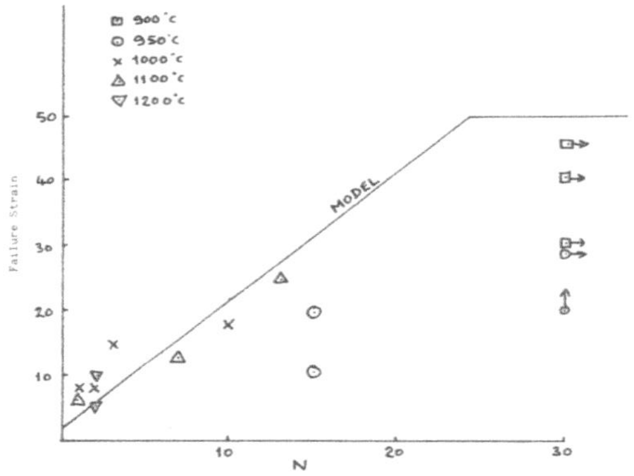

FIG. 12 RELATIONSHIP BETWEEN NUMBER OF "ACTIVE" OXIDE CRACKS AND FAILURE STRAIN

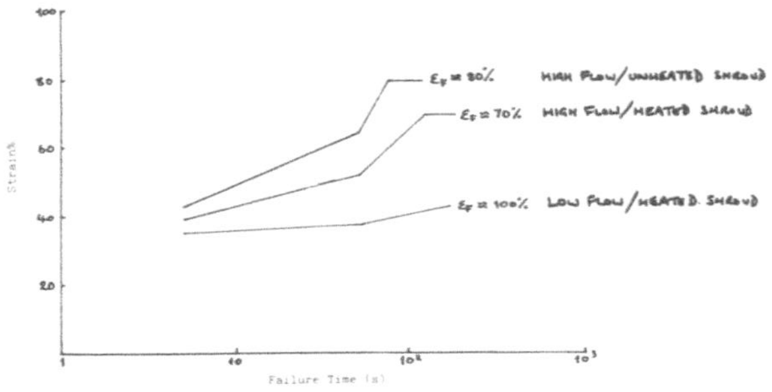

Fig. 13 EXAMPLE OF BALLOON STABILISATION PROCESS - STRAIN IN STRONGER PORTION OF TUBE INCREASES WITH TIME AND HEAT FLUX.

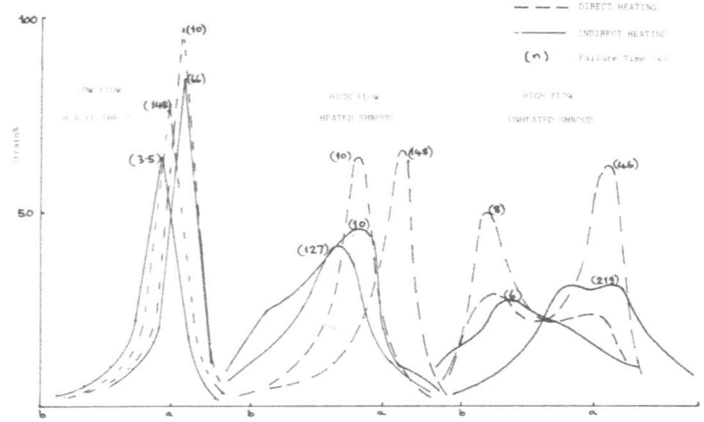

FIG. 14 AXIAL STRAIN DISTRIBUTIONS FOR THE THREE HEAT TRANSFER CONFIGURATIONS

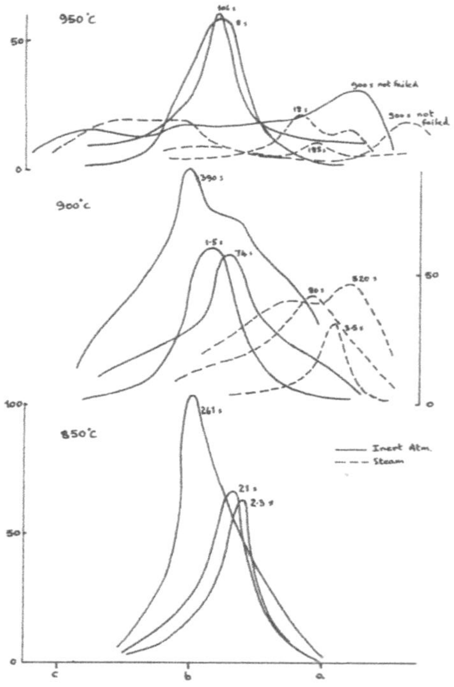

FIG. 15 AXIAL STRAIN DISTRIBUTIONS FOR MIXED-PHASE CREEP

Discussion

Question: J.L. Nigon, CEA/CEN-G

Let us suppose that you are in charge of code development and of modelling of clad ballooning and bursting. Which parameters would you recommend to take into account as the most important, since we seem to be in a rather complex situation.

Answer: P.M. Jones - CEC/JRC

The first priority of thermohydraulic modelling must be to predict the temperature history since this determines whether or not ballooning can occur within the timescale of the transient.
However, to predict the clad failure strain and resultant blockage we need also the cladding surface heat fluxes. Ideally speaking, the heat transfer coefficients and coolant temperatures should be known as well, but only fairly approximate, time-averaged values of these are adequate.

Question: K. Wolfert - GRS

Is it right, when I interprete one of your conclusions that : "co-planar ballooning" can be excluded, if we take into account, a realistic scattering of the parameters influencing the ballooning process".

Answer: P. Jones, CEC/JRC

Unfortunately, this would be going too far! What is shown is that the heat transfer conditions favouring large strains do not favour axial extension of ballooning, hence co-planarity. Therefore, very high blockages in a LB-Loca are not very likely, but they cannot be completely excluded.

Question: R. Deruaz, CEA/CEN-G

1) Did you perform sensitivity studies with the RODSWELL code concerning the effect of pellet eccentricity?

2) How, if the effect is significant, do you manage the comparison of RODSWELL calculations, where you are obliged to introduce assumptions on this parameter, with experimental results for which you do not know the pellet eccentricity?

Answer: P. Jones, CEC/JRC

1) Sensitivity studies showed pellet eccentricity and circumterential variations in heat transfer coefficients, coolant temperature, rod power and clad wall thickness all to be potential sources of significant non-uniformity of strain rate. However, once non-uniform ballooning starts, the pellet eccentricity caused by the "hot-side-straight" mechanism (assuming that the pellets do not move) dominates the situation completely, and the eccentricity is approximately one.

2) The "hot-side-straight" mechanism is a real physical phenomenon caused by anisotropic creep shortening of a ballooning tube. It is expected to approximate quite well the behaviour of alpha-phase Zircaloy in indirectly

heated single rod tests.
In a bundle other stochastic variables play a larger role but the model can still predict the minimum failure strain which is a useful parameter in identifying conditions leading to high blockages.

Question: Dr. Karwat, TU/München

1) Which flow regimes have been applied during the tests and which measured values were taken to derive HTC and to back-up the conclusion that HTC is dominating?

2) Wich method has been applied (measurements?) to estimate the magnitude of azimuthal temperature differences?

Answer: P.M. Jones, CEC/JRC

1) Steam flows range from laminar to turbulent regime (RE Number <1000 to >4000). Attempts were made to derive heat transfer coefficients and coolant temperatures from thermocouple measurements on heater rods, clad and shroud surfaces and in the steam itself. These were generally unsuccessful, so power measurements in the steady state were used to determine the "nominal surface flux". This has been predicted to determine the clad strain, variations of H.T.C., and coolant temperatures independently being secondary effects.

2) Measurements of azimuthal temperature distributions were made by means of two or three thermocouples welded to the clad outer surface at two axial levels. These were in general near to, but not at the failure position.

General discussion concerning on Sessions A1 and A2

Note: The acting chairman, R. Deruaz, proposed and the participants agreed to concentrate the discussion on the problem of ballooning. The discussion was introduced by R. Deruaz as follows:

In the light of experimental and analytical works performed so far (including the work done at Westinghouse and KfK), with different

- balloon shapes
- subchannel flow-area blockages
- axial distributions of balloons with
- coplanarity or not,

it is suggested to discuss particularly the following issues:

1. Modelling:

Are we presently able to define the kind of modelling which is appropriate for introduction into safety codes, keeping in mind that:

- all known experimental results indicate no major problem with blockage cooling,
- one needs for obvious reasons of tractability and costs a modeling as simple as possible.

2. Experiments:

Can we estimate the content of further future work in the field:

- considering again, on one hand, that no significant heat transfer degradation was found so far,
- on the other hand taking into account for example the following aspects:

- test section effects such as size, influence of more or less realistic spacer grids at the entry of the blocked region are always existing in such experiments, and we must be sure that they do not modify significantly the studied phenomena.

- possibly other shapes of balloons, other blockage ratios (more than 90%), or less than 60%) could constitute worse cases than the ones already studied.

Contractors from Berkeley and Winfrith as well as code developpers from Germany and France are especially invited to give their opinion on these points.

Comments: S.A. Fairbairn CEGB/BNL

We cannot say that large balloons will never occur only that they are unlikely. For example large rod strains, albeit of short axial extent, have been observed in the NRU-MT3 test. Thefore, the U.K. approach has been to use a conservative code like BART and to carry out experimental work which gives information on the degree of conservatism in such a code and verifies that the models used are physically reasonable. This is the background to the work at BNL which has been presented to the seminar. We can, in addition, identify an area of considerable interest to us which has not been mentioned in this seminar, namely pellet stack behaviour which has a strong influence on the ballooning process. I think also that it would be useful to develop a model for the heat transfer occuring with actual or near surface contacts with water in a blockage throat as this will help us to model early rewetting of ballooned clad especially for the case when flaps of ruptured clad project into the subchannel flow area.

Comments: Dr. Karwat - TU/München

We should remind the different objectives of blockage cooling and fuel behaviour experiments. Blockage cooling experiments were mainly directed towards studying the coolability of artificially blocked or constrained fuel rod areas. Main aim was to demonstrate that the core cooling acceptance criteria can be fulfilled even for such situations (T < 1200°C, 17% oxidation limit). Fuel behaviour experiments, however, involve the reality of coupling transient fluiddynamics with transient thermomechanics of the fuel clad. Main aim here is to understand and eventually predict fuel ballooning and/or burst events. For both objectives of corresponding tests there is no unique answer to what is a conservative approach to perform tests or interprete obtained results.

Comments: Dr. K. Wolfert, GRS

As concerns the modelling of fuel behaviour, I see no chance, to predict with a thermohydraulic system code the local clad temperature over the whole core with an accuracy of 10-15K, which is required for a realistic description of ballooning processes.

Comments: J.L. Nigon CEA/CEN-G

I express my general agreement with the above comments by Dr. Wolfert and Dr. Karwat. I just would like to add that, according to this fact: - Fuel deformation has no incidence on the overall thermalhydraulic behaviour of the primary circuit-, the Cathare code is not aimed at the modelling but only at the definition of thermalhydraulic conditions at the boundary of the core channel, and in the core channel as long as degradation is not under way.

Question: E. Skupinski - CEC/BXL

I have understood until now, that in choosing in various experimental research programmes test sections with very strong blockages - in the case of the THETIS - programme coplanar blockages of 80 and 90%, with a by-pass-area - this was considered as a "worst case approach" or as a very conservative assumption, and that the aim was to show that cooling of even such "worst-case" blockages is possible. Now, in a number of experiments, including those of this research programme, it has been demonstrated that such strong blockages are coolable.

Can we consider now that the blockage problem is solved as a safety issue, or and why do we need further experiments?

Answer: C.A. Cooper, UKAEA/AEE-W

The THETIS experiments have investigated the cooling of blockages more severe than those arising in the UK PWR Safety case and, as you have suggested, the THETIS blockages were coolable. However, this only shows that the THETIS fuel rod simulators were coolable under the conditions tested in the THETIS experiments, and does not provide a direct commentary on the coolability of fuel assemblies containing blockages in a reactor LOCA. This is primarily because the fuel rod simulators used in the THETIS experiments, in common with those used in the previous FLECHT and FEBA blockage heat transfer experiments, do not give an excellent simulation of the thermal response of a reactor fuel rod in a reflood transient, having no simulation of the fuel-to-cladding gap and being clad in Inconel rather than Zircaloy.

Thus, in order to comment on blockage coolability for the reactor we must use computer codes and a two-stage process is involved in demonstrating the acceptability of the clad ballooning safety case. The first stage is to demonstrate that the codes can adequately calculate the results of both in-pile and out-of-pile balloning and blockage heat transfer experiments. When this point has been demonstrated the codes can be used to predict the response of reactor fuel assemblies in a LOCA.

The current position in the UK is that the licensing authority accepts that ballooning leading to a clad temperature runaway is unlikely to occur but has indicated that it requires a programme of future work in support of the clad ballooning safety case to be mounted. This programme includes code validation and a further blockage heat transfer experiment, in the ACHILLES rig at Winfrith, is planned to provide data for code validation.

Comments: D. Mewes - U/Hannover

I think the gained physical understanding of ballooning effects should not be judged by the criteria how to implement it into a code system.
The present results on "ballooning" should be considered in the sense that the engineer knows now more about the fluiddynamics in the blocked area. He learned to know that the fluiddynamics obviously changes, but that it does not cause the heat transfer to break down.

Comments: Dr. S.N. Aksan - EIR

It is rather difficult to define experiments which will be representing real reactor cases in relation to ballooned fuel rods causing blockages in the bundle. Even under relatively severe conditions defined in LOFT (Loss-fluid-test) double-ended rupture LOCA experiments, could not be seen really significant fuel deformations and blockages of fuel bundles. LOFT is among the existing test facilities probably the best simulation of large PWR's. Any defined blockage experiments has to be considered as hypothetical.

Parallel Sessions Research Area A

The loss of coolant accident (LOCA) and the functioning and performance of the emergency core cooling system (ECCS)

Session A3

Chairman: G.L. Shires (UKAEA/AEE-W)

Session secretary: E. Skupinski (CEC/BXL)

2D EFFECTS IN THE CORE DURING THE REFLOODING PHASE OF A LOCA

R. DERUAZ, P. CLEMENT, J.M. VETEAU
CEA-CEN GRENOBLE - FRANCE -

SUMMARY

In a PWR core the power is not identical from an assembly to another and this may result in 2D phenomena which have to be studied in order to know whether it is necessary to take them into account in safety codes or not and, if yes, to define the simplest models suitable for that. An experimental programme has been conducted during which numerous aspects were investigated. Boil up and reflood tests (constant and oscillating flow rate) have been performed which show the effect of various parameters and somewhat clarify the problem of fall back and cross flows as well as their consequences on quench front propagation and heat transfer in the dry region.

1. INTRODUCTION

Complex phenomena occur during the reflooding phase of a PWR core and resulting modeling difficulties explain that a lot of reflood tests have been/are performed in many laboratories. However these experimental works especially make use of test sections like tubes, annuli, or heater rod bundles of small radial dimensions where phenomena are mainly axial. In an actual PWR, additional phenomena are expected to occur due to the non uniform radial power distribution in the core : cross-flows of water and steam between hot and cold assemblies may take place as well as fall back from the upper plenum to the core in particular along cold assemblies where, if radial mixing of steam is not complete, the steam velocity is smaller than in hot ones.

These 2D phenomena could significantly modify the thermal behaviour of fuel rods in comparison with predictions of codes which include models based on experiments where 1D phenomena only are occuring.

The PERICLES programme (a joint CEA-EDF-FRAMATOME programme) - part one - aims at studying these 2D effects in well defined conditions and providing the needed data base to help writing and adjusting suitable models for safety codes.

2. TEST FACILITY [1] [2]

The test section which is used here is a rectangular one : three 7x17 full length heater rod assemblies are arranged side by side in order to form a 7x51 bundle which is placed in an electrically heat traced housing. Each fuel pin simulator consists of 3 helical Nichrome V wires which are star connected at the top in order to be able to well represent the geometry of the upper part of fuel rods, and are embedded in boron nitride. The pitch of the heater wires is varied to get a chopped cosine axial power profile (11 steps), the axial peaking factor being equal to 1.6. The cladding (9.5 mm O.D. and .6 mm thick) is made of stainless steel.

The assemblies are supplied at their bottom only with two independent electrical sources which power is adjusted to get the desired radial peaking factor and can be controlled to represent the residual power decay. In a given assembly the radial power distribution is uniform and the hot assembly is generally placed between two cold ones.

The whole bundle includes 85 instrumented pins, six thermocouples each, and four types of pins provide 24 instrumented elevations. In addition, in each assembly, fluid temperature and pressure drop are measured at 6 elevations.

The upper plenum corresponding to each assembly is equipped with a steam-water separator : the steam flow rate is measured at each outlet, while water is either collected (no fall back) and its flowrate measured, or is drained on the upper core plate where it may either accumulate or fall back in the core through the upper tie plates.

Boil up and reflood tests have been carried out.

3. BOIL UP EXPERIMENTS

In that case, the liquid inventory is held constant all along the test by equalizing inlet and outlet flow rates. Tests are performed at 3 and 6 bar and the main parameters are the radial peaking factor which is varied from 1 to 1.85, the liquid inventory and the power level. Liquid entrainment is negligible.

It was found, in all cases, that [3] :

- the two-phase level can be considered as identical in cold assemblies (CA) and in the hot one (HA).
- this level (at given pressure, liquid inventory and inlet subcooling) depends upon the average power and not upon the radial power distribution, and can be calculated by making use of void fraction models adjusted from tests at fxy=1.
- a simple model assuming perfect radial mixing of steam in the wet region allows a reasonably good prediction of heat transfer in the dry region (figure 1).

4. FORCED REFLOOD TESTS AT CONSTANT INLET FLOW RATE

Starting with tests at uniform radial power for reference and the base case (P=3 bar, G=3.6 g/cm^{2}s, ΔT_{sub}=60°C, T_{wi}(HA)=600°C and fxy=1.43) numerous aspects were then investigated at constant or/and decaying power, including in particular the effect of :

- classical external parameters like pressure P, inlet mass velocity G, inlet subcooling ΔT_{sub}, initial wall temperature in the hot assembly T_{wi}(HA)
- the radial peaking factor fxy for unchanged conditions regarding the HA power and T_{wi}(HA)
- fall back from the upper plenum to the core
- guide thimbles
- the relative position of HA and CA

In addition, and so as to help analysis of cross-flows, some tests were performed at fxy $\neq$ 1 and T_{wi}(HA) = T_{wi}(CA) as well as fxy=1 and T_{wi}(HA)$\neq T_{wi}$(CA) - investigation of T_{wi} and fxy separate effects -
There is no place here to present all the results which were got and we'll focus our attention on some aspects which seem of particular interest.

4.1. Common features to all results

Whatever the test conditions may be it appears that, as soon as fxy$\neq$1, the comparative behaviour of the assemblies is such that (figure 2) :

- the maximum temperature rise is larger in the HA than in the cold ones

- quench front progression is slower in the HA in, roughly, the lower third of the test section, and then becomes very close to the CA one.

In addition, it is to be mentionned that pressure drop differences between HA and CA are very small, practically not detectable ; steam flow rates at the outlet of every assembly are close together ; but liquid entrainment is, after a time, much larger in the hot assembly.

When modifying P, G, ΔT_{sub}, the classical overall influence of these parameters is found, but the comparative behaviour of assemblies is not significantly modified.

These results still stand for other relative positions of HA and CA while differences in cold and hot assemblies behaviour are found to be somewhat reduced when guide thimbles are represented [4] [5].

4.2. Effect of the radial peaking factor

Increasing fxy for given initial conditions and proper power in the HA (so placing the HA in colder surroundings) from 1 to 1.85 leads to results which are presented on figure 3 :

- during the major part of the transient, quench front progression and heat transfer in the dry region of the HA are enhanced.

- however, very early in the transient, it is just the opposite which occurs : both the quench front velocity and heat transfer are reduced in the HA and, finally, contrary to what could be expected, the turnaround temperatures have the same order of magnitude.

4.3. Fall-back

The comparison of test results with and whitout possibility of fall-back from the upper plenum to the core (figure 4) shows that, if fall-back occurs, it's effect on the thermal behaviour of heater rods is not significant.

The flow rate of fall-back is not really measurable and it is easy to show [4] that there is no way to get it from mass balances but by making some simplifying assumptions.

Other useful comparisons regarding effects of possible fall-back can be made which contribute to clarify the problem. So it appears that :

- water accumulation in the assemblies is not modified

- liquid entrainment, as measured at the outlet of the test section is very close to the value calculated from water accumulation on the upper core plate.

- the total steam flow rate at the outlet remains unchanged.

Taken as a whole these results show that no significant fall-back is occuring in the test conditions chosen for Pericles programme.

Experiments were carried out on Serops facility with air and water at atmospheric pressure, with the same geometry, in order to determine the maximum gas velocity which allows fall back to occur. The transposition of results to steam and water at 3 bar gives a value close to 7m/s. It is seen on figure 5 that steam velocities smaller than that at the outlet are only obtained early in the transient, at a stage where liquid entrainment is very small. This is partly due to the complete radial mixing of steam which increases steam velocity in cold assemblies where, a priori, there is the largest probability for fall back to take place.

We failed, despite several attempts including the base case with no residual power, in finding test conditions such that fall back could occur.

4.4. Cross-flows

Fall back not occuring significantly, cross-flows become the main mechanism to be considered in case of a 2D reflooding process. This simplifies somewhat the problem which however remains complex as, in principle, cross-flows of both steam and water can take place and this either in the wet or in the dry region - or both - Simple calculations involving energy and mass balances lead however to interesting conclusions [4].

4.4.1. Cross-flows of steam

Calculations of steam flow rate at the outlet under the assumption of either perfect or no radial mixing in the rewetted region and no radial mixing in the dry region lead to $M_{v2}(CA) < M_{v2}(HA)$ due to less steam generation in the dry region of the CA, while measured values are such that $M_{v2}(CA) > M_{v2}(HA)$. This indicates clearly that steam cross-flows ΔM_v occur in the dry region and from HA to CA. An example of calculated ΔM_v in case of large fxy is given figure 6.

4.4.2. Cross-flows of water

It is suggested here that liquid cross-flows (ΔM_L) mainly take place in the wet region where differences in steam generation between HA and CA as soon as $T_\ell = T_{sat}$ can become significant and result in driving forces responsible of liquid cross-flows.

Another possible mechanism is liquid overflowing from CA to HA due to the fact that $z_{quench}(CA) > z_q(Ha)$

ΔM_L was calculated using again energy and mass balances and it is seen on figure 6 that ΔM_L can reach very large values at high fxy, increasing significantly the flow rate entering the hot assembly in comparison with values imposed at the very inlet of every assembly. The fact that liquid cross-flows start before $T_\ell = T_{sat}$ at the quench fronts, at a time where liquid density differences could produce only small liquid cross flows, shows there is a possibility that the second mechanism postulated above actually takes place.

The existence of such ΔM_L from CA to HA explains now some results presented above :

- ΔM_L influences quench front progression which rate is not very different in HA and CA despite the differences regarding power and initial cladding temperature
- ΔM_L results in a large liquid entrainment flow rate in the HA which has two consequences :
  . the heat transfer coefficient in the dry region of the HA increases with fxy (if we except the very beginning of the transient), and this despite the loss of steam to CA.
  . this increased precursory cooling influences also the quench front velocity on the HA which was found to increase with fxy at the same stage of the transient.

In addition the presence of much more liquid downstream of the HA quench front results in an increase of pressure due to interfacial shear and makes possible ΔM_v in the dry region from HA to CA.

Finally, early in the transient, at a time where $\Delta M_v \approx 0$, it seems that ΔM_v already exists due to vapour escaping the HA at the level of the quench front : heat transfer in the dry region is degraded compared with the no radial mixing case, reducing quench front progression which, in turn, contributes to the degradation of heat transfer in the dry region of HA. This trend is stopped as soon as ΔM_L is sufficiently large to compensate the steam loss.

4.4.3. Quench front progression

Such phenomena may have a complex effect on quench front progression. However, and as it could be expected, it has been shown that quench front velocities are such that power release at the quench fronts are identical in HA and CA [4]. This relationship is clearly of interest in the view of quench front progression calculations in 2D situations with 1D tools.

5. FORCED OSCILLATING FLOW RATE REFLOOD TESTS

Codes generally predict oscillations of the flow rate at the inlet of the core during reflooding. These oscillations are however not found on all system loops under operation. It was nevertheless decided to check the influence of inlet flow rate oscillations upon 2D phenomena : are cross-flows modified and, especially, do these oscillations result in oscillations of the outlet steam velocity such that, in the decreasing phase, the steam velocity allowing fall back to occur could be reached ?

Oscillations were achieved making use of a special device allowing reverse flow (figure 7). The choosen period for oscillations was 6 seconds and the total amplitude was varied from 20 to 26 g/cm^2s, the negative alternation corresponding to 8 g/cm^2s. These tests were conducted at atmospheric pressure in order to avoid perturbations due to the feed back of the pressure control system of the loop.

Comparisons of the results with the cases where are imposed the average, non oscillating, flow rates at the inlet show neither an effect on quench front progression which remains very smooth as it was already found in the 1D case [6], nor on cladding temperature evolution (figure7) Moreover oscillations of the outlet steam flow rate are not sufficient to allow fall back to occur.

However, a question must be asked : would the results be the same with actual fuel rods for which potentialities exist for getting oscillations of the quench front in the same hydraulic boundary conditions ?

6. CONCLUSION

An effort has been made to try to clarify the problem of 2D phenomena in the core during reflooding. The effect of the main parameters has been investigated, physical mechanisms like cross-flows of steam and water as well as fall back of water from the upper plenum to the core have been studied and an interesting relationship between quench front propagation in hot and cold assemblies was found. Work is going on trying, in the light of the present results, to select the simplest modeling which would be able to take into account the major aspects of 2D phenomena in the core during the reflooding phase.

REFERENCES

[1] J.M. VETEAU, P. CLEMENT, R. DERUAZ
Reflooding of a PWR bundle. Pericles programme and first results
European Two Phase Flow Group Meeting. Zurich 1983.

[2] Indirect Action Programme on the Safety of Thermal Water Reactors -Project 3A-
Progress Report n°7. April-September 1983.

[3] J.M. VETEAU, P. CLEMENT, R. DERUAZ
Reflooding of a PWR bundle. Pericles Programme and first results
11th Water Reactor Safety Research Information Meeting.
Gaithersburg 1983.

[4] Indirect Action Programme on the Safety of Thermal Water Reactors.
-Project 3A-
Progress Report n°8. October 83-March 84.

[5] J.M. VETEAU, P. CLEMENT, R. DERUAZ
2D effects during reflooding of a PWR : forced reflood tests of Pericles programme.
European Two Phase Flow Group Meeting. Rome 1984.

[6] P. CLEMENT, R. DERUAZ, J.M. VETEAU
Reflooding of a PWR bundle. Effect of inlet flow rate oscillations and spacer grids.
Intl Meeting on Thermal Nuclear Reactor Safety. Chicago 1982.

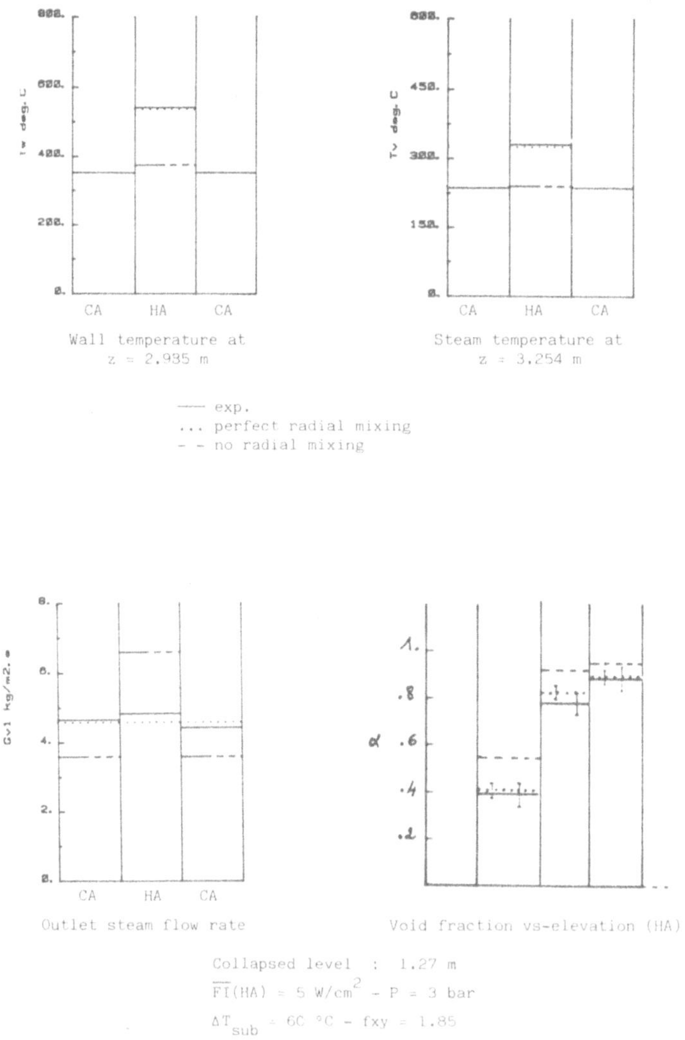

Figure 1 : BOIL-UP TESTS. Comparison between calculations and experimental data.

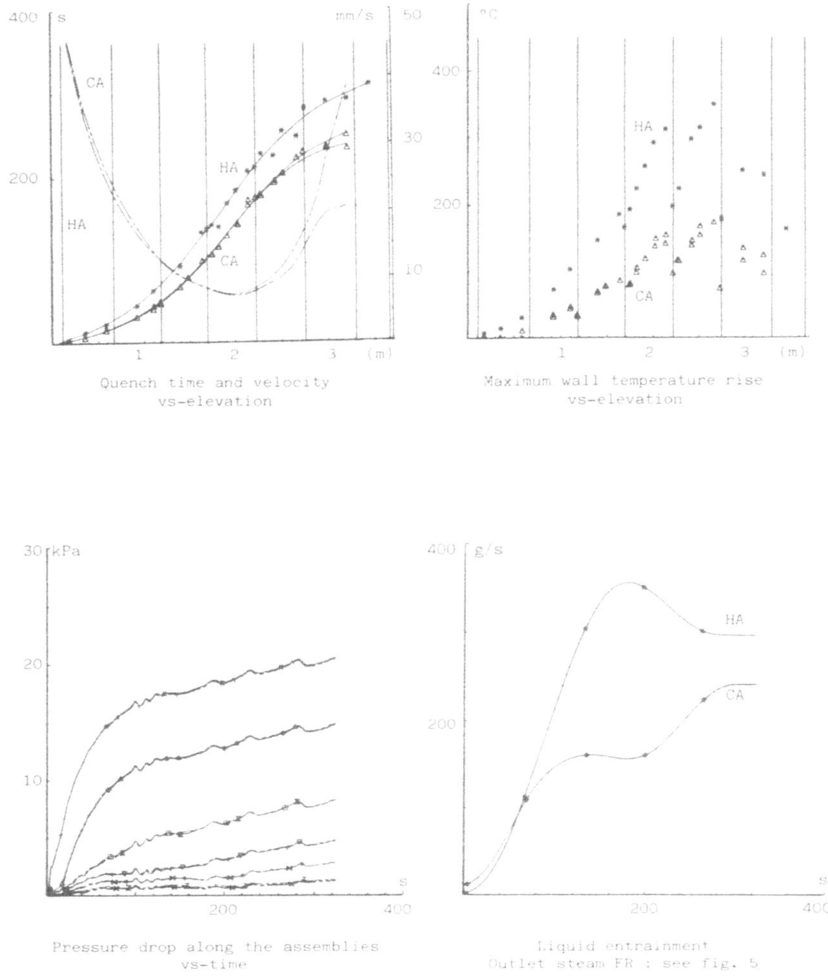

P = 3 bar - G = 3.5 $g/cm^2 s$ - $\overline{FI}$(HA) = 4.2 W/cm^2
ΔT_{sub} = 60°C - T_{wi}(Ha) = 600°C - fxy = 1.85

Figure 2 : COMPARATIVE BEHAVIOUR OF COLD AND HOT ASSEMBLIES

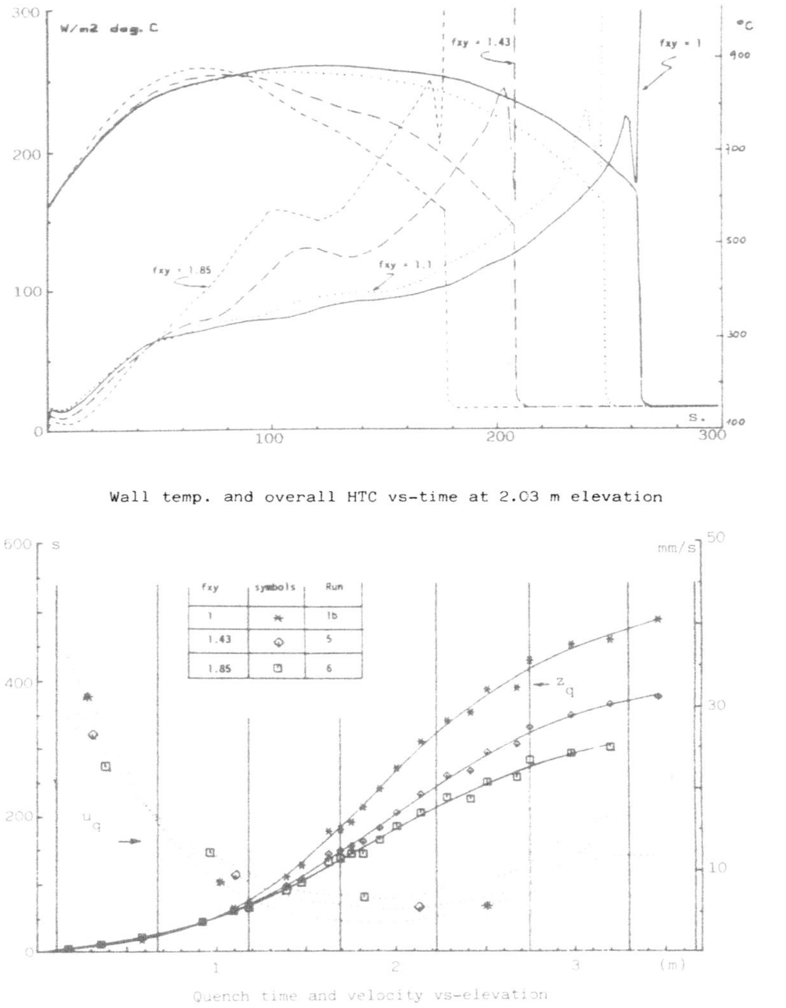

P = 3 bar - G = 3.6 $g/cm^2 s$ - $\overline{FI}$(HA) = 4.2 W/cm^2
ΔT_{sub} = 60°C - T_{wi}(HA) = 600°C

Figure 3 : INFLUENCE OF THE RADIAL PEAKING FACTOR ON HOT ASSEMBLY BEHAVIOUR

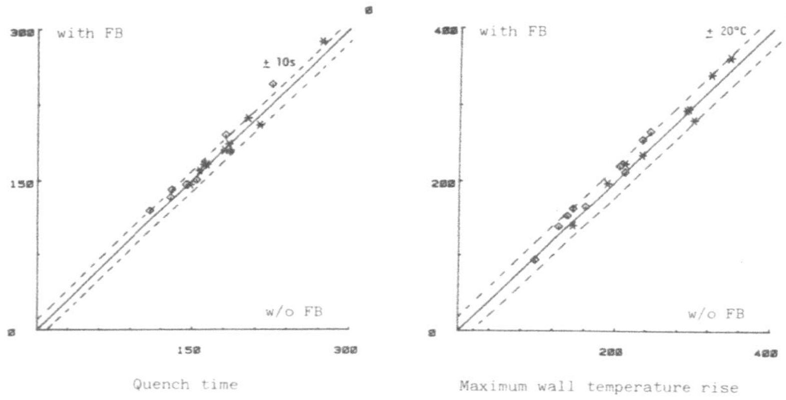

Figure 4 : EFFECT OF FALL-BACK at 2.03 m-elevation (8 different test conditions)

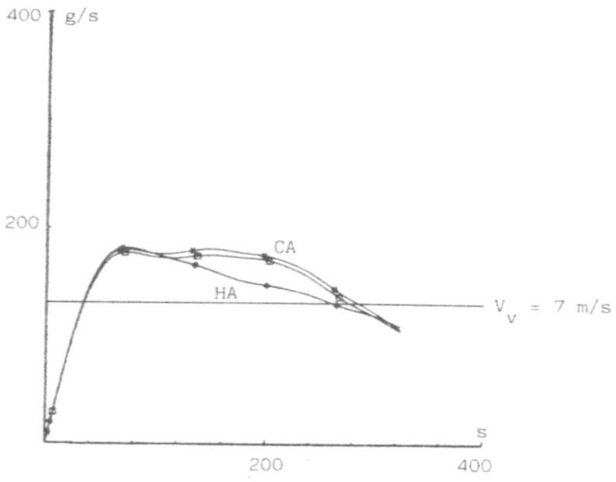

P = 3 bar - G = 3.6 g/cm^{2}s - $\overline{FI}$(HA) = 4.2 W/cm^{2}s

ΔT_{sub} = 60°C - T_{wi}(HA) = 600°C - fxy = 1.85

Figure 5 : OUTLET STEAM FLOW RATE vs.TIME

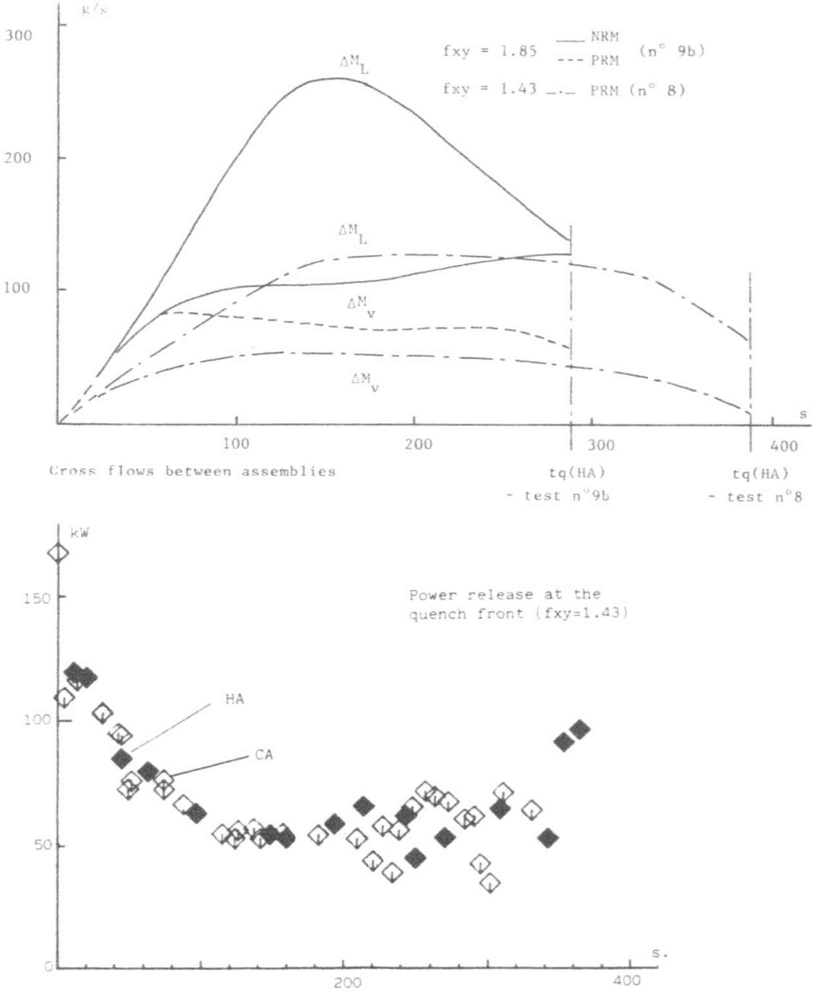

P = 3 bar - G = 3.6 g/cm^{2}s - $\overline{FI}$(HA) = 4.2 W/cm^2

ΔT_{sub} = 60°C - T_{wi}(HA) = 600°C

Figure 6

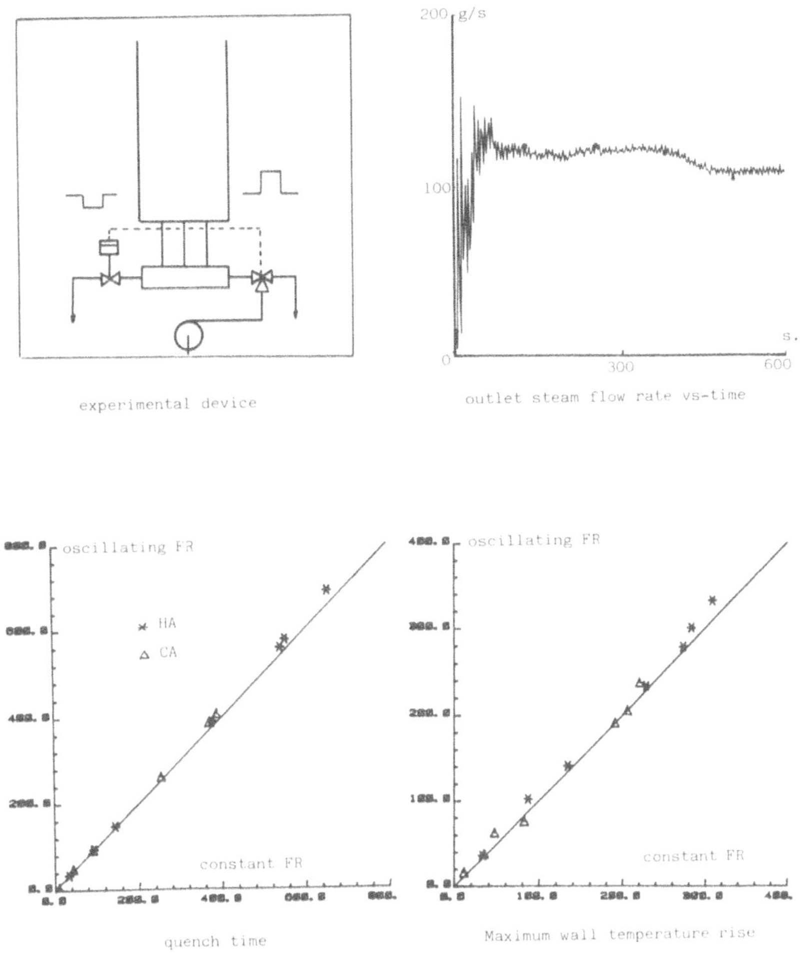

$P = 1$ bar - $G = \overset{+13}{-8}$ g/cm^{2}s - $\overline{FI}(HA) = 3.2$ W/cm^2

$T_{sub} = 20°C$ - $T_{wi}(HA) = 600°C$ - $fxy = 1.85$

Figure 7 : OSCILLATING FLOW RATE TESTS

Discussion

Question: S.A. Fairbairn - CEGB/BNL

In our project we have found that a comparison of calculated and measured values of test section outlet steam flows suggests that the measured value is overestimated because of the evaporation of water by superheated steam in the zone of phase separation. Do similar considerations apply to Pericles?

Answer: R. Deruaz - CEA/CEN-G

In the Pericles facility is made use of spacer grids which are equipped with mixing vanes. This kind of grids is very efficient in improving cooling rates and then significant superheating of steam does not occur at the outlet of the test section.

Question: E. Skupinski CEC/BXL

In which respect do your results agree with results obtained with SCTF experiments?

Answer: R. Deruaz - CEA/CEN

SCTF and Pericles are both dealing with the investigation of 2D phenomena in the core during reflooding, but important differences exist however between these two test facilities, Pericles being smaller, allows for a more analytical approach of the problem. As far as I know (France is not in the 2D/3D agreement) following statements may be made:

- similarities with SCTF:
  . quench front velocities are larger or equal in the low power sub-assemblies, but the collapsed liquid level is identical in the whole bundle.
  . fall back, if any existing, is not claimed to have a significant effect.
  . an enhancement of heat transfer in the high power bundles is observed, which seems to compensate roughly the difference in power supply.
- differences with SCTF: this heat transfer enhancement:
  . is not obtained with Pericles in the early phases of the transient.
  . is explained by the SCTF team by the so-called "chimney effect", which seems to correspond both to more vapour (due to larger steam generation) and to more liquid (cross-flows of water) in the high power sub-assemblies. We agree with the strong cross flows of water, whereas, as was shown by Pericles data, cross flows of steam take place in the direction from hot to cold assemblies. SCTF, using horizontal Δp-measurements and homegeneous flow modelling, nevertheless agrees with Pericles by stating that cross flows of water represent the most significant contribution to heat transfer enhancement.

Question: G.L. Shires - UKAEA/AEEW

What conclusions do you draw from your experiments as to the potential use of 1-D codes for assessing reflooding in a real reactor?

Answer: R. Deruaz - CEA/CEN

The conclusion of the work is that we must find a way to take into account 2D effects. This does not mean, however, that a 2D treatment is necessary. Possibly a solution involving parallel channels with connections could be sufficient. Anyway, an attempt in this direction is suitable, keeping in mind the need for simplicity of safety codes.

FLUID DYNAMIC EFFECTS IN THE FUELELEMENT TOP NOZZLE AREA DURING REFILLING AND REFLOODING

A. HAWIGHORST, R. SPATZ, H.KRÖNING, D. MEWES
Institut für Verfahrenstechnik, Universität Hannover,
Bundesrepublik Deutschland
F. MAYINGER
Lehrstuhl A für Thermodynamik, Technische Universität München,
Bundesrepublik Deutschland

Summary

Counter-current flow of a gas and a liquid occurs during emergency cooling conditions in the region of downcomer and the core of ligthwater cooled nuclear reactors. In order to identify the flooding mechanisms, investigations of counter-current flow were performed under a large variety of test conditions. As internals to the flow-duct, perforated plates, a 4 x 4 rod bundle with tie plate and a complete fuel element dummy, were used to simulate hydraulically a section of the upper part of a PWR fuel element. The selected fluid systems were air-water and steam-water. During the steam-water tests the effects of subcooled water injection were also investigated.

The results of the experiments indicate similar mechanisms for counter-current flow in tubes and square ducts with and without internals. In all cases there is a distinct gradient in pressure drop and a remarkable apearance of flow instabilities near the stagnation point.

The mathematical methods leading to a flooding model are outlined. The model is based on the stability condition of the wavy interfaces. The own experimental data and those from the literature are well represented by the derived model.

1. INTRODUCTION

The knowledge about the fluiddynamic behavior of gas-liquid mixtures is an important basis of a great number of technical processes. In counter-current flow the handling of the two phases enables very often an improvement of heat and mass transfer. In general the liquid flows downwards and the gas flow is directed upwards. In the process industry the counter-current flow is used in packed columns, perforated tray columns or film evaporators. Counter-current flow of steam and water can occur also during a hypothetical emergency core cooling of light water cooled nuclear reactors. In the pressurized water reactors of the Federal Republic of Germany (FRG-PWR), two separate emergency core cooling systems are provided, which reflood the core. These are the bottom injection (cold leg injection) and the top injection (hot leg injection). From the top injection the water must pass through the upper plenum and the end-box into the core region (Fig 1). Furthermore, steam will be generated from the bottom and the top. Both coolant and steam will flow upwards through the core into the upper plenum region. The action of steam flowing upward while water flows downward can lead to a condition defined as counter-current flow limiting (CCFL), which tends to restrict the water downflow. Furthermore, water droplets will be entrained and transported by the steam flow through the core and can deentrain in the upper plenum region. This deentrained water and the upward flowing steam can lead to the same fluid behavior discussed previously.

The purpose of the project is to investigate these mechanisms at countercurrent flow conditions in order to improve the understanding of the fluid dynamic effects during refilling and reflooding. The results should make a contribution for the improvement of the reactor computer codes, for example TRAC. In order to investigate the mentioned problem, the experimental work is split in different stages:

- Test facilities:
  - Air/Water
  - Steam/Water
- Internals:
  - Orifices and perforated plates
  - 1/16 Fuelelement (4 x 4)
  - Complete fuelelement (16 x 16)
- Optical observations of flow conditions
- Investigations of droplet entrainment
- Determination of flooding-curves
- Measurement of pressure drop

2. PHENOMENOLOGICAL DESCRIPTION OF COUNTER-CURRENT FLOW

The principle phenomenons at counter-current flow conditions are shown in Fig. 2. The lower part of the diagram gives informations about the carry-over ratio of the injected water dependent on the superficial gas velocity. At lower gas velocities all injected water flows downwards as an undisturbed annular flow. With increasing gas velocity there are greater interactions between the two phases. Waves begin to grow and become instable. Some droplets are entrained and transported upwards by the gas flow (b) . At the flooding point a considerable amount of liquid is carried out. At further increasing gas flows the carry-over ratio raises from 0 % up to 100 % (c). If the gas velocity reaches a definite value a hanging wavy film moves upwards through the flow duct and all injected water is carried out (d). At higher gas velocities there is a concurrent two phase flow above the injection and a single phase gas flow below the injection (e).

The previous discription is valid for a distinct constant water injection rate. At increasing water rates the film thickness is growing and the flow area of gas is decreasing. Therefore flooding occurs at lower gas velocities. The onset of flooding is depending on the superficial gas and water velocities. In the flooding diagram (Fig. 3) the downwards penetrating liquid is plotted versus the upwards flowing gas. The II. and IV. quadrant represent the downwards or upwards directed concurrent flow. Flow conditions in the III. quadrant, i.e. water upflow and gas downflow are physically impossible. In the I. quadrant the possible counter-current flow conditions are limited by the flooding curve. This line indicates the maximum water flow, which can penetrate downwards against a given gas flow.

3. HYDRODYNAMIC STABILITY MODEL OF FILM FLOW AT COUNTER-CURRENT FLOW

In the literature a lot of flooding models were presented. In general they can be devided in three groups, which are based on

- separated film flow models (1, 2, 3)
- instability models of surface waves (4, 5, 6, 7) and
- dimensionless numbers formed by superficial velocities (8, 9, 10, 11).

A comparison of the different models resulted, that all models do not consider the complex flow conditions in the region of the fuel element top nozzle. The parameters of the existing empirical correlations have to be adjusted to the experimental data related to different geometries.

The theory of hydrodynamic stability (12, 13) opens the possibility to extend the model conception of counter-current flow. The mathematical method is based on the stability of wavy interfaces (Fig. 4). The maximum differential velocity between the phases for a disturbance not to change the wavy boundary layer, results from the Orr-Sommerfeld equation of a frictionless flow. In order to calculate the superficial gas velocities the volumetric gas fraction is used as a parameter:

$$\frac{w_{OG}}{C\cdot\sqrt{\frac{k\cdot\sigma}{\rho_G}}}\cdot\frac{1}{\varepsilon}+\frac{w_{OL}}{C\cdot\sqrt{\frac{k\cdot\sigma}{\rho_G}}}\cdot\frac{1}{1-\varepsilon}-1=0 \qquad (1)$$

The integration constant C depends on the flow duct geometry. Fig. 5 shows a band of lines with various gas fractions. The solutions of the variation problem results in the basis equation of the flooding model, which is marked as envelope in the diagram:

$$\left(\frac{w_{OG}^2}{g}\cdot\frac{w_{OG}^2\cdot\rho_G}{\sigma}\cdot\frac{\rho_G}{(\rho_L-\rho_G)}\right)^{0{,}125}+M\cdot\left(\frac{w_{OL}^2}{g}\cdot\frac{w_{OL}^2\cdot\rho_L}{\sigma}\cdot\frac{\rho_L}{(\rho_L-\rho_G)}\right)^{0{,}125}=C \qquad (2)$$

As dimensionless numbers for both phases the Froude number, the Weber number and the density ratio are used. The equation considers the inertial, the weight and the surface forces. The influence of friction and the geometry are comprehended by parameters M and C. This dependencies are found by mean of own experiments and literature data.

The geometric parameter C is given by a hyperbolic function:

$$C = 1 \div \tanh\,(Bo^{*}/80) \qquad (3)$$

The bond number Bo^{*} is defined as followed:

$$Bo^{*} = L\cdot\left(g\cdot\frac{(\rho_L-\rho_G)}{\sigma}\right)^{1/2} \qquad (4)$$

L is wetable parameter, an important value for the flooding behavior. A comparison in Fig. 6 shows, that the experimental data are well represented by equation 3.

The influence of friction is comprehended by an exponential equation with the viscosity ratio of the two phases.

$$\frac{M}{C} = 0{,}81\cdot\left(\frac{\nu_L}{\nu_G}\right)^{0{,}123} \qquad (5)$$

A comparison with experimental data is given in Fig. 7.

4. AIR WATER EXPERIMENTS

The test setup for the air-water experiments (Fig. 8) consists of an air loop and a water loop. The air is injected into the bottom of the test section after passing one of the flow element flow meters. The two phases leaving the test section at the top are separated and the air is vented through a valve, which controls the system pressure. In the counter-current flow investigations the water is pumped into the upper part of the test section through a nozzle or porous sinter metal. The outlet mass flows - fall back and carry over - are measured by collecting tanks.

For the entrainment investigations an optical measuring device is developed, as shown in Fig. 9. The shadow photography arrangement consists of a gas flash ligth and a normal camera with an extension tube for the magnification. Weir elements near the holes of the flow duct separate the liquid film from the aperture. Droplets, entering the separating volume and wetting the windows, are removed by a small air jet and drained.

Entrainment photographs at different flow conditions are shown in the upper part of Fig. 10. The flow conditions are indicated in the carry over ratio diagram in the lower part of the figure. Increasing drop sizes with increasing air velocities can be observed. At the evalution of the droplet photographs a digital image analysing system is developed in order to get more objectiv and reproducable results of drop size distributions.

Fig. 11 shows a comparison of drop size distribution gained from single tube and bundle tests at the same air and water superficial velocities. The droplets above the tie-plate are of greater diameter. Their sizes can be explained by two quite different generating mechanisms. Typical results of the experiments with counter-current mass flow are plotted in Fig. 12. The experimental data are well represented by the derived model (Eq. 2).

5. STEAM-WATER EXPERIMENTS

An important task of the project was to investigate the fluid dynamic behavior at nearly the same flow conditions as during refilling and reflooding. Therefore a steam-water test facility was built. In the test section internals were installed up to a complete fuel element dummy. The piping and instrumentation flow sheet is shown in Fig. 13. In principal it is a similar arrangement as at the air water experiments. The maximum steam flow of 1900 kg/h can be recirculated in the test setup by a steam blower. The inlet mass flow rates are measured by different orifices and the water outlet mass flow rates are measured by collecting tanks. The system pressure is 4.8 bar (abs) corresponding to ECC-condition after the end of blow down. Fig. 14 shows the experimental results of test series with a 4 x 4 rod bundle and with the complete end box (16 x 16 rod bundle) as internal. The water is injected at saturation temperature.

Flooding can be observed at lower steam velocities in the case of the complete fuel element. At the same superficial steam velocities much less water can penetrate downwards. The main reason for this behavior is the different flow distribution in both cases. The main part of water fall back flows as a film along the flow duct wall, especially along the corners, where the local steam velocity is lower than on an average. This phenomena occurs at both geometries, but the perimeter to cross section ratio decreases with increasing sizes. Therefore the influence of the wall to the flooding behavior is much greater in smaller test sections.

With subcooled water injection an additional effect by phase changes must be considered. A part of steam flow condensates until the water attains the saturation temperature. Therefore the flooding model is extended by a term, which comprehends the steam reduction caused by condensation:

$$\left[\left(\frac{w_{0G}^2}{g}\cdot\frac{w_{0G}^2\cdot\rho_G}{\sigma}\cdot\frac{\rho_G}{(\rho_L-\rho_G)}\right)^{0.25} - f\cdot Ja\cdot\left(\frac{w_{0L,in}^2}{g}\cdot\frac{w_{0L,in}^2\,\rho_{L,in}}{\sigma}\cdot\frac{\rho_{L,in}}{(\rho_{L,in}-\rho_G)}\right)^{0.25}\right]^{0.5} + M\cdot\left(\frac{w_{0L}^2}{g}\cdot\frac{w_{0L}^2\cdot\rho_L}{\sigma}\cdot\frac{\rho_L}{(\rho_L-\rho_G)}\right)^{0.125} = C \qquad (6)$$

M and C are the already mentioned flooding parameters, Ja is the Jakob number, $Ja = C_p\cdot\Delta T_{sub}/r$ and f gives an information about the degree of equilibrium between the phases. It depends on flow turbulences and phase interactions, on the residence time and on the flow duct geometry. The value of f can come up to a maximum of 1, i.e. all water has saturation temperature.

The flooding curves of tests with the 4 x 4 rod bundle and tie-plate as internals are presented in Fig. 15. With the same steam flow rate more water can penetrate downwards, if the injected water is subcooled. A part of the injected steam is condensed below the smallest free flow area. So the onset of flooding in this region occurs at higher steam inlet flow. The water subcooling temperature is measured in the injection tube and not at the tie-plate. Therefore the difference in flooding behavior between saturated or subcooled water injection is slight, because the water is partly heated up on its way from the injection to the smallest free flow area, which turned out to be the most important section for flooding. Only subcooled water, which attains this area, can influence the flooding behavior.

Fig. 16 shows the results of pressure drop measurements at different water injection rates. It can be observed, that with increasing steam velocities the pressure drop shows a marked maximum and approaches to the value of the single phase pressure drop at higher steam flow rates. In consequence of the mutual restrictions of gas and liquid at the onset of flooding a turbulent two phase mixture arises and the pressure drop reaches a maximum. At increasing steam velocities the considerable amount of the water is carried out and the pressure drop slightly decreases. With increasing water injection the pressure drop becomes greater and flooding takes place at lower steam velocities.

REFERENCES

1. Brauer, H.: Strömung dünner Flüssigkeitsschichten unter der Einwirkung eines angrenzenden Gasstromes. Allg. Wärmetechnik (1961)
2. Feind, K.: Strömungsuntersuchungen bei Gegenstrom von Rieselfilmen und Gas in Lotrechten Rohren. VDI-Forschungsheft 481 (1960)
3. Taitel, Y.; Bornea, D.; Dukler, A.E.: A film model for the prediction of flooding and flow reversal for gas-liquid flow in vertical tubes. Int. J. Muliphase Flow 8 (1982) 1, 1/10
4. Shearer,C.J.; Davidson, J.F.: The investigation of a standing wave due to gas blowing upwards over a liquid film; its relation to flooding in wetted-wall columns. J. Fluid Mech. 22 (1965) 2, 321/335
5. Cetinbudakler, A.G; Jameson, G.J.: The mechanism of flooding in vertical countercurrent two-phase flow. Chem. Eng. Sci. 24 (1969), 1669/1680
6. Imura, H.; Kusuda, H.; Funatsu, S.: Flooding velocity in a counter-current annular two-phase flow. Chem. Eng. Sci. 32 (1977), 79/87

7. Richter, H.J.: Flooding in tubes and annuli. Int. J. Multiphase Flow 7 (1981) 6, 647/658
8. Kutateladze, S; Sorokin, L.: The hydrodynamic stability of vapour-liquid systems. In: Problems of heat transfer and hydraulic of two-phase media. Ed. by Kutateladze, Pergamon Press, Oxford 1969, 385/395
9. Ruff, K; Pilhofer, T.; Mersmann, A.: Vollständige Durchströmung von Lochboden bei der Fluid-Dispergierung. Chem.-Ing.-Tech. 48 (1976) 9
10. Suzuki, S.; Ueda, T.: Behaviour of liquid film and flooding in counter-current two-phase flow; Part 1.Flow in circular tubes. Int. J. Multiphase Flow 3 (1977), 517/532
11. Wallis, G.B. et al.: Countercurrent gas-liquid flow in parallel vertical tubes. Int. J. Multiphase Flow 7 (1980), 1/19
12. Schlichting, H.: Grenzschichttheorie. G. Braun-Verlag, Karlsruhe (1965)
13. Wieghard, K.: Theoretische Strömungslehre. Teubner, Stuttgart (1965)

Discussion

Question: K. Wolfert - GRS

Mr. Deruaz has shown that counter current flow limitation (CCFL) occured at the upper grid plate in Pericles at a steam velocity of about 9-10 m/s (corresponding to 7-8 m/s in the core).

Did you check your model on the basis of the Pericles data, do you find any agreement?

Answer: R. Spatz - U/Hannover

We checked our model with the Pericles data.
In Pericles the counter current flow limitation is associated to a superficial steam velocity of 7-8 m/s prevailing in the core region. In our tests the superficial velocities are related to the smallest free flow area, which is the tie-plate area. In this cross section the steam velocity is nearly 20% greater. This means with respect to the Pericles data that the steam velocities are about 10 m/s.
Such velocities correspond nearly to values calculated with the model developed by us.

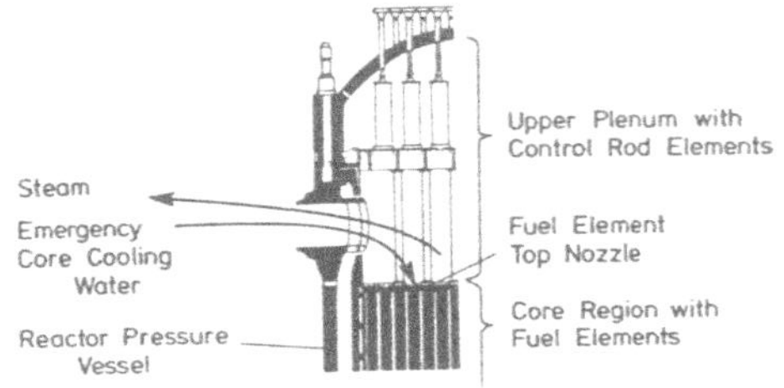

Fig. 1 Upper plenum of the reactor vessel of a PWR

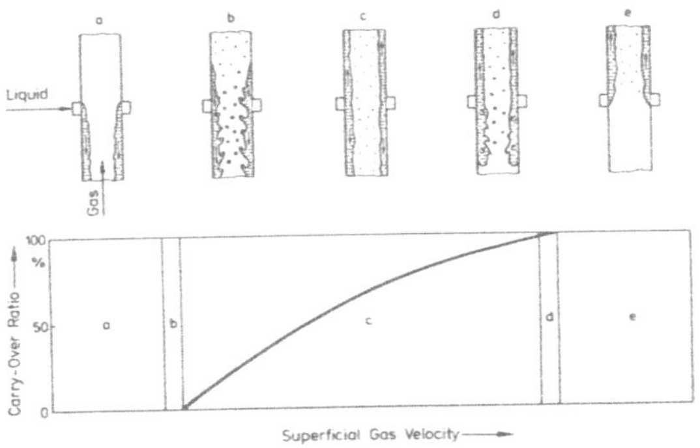

Fig. 2 Scheme of changing from concurrent to counter-current flow
a) undisturbed counter-current flow
b) onset of flooding
c) splitting of the liquid flow
d) stagnation point
e) concurrent flow upwards

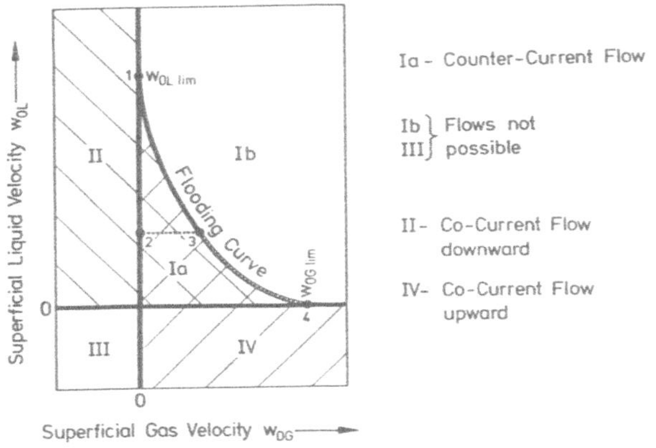

Fig. 3 Flow diagram (liquid downwards flow positiv)

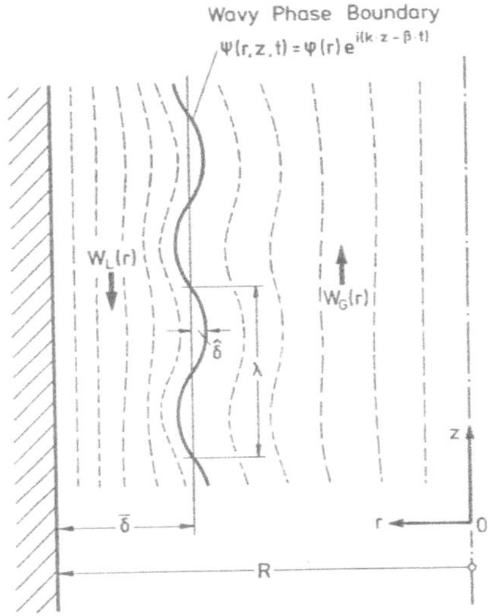

Fig. 4 Wavy interface at counter-current flow

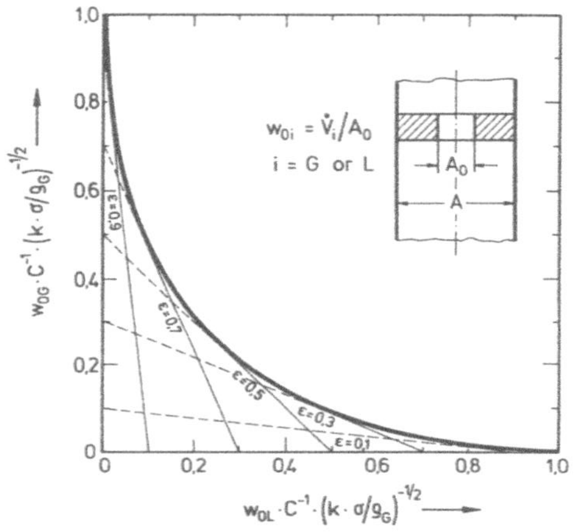

Fig. 5 Flooding function (Eq. 1)

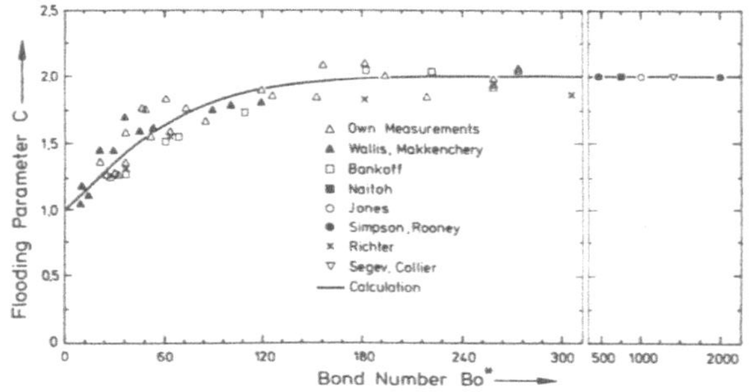

Fig. 6 Dependence of the flooding parameter C on the Bond-number (Eq. 3 ÷ 4)

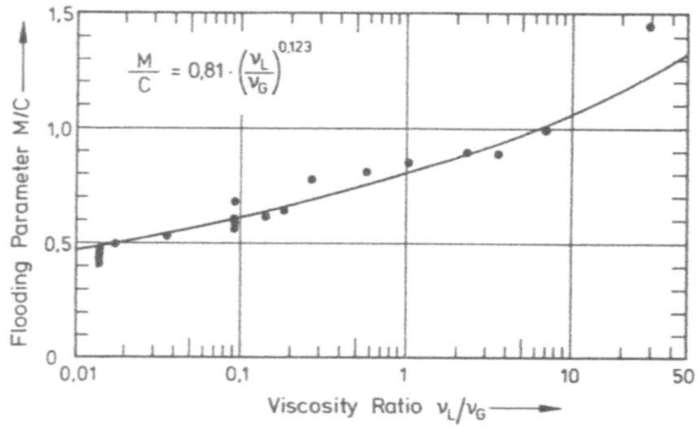

Fig. 7 Dependence of the flooding parameter M/C on the viscosity

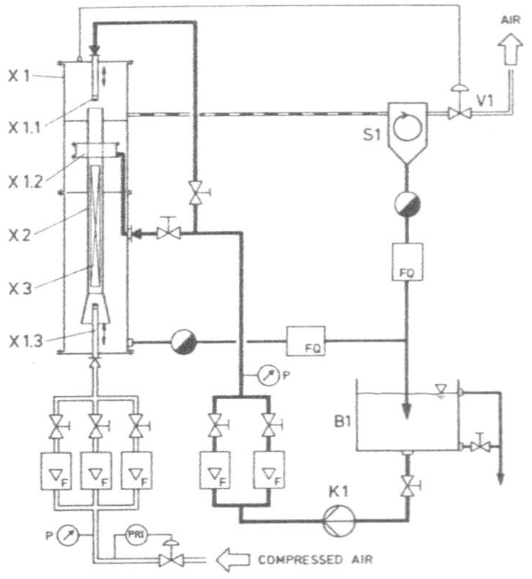

X1	Test Vessel
X1.1	Water Injection Through a Nozzle
X1.2	Water Injection Through Porous Sinter Metal
X1.3	Air Injection
X2	Flow Duct
X3	Internals
V1	Pressure Controlling System
S1	Twist Separator
B1	Water Storage Tank
K1	Centrifugal Pump

Fig. 8 Air-water test facility

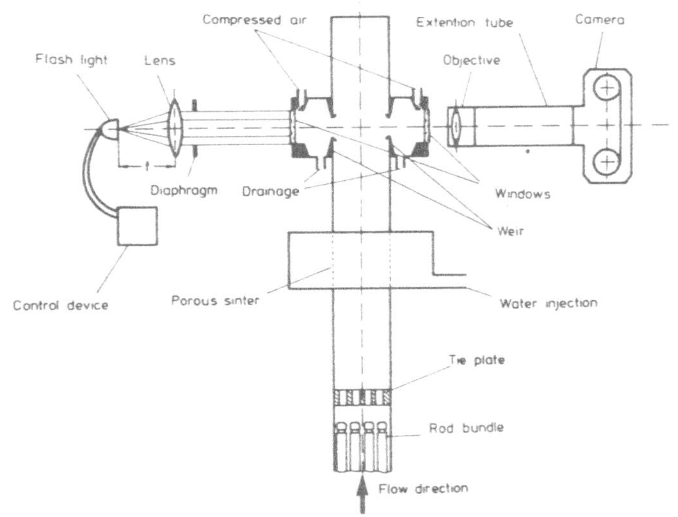

Fig. 9 Arrangement for entrainment photography

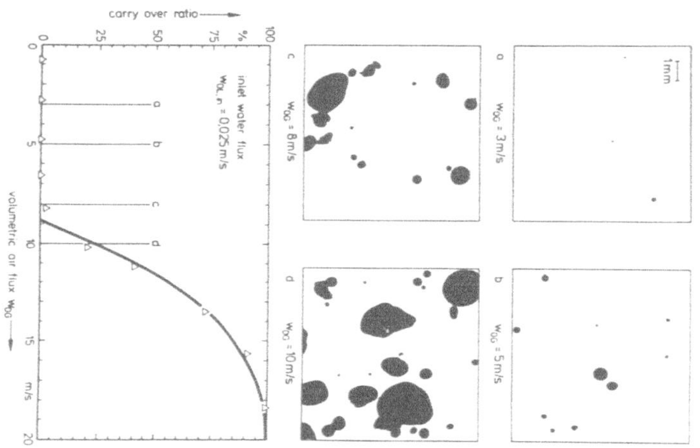

Fig. 10 Droplet photographies above the tie-plate at different superficial velocities

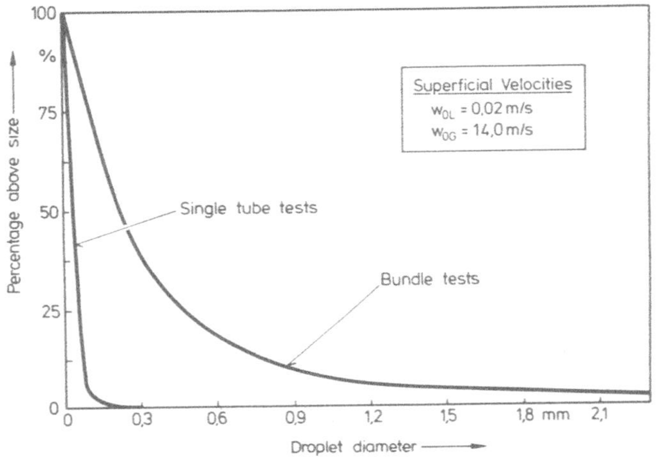

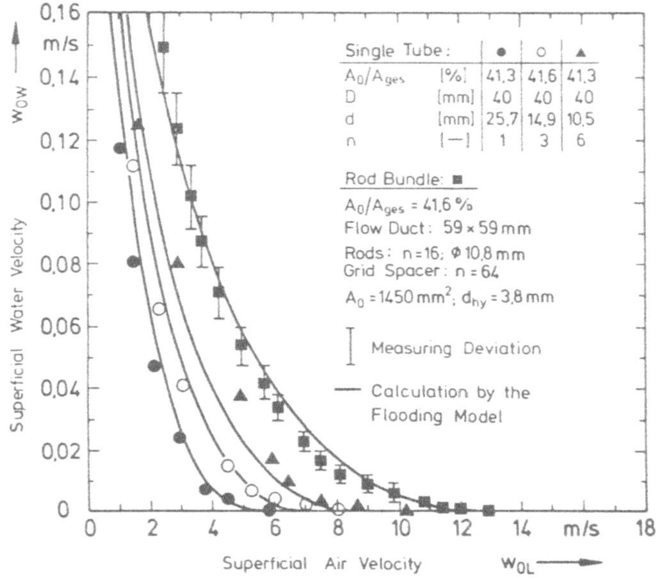

Fig. 12 Flooding behavior at different internals in the test section

Fig. 11 Drop size distribution in a single tube and above the tie-plate

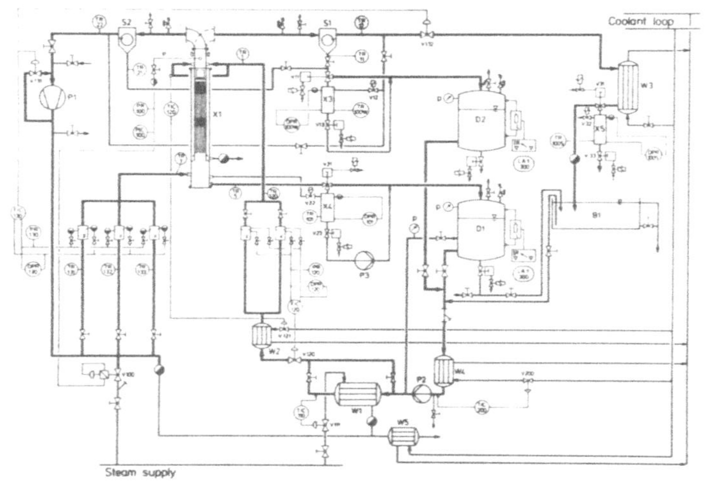

Fig. 13 Piping and instrumentation flow sheet of the steam-water loop

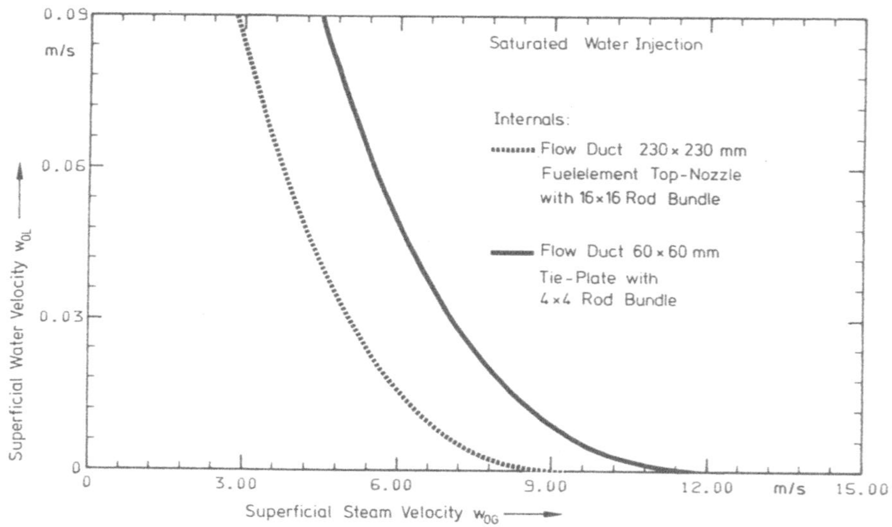

Fig. 14 Comparison of the flooding behavior for tests with the 4 x 4 and the 16 x 16 rod bundle

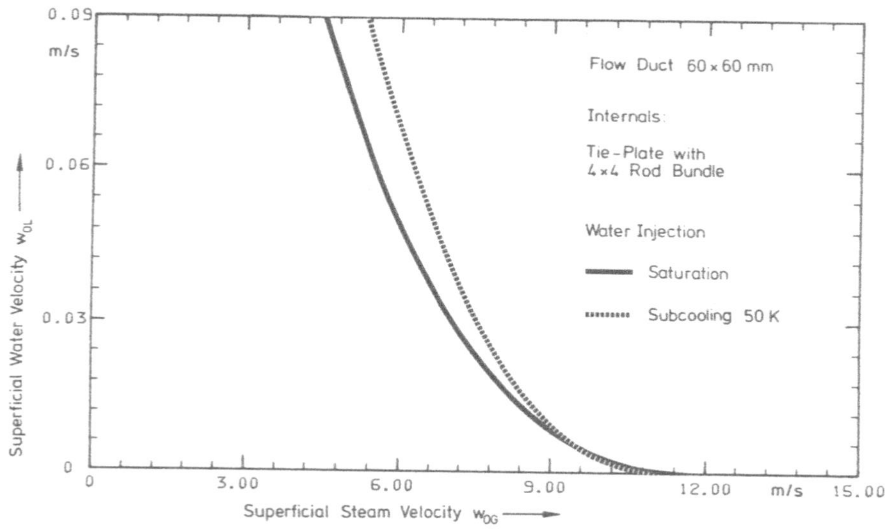

Fig. 15 Comparison of the flooding behavior at saturated and subcooled water injection

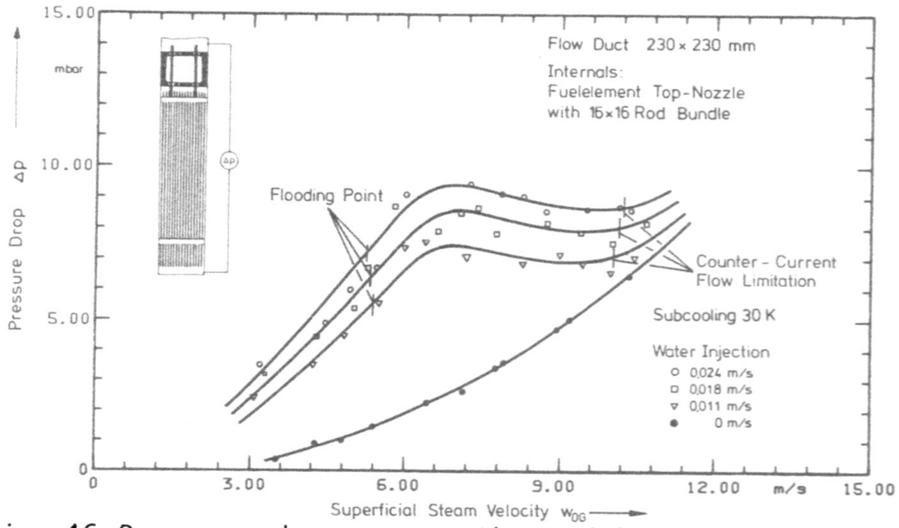

Fig. 16 Pressure drop across the rod bundle at different water injection rates

STUDY OF REWETTING AND QUENCH PHENOMENA BY SINGLE PIN OUT-OF- PILE EXPERIMENTS, WITH SPECIAL EMPHASIS ON THE EFFECT OF PIN COMPOSITION

N .E. Kaiser and O. Rathmann
Risø National Laboratory, Denmark

Summary

A series of comparative experiments with single electrically heated fuel pin simulators of different compositions is performed under identical boundary conditions simulating emergency core cooling by reflooding.

Three types of pins are used. A hollow tube heated directly by Joule heating, an indirectly heated rod with an internal heater and the space between the heater and the cladding filled with alumina-oxide powder, and finally another indirectly heated rod with the space filled with UO_2 ring pellets, simulating very closely a real fuel rod.

The pins are tested in the same test rig. The outer dimensions of the pins are identical for all three pins, 12 mm diameter and a heated length of 4 m. The pins are placled in a circular shroud with an internal diameter of 22 mm, leaving an annular space of a width of 5 mm for the coolant flow. The shroud can be heated to simulate an array of surrounding pins and to avoid cold wall effects.

The test parameters are: initial temperature 600-800 ^{o}C, liquid flow rate at the inlet 1-10 cm/s, system pressure 1-10 bar.

The experimental results have been compared to calculations with the computer code NORCOOL-II. A rather good agreement have been achieved between experiments and calculations for a wide range of parameters, i.e. rod composition, flow rate, system pressure.

INTRODUCTION

The project is aimed at increasing our knowledge concerning the rewetting and quench behaviour of nuclear reactor cores during the emergency core cooling phase of a loss-of-coolant accident. This knowledge is of major importance in reactor accident analysis for the prediction of cladding temperatures, metal water reaction, ballooning etc.

Our present knowledge concerning the quenching behaviour of heated pins is mainly based on experiments with electrically heated tubes with internal flow and with electrically heated pins of two types, i.e. direcly and indirectly heated pins. In all cases the composition is very different from that of nuclear pins as regards heat capacities and heat conduction characteristics.

The thermal behaviour of the experimental pins during a temperature transient can therefore be expected to deviate from each other and from the behaviour of nuclear pins under corresponding boundary conditions.

As a result of this difference and other factors, predictions concerning the behaviour of nuclear reactors, based mainly on small scale experiments with electrically heated pins may not be too reliable, as was in fact demonstrated in connection with the initial LOFT tests where maximum cladding temperatures were overestimated quite considerably.

The main task of this project is the performance of a series of comparative experiments simulating emergency core cooling, ECC, conditions with electrically heated single pins of different compositions under identical boundary conditions.

The pins consist of directly heated pins (tubes), conventional indirect ly heated pins and a more advanced type of indirectly heated pins. This last type is based on a central heater surrounded by UO_2 ring pellets and with inconel cladding, simulating very closely a real fuel pin.

The pins are placed in a circular shroud, which may be unheated or heated by radiation from an outer shell. The annular space between pin and shroud have a hydraulic diameter corresponding to a real reactor core geometry. The pin and the shroud are cooled by a flow of cooling water through the annular space. The movement of the rewetting front and temperatures of canning and shroud are measured by thermocouples.

The second task of the project consists of an analysis of the experimental results. This analysis includes a series of comparative calculations using the computer code NOORCOOL-II, which is developed at Risø National Laboratory.

EXPERIMENTAL PROGRAMME

Geometry

The quenching phenomena have been investigated in two different geometries. The first one was quenching inside a circular tube and the other one was quenching of a single rod placed concentric in a circular shroud with the coolant flowing in the annular space between the shroud and the rod. The shroud could be heated to simulate an array of surrounding hot fuel rods.

The length of the test section was the same in both geometries. The heated length was 4 m., which is a typical length of real fuel channels in nuclear reactors. The hydraulic diameter is the same in the two geometries, 10 mm, very similar to that of a real fuel rod bundle.

Fuel rod simulators

In the annular geometry 3 different types of rods have been tested. The 3 types are shown in fig. 1, together with a typical real fuel rod for comparison. The outer diameter is equal for all 3 rods. The cladding in the 3 test rods are identical. The cladding material is inconel 600 with a thickness of 1 mm.

The interior of the 3 rods are very different. In the first test rod the space inside the cladding is empty except for a few thermocouple leads. The heat is generated in the cladding by Joule effect using the cladding direct as an electric resistor.

In the second test rod the inner space is filled with alumina-oxide powder vibrated in the space between a central heater, the thermocouple leads and the cladding. The heat is generated in the central heater, which is made of mineral insulated thermocoax cable.

In the third test rod the space between the central heater and the cladding is filled with ring pellets made of uraniumdioxide and thus simulating rather closely a real fuel rod with solid pellets.

In a real fuel rod the cladding is zircalloy with a thickness of typical 0.8mm. The rod is filled with solid uraniumdioxide pellets.

In the real fuel rod the heat generation in the uranium is transferred from the pellets to the cladding across a small gap filled with gas, originally pure helium, but gradually contaminated by fission products during the lifetime of the rod. This heat transfer is a combination of conduction in the gas, radiation between pellets and cladding and direct contact at some spots between the fuel and the cladding.

In the test rod with uraniumdioxide ring pellets the conditions in the gap are intended to be the same as in a new fuel rod. The gas gap has the same size as in a real rod and the gap is filled with helium.

Test section

A cross section of the test section is shown in fig. 2. The annular space between the rod and the shroud forms the coolant channel. The shroud is not directly heated, but can be heated by radiation, from an outer heater made of a stainless steel tube.

Test rig

A diagram of the test rig is shown in fig. 3. The coolant water is stored in a supply tant, which can be pressurized by nitrogen pressure. The pressure is used to maintain the coolant flow in the test section. The flow rate is regulated manually by a regulating valve. The inlet temperature of the water is maintained using either a heat exchanger or an electrical preheater or both.

At the high pressure test runs, the whole test section, the preheater, the steam separator, the condenser and the collecting tank are surrounded by the pressure vessel. The vessel is pressurized by nitrogen. The pressure system is indicated by the dotted line on the diagram.

Instruments

The measurement of the flow is made using a load cell. The water supply tank is hanging in the load cell. The weigth of the tank and the remaining amount of water is continously recorded. The slope of the curve indicating the weight as function of time is the flow rate.

All temperatures were measured using mineral insulated thermocouples sheated by inconel. Type K, Cromel-Alumel, thermocouples were used.

The surface temperatures on the rod and the shroud were measured in several different levels above the inlet. The thermocouples for these measurements were mounted so that neither the hot junctions nor the leads disturbed the coolant flow.

The collapsed liquid level was measured using a differential pressure transducer with electrical output.

Void is measured using γ-absorption technique. The γ -rays are absorbed in the test section dependent of the amount of water. In the experiments a 1 mCi Cs-137 γsourse is used. The sampling time is 10 sec, which gives about 42000 counts. The difference between 0 and 100 % void is about 4000 counts.

All the experimental data were collected by a PDP-11 computer system at fixed time intervals. A number of selected temperatures were furthermore recorded together with the load cell output continously on 2 6-pen recorders.

Experimental results

In all fully 300 test runs have been made covering variations of all test parameters, but of course it has not been possible to make experiments with all imaginable combinations of all parameters.

From the temperature transients the time for the passage of the quench front at different levels can be detected. At the annular test section in fact two different quench front positions versus time can be detected, one at the rod and the other at the shroud as shown in fig. 4. In this figure also the collapsed liquid level is indicated.

From the temperature transients the quench time for differents levels can be plotted versus inlet velocity as shown in fig. 5. This type of plot is very suitable to make comparisons between tests with different sets of parameters.

A typical plot of void fraction is shown in fig. 6. The corresponding time for the passage of the quench front detected from the temperature transient is indicated on the diagram and shows good agreement with the rapid decrease in void fraction when the front passes.

Effect of parameter variations on quench time

The effect of parameter variations is summarized below.

Increasing quench time (decreasing quench front velocitty) was obtained by increasing initial temperature, increasing coolant inlet temperature and increasing power supply during the quenching, or by decreasing coolant flow rate and decreasing pressure.

A comparison between the 4 different test sections, the tubular one and the annular one 3 different rod types, at atmospheric pressure is shown in fig. 7. A similar plot at 5 bar is shown in fig. 8.

Rather big differences can be seen between the tubular and annular test sections. The annular section quenches significantly faster than the tube. This may be due to some cold wall effect from the shroud.

The difference between the 3 types of rods are moderate. The directly heated rod quenches somewhat faster than the indirectly heated rods. But there is no significant difference between the two indirectly heated rods.

The higher thermal capacity of the filler material in the UO_2 rod is apparently compensated by the poorer heat transfer from the inner area of the rod across the gas gap to the cladding.

THEORY

Description of NORCOOL-II.

NORCOOL-II (1,2,3) is a best estimate transient thermohydraulic computer simulation code with 3-fluid, 1-D network hydraulics and 2-D representation of "Heat Components" (i.e. pipe walls, fuel rods etc.). "1-D network hydraulics" means that transients in a hydraulic network can be simulated but with 1-D description of each of the branches of the network.

The 3-fluid hydraulics has separate description of each of the three fluids (1) liquid film or continous liquid, (2) droplets and (3) gas in terms of seperate volume fractions, temperatures, velocities and a pressure common to all three fluids. Each of these fluids is described by separate continuity, momentum- and energy equations.

The flow equations are integrated by a finite difference technique which is implicit in time and which has donor-cell (upstream) differenced in space.

The thermal behaviour of "Heat Components" (fuel rod simulators, pipe walls etc.) is described by a 2-D finite element model in r-z geometry, i.e. with rotational symmetry. Close to large axial temperature gradients this model automatically inserts additional axial fine mesh nodes which "moves" (i.e. are inserted and removed again when no longer necessary) along with the temperature gradients, e.g. quench fronts.

Modelling of fuel rods and fuel rod simulators include gap heat transfer (gas conduction + radiation) with thermal expansion effects estimated from pellet and cladding mean temperature.

NORCOOL-II contains a package of constitutive equations for friction, heat transfer and mass transfer for wall-liquid and wall-gas, and for the interfacecial intertractions liquid film-gas and droplet-gas. These constitutive equations depend on the flow regime which is assumed controlled by the void fraction α:

$\alpha<0.6$:	Bubble flow, incl. single phase liquid.
$\alpha>0.8$:	Ann. disp. flow: annular flow with both liquid film and liquid droplets.
$0.6\leq\alpha\leq0.8$:	Transition with interpolation between the above two regimes.

Of special importance for quenching simulation are heat transfer and droplet deposition/entrainment.

The wall heat transfer model is roughly composed of three parts: liquid forced convection boiling up to the critical heat flux point, film boiling or gas convection for wall temperatures above minimum film boiling temperature and transition boiling (interpolation between the two) in the intermidiate range. The wall heat transfer does not include radiation from wall to liquid or gas. However, for wall temperatures up to 700 ^{o}C such radiation heat transfer is of minor inportance.

The entrainment/deposition model (4) is experimentally based and describes the entrainment/deposition process in annular dispersed flow.

For quenching it is important that the interfacial constitutive equations depend on the droplet diameter, which in NORCOOL-II is determind from a constant Weber number. A value of We_d = 0.72 is used in correspondance with the analysis of Kocamustafaogullari et al. (5).

Simulation of Selected Annulus Experiments

A series of quench experiments with indirectly heated rods were simulated with NORCOOL-II experiments with "simple" rods (with Al_2O_3 filler) as well as with "advanved" rods with UO_2 pellets.

The 4 m long annulus was divided in 13 calculated cells with max. cell length 0.5 m for the hydraulic calculation.

For heat conduction calculation in rod and shroud etc. 2 fixed axial nodes per hydraulic cell and fine mesh nodes down to 1 mm, were used. In the radial direction 8 and 11 nodes were used for the tywo rod types, respectively, and 5 radial nodes for shroud and radiator. Included here were 3 radial nodes for the rod cladding and 3 for the shroud.

An example of the agreemtn between calculated and experimental results are shown in figs. 9 - 12, a low power, low flooding rate, 1 bar experiment with the advanced rod. As seen the agreement in the progression of the quench front is fine while the calculated results show some overprediction of the wall cooling by steam with droplet above the quench front.

The entire set of calculation shows that NORCOOL-II predicts the effects of increased rod power, increased flooding rate and elevated pressure qualitatively, and mostly also quantitatively, correct for both rod types. However, discrepancies in comparison with the experimental results are seen due to inadequate modelling of flow regime effects and gap heat transfer and due to the lack of a rod-to-shroud radiation model.

However, because quenching of simple and advanced fuel rod simulators behave qualitatively in the same way it is believed that by means of a computer model the NORCOOL-II experience regarding quenching gained from simple fuel simulators may be transferred to real power reactor fuel rods.

REFERENCES

1. Rathmann, O. Rosdahl, Ö. and Astrup, P. NORCOOL-II: An advanced Computer Code for thermohydraulic LOCA Analysis of a BWR. Part I: Model Description. NORHAV-D-79, Risø National Laboratory (1979)

2. Rathmann, O. NORCOOL-II. Part III: Numerical Methods and Program Description of the Heat Component Part. NORHAV-D-96, Risø National Laboratory (1982)

3. Rathmann, O. Global solution for NORCOOL-II. NORHAV-D-97, Risø National Laboratory (1982)

4. Würtz, J. An Experimental and Theoretical Investigation of Annular Steam-Water Flow in Tubes and Annuli at 30 to 90 bar. Risø Report No. 372, Risø National Laboratory (1978)

5. Kocamustafaogullari, G., De Jarlais, G., Ishii, M. Droplet Generation During Core Reflood. CONF-831047-60, Argonne Nat. Lab., 1983

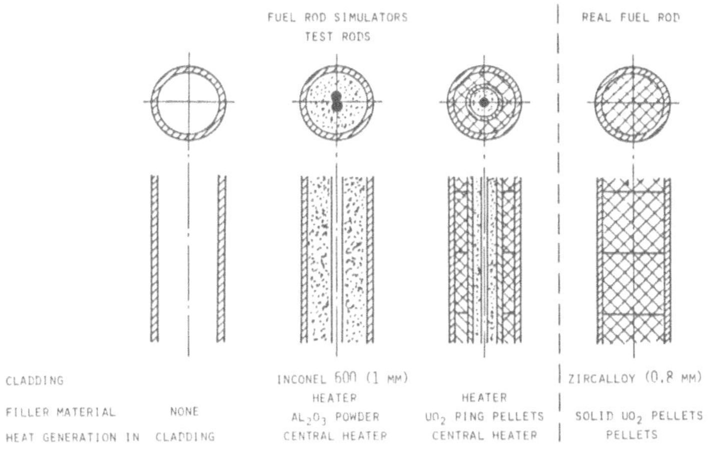

COMPARISON BETWEEN FUEL ROD SIMULATORS AND REAL FUEL ROD

Fig. 1

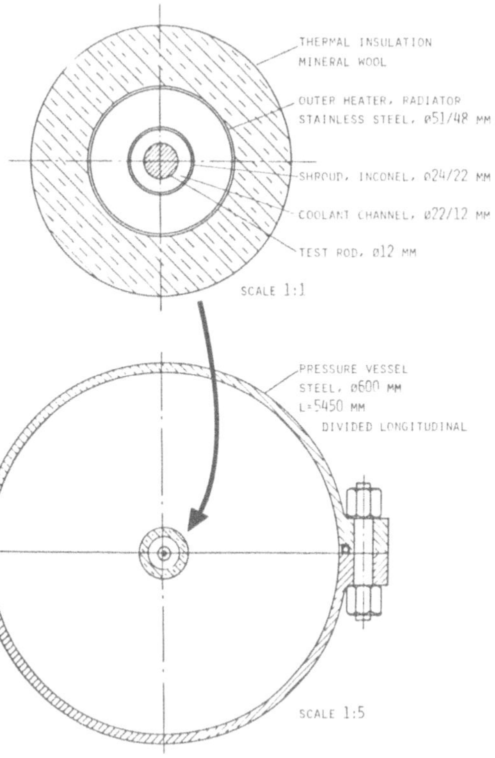

TEST SECTION, CROSS SECTION

Fig. 2

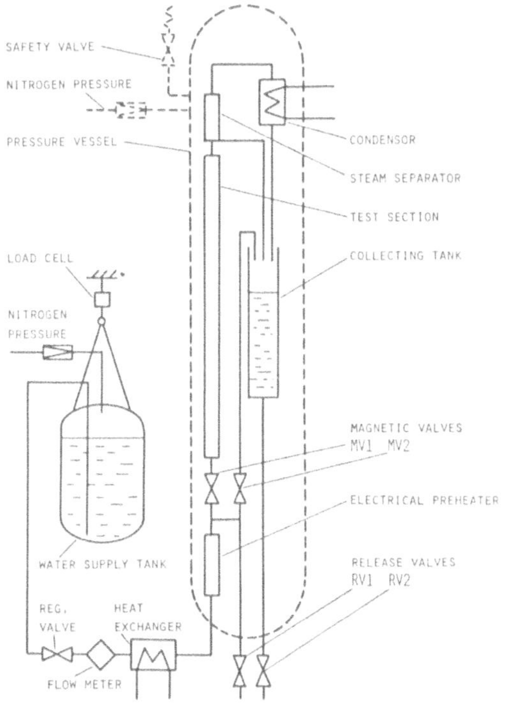

TEST RIG, DIAGRAM

Fig. 3

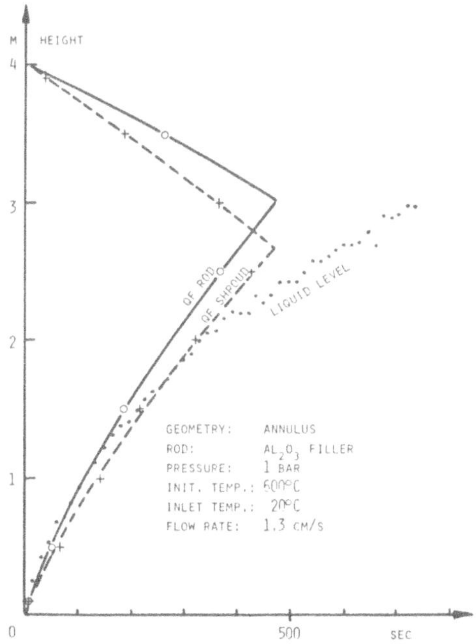

COLLAPSED LIQUID LEVEL AND QUENCH FRONTS

Fig. 4

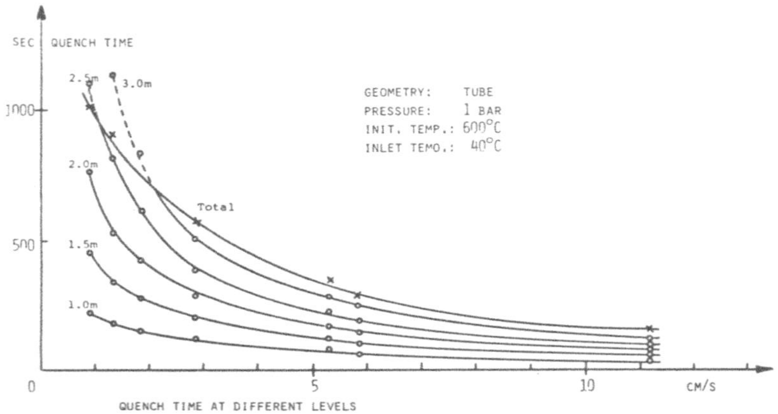

Fig. 5

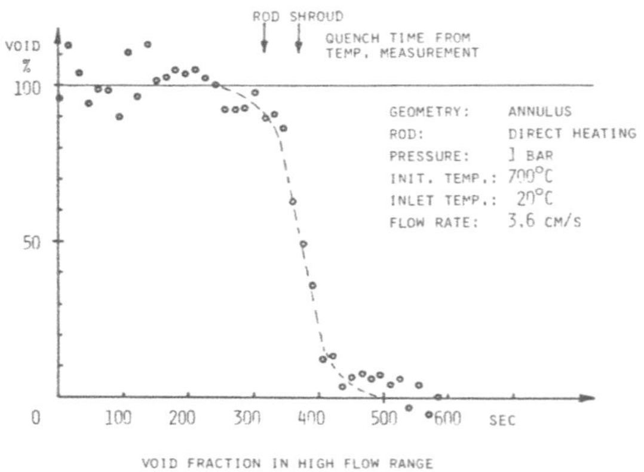

Fig. 6

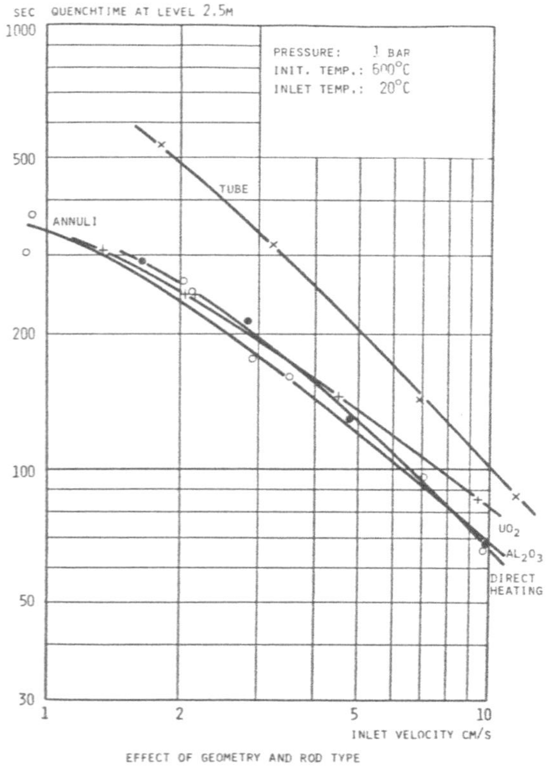

Fig. 7

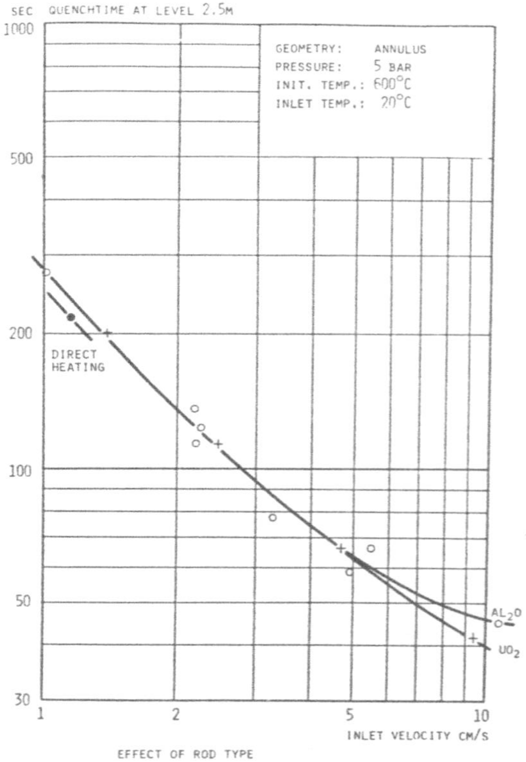

Fig. 8

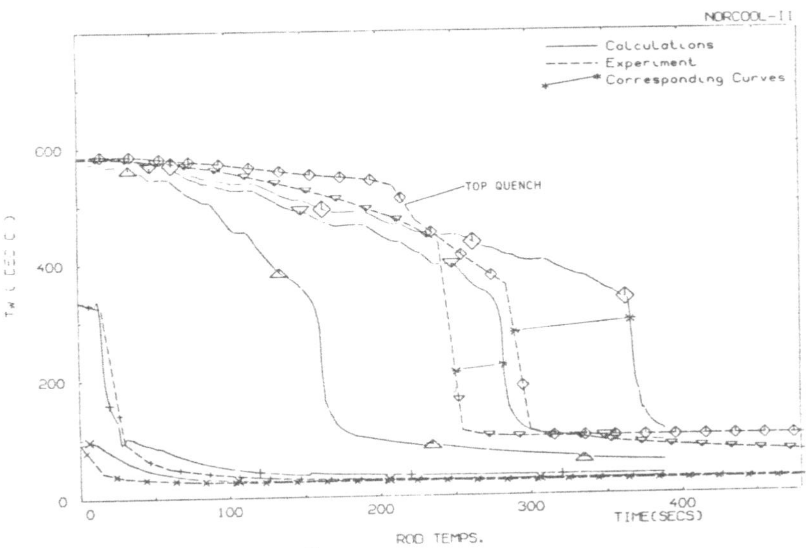

Fig. 9

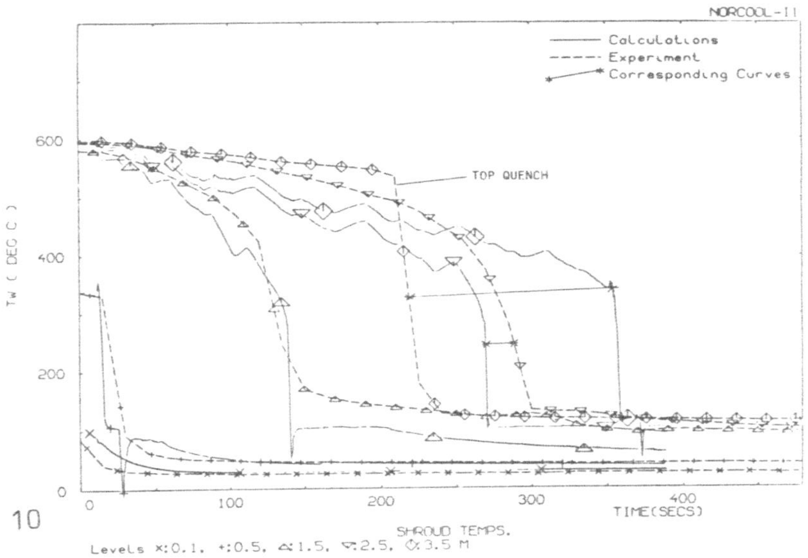

Fig. 10

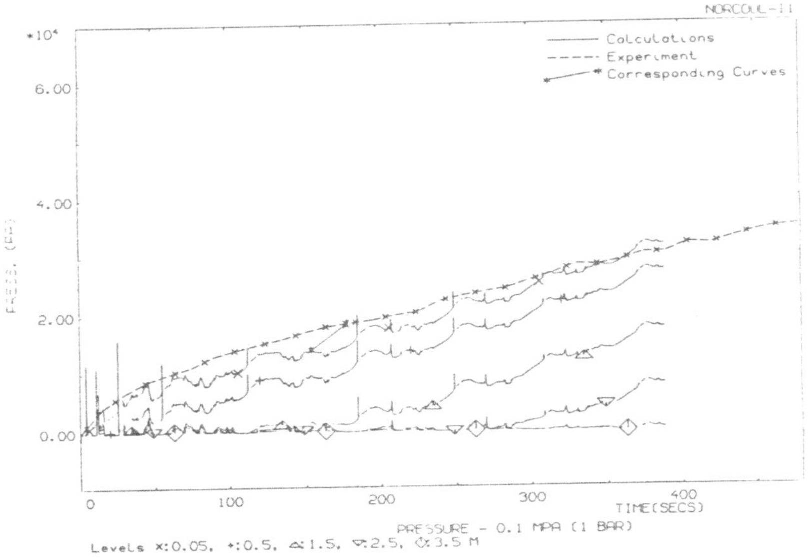

Fig. 11

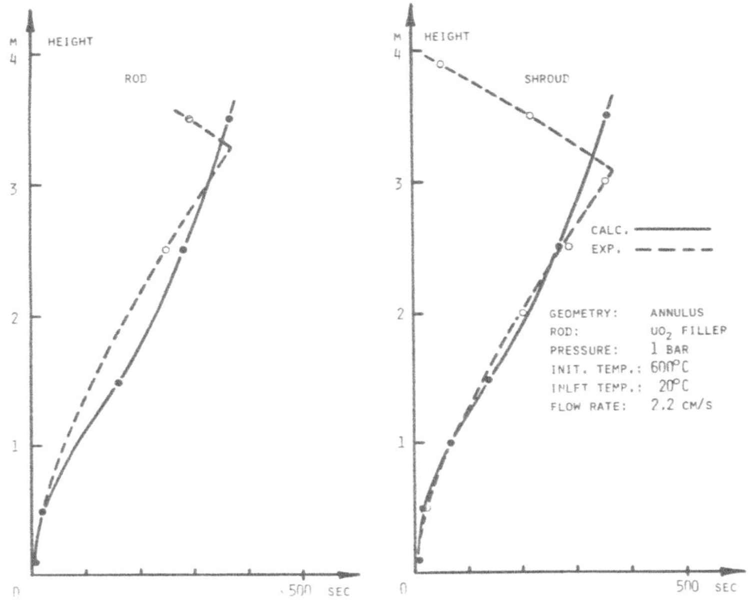

Fig. 12

Discussion

Question: G.L. Shires - UKAEA/AEEW

The use of a heated shroud is necessary in cluster reflood experiments and it is assumed that if the quench rate of the shroud equals that of the rods this is a good representation of surrounding fuel. Perfect matching is often not achieved. Does your work indicate how significantly the rod quenching is affected by a mismatch of the shroud quench rate, i.e. is the rewetting of rods very sensitive to the quenching of the shroud tube?

Answer: N.E. Kaiser - Risø N.L./Roskilde

No special investigations have been done to find out the influence of a mismatch between the two quench rates on the rewetting of the rod.

Question: E. Skupinski - CEC/BXL

Did you consider for the comparison of the quench-behaviour of the different heater rods the same boundary conditions of all relevant thermalhydraulic parameters?

Answer: N.E. Kaiser - Risø N.L./Roskilde

The three heater rod types were tested in the same test rig with all other parameters (i.e. temperatures, flow rate, pressure, etc.) as identical as possible. Even the shroud of the annulus was the same in all tests, only the heater rods were changed.

Question: P. Clement - CEA/CEN G

You have checked the effect of the filling material of the rod, but you have not checked the effect of the cladding material, which may be important. Do you intend to perform tests with Zircaloy claddings?

Answer: N.E. Kaiser

We do not intend to make experiments with Zircaloy cladding mainly due to problems with oxidation of the Zircaloy during repeated quenching from high temperatures.

The effect of Zircaloy oxidation have been investigated by M.K. Denham et al. from Winfrith.

Question: M.K. Denham, UKAEA

In the experiments, the Al_2O_3 swaged pin quenched at the same rate as the UO_2 pin with a gas gap. It was suggested that the resistance of the gas gap was compensated by the higher thermal capacity of the UO_2. Was this interpretation confirmed by calculations using the NORCOOL code?

Answer: N.E. Kaiser - Risø N.L.

The NORCOOL code has successfully calculated the quench speeds for both types of rods.

HEAT TRANSFER TO A DISPERSED TWO-PHASE FLOW AND DETAILED QUENCH FRONT VELOCITY RESEARCH

T.C. DE BOER, S.B. VAN DER MOLEN, et al.
Netherlands Energy Research Foundation ECN
Petten, The Netherlands

Summary

During the blow-down phase of a loss-off coolant accident (LOCA) in a large pressurized water reactor (LWPR) the core will heat up dramatically. Water will be injected in the system, and by bottom flooding the core will be cooled.
The objectives of this experimental programme are, to obtain a data base for "Boildown and Reflood" computer code development and to obtain information concerning the influence of non uniform temperature and/or power profile on the quench front velocity and prequench heat transfer, including unheated wall and grid effects.
This programme can be divided into two parts: firstly the tube, annulus and a 4-rod bundle experimental programme and secondly the 36-rod bundle reflood test programme.
The experiments performed in the first part of the programme show an early wetting of the unheated shroud, which leads to a quenchfront velocity increasement and an increase in liquid transport downstream the quenchfront. Furthermore an increased efficiency of inverted annular flow regime in the cooling of the heater pin is found. For the inverted annular flow regime the extended Bromley correlation gives rather good agreement with the experimental data. In the second part of the programme, boildown and reflood experiments have been performed, at 0.2 MPa pressure in the ECN 36-rod bundle test facility with axial uniform heated rods (3m length) The experiments show the following:
The considerable wall-temperature differences in the radial direction either present initially or obtained during the quench process deform the quench front. Under these process conditions high increase of prequench heat transfer is observed on the hotter surfaces, suggesting an inverted annular or transient flow regime to be induced. This intensive heat transfer considerably levels the wall temperatures just before quenching and helps to smooth out the quench-front velocity differences. Heat transfer improvement induced by the presence of spacer grids is observed.

1. INTRODUCTION

The non-proprietary data base for reflood heat transfer and quench front velocity research in a simulated pressurized water reactor (PWR) is available for heater rod bundles that are typified by the Westinghouse 15 x 15 design and resulted from FLECHT tests [1-4]. This study aimed to result in a better understanding of the prequench heat transfer and quench front propagation, influenced by grids, cold walls, a radial non-uniform power and/or temperature profile. Furthermore, the use of one-dimensional computer models for the calculation of the reflood process in a bundle needs a better justification. In the framework of the CEC Indirect Action Research Programme "Safety of Thermal Water Reactors 1979-1983", the ECN-programme "Heat Transfer to a Dispersed Two-Phase Flow and Detailed Quench Front Velocity Research" was defined. The ECN "Prequench Heat Transfer" research as far as the experiments in a tube, an annulus and a 4-rod bundle, the optical analysis of dispersed two-phase flow and the ECN 36-rod bundle experiments is concerned, has been reported [5,6,7]. The first part is shortly discussed in chapter 2.1, the second part is the major content of this paper.

2. EXPERIMENTAL PROGRAMME

2.1 The tube, annulus and 4-rod bundle prequench heat transfer research programme [5,6]

Prequench heat transfer data, obtained from small test sections are obscured by shroud effects. The objectives of the first part of this project are to investigate the influence of an unheated wall on the prequench heat transfer and to compare heat transfer coefficients derived from experimental data with values calculated using existing heat transfer correlations. Experiments have been performed in three well instrumented test sections, a tube, an annulus and a 4-rod bundle (heated length 1.5 m), with the following conditions: inlet velocity 2-20 cm/s, inlet temperature 20-80 °C, initial wall temperature 400-800 °C, axial uniform power distribution, at atmospheric pressure.

2.2 The 36-rod bundle reflood programme [7]

The second part of this project is defined as an ECN 36-rod bundle reflood test programme. The objectives are, to develop a data base to aid in computer code development by others and to obtain information concerning the influences of core inlet conditions, spacer grids, radial non-uniform power and/or temperature profile on the reflood process and especially on the prequench heat transfer.

The test section comprises a 36-rod bundle within a rectangular shroud. The bundle consists of 32 axial uniform indirectly heated rods (3 m heated length) and 4 unheated corner rods, bearing the instrumentation. Among other parameters, coolant and heater rod temperatures at different levels, pressure differences along the shroud, as well as the outlet steam flow and the water carry-over have been measured.

The experiments (table 1) performed in the 36-rod bundle test facility Figs. 1a, b, c, are mainly reflood experiments with the following conditions: pressure 2-6 bar; heat fluxes 1.6-5.0 W/cm^2; inlet subcooling 20, 80 °C; initial wall temperatures 200-800 °C flooding rates 1.4-8.0 W/cm^2.

3. EVALUATION OF THE EXPERIMENTS

Three aspects of the experimental results will be discussed:

- the influence of the radial non uniformity of wall temperatures and heat production on the quench front progression;
- the influence of the radial non uniformity of wall temperature and heat production on the prequench heat transfer;
- the effect of spacer grids on the prequench heat transfer.

3.1 Quench front progression (QF)

During a reflood process three different regions in a coolant subchannel can be distinguished:

- the prequench region where the heated wall has to be cooled down to the sputtering or rewetting temperature;
- the quench front region were the ultimate cooling takes place;
- the post quench region where the wall is cooled by subcooled or saturated liquid.

The relevant parameters determining the progress of the quench front are shown by the equation obtained by combining the one-dimensional approximation of Semeria and Martinet [8] and Yamanuchi [9] combined with the Yu correlation [10].
The combination is given by:

$$V_q = \frac{1}{\rho c}\sqrt{\frac{k}{d}}\;\frac{4.24 * 10^4 A V^{0.15}(1+V\Delta T_c^{\,2})^a}{\sqrt{(T_w-T_q)(T_w-T_s)}} \qquad (5.1)$$

The formula shows as the relevant parameters:

- the prequench wall temperature T_w;
- the liquid velocity at quench front V;
- the liquid cubcooling at the quench front ΔT_c.

$A = 1$; $a = 0.13$ if $(1+V\Delta T_c^{\,2}) < 40$
$A = 0.4839$; $a = 0.346$ if $(1+V\Delta T_c^{\,2}) > 40$

3.1.1 Influence of the initial wall temperature

The initial wall temperatures determines the quench-front progression for the first period of the reflood process. One has to distinguish between the initial wall temprature lower than the rewetting temperature and higher than the rewetting temperature. For the initial wall temperature of 200 °C (lower than the rewetting temperature) the initial front velocity is equal to the inlet velocity up to that height in the bundle where the increase of the wall temperature is high enough to exceed the rewetting temperature. The deviation of the quench front velocity from the inlet velocity is obtained for prequench wall temperatures between 250 °C and 300 °C.

The example is of the quench front progression for the low initial wall temperature is shown in Fig. 2. It can be concluded that the initial quench front velocity is completely determined by the inlet velocity in case of prequench wall temperature smaller than rewetting temperature. At higher bundle positions the influence of the heat production becomes the leading parameter. In case of experiments with uniform initial wall temperatures at temperature levels of 600 °C and 850 °C, higher than the rewetting temperature, the initial quench front velocity is completely different from the inlet velocity. Comparison of both experiments (Fig. 3) shows the initial quench front progression difference. After sufficient prequench cooling the progression becomes dominated by the heat production and inlet coolant conditions (V, ΔT_c) which are the same for both experiments (indicated by an equal slope of the lines in Fig. 3).

3.1.2 Influence of the heat production level

As far as the heat production level is concerned the influence of the non uniform heat production in case of the initial wall temperatures of 600 °C is illustrated by Fig. 4 where the quench front progression of two uniform heat production experiments are compared to a non uniform heat production experiment.

- 4113 Uniform heat production 3.6; 3.6; 3.6 W/cm^2 $T_w$init = 600 °C
- 4115 Non uniform heat production 3.6; 2.2; 1.4 W/cm^2 $T_w$init = 600 °C
- 4122 Uniform heat production 1.8; 1.8; 1.8 W/cm^2 $T_w$init = 600 °C

The total bundle power of 4115 and 4122 experiments is the same within 10% and half of the bundle power of experiment 4113. Fig. 4 shows that the quench front velocities of experiments 4115 and 4112 are of the same magnitude and about double than that of experiment 4113 after the first meter of the bundle. There are two reasons to explain it. On one hand the prequench wall temperatures T_w just downstream the quench front of Exp. 4115 and 4122 are nearly at the same level of around 300 °C whereas T_w for experiment 4113 is around 350 °C. On the other hand the subcooling at the quench front position is smaller for experiment 4113 because of the higher total power.

3.1.3 Influence of the heat production profile

In the following the influence of the heat production profile will be discussed. Looking to the plots of wall temperature versus time it is observed that the low powered rod quench systematically earlier than the high powered ones. This implies that the quench position of the low powered rods is displaced downstream the quench front position of high powered rods for high inlet velocity and high inlet subcooling, once the differences are established at the beginning of the test section they remain almost the same throughout the rest of the test section. This means the same velocity of the quench front on the individual rods (Fig. 5). In case of void fractions at the quench front the differences increases during the process with an extreme situation in case of high inlet velocity and low subcooling Fig. 6.

3.1.4 Influence of initial wall temperature profile

Two types of experiments with a radial non uniform initial wall temperature distribution combined with a radial non uniform heat production have been carried out. For both types the temperature distribution is a consequence of the radial power distribution in the period preceeding to the start of the reflood process. In the first type the initial wall temperature distribution is obtained by a radial power distribution comparable to the experiments described in the foregoing section with values of 3.6; 2.2 and 1.4 W/cm^2 for the central, inner and outer ring elements respectively (approximately radial cosine distribution). The initial temperature distribution of the second type is caused by an extreme power distribution of the second type is caused by an extreme power distribution (0.0; 3.6 and 3.6 W/cm^2 in the central, inner ring and outer ring elements) simulating a guide tube situation. The initial wall temperature distribution is very important for these experiments. In the first type the power distribution provides for a radial "cosine type" wall temperature distribution with the high wall temperatures in the centre and the low wall temperatures for the outer ring elements. In the second type experiments the power on the inner ring and outer ring rods provides for an initial temperature distribution with lower wall temperatures for both the shroud and the four centre elements ("guide tubes").

The early wetted walls decrease the prequench wall temperature of the heated elements downstream the quench front and there fore influences the quench front behaviour. The effect of the initial wall temperatures and "cold wall" effects is demonstrated most clearly by comparing the initial quench front velocities for the differenct experiments (Fig. 7,8).

3.2. Prequench heat transfer

The prequench heat transfer determines the wall temperature T_w just downstream of the quench front which is one of the governing parameters of the quench front behaviour. The prequench heat transfer is strongly dependent on the flow regime in the prequench region. For subcooled conditions at the quench front the flow regimes downstream of the front are successively: the inverted annular flow regime; the transition region; the dispersed two-phase flow region; the one phase steam flow, however in case of saturated condition, no inverted annular flow regime occurs.

The saturated conditions are reached quite early in the reflood process for low inlet velocities (1.4 and 2.4 cm/s) and low inlet subcooling (20 °C) and high initial wall temperatures (600 °C, 850 °C). For low initial wall temperatures (200 °C) the steam production starts later in the reflood process due to the low heat transfer rate for wall temperatures below the wetting temperature. The steam production starts after reaching the saturation condition or after the wall temperature reaches the wetting temperature. The subcooled conditions at the quench front remains for a considerable time of the reflood process in experiments for high inlet velocities (8.0 cm/s) and high inlet subcooling (80 °C). The combination of a high inlet velocity (8 cm/s) and a low inlet subcooling of 20 °C results in a situation between those two extreme cases of saturated and subcooled conditions.

The prequench heat transfer is a combination of:

- the contribution of the convective heat transfer by the steam flow or dispersed two phase flow;
- the contribution of the radiation heat transfer from wall to the droplets, from wall to the vapour and from vapour to the droplets in the dispersed two-phase flow;
- the contribution of the radiation between element of different wall temperatures;
- the "Bromley type" heat transfer obtained for inverted annular flow regimes caused by the heat conduction in the vapour film between the wall and liquid and by the radiation of the high-wall temperture to the liquid.

  In order to account for the flow regime the contributions should be weighted by the void fraction as given in [5]. The above mentioned contribution to prequench heat transfer is changing during the reflood process because of the change in wall temperatures in the bundle throughout the experiments.

The actual flow regimes cannot be detected in the bundle. Calculating the heat transfer coefficients form the coolant parameters, the void fraction α can be used as the weighting function for the convective heat transfer and $(1-\alpha)^{0.5}$ as the weighting function for the "Bromley type" heat transfer coefficient. The radiation heat transfer coefficient consists of the radiation component to the coolant as well as the radiation component to the neighbouring elements if wall temperatures differ. The combined heat transfer coefficient can be written as:

$$h_{tot} = \alpha\, h_{conv} + h_{rod} + (1-\alpha)^{0.5}\, h_{Br} \,. \qquad (5.2)$$

5.2.1 Low flow reflood experiments with low initial wall temperatures

In the first 100 seconds of the reflood process the steam production in the wetted region of the bundle is low. Therefore, the prequench heat flux from the element to the coolant is low. Later in the reflood process the heat flux increases to values above the heat production. In case of the uniform power experiments 3216 and 3224 the calculated heat fluxes to the coolant on the basis of the production and the change of the accumulated heat are almost the same which leads to comparable heat transfer coefficients. For a non uniform power profile experiments 3221 and 3229 with a radial cosine type profile the calculated heat fluxes are different for the different type of heater leements, the highest in case of the high pwered rods and the lowest in case of the low powered rods.

The prequench heat fluxes and heat transfer coefficients (Figs. 9) for the uniform heat production experiments show comparable values for the representative rods. Comparing these values with values from the non uniform heat production experiment for the same heater rods differences are observed. Three effects can explain these differences:

- different raidation heat transfer between neighbouring elements;
- different radiation heat transfer to the dispersed two phase flow;
- different convective heat transfer due to imperfect mixing in the dry area.

5.2.2. Low flow reflood experiments with high initial wall temperatures
Inlet velocity 2.4 cm/s

For these experiments the reflood process starts with the initial wall temperature above the rewetting temperature. Therefore the steam production and heat transfer in the dry region starts immediately at the beginning of the reflood process. The group of experiments in question consists of four types of experiments:

- the radial uniform power and radial uniform initial wall temperature experiments;
- the radial non uniform power and radial uniform initial wall temperature experiments with a radial "cosine type" heat production;
- the radial non uniform power and radial non uniform initial wall temeprature experiments with a radial "cosine type" heat production;
- the radial non uniform power nad radial non uniform initial wall temperature experiments with a "guid tube" type heat production. The influence of the following parameters on the prequench heat transfer will beshown in Fig. 10.
- the influence of the non-uniform heat production;
- the influence of the initial wall temperature (profile and level);
- the influence of the power level;
- the influence of pressure.

5.2.3 High flow reflood experiments with high initial wall temperatures

(V_{inl} = 8 cm/s, T_{init} = 600 °C respectively 850 °C)

In case of a high inlet subcooling (80 °C) combined with the high flooding rate the local conditions at the quench front are subcooled. This will influence the void fraction downstream the quench front and therefore the heat transfer coefficient, see Eq. 5.2.

* The influences of power and temperature profile, the influence of the power level and the influence of pressure are shown in Fig 11 of the heat transfer coefficients versus time and versus distance to the quench front. The case of a low inlet subcooling (20 °C), combines with the high flooding rate Fig. 12, results in higher void fraction downstream the quench front and consequently in lower heat transfer coefficient just downstream the quenchfront. From these figures it appears that the level of the prequench heat transfer coefficients based on heat fluw calculations and saturated coolant conditions for the uniform power experiments is around 100 W/m^{2}K.

6. CONCLUSIONS

6.1 Conclusions from the tube, annulus and 4-rod bundle experiments [5]

* Effects of the unheated wall:

- the quench front velocity V_q increasement, due to the early wetting of the shroud;
- the early wetting of the shroud contributes to the initial steam production;
- additional contribution to the cooling of the heater pin by the radiation heat transfer before and after shroud wettting;
- the early wetting of the shroud influences liquid transport downstream the quench front;
- increased efficiency of inverted annular flow regime in the cooling of the heater pin.

* The difference between the experimental heat transfer coefficients for the single phase and the dispersed two-phase flow regime, and the calculated values of convective heat transfer by the Heinemann correlation and radiation heat transfer indicates the presence of superheat in the very beginning of the reflood process.

* For the inverted annular flow regime the extended Bromley correlation gives rather good agreement with the experimental date.

6.2 Conclusions from ECN 36-rod bundle experiments [7]

* The boildown and reflood tests are executed with the ECN 36-rod bundle test facility. The objectives of the experimental test programme have been met. They were aimed to obtain

- a valuable data base for computer programme development and validation;
- information on the influence of radial non uniform temperature and for power profile on quench front propagation and the prequench heat transfer;
- information about the influence of grids and unheated rods (guid tube simulation) on the prequench heat transfer.

* The performed experiments confirm that, the local coolant velocity and subcooling and the prequench wall temperature determine mainly the quench front velocity. In case of saturation at the quench front the local void fraction becomes also a significant parameter. Increase of coolant velocity or subcooling and decrease of prequench wall temperature accelerate the quench front propagation. During the first period of the reflood process the initial wall temperature determines the quench front velocity. Later on in this process the power becomes dominant, such that the mean quench front propagation depends on the bundle mean power but not on the radial power distribution.

* Radial non-uniform temperature and/or power profile, unheated walls such as shroud or guide tubes produces wall temperature differences in the radial direction, which, in turn, induce differences of quench front position. The differences in quench front position or smaller for high subcooling at the quench front and larger for increased void fraction at the quench front.

* If the quench front is deformed due to induced radial temperature differences, i.e. colder surfaces are quenched already at a certain distance downstream relative to hotter surfaces, a high increase of the prequench heat transfer is observed on hotter surfaces along this distance, suggesting an inverted annular or transient flow regime being present here. The intensive heat transfer considerably levels the wall temperatures just before quenching and aids to smooth out the quench front velocity differences.

* The heat transfer coefficients for the dispersed two-phase flow regime evaluated for experiments with the uniform and correspondent non-uniform power distribution show hardly any significant differences other than the contribution due to wall-to-wall radiation. This result suggests an intensive mixing in the prequench region of the radial non-uniform power experiments. Moreover, in these non-uniform power experiments the prequench wall temperature and the quench front velocity in individual subchannels are almost the same, at least for the cases of small inlet velocities and subcooling. It can be concluded that also the mixing in the wetted region is intensive and the void fraction at the quench front is rather uniformly distributed among subchannels.

* Grids improve the heat transfer at their position and downstream. Consequently, the grid temperature is decreased, which eventually leads to an early wetting of the grid surface.

REFERENCES

1. CADEK, F.F. et al., PWR FLECHT (Full Length Emergency Cooling Heat Transfer), Final Report, WCAP-7665, April 1971.
2. CADEK, F.F., et al., PWR FLECHT Final Report Supplement, WCAP-7931, October 1972.
3. LILLY, G.P., et al., PWR FLECHT Cosine Low Flooding Rate Test Series Evaluation Report, WCAP-8838, March 1977.
4. LILLY, G.P. et al., PWR FLECHT Skewed Profile Low Flooding Rate Test Series, Evaluation Report, WCAP-9183, November 1977.
5. MOLEN, S.B. van der, BOGAARD, P.P.A. van den, VONKA, V., BOER, T.C. de, Prequench Heat Transfer in a Tube Annulus and a Four-rod Bundle, Final report ECN-83-12, July 1983.
6. GALJEE, F.W.B.M., VONKA, V., MOLEN, S.B. van der, BOER, T.C. de, Optical Analysis of the Dispersed Two-Phase Flow Regime, Final Report ECN-83-11, July 1983.
7. BOER, T.C. de, MOLEN, S.B. van der, MAARLEVELD, J.R.C., HOOGLAND, H, Final Report "ECN 36-Rod Bundle Boildown and reflood experiments data and evaluation", Volume I, II and III, ECN-84-25, September 1984.
8. SEMERIA, R. and MARTINET, B., Calefaction Spots on a Heating Wall: Temperature Distribution and Resorption. Proc. IME, Vol. 180 part 3C, pp 192-205, 1965.
9. YAMANOUCHI, A, Effect of Core Spray Cooling in Transient State after loss-of-coolant Accident, J. Nucl. Sci. Technol. Vol. 5 pp 547-558, 1968.
10. YU, S.K.W., FARME, P.R. and CONEY, M.W.E. Methods and correlations for the prediction of quenching rates on hot surfaces. Int. J. Multiphase Flow, Vol 3, pp 4, 5.443, 1977.

	Exp. No.	p (MPA)	T wall initial, lev. 4 (°C)			Power (W/cm^2)			V_{in} (cm/sec)	T_{in} (°C)
			c.r.	i.r.	o.r.	c.r.	i.r.	o.r.		
	3215	.6	200	200	200	3.3	3.6	3.6	1.5	80
*	3216	.2	-	-	-	3.4	3.7	3.6	-	40
	3218	.6	-	-	-	3.3	3.6	3.6	-	140
	3220	.2	-	-	-	1.7	1.8	1.8	.7	100
*	3221	-	-	-	-	4.0	2.2	1.8	1.3	-
*	3224	-	-	-	-	3.4	3.7	3.6	1.5	-
*	3229	-	-	-	-	4.1	2.9	2.2	1.4	-
	3230	.2	200	200	200	3.4	3.7	3.6	1.1-0.9	100
	3231	-	-	-	-	3.4	3.7	3.6	1.7-0.9	40
*	4100	.2	600	600	600	3.3	3.6	3.5	8.0	100
	4102	-	-	-	-	3.3	3.6	3.5	5.0	-
*	4106	-	-	-	-	3.6	3.6	3.5	2.4	-
	4113	-	-	-	-	3.8	3.6	3.5	-	40
	4114	-	-	-	-	3.8	3.0	2.2	-	-
	4115	-	-	-	-	3.8	2.2	1.4	-	-
	4116	-	-	-	-	3.8	3.0	2.1	-	100
*	4117	-	-	-	-	3.7	2.2	1.4	-	-
	4118	-	400	400	400	3.6	3.6	3.5	-	-
	4119	-	600	600	600	3.6	3.6	3.5	5.0	40
*	4120	-	-	-	-	3.6	3.6	3.6	8.0	-
	4121	-	400	400	400	3.6	3.6	3.5	2.4	-
	4122	-	600	600	600	1.9	1.7	1.7	-	-
	4124	-	600	<	<<	3.6	2.2	1.4	-	-
	4125	-	<<	600	600	0	3.6	3.5	-	-
*	4126	-	600	<	<<	3.6	2.2	1.4	8	-
*	4127	-	<<	600	600	0	3.6	3.5	-	-
*	4128	-	600	600	600	5.2	5.0	4.8	2.4	-
*	4129	-	-	-	-	4.8	5.0	4.8	8	-
*	4130	-	-	-	-	3.6	2.1	1.3	-	-
*	4131	-	600	<	<<	3.6	2.2	1.4	2.4	100
*	4132	-	<<	600	600	0	3.6	3.5	-	-
*	4133	-	-	-	-	0	3.6	3.5	8	100
*	4134	-	600	<	<<	3.6	2.2	1.4	-	-
*	4135	-	600	600	600	3.6	2.1	1.3	-	-
*	4136	-	-	-	-	5	5	5	2.4	100
*	4137	-	-	-	-	5	5	5	8	-
*	4138	-	800	800	800	3.6	3.6	3.6	2.4	-
*	4139	-	<<	800	800	0	3.6	3.6	-	-
*	4140	-	800	800	800	3.6	3.6	3.6	8	-
*	4141	-	<<	800	800	0	3.6	3.6	-	-
	4142	-	800	800	800	3.6	3.6	3.6	2.4	40
	4143	-	<<	800	800	0	3.6	3.6	-	-
	4144	-	800	800	800	3.6	3.6	3.6	8	-
	4145	-	<<	800	800	0	3.6	3.6	-	-
	4146	.4	800	800	800	3.6	3.6	3.6	2.4	60
	4147	-	-	-	-	3.6	3.6	3.6	8	-
*	4149	-	-	-	-	3.6	3.6	3.6	2.4	120
*	4150	-	-	-	-	3.6	3.6	3.6	8	-

Table 1: Initial run conditions for the performed Boildown and Reflood experiments, as reported in Volume II and III [7] and evaluated (*).

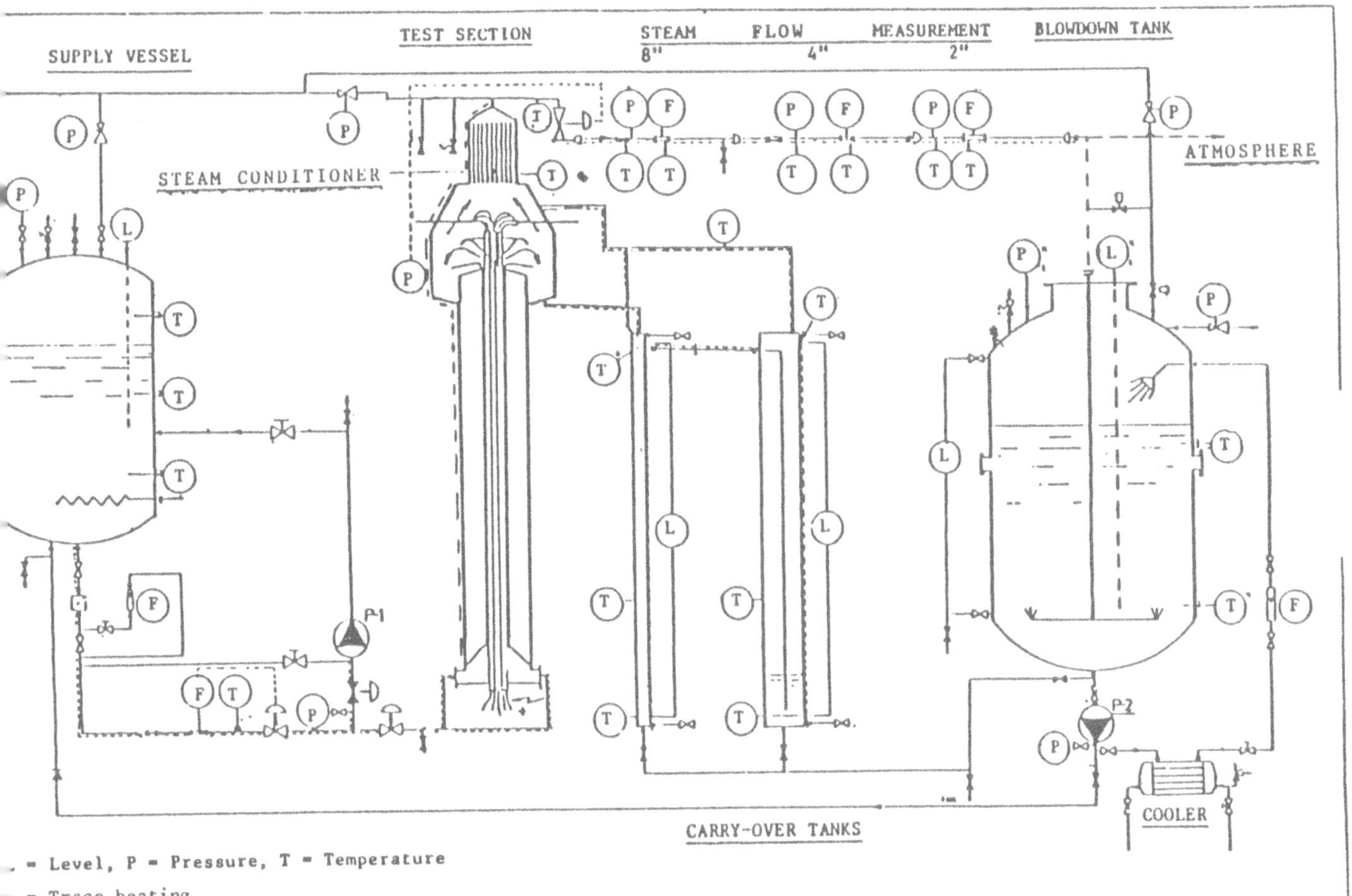

FIG.1A Scheme "ECN Reflood Facility"

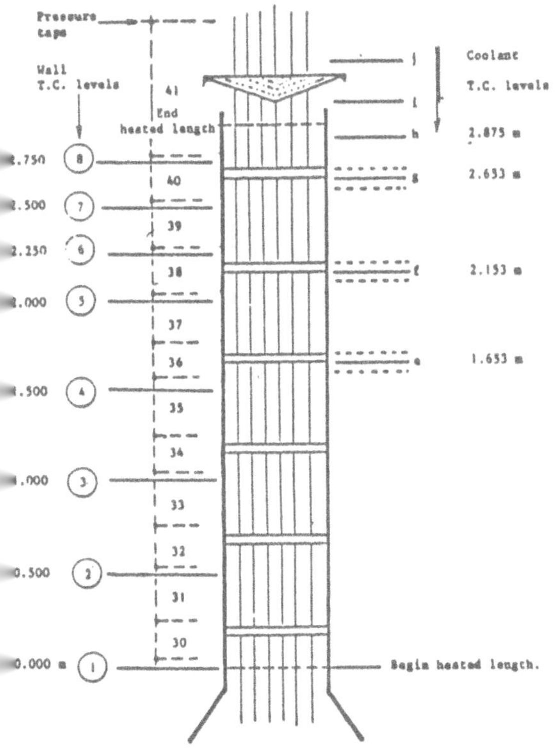

Fig.1B Thermocouple and DP-cell indications in bundle.

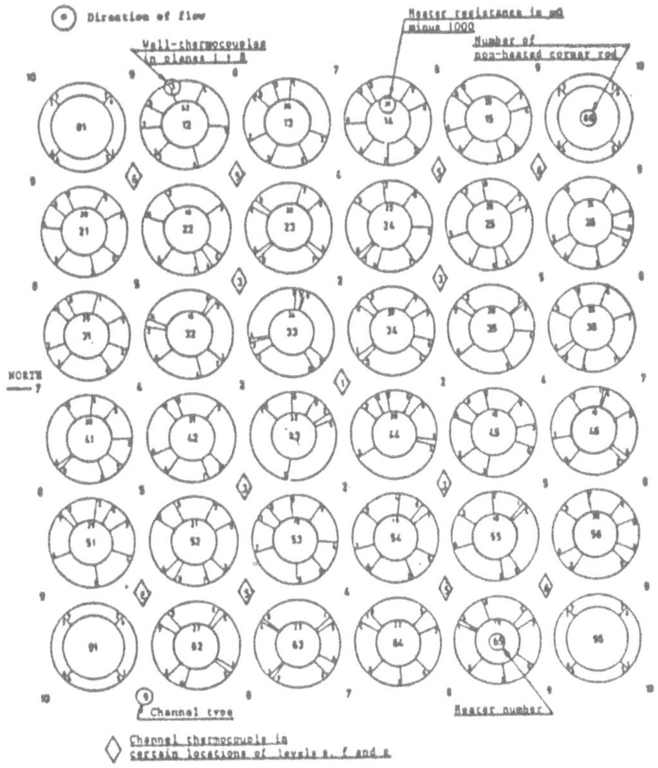

cross section area	47.8 cm^2	subchannel hydraulic diameter	13.6 mm
hydraulic diameter	12.3 mm	rod diameter	10.7 mm

FIG.1C Cross-section of the bundle with thermocouple positions.

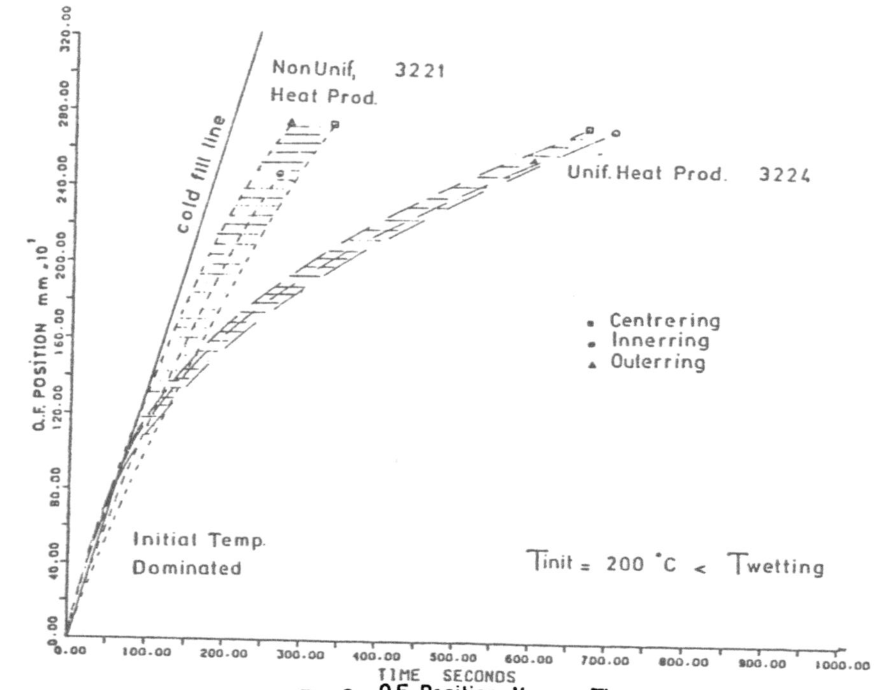

Fig. 2 QF Position Versus Time
For Low Initial Wall Temp.

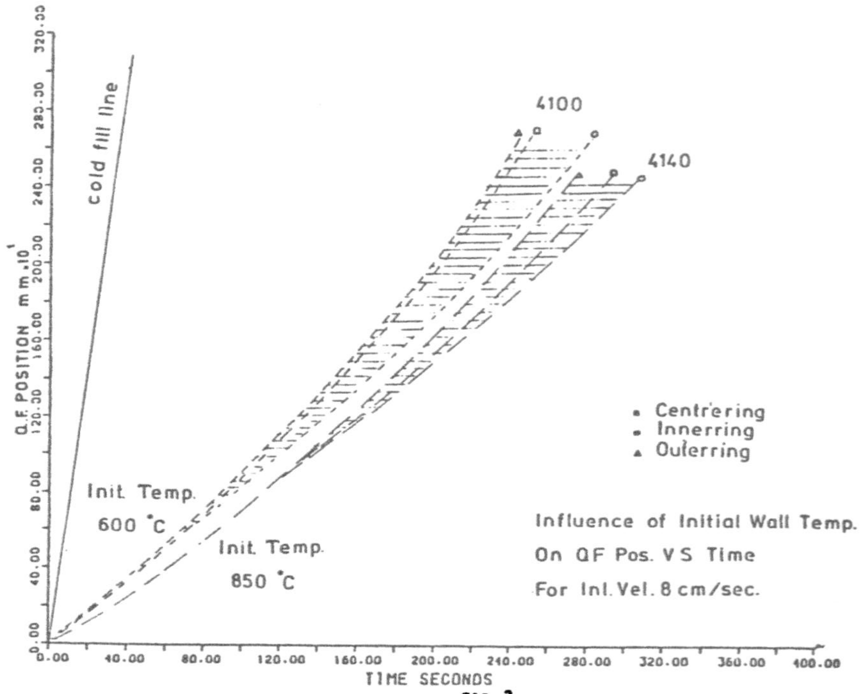

FIG. 3

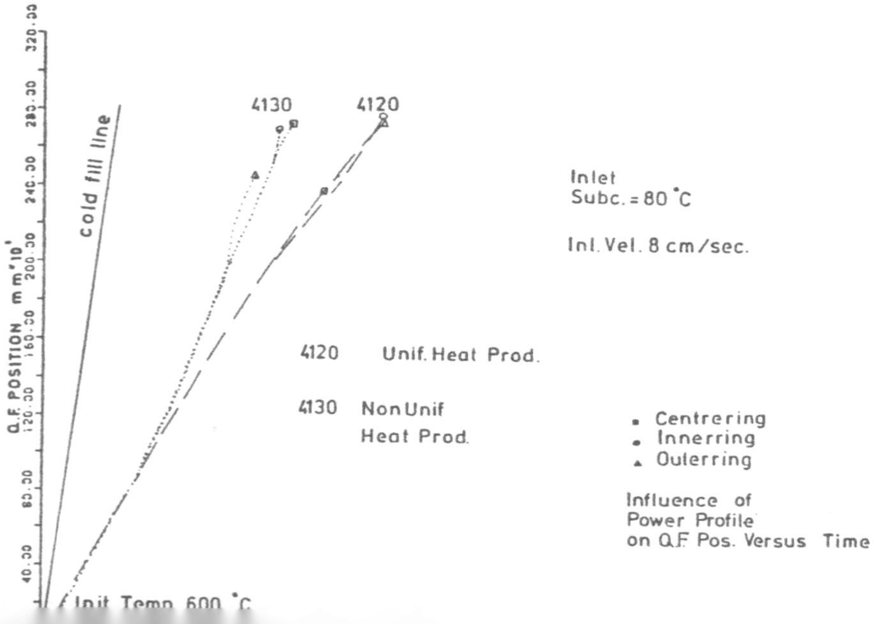

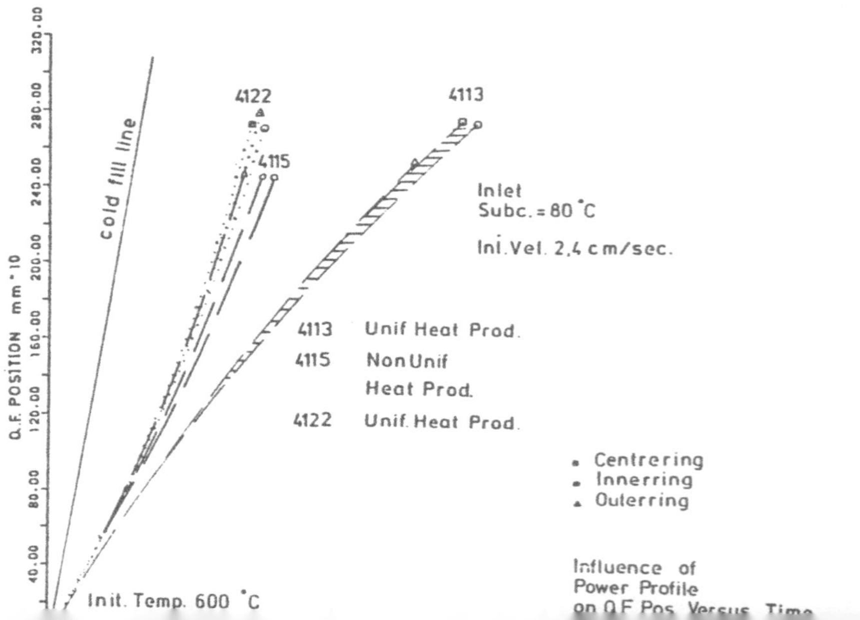

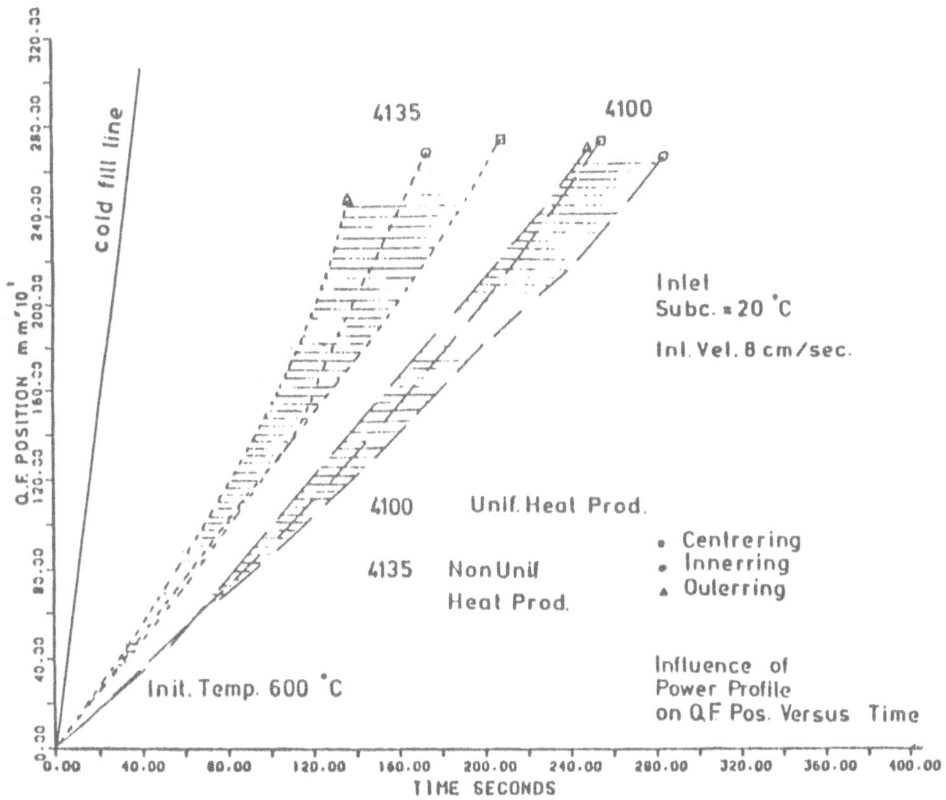

FIG. 6

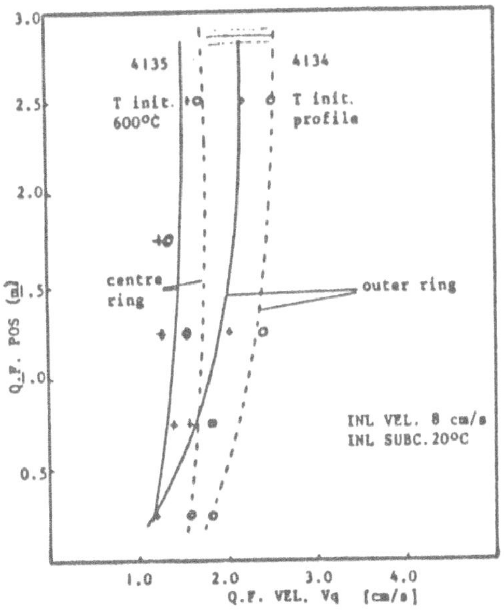

Fig. 7 INFLUENCE OF INITIAL WALL TEMP. PROFILE ON QUENCH FRONT VEL.

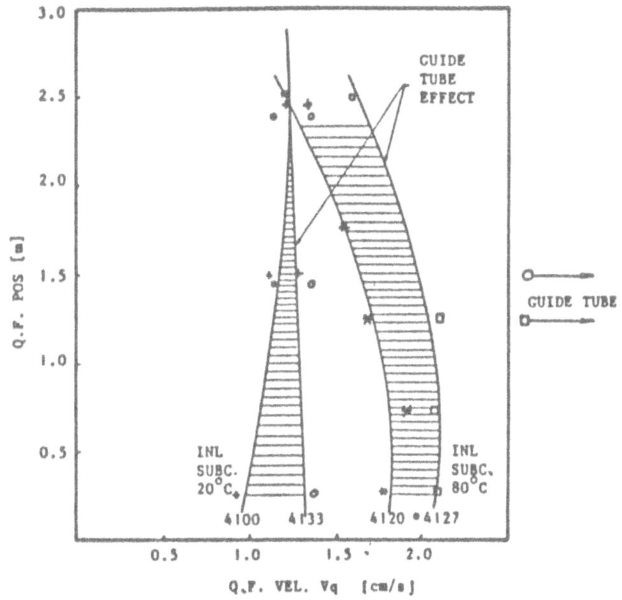

Fig. 8 INFLUENCE OF HEAT AND TEMP PROFILE ON QUENCH FRONT VEL. FOR INL. VEL. 8 cm/s

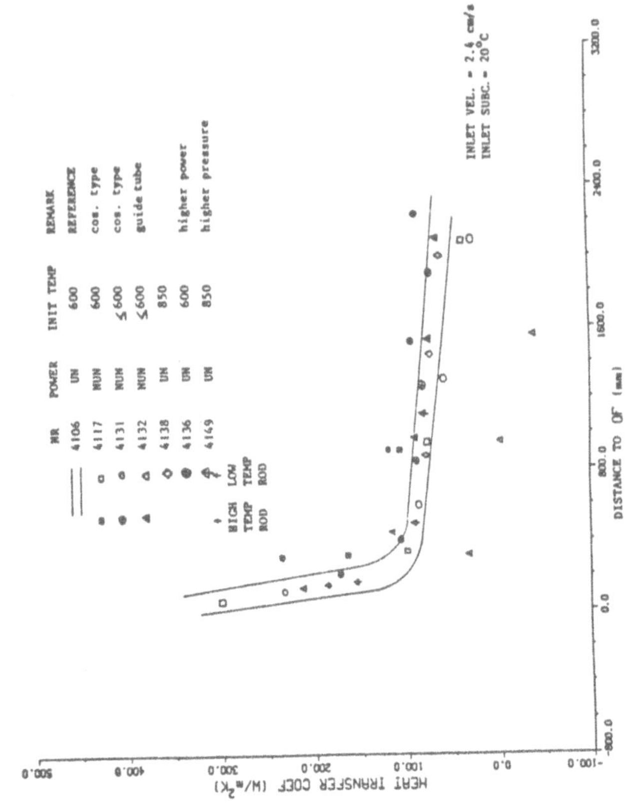

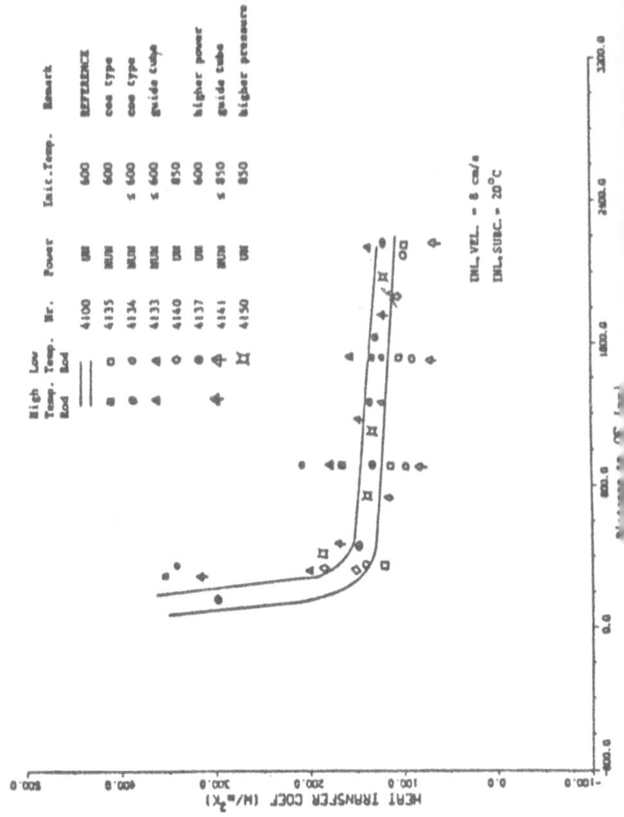

Fig. 10 HEAT TRANSFER COEFF. FOR LOW FLOW, HIGH INITIAL TEMP. EXPS.

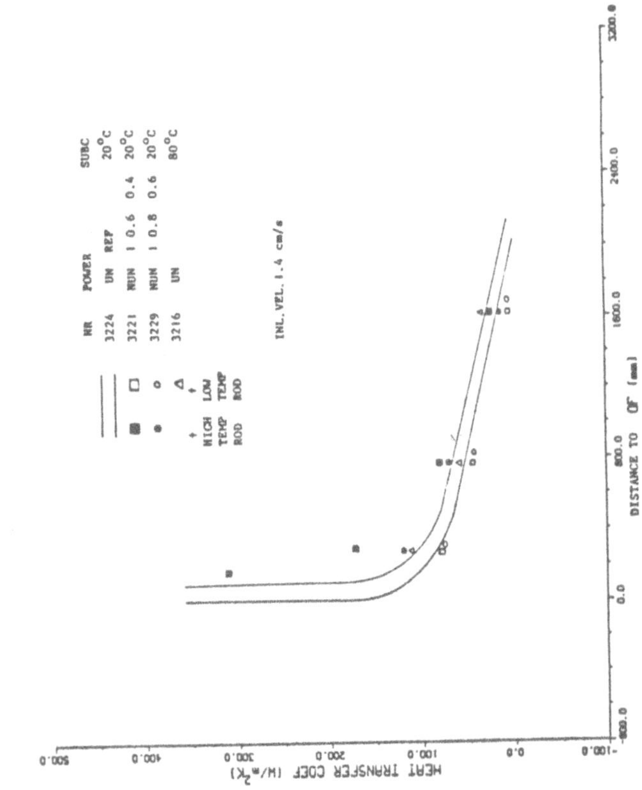

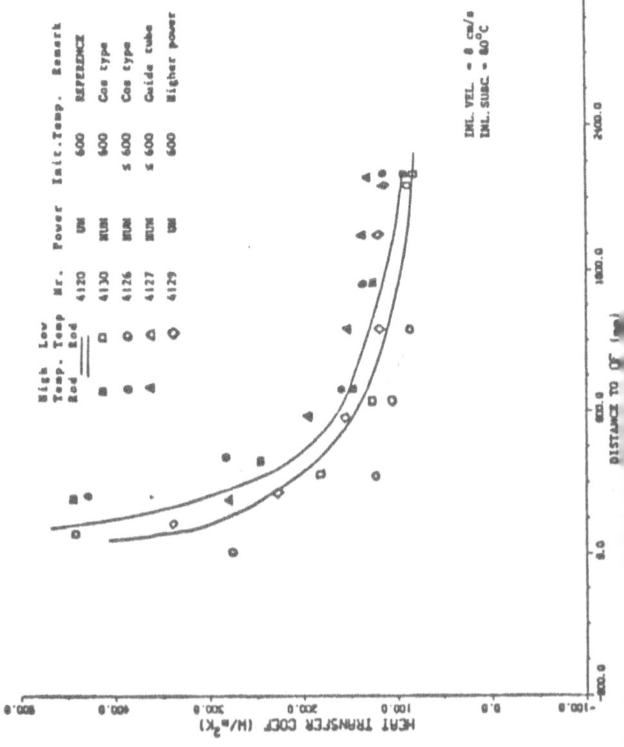

Fig. 9 HEAT TRANSFER COEFF. FOR LOW FLOW, LOW INITIAL TEMP. EXPS.

Discussion

Question: G.L. Shires - UKAEA/AEE-Winfrith

The form of your wetting front versus time profiles does not appear to be the same as shown by other contributors. They diverge earlier from the collapsed liquid level. Can you explain this, please?

Answer: S.B. vander Molen - ECN Petten

The difference with respect to the divergence of the E.C.N. experiments and those of other contributors is most probably due to the axial uniform heat production in case of the E.C.N. heater elements.

Question: R. Deruaz - CEA/CEN

Regarding point 7 of your conclusions: do you mean that the difference in quench front velocities on a hot and a cold rod is due:

- early in time to differences in initial wall temperature?
- later in time to differences in power?

Answer: T.C. de Boer - ECN

Conclusion 7 is given with respect to two types of experiments.

- Firstly, in case of a radial uniform initial temperature, uniform power experiment, the level of the uniform initial wall temperature determines the initial quench front velocity, whereas the power level determines the quench front velocity later on.

- Secondly, in case of a non uniform initial wall temperature and a non uniform heat production we think that the differences in quench front velocities on a hot and on a cold rod are due to initial wall temperature differences early in time and later on to the differences in power.

Parallel Sessions Research Area A

The loss of coolant accident (LOCA) and the functioning and performance of the emergency core cooling system (ECCS)

Session A4

Chairman: K. Wolfert (GRS-München)

Session secretary: E. Skupinski (CEC/BXL)

REWETTING PROPAGATION OVER ZIRCALOY UNDER BOTTOM FLOODING CONDITIONS

M. K. DENHAM and D. BLACKBURN
UKAEA, AEE Winfrith, Dorchester, Dorset, UK

SUMMARY

An experiment has been performed in which a hot, electrically heated, Zircaloy tube was internally reflooded with water from below. The object was to learn about the rewetting of the Zircaloy cladding of PWR fuel rods during the reflood phase of a loss of coolant accident. It was found, as previous work had shown, that the Zircaloy tube rewetted approximately twice as quickly as a similar tube of Inconel, because of its lower thermal capacity.

The effect of an oxide layer on the Zircaloy was systematically investigated. Layers of average thickness from 1 to 75 μm were studied, and the nature of the surface was carefully recorded. It was found that oxidation had a small effect on the quench front speed compared with the effect of other variables such as flooding rate and pressure. The speed increased by about 40% at the highest oxide thickness.

Calculations indicated that oxidation would have a similar effect on the rewetting of a PWR fuel rod.

1 INTRODUCTION

A study of the reflooding of a hot Zircaloy tube has been carried out at the Winfrith Establishment of the UKAEA. The study forms part of the Winfrith programme of experiments aimed at a better understanding of the thermal hydraulics of a PWR loss of coolant accident (LOCA). This programme includes the reflooding of clusters of electrically heated fuel rod simulators in the THETIS rig and internal reflooding of single tubes in the REFLEX Rig.

The Zircaloy cladding of reactor fuel is susceptible to oxidation at the temperatures which may be reached during a LOCA. To avoid this problem, experimental fuel rod simulators are usually clad in stainless steel or a nickel-chromium alloy. However, this does mean that, in order to relate such experiments to a reactor, it is necessary to understand the influence of the Zircaloy cladding. In a general review of rewetting published in 1978 [1] the need for more information on the rewetting behaviour of Zircaloy was pointed out, and it is still the case that little information is available in the open literature. More recent work by Andreoni et al [2] has shown that Zircaloy rewets more quickly than Inconel, in inverse proportion to its thermal capacity. Dhir et al [3] showed that this increase was tempered by the influence of the fuel pellets. Both workers found that oxidation of the Zircaloy tended to decrease the time required to quench their test sections.

The present work was carried out with the aim of carefully characterising the oxide layer and measuring its influence on rewetting. The results were then interpreted in terms of a model of rewetting and the model used to calculate the influence of oxide on the rewetting of reactor fuel.

2 DESCRIPTION OF THE EXPERIMENTS

The reflooding experiments were performed in the REFLEX Rig [4], which is illustrated in Figure 1. The test section was a 4.2 m long Zircaloy-2 tube, 12.7 mm inside diameter and 1.48 mm wall thickness. It was instrumented along its length with 21 0.5 mm thermocouples inserted into the wall. The assembly was mounted in a vacuum vessel so that heat losses would be small and calculable.

A reflooding experiment was carried out by first heating the test section by passing electrical current through it. When the temperature reached 600°C, water was made to flood into the tube from the bottom. A quench front progressed upwards and eventually the whole tube quenched. The test section temperatures were recorded at 0.5 s intervals. Measurements of quench front speed and rates of heat transfer to the coolant were deduced from the temperature records.

The experimental programme started with the test section in a clean, unoxidised condition and a series of 15 refloods was performed. This covered a range of pressure (1 - 4 bar) flooding rate (1 - 16 cm s^{-1}) and inlet subcooling (0 - 80 K) appropriate to a large break LOCA. Then, after examining the tube using an intrascope and a specially developed oxide thickness probe, the test section was oxidised in steam at 650°C until a layer of suitable thickness had grown. The series of 15 refloods was then repeated. In this way, layers of 1, 15, 25, 58 and 75 μm average thickness were studied in turn. Finally, the test section was cut up for detailed examination.

3 NATURE OF THE OXIDE LAYER

In a separate but related study, samples of the same Zircaloy tubing were oxidised in steam in a rig designed to reproduce the REFLEX rig conditions. The samples were held at 700 ± 2°C. Subsequently they were carefully examined to establish the physical structure of oxide layers of various thicknesses. Information on the oxidation rate of Zircaloy at 700°C, where previous data were sparse, was also obtained and used to select the oxidation times for the REFLEX rig.

The clean, unoxidised tube had a dull, metallic appearance, with a few longitudinal marks probably made during its manufacture. The 20 μm oxide layer was black with light coloured flecks; it had a smooth appearance. The 34 μm layer was a cream colour, again with some lighter patches. It also appeared smooth. The 56 μm layer was a more uniform, paler colour, it had evident cracks running circumferentially, and in a few places (perhaps one per cm^2) a piece had spalled off. The 94 μm oxide layer looked similar to the 56 μm layer.

The layers were quite uniform in thickness over distances comparable to their thickness. However, the thickness varied considerably over distances of 1 mm or more. Cracks running parallel to the surface could be clearly seen when the layers were sectioned. The oxidised surfaces had roughness averages of about 1.2 μm, twice that of the clean tube. All these features are consistent with the Zircaloy oxidation literature.

Intrascope inspection and post-test examination of the REFLEX test section showed that the oxide layers formed were similar to those described above, except that there was more spalling and regrowth of oxide. The oxide thickness was not very uniform, varying from place to place by a factor of two.

4 RESULTS OF REFLOODING EXPERIMENTS

4.1 Typical Results

Figure 2 shows results from one of the reflooding experiments. In the upper graph, temperatures at several heights on the test section are shown varying with time. The lower graph shows the position of the quench front which advanced up the test section, and also the collapsed liquid level in the test section, deduced from pressure measurements. It is noticeable that the collapsed liquid level stopped rising after about 115 s. Until shortly before this time, the water arriving at the quench front had been subcooled. Once the water became saturated, sufficient steam was generated below the quench front to entrain nearly all the water, and little further water accumulated in the test section. A corresponding change in the flow regime above the quench front, from inverted annular to dispersed flow, is believed to have accompanied this change. One effect of this was to slow down the quench front, as clearly shown in Figure 2. In the following paragraphs we distinguish between the initial 'liquid level rising' phase of the transient, and the subsequent 'liquid level constant' phase.

4.2 Quench Front Speed

Figure 3 shows how increasing the oxide thickness affected the quench front speed. In this Figure, the quench front speeds measured at particular conditions (flow rate, pressure, subcooling and axial position) and various oxide thicknesses, have been normalised by the corresponding speed measured at 1 μm thickness. It can be seen that while the liquid level was rising the speed varied only a little. Oxidation had a greater effect once the liquid level became constant, increasing the quench front speed by an average of 40% as the thickness rose to 75 μm.

The quench front speed did not vary steadily with oxide thickness in either case, probably because of significant spalling which occurred during and after the 25 μm test series. As a result of the spalling, some areas of the test section were covered in a thinner than average layer of oxide. The effect of oxide was relatively small compared with the effect of variations in the other experimental parameters (flow rate, for example) and this is another reason for the scatter in the experimental results.

4.3 Apparent Rewetting Temperature

The temperature of the wall just before quenching occurs is known as the apparent rewetting temperature, and plays an important role in most theories of quench front propagation. This measured temperature is generally higher than the true rewetting temperature because of axial and radial temperature gradients in the tube wall and oxide layer. Figure 4 shows the effect of oxide thickness on the apparent rewetting temperature. These results cover a range of flow rate, pressure and subcooling.

When the liquid level was rising, the rewetting temperature increased with oxide thickness by as much as 90 K at 75 μm. A similar increase was reported by Dhir et al [3] who attributed it to an increase in the wettability of the surface. However, as Figure 4 shows, the increase is also consistent with the difference in thermal effusivity between Zircaloy and Zircaloy oxide [5]. In other words the surface of the oxide layer may have been able to cool to the rewetting temperature while the bulk, measured, wall temperature remained higher.

The lower graph in Figure 4 shows that once the liquid level became constant, the apparent rewetting temperature became insensitive to the oxide thickness.

It is noted that the experimental apparent rewetting temperature increased with pressure and flow rate. This is why Figure 4 reveals a range of temperature at each oxide thickness. The homoeneous nucleation temperature theory [5] does not account for this, but an instability model, though not strictly applicable to the vertical tube geometry, does show an appropriate sensitivity to pressure.

It is interesting that while the liquid level was rising, oxidation increased the apparent rewetting temperature without very much change in the quench front speed. In contrast, when the liquid level was constant the quench front speed increased but the apparent rewetting temperature remained constant. In both cases, the rate at which heat was released at the quench front increased with oxide thickness by about the same amount, 40% at 70 μm oxide thickness.

5 ANALYSIS AND INTERPRETATION OF THE RESULTS

The results have been studied in order to answer the following questions concerning the quench front speed:

- how do the measured quench front speeds compare with previous results?
- can the observed effects of oxidation be explained?
- what are the implications for the rewetting of reactor fuel?

5.1 Comparison with Previous Results

A straightforward comparison of the Zircaloy tube results with our earlier REFLEX results for an Inconel tube of similar dimensions is shown in Figure 5. The speeds were in inverse proportion to the thermal capacities of the two tubes, a ratio of 2.18 in this case. This agreed with previous workers' findings.

In order to go further than this, it was necessary to analyse the results in terms of a model of rewetting. The Clement-Regnier k_2 model [6], which provides a way of specifying the surface heat flux during quenching using just one empirical parameter k_2, was chosen. Tools for carrying out two-dimensional conduction calculations for various geometries, including tubes and reactor fuel rods with oxide layers, were developed and used for this work.

As a precursor to the analysis of the Zircaloy results, our earlier REFLEX Inconel tube data were compared with the Clement-Regnier k_2 correlation. As Figure 6A shows, the results were in broad agreement. Closer examination of the results showed that the rewetting temperature T_w had an effect which was not accounted for by the correlation. It was found that the parameter $q^* = \rho c \beta u \, (T_w - T_o)$, plotted in Figure 6B, was effective in correlating this dependence. Here $\rho c \beta$ is the thermal capacity per unit length of the test section and u is the quench front speed.

Clement and Regnier discovered that the same values of k_2 correlated results both from Inconel and from Zircaloy tubes. Our results support this. Firstly, the rewetting temperature behaved similarly for Inconel and Zircaloy (without oxide). Secondly, we found that k_2 was actually nearly proportional to the product of the thermal capacity of the tube and the quench front speed, $\rho c \beta u$, and that the constant of proportionality was nearly the same for both our test sections (because of their similar thickness and thermal conductivity). Therefore, since both test sections yielded the same values of the product $\rho c \beta u$ (Figure 5), their k_2 values must also have been similar.

5.2 Effect of Oxidation

Calculations of rewetting speed were carried out for the Zircaloy test section for a range of oxide thicknesses. First, k_2 was calculated for a typical quench front speed at low (1 μm) oxide thickness. k_2 was then kept constant as the oxide thickness was increased, and T_w was also kept constant as suggested by the experimental observations for the 'liquid level constant' phase. The calculated rewetting speed rose with oxide thickness in accordance with the experimental measurements. Similarly, when T_w was made to rise in a way consistent with the experimental observations for the 'liquid level rising' phase, and k_2 was allowed to vary as implied by the q^* correlation described above, the calculated rewetting speed remained approximately constant, again in agreement with experiment. These encouraging results are shown in Figure 3.

It is noted that if k_2 had been kept constant in the latter calculations, according to the Clement-Regnier correlation, the results would have been insensitive to T_w. So the effect of oxide in the 'liquid level rising' phase, when T_w increased with oxide thickness, would not have been correctly reproduced.

In all these calculations, it was assumed that the oxide had a thermal conductivity of 0.75 W m^{-1} K^{-1}. This is a plausible value for this rather uncertain quantity [7]. The change in quench front speed caused by the oxide layer was predicted to be inversely proportional to the conductivity of the oxide.

5.3 Implications for Rewetting of Reactor Fuel

The rewetting speed for reactor fuel was calculated using the k_2 model, for two cases, one giving fairly high speeds and the other, low speeds. Two oxide thicknesses, zero and 75 μm, were examined and the calculations covered a range of values for the thermal conductance of the gas gap between the fuel and the cladding.

The results are shown in Figure 7. The effect of oxide, at a typical gas gap conductance of 5 000 W m^{-2} K^{-1}, was to increase the quench front speed by 65 to 70%. This increase was therefore of the same order as that calculated (and observed) for the Zircaloy tubular test section. At higher values of gas gap conductance, the predicted effect of oxide was similar, whereas at lower values when the quench front speed was naturally higher, the predicted effect of oxide was relatively small.

6 FINAL REMARKS

This work has confirmed that a Zircaloy tube rewets approximately twice as fast as a similar tube of Inconel or Stainless Steel, because of its lower thermal capacity. The systematic study of the effect of oxidation of the Zircaloy has shown that oxidation has an effect (over the range of oxide thickness likely during a loss of coolant accident) which is relatively small compared with the effect of flooding rate or pressure. Furthermore, calculations have indicated that the effect of oxidation on the rewetting of a fuel rod would be equally small. Under conditions of practical importance, a small increase in quench front speed can be expected, and this should be beneficial from the point of view of reactor safety.

REFERENCES

[1] ELIAS, E. and YADIGAROGLU, G. (1978). The Reflooding Phase of the LOCA in PWRs Part II: Rewetting and Liquid Entrainment. Nucl Safety 19, 160-175.

[2] ANDREONI, D., CLEMENT, P., DERUAZ, R., PORRACCHIA, A. and REGNIER, P. (1980). Calculation of Single Tube Reflooding with the HEXECO Code and Analysis of ERSEC Experiments. Paper B6 presented to the European Two Phase Flow Group, Glasgow.

[3] DHIR, V. K., DUFFEY, R. B. and CATON, I. (1981). Quenching Studies on a Zircaloy Rod Bundle. J Heat Transfer Trans ASME 103, 293-299.

[4] DENHAM, M. K., ELLIOTT, D. F. and BRITTON-JONES, K. A. (1984). A Study of Rewetting Propagation over Zircaloy under Bottom Flooding Conditions. European Applied Research Reports - Nucl Sci Tech. Vol 5 No 5, 917-1100. (Also EUR 8894 EN and AEEW - R 1600).

[5] YAO, S. C., HENRY, R. E. (1978). An Investigation of the Minimum Film Boiling Temperature on Horizontal Surfaces. Trans ASME J Heat Transfer 100, 260-267.

[6] CLEMENT, P. and REGNIER, P. (1978). Heat Transfer Modelling at the Quench Front During Reflooding Phase of a LOCA. Paper presented to the European Two Phase Flow Group, Stockholm.

[7] GILCHRIST, K. E. (1979). Thermal Conductivity of Oxide Deposited on Zircaloy Fuel Tube Material. J Nucl Materials 82, 193-194.

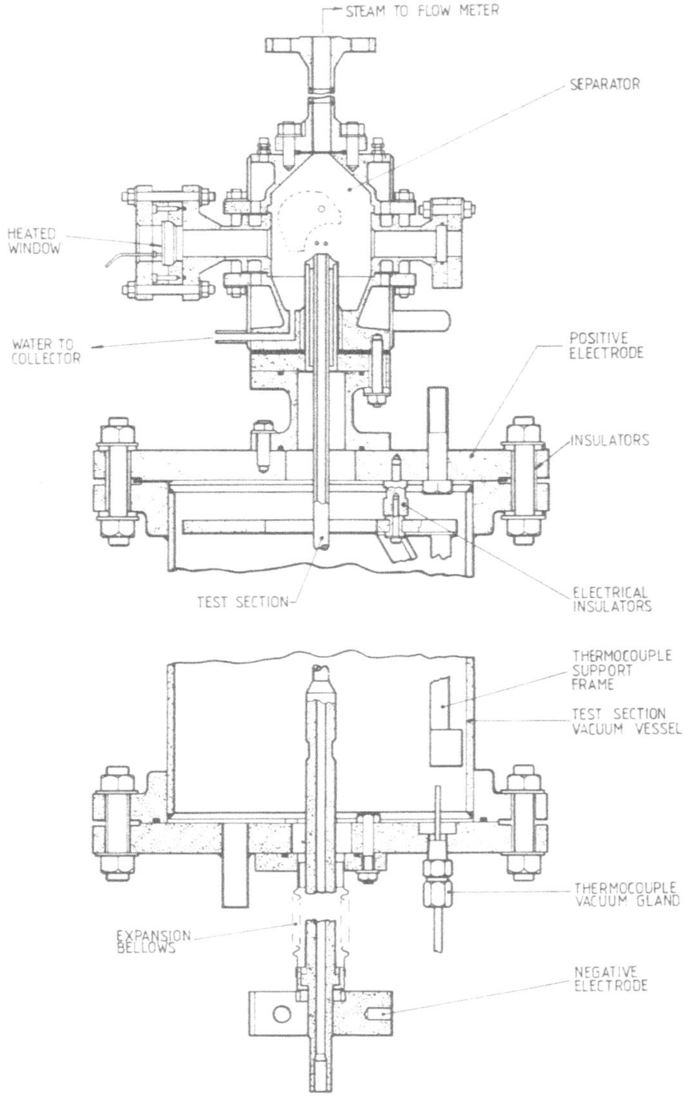

1 **REFLEX TEST SECTION ASSEMBLY DRAWING**

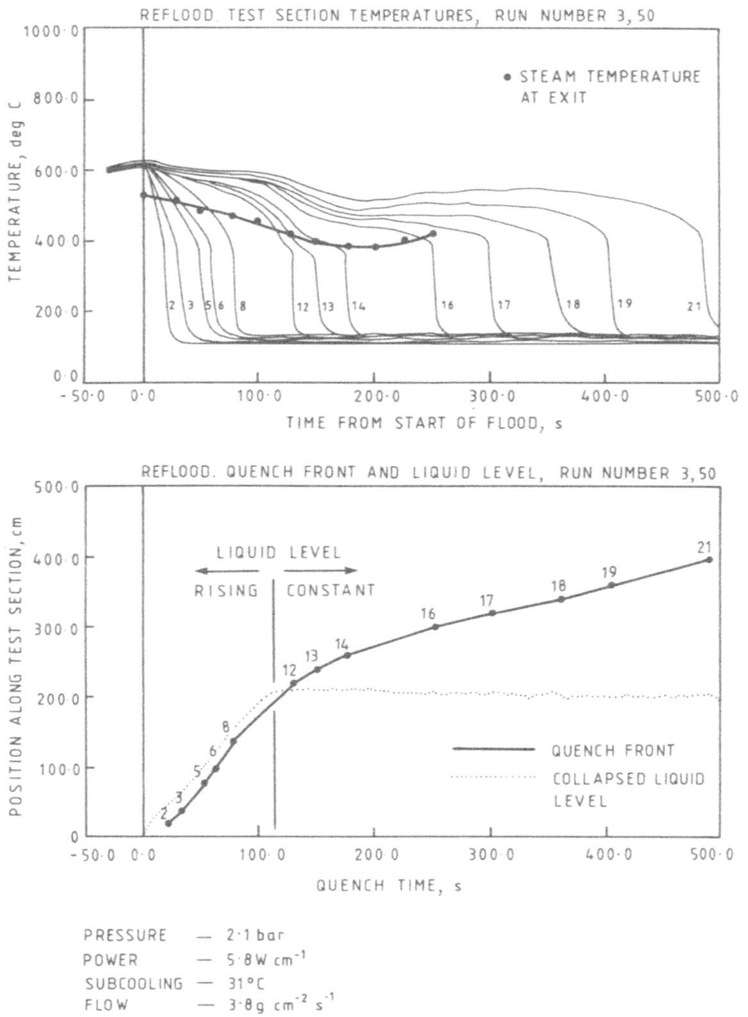

2 **ILLUSTRATIVE RESULTS OF A REFLOOD EXPERIMENT**

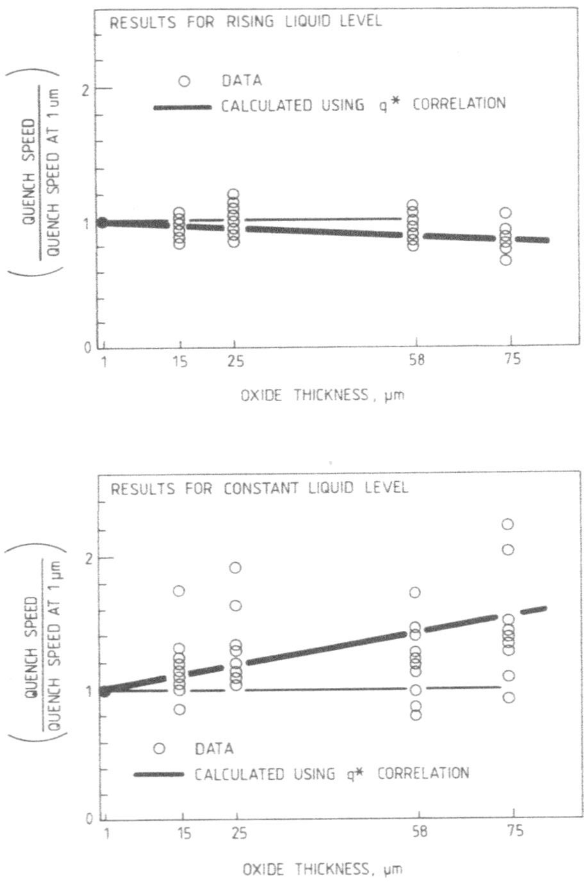

3 EFFECT OF OXIDE THICKNESS ON QUENCH FRONT SPEED

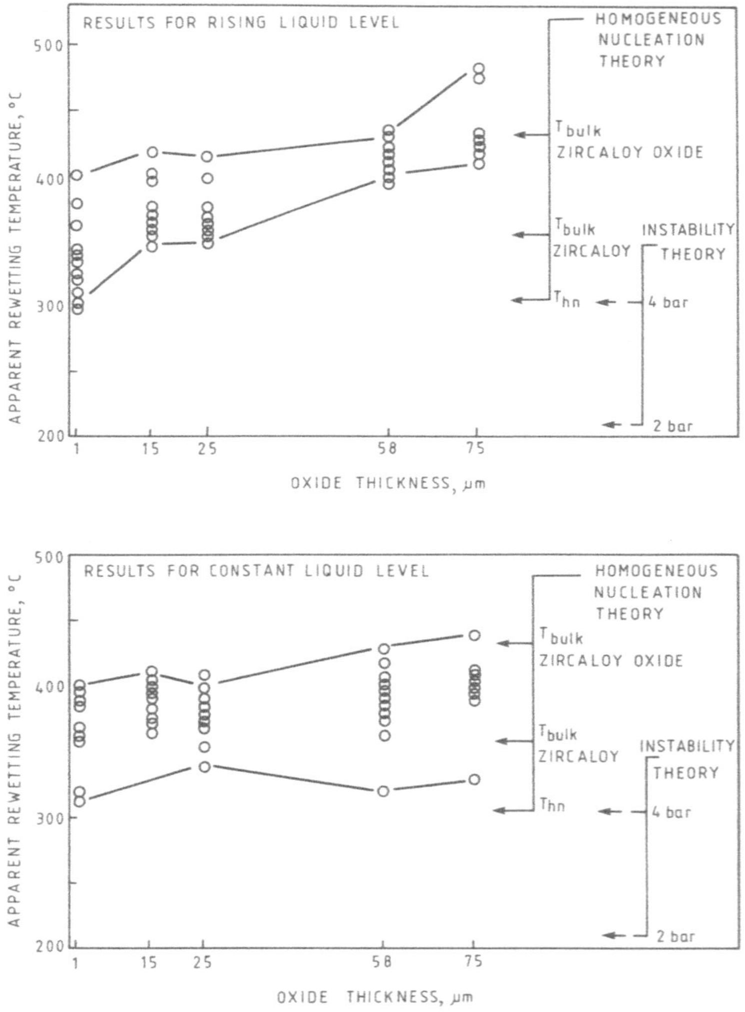

4 EFFECT OF OXIDE THICKNESS ON REWETTING TEMPERATURE

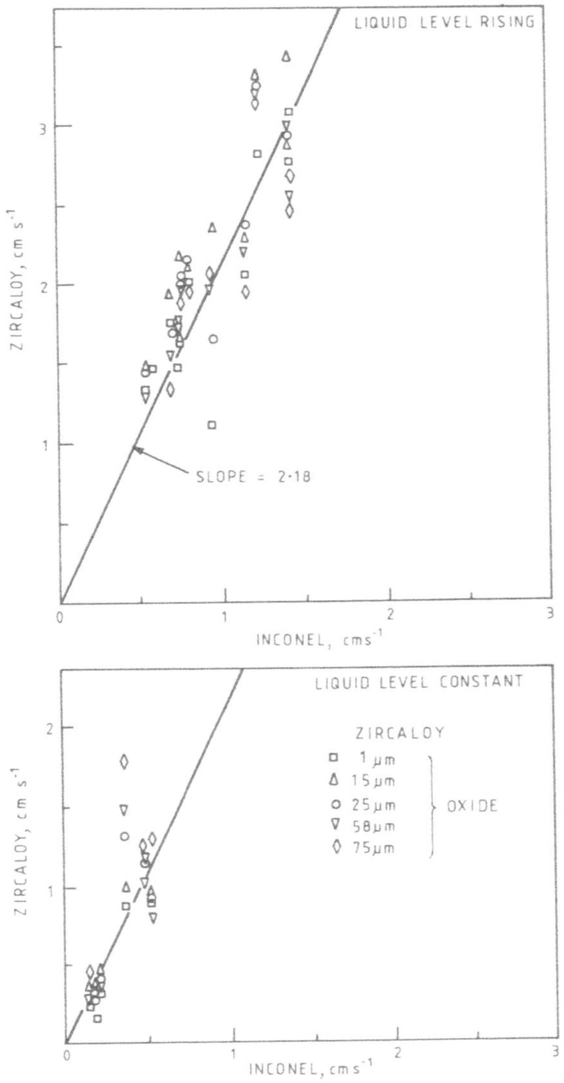

5 COMPARISON OF QUENCH FRONT SPEEDS OVER INCONEL AND ZIRCALOY

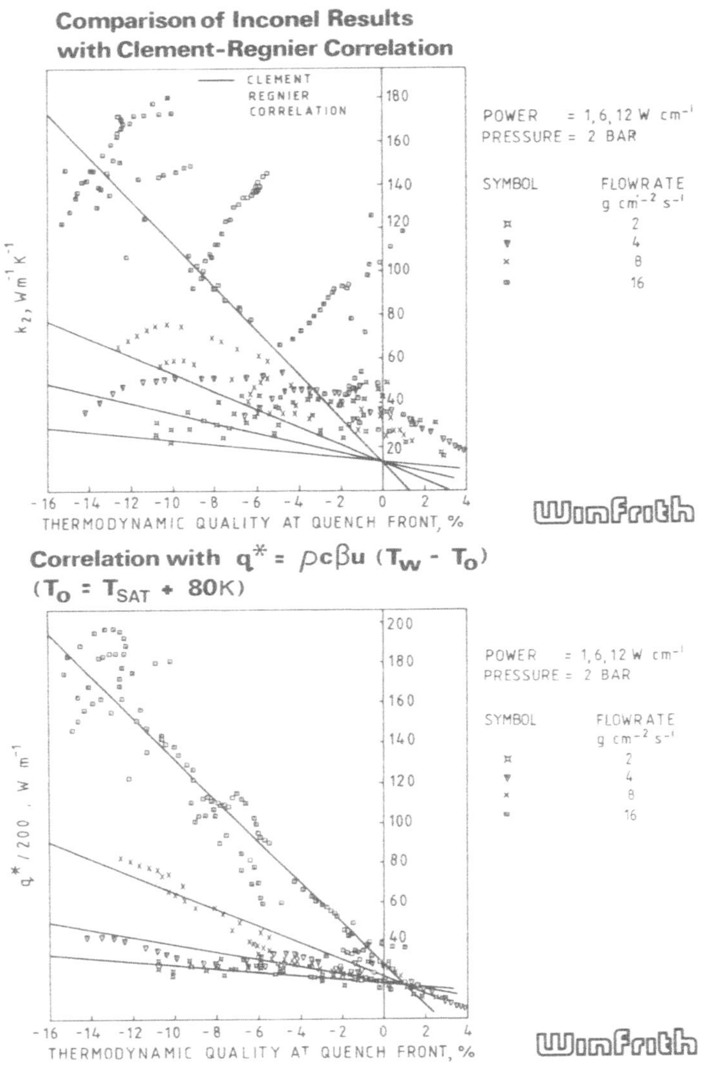

6 CORRELATION OF INCONEL TUBE RESULTS

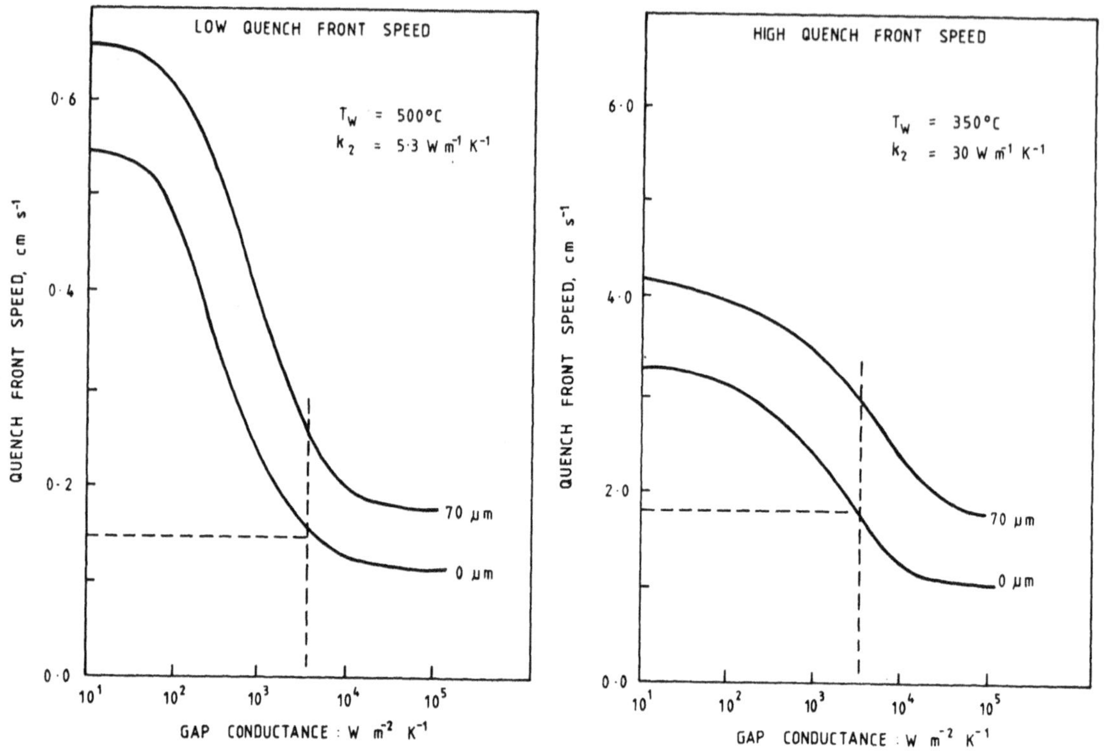

7 **CALCULATED EFFECT OF CLADDING OXIDATION ON THE REWETTING OF REACTOR FUEL**

Discussion

Question: K. Wolfert - GRS

What would be the difference in the results of a PWR calculation using the modified or the original k2 factor?

Answer: M.K. Denham - UKAEA

This question can certainly not be answered with without performing a complete reflood calculation using the two different factors. This is because the apparent rewetting temperature, on which the modified k2 depends, is not an independant variable; it depends on the quench front speed which in turn depends on k2.

Our calculations have been confined to the region of the quench front. Here the effect of modifying k2 was to decrease the quench front speed calculated for our Zircaloy test section by up to 50% under some conditions, as described in the paper. As discussed, the quench front speed for reactor fuel would probably be affected to a similar extent.

Question: R. Deruaz - CEA/CEN

In your modified version of k2, what is the meaning of T_o? Is it a physical quantity or a correlating parameter?

Answer: M.K. Denham - UKAEA

T_o is simply a correlating parameter, chosen to give the optimum correlation of the effect of apparent rewetting temperature, T_w, on the quench front speed.

Question: S.A. Fairbairn CEGB/BNL

Can you comment on the likely effect on quench rate of replacing the Inconel clad of the heaters in Mr. Kaiser's project with a Zircaloy clad.

Answer: M.K. Denham - UKAEA/AEE-W

Since Zircaloy has half the thermal capacity of Inconel, using cladding of Zircaloy would increase the quench front speed, possibly by a factor of two, though probably somewhat less than this because the increase would be tempered by the presence of the simulated fuel pellets. It should also be noted that reducing the cladding thickness from Mr. Kaiser's 1 mm to 0.5 mm thickness of PWR fuel cladding would tend to increase the speed, again by a maximum of a factor of two. This could be quantified using the methods described in my paper.

THE ROLE OF EXPERIMENTS IN THE DEVELOPMENT AND QUALIFICATION OF THERMALHYDRAULIC CODES, AS EXPERIENCED IN THE DEVELOPMENT OF THE CATHARE CODE

J.L. NIGON (CEA/CEN-GRENOBLE)

SUMMARY

On one hand, analytical tests are used to develop and to qualify physical laws ; on the other hand integral tests are used for global verification . This integral tests should never be used for adjustment of correlations, rather, only in order to point out eventual needs for improvement.

A stepwise approach is recommended for the development work. The mechanical laws should be derived first. Using them as a basis for further interpretation of experimental results, the mass transfer laws should then be established. The wall heat transfer correlations are the last ones to be fixed. Of course the correlations required for the description of component behaviour are developed (or adjusted) only after the physical laws of pipe flow are fixed. A standard series of analytical tests which covers a very wide range of parameters is designed to be used as the basis of the systematic qualification process. This methodology leads to a complete consistent set of physical laws (named a Revision in the CATHARE code).

The qualification work together with the first verification calculations reveal needs for improvements, particularly in the areas of condensation and of reflood (grid effect, multidimensional effects), which we hope could be achieved through on-going or planned experiments.

INTRODUCTION

A rigourous methodology, which is though to be a successful way towards best estimate reactor calculations, has been applied in the best of possible ways to the development and assessment of the CATHARE code.

The basic concept is to go from simple phenomena to more complex ones. Then according to a stepwise "no return" strategy correlations should be deduced from separate effect tests only, whereas integral test calculations are devoted to verification and identification of new needs.

We will first present the development strategy and its experimental basis. Then the qualification process and the definition of a "Revision" of the physical laws will be explained. A few words about verification will precede the presentation of some important needs and of the subsequent aspects of the future work.

1. DEVELOPMENT STRATEGY

1.1. Once the model, namely the 1D-6 equation model, is chosen, the differential terms (if any) of the physical laws have to be fixed. In our case, added mass term and pressure difference correction are based on theoretical considerations. They have been verified against existing experiences and/or against the most commonly accepted theories (Wallis, Ishii, Taitel and Duckler for stratification criterion).

1.2. The mechanical laws are derived from a large data base, which consists of experimental results and of previously qualified models and correlations.

It is the case, for example, of the Drift Flux Model which was extensively used to establish the interfacial friction factor of CATHARE in the range of bubbly flow. The experiments used for the derivation of these mechanical laws are either air water tests or steam water tests carried out in thermal equilibrium. They are listed in table I. For most of these experiments the measured values of mass flow rate, void fraction and pressure along the test section, allow a direct estimation of the interfacial friction factor. For the other experiments (such as REBECA) the use of the CATHARE model itself [through an "inverse" calculation in which pressure values are used as data] leads to an evaluation of τ_i assuming that τ_i can be written in the form :

$$\tau_i = \xi(\alpha, P, V_\ell, V_v, D)\, \rho_\ell \frac{(V_v - V_\ell)^2}{D} \tag{1}$$

The interpretation of the previous experiments leads to a data bank $[\xi; \alpha, P, V_\ell, V_v, D]$ from which the correlations are derived. The interfacial friction factor ξ introduced in the current version of CATHARE (Revision2) is represented on figure 1.

1.3. Because of its major importance in critical flow estimation, the vaporization of liquid has been given emphasis during the development of the mass transfer laws. The mass transfer is written in the form :

$$\Gamma = -\frac{q_{\ell i} + q_{vi}}{h_v - h_\ell} \tag{2}$$

Assuming saturation conditions at the interface, the liquid heat transfer is supposed to be proportionnal to $h_\ell - h_{\ell sat}$

$$q_{\ell i} = f(\alpha, P, V_\ell, V_v)\,(h_\ell - h_{\ell sat}) \tag{3}$$

The mass transfer experiments used for the development of CATHARE are listed in table II [we will see in the next chapter another set of experiments used for qualification of both momentum and mass transfer laws]. Assuming that mechanical laws are still applicable in the case of mass and energy transfer between phases, the use of the CATHARE model itself, thanks to an "inverse" calculation, provides a data bank for $q_{\ell i}[\alpha, P, V_v, V_\ell]$. This inverse calculation is possible in the case of steady state conditions, under which the pressure is known (measured) along the flow line and all other conditions are known at inlet. However, since the numerical scheme is slightly different from that of the "direct" calculation, the consistency of the result is checked by a reconstitution of each test used for the data base.

1.4. Finally wall heat transfer correlations are developed from non adiabatic experiments. Again, assuming that the transfer laws previously established are applicable, we use them to evaluate the local state of the fluid (α, h_v, h_ℓ, V_v, V_ℓ) which is to be introduced in the correlations. The heat transfer experiments presently used for CATHARE are listed in table III. Inverse calculations are no longer applicable, except for the case of reflood in tubular test section, where the wall heat transfer downstream of the quench front may be well represented by a steady state calculation. In the other cases, the complexity of the situation resulting from fast transient conditions and/or from the rod bundle and mixing grid effects, leads to a trial and error strategy, and the search of coefficients in correlations of given form.

1.5. In the present version of CATHARE the specific development for components has been limited to capacities and pumps ; further development will be performed in the next version.
The internal laws inside the capacities are based on a bubble rise velocity for the lower plenum and a droplet velocity for the upper plenum. Up to now emphasis has been put on the correlations to be used at the junctions between pipes and capacity. Special attention has been given to the lower plenum the behaviour of which is studied in the PIERO experiment (1/4 scale, air water tests at atmospheric pressure, including CCFL in the downcomer, entrainment of water in the bottom of the lower plenum by air flowing out of the core...). Regarding the primary pumps, head and torque correlations including two phase degradation, fit the experimental results of EVA and EPOPEE (1/3 and 1/10 scale respectively).

1.6. Finally the results of clad deformation and burst tests obtained during the EDGAR programme are included in the fuel model.

2. QUALIFICATION

Whereas development of physical laws relies on a rather limited number of experiments, most of them being steady state tests, the most wide possible range of parameters should be covered by the qualification tests. Additionnal separate effect tests used for qualification are listed in Table IV.

As soon as a formulation is established for a given law, many calculations are performed by the developer himself, using series of tests already used for the derivation of the correlation as well as tests not yet used in the development work. After some adjustment, if needed, the correlation reaches its final form.

After representation of all physical phenomena is completed, the whole set of models and correlations is proposed to be the next "REVISION" of CATHARE.

Before the revision is accepted in the code, one has to check by means of systematic computation of a series of analytical tests that no major fault or inconsistency has been introduced in the package. This step is a kind of "quality insurance". Among the many experiments used for qualification, about seventy tests are selected for this systematic check ; the selection criteria are the following :

(i) Thermohydraulic conditions of the test should be as close as possible to the anticipated conditions during a reactor accident.

(ii) The range of parameters should be as wide as possible. Thus, the values of the parameters of all the tests of a facility should be inside the range of the parameters of the selected tests.

(iii) Tests should have significant features. They should illustrate some special effects such as countercurrent, level formation, level migration..., grid effect, etc...

(iv) Experimental results, i.e. measurements, should be accurate enough, and reliable.

(v) Tests should be chosen in such a way that as few tests as possible are needed to satisfy the above goals.

During this qualification phase, we had the oportunity to compute experiments which were not yet used, some of them covering a new range of parameters. It was the case for example of the VERTICAL CANON, TAPIOCA, SUPER MOBY DICK (vertical large diameter pipe), G2 LOOP (Westinghouse). Some results are a remarkable confirmation of the quality of the mechanical laws. Others, namely the G2 level swell experiments, reveal an important need for improvement, which is expected in the future work.

The overall development and qualification process is reported on figure 2

3. VERIFICATION

As it has been previously said, the objective of integral test calculations is strictly limited to the verification of the overall behaviour of the code ; major trends are manifested by sensitivity studies [influence of the modification of basic physical correlations, inside the range of uncertainty, as well as the influence of numerical approximation, or of component description]. The major discrepancies, and the high sensitivity of the final results to some physical parameters, lead to the identification of needs for further developments or improvements.

Up to now our own experience in this verification process is limited. Only three integral tests LOFT L3-5 and L3-6 and LOBI SDSL 03 are completed . Altogether, the CATHARE results are comparable to the results of other codes - not better, not worse -. The main discrepancy seems to be the evaluation of the mass flow rate at the break... The reason for this discrepancy is not clear : the measured conditions upstream of the break are correctly evaluated, and the critical flow measured on Super Moby Dick for similar upstream conditions is accurately calculated by CATHARE (as well as by most other codes).

Besides this major concern, the pump behaviour in two phase flow had to be improved : we had an improper split of the pump head between the impulsions of liquid and vapor !.

Further verification on LOFT, LOBI, PKL and, in the future, BETHSY, will provide us with much more essential information in order to use the code for safety evaluation and to define the priorities for next Revision [see figure 3].

4. NEEDS AND FUTURE WORK

4.1. As far as the mechanical laws are concerned, the most important need is identified as the interfacial friction factor in rod bundle geometry (core channels). As a matter of fact it is inappropriate to describe the swelled level in boil off conditions in rod bundles with the interfacial friction factor of pipe flow. Let us summarize the results which we obtained on this subject :

(i) The correlation of τ_i has been derived from the Drift Flux Model and from Moby Dick ;

(ii) The qualification of τ_i , including the validation of the diameter effect (0.014 m to 0.300 m) is based on Super Moby Dick, Vertical Canon and Tapioca ;

(iii) The calculated swelled level in Vertical Canon and Tapioca was quite perfect in most cases ;

(iv) The calculated swelled levels in G2 loop (of Westinghouse - rod bundle boil off tests) were not comparable to experimental data ;

(v) The calculated swelled level in boil off experiment, tubular test section of 12 mm inside diameter (ERSEC) was correct.

From this five established facts, we conclude that the low heat flux in boil off experiments is not the reason of the nonapplicability of τ_i, that there is no reason to modify the τ_i correlation in tubular geometry and that a new correlation has to be derived for rod bundle geometry ; this is under way.

4.2. Another indentified important need is related to rapid condensation of steam on undersaturated water flow in pipes and in capacities. Significant amount of work has still to be done on this subject, in the domains of experiments, modelling and numerics. A new separate effect experiment (COSI) is planned in the Heat Transfer Lab. in Grenoble.

4.3. Regarding the wall heat transfer, many modifications are planned ; they will be implemented in the next Revision (Revision 3). The most important expected modifications deal with reflooding ; in the present status of the correlations, heat transfer downstream of the quench front is accurately predicted for a tubular test section ; the development of the modelling of grid effects in under way ; it is based on the interpretation of ERSEC results, some FEBA results, and of PERICLES both in rectangular geometry and in the future cylindrical geometry. Multidimensional effects during reflood will be modelled in the next version of CATHARE ; we intend to use the results of boil off and reflood tests obtained on PERICLES (rectangular test section) ; no real multidimensional calculation of the core is planned ; we intend to model cross-flows through core channels.

4.4. Future work includes also the development of new models and correlations for components. Phase separation in tees will be one of the first efforts ; experiments on SUPER MOBY DICK are underway ; they will be completed by mid 85.

Two new modules will be developed in the near future : a 2D downcomer and a 1D pump ; existing experimental results have to be exhaustively used before some new need can be indentified in this area.

LIST OF EXPERIMENTS USED FOR DEVELOPMENT AND QUALIFICATION OF THE MOMENTUM TRANSFER LAW

	Range of Parameters	Comments
Air Water Tests		
MOBY DICK	$P = .2$ MPa $\alpha < .6$ vertical tube 0.014 m I.D.	used for derivation of τ_i
ECTHOR	$P = 0.1$ MPa horizontal tube 0.25 m I.D. low liquid velocity stratification limit	used for derivation of τ_i in stratified flow
Steam Water Tests		
SUPER MOBY DICK	$2 \leq P \leq 12$ MPa low velocity vertical tube 0.135 m I.D. $.5 \leq \alpha \leq .9$	used for qualification of τ_i
SUPER MOBY DICK	$2 \leq P \leq 12$ MPa horizontal tube 0.086 m I.D. stratification limit	- id -
DADINE	$P \leq .6$ MPa $\alpha > .6$ vertical heated tube 0.012 m I.D. $Q_{wall} \leq 5.5\ 10^4$ W/m^2	used for derivation of τ_i (slug flow to annular)
REBECA	$P \leq 1.$ MPa $\alpha > .85$ $V_v \leq 200$ m/s Nozzles and Diaphragms 0.03 to 0.2 m I.D.	used for development of τ_i (mist flow)

TABLE I

LIST OF EXPERIMENTS USED FOR DEVELOPMENT OF THE MASS TRANSFER LAW

Critical Flow Tests		
MOBY DICK	$P \leq .7$ MPa Vertical tube 0.014 m I.D. and 0.020 m I.D.	Data base $q_{li}(\alpha,P,V_v,V_l)$ provided by "inverse" calculations
SUPER MOBY DICK	$2 \leq P \leq 12$ MPa vertical tube 0.020 m I.D.	
MARVIKEN CFT	short and Long Nozzle	used only to ckeck the validity of the diameter dependance

TABLE II

LIST OF EXPERIMENTS USED FOR DEVELOPMENT AND QUALIFICATION
OF THE WALL HEAT TRANSFER CORRELATIONS

Blowdown Heat Transfer - **OMEGA**

Tubular Test section	Initial conditions = Nominal conditions of a PWR tube 0.012 m I.D., L = 3,65 m, uniform axial heat flux
Rod Bundle Test section	Initial conditions = Nominal conditions of a PWR 36 rod bundle (PWR 17x17), L = 3,65 m cosine axial heat flux

Reflood Heat Transfer - **ERSEC**

Tubular Test section	Vertical tube Ø 0.012 m, L = 1 to 3.3 m $.1 \leq P \leq .6$ MPa, $q_{wall} = 3$ and $5.\ 10^4$ W/m^2 imposed mass flow rates 20, 50 and 120 kg/m^2.s.
Rod Bundle Test Section	5x5 (PWR 15x15) and 6x6 (PWR 17x17) rod bundle L = 3,65 m cosine axial heat flux $.1 \leq P \leq .6$ MPa, $q_{wall} = 3$ and $4.5\ 10^4$ W/m^2 imposed mass flow rates 20, 60 and 100 kg/m^2.s

Condensation Heat Transfer (in steam generator) - **PATRICIA S.G.I.**

Single, full scale, model 51 Steam Generator tube, with controlled (organic liquid) secondary side, operated in steady state condition with imposed pressure difference at tube inlet and outlet.

TABLE III

ADDITIONNAL LIST OF EXPERIMENTS USED FOR THE QUALIFICATION OF CATHARE

FLOW TESTS

CANON SUPER CANON	Blowdown from 3.2 and 15.5 MPa of an horizontal tube 4 m long, 0.10 m I.D. break from 0.1 m to 0.01 m diameter
VERTICAL CANON	Blowdown from 15.5 MPa of a vertical tube 4 m long, 0.10 m I.D. through breaks 3 to 15 mm diameter
TAPIOCA	Blowdown from 15. MPa of a vertical capacity 2.5 m long, 0.30 m I.D. through breaks 1 to 10 mm diameter

HEAT TRANSFER TESTS

PERICLES / Rectangular Test Section
357 [7x(3x17)] heated rods (17x17 geometry)
P < .6 MPa
Boil off, constant head Reflood, constant or oscillating flow rate Reflood, axial cosine heat flux, non uniform radial power (allowing 2D effects).
[Experimental results to be used for Cathare in the near future]

COMPONENT TESTS

Capacities

PIERO	1/4 scale, air water at atmospheric pressure, simulating the vessel lower plenum
SEROPS I and SEROPS II	Full scale, air water at atmospheric pressure, Phase separation on the various structures in the upper plenum

[Results of SEROPS not yet used for the modelling of the capacities in Cathare]

Pumps

EPOPEE	1/10 scale of a 93D pump model, rotating speed up to 32 000 rpm - Flow blockage in the pump
EVA	1/3 scale of a 93A pump model, pump head and torque under degraded conditions in the quadrants 1, 2 and 3.

[Experimental results are proprietary]

TABLE IV

Steam Generator

PATRICIA S.G.II — P = 7.5 MPa
Heat Transfer and secondary side behaviour in full scale. 9 channels of the tube region [model 51 S.G.]
Organic Liquid in primary side.

ECCS Injection Zone

EPIS I — 1/11 scale, air water P = 0.1 MPa
Momentum transfer at high velocity

EPIS II — 1/25 scale, steam water P = 0.6 MPa
Flow regime near the injection.

FUTURE PLANNED TESTS

ECCS Injection and Condensation

COSI — P = 7. MPa, steam water, primary pipes 0.12 cm I.D., flow behaviour and condensation (small break LOCA conditions, same geometrical scale as BETHSY).

Tee

SUPER MOBY DICK — $P \leq 12$ MPa
main pipe { 0.135 m I.D. / 0.086 m I.D. }
secondary pipe 0.020 m I.D.
horizontal or vertical, 45° and 90° angle.

Reflood Heat Transfer

PERICLES — Cylindrical Test Section
357 rods (17x17 bundle geometry)
$P \leq 7$ MPa
3,65 m axial cosine heat flux
Boil off, Core Uncovery and Reflood Tests with neglectible housing effects.

TABLE IV (End)

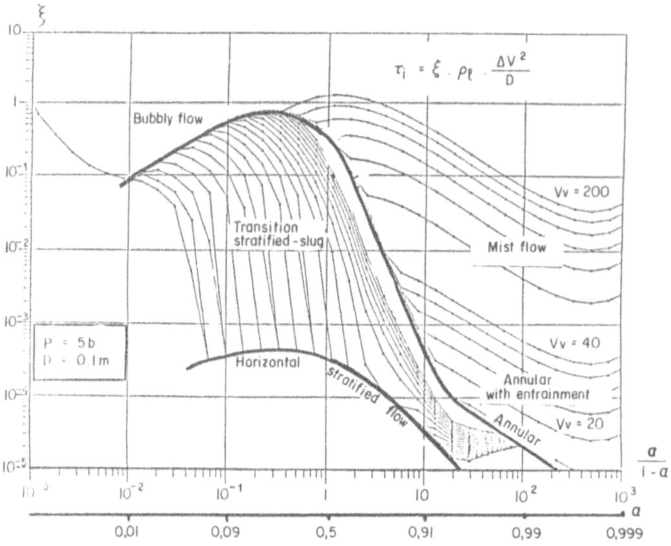

Figure 1

DEVELOPMENT AND QUALIFICATION OF CATHARE

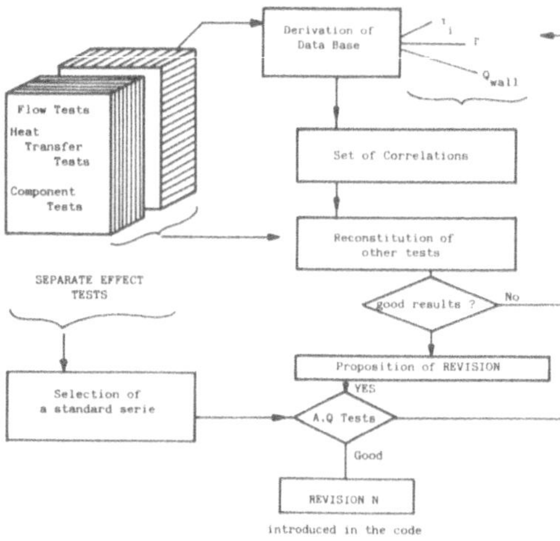

Figure 2

ASSESSMENT STRATEGY

QUALIFICATION AND VERIFICATION

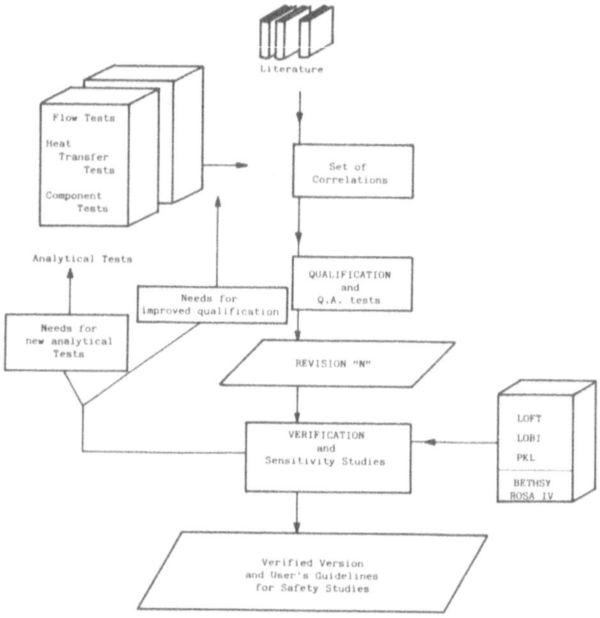

Figure 3

Discussion

Question: G.L. Shires - UKAEA/AEE-W

I was most impressed by the presentation on the development, qualification and verification of Cathare, which is clearly a logically planned programme. I ask the general question concerning time and effort required. When do you expect to reach the level of perfection which you believe will be necessary before the code can be applied with confidence?

Answer: J.L. Nigon - CEA/CEN

At present time, the "Revision 2" (package of physical laws) in Cathare still needs some improvements for:

- interfacial friction factor in rod bundles;
- condensation rate (mass transfer during condensation);
- heat transfer in rod bundles (blowdown and reflood).

We do not want "perfection" at this stage. The general strategy applied to the development and assessment of Cathare should only, after a given period of time, avoid the persistance of "stupid parts" in the package. We are confident that the next "Revision 3" version of physical laws (to be released mid'85) will be good enough to allow the use of Cathare for safety studies.

Even now, some safety studies are under way with Cathare, by people who participated to the development of the present version. Of course, in the present state of the code, this requires a great care.

Question: R. Deruaz - CEA/CEN-G

Regarding void fraction in rod bundles for level swell experiments where you seem to get difficulties with the Cathare code: did you try to use the Winfrith correlation based on the drift flux model and issued from boil up and boil down experiments in Thetis rig?

Answer: J.L. Nigon - CEA/CEN-G

The results of qualification work dealing with level swell experiments G.2 are recent. During this short period of time only three Vgj correlations issued from tube-slug flow/tube bubbly flow/vessel bubbly flow have been compared to experiments (including Thetis). The first effort has been to prove the absolute necessity of 2 different correlations for tubes and for rod bundles. This objective is reached. Now we have to choose the appropriate correlation; in any case, the work performed at Winfrith will be considered for this.

You should consider that:

1) before a new correlation is introduced, it has to be tested for all parameters of future use. In the case under consideration, we had to develop (or to choose) the appropriate extension of the τ_i correlation for rod bundles valid for both low and high velocity, between 1 bar and $\simeq$ 100 bar, and for the whole range of void fraction.

2) the numerical scheme is sensitive to the numerical properties of

right hand side terms. A correlation of τ_i which is proportional to ΔV_i^4 might be "dangerous".

Parallel Sessions Research Area B

The protection of nuclear power plants against external gas cloud explosions

Session B1

Chairman: A.R. Edwards (UKAEA/SRD-C)

Session secretary: J. Vilain (CEC/BXL)

Large-Scale Field Trials on Dense Vapour Dispersion

by

J McQuaid, B Roebuck and D G Wilde
Safety Engineering Laboratory
Health and Safety Executive
Broad Lane
Sheffield S3 7HQ

1 Introduction

During 1982/83, the Health and Safety Executive (HSE) organised a series of large-scale experiments on the dispersion of heavy gas clouds. The experiments were funded on a cooperative basis by 39 organisations from 10 countries including the CEC. The field trials were carried out under contract by the National Maritime Institute (now NMI Ltd) at Thorney Island on the south coast of England. This paper describes the purpose of the experiments, their design and the results achieved.

2 Purpose of the Experiments

In 1976 HSE instituted a programme of research on the dispersion of heavy gases. The principal theme of the programme is the study of the dispersion of fixed-volume clouds, initially formed in a container at atmospheric pressure and temperature and then suddenly released. Experiments have been conducted at three different scales, the Thorney Island experiments being the largest.

The motivation for the programme was the absence of a reliable body of data with which to distinguish between the different predictive models that were available. Furthermore, there was little information on the detailed structure of a dispersing cloud of a kind useful for the construction of physical models. Predictions of dispersion provided by the available models were known to differ widely. Indeed, during the planning of the Thorney Island trials, dispersion predictions were invited for the conditions of the trials and these differed by up to a factor of 100 in the concentration at a given downwind distance under identical conditions.

The identification of possible effects of the scale of release was also an important consideration in undertaking large-scale field experiments. This question is discussed in detail in McQuaid (1982). The Thorney Island experiments represent the largest experiments that were considered feasible, from the points of view of available sites in the UK and of the cost of conducting field experiments.

The trials were organised as a multi-sponsored project, co-ordinated by HSE. They were similar in concept to two other sets of experiments in the HSE programme. These were a series of medium-scale field experiments reported by Picknett (1981) and a series of wind tunnel experiments reported by Hall et al (1982). The list of sponsors of the Thorney Island trials is given in the Annex.

3 The Trials Site

The trials site was a disused airfield on Thorney Island in Chichester Harbour on the south coast of England. A map of the area, with the location of the site indicated, is shown in Fig 1. The trials area comprised an upwind fetch of 1.0 km and an instrumented area of 900 m by 500 m. It was essentially flat (terrain gradients less than 1 in 100), unobstructed and of uniform roughness. Meteorological records, comprising wind speed, wind direction and Pasquill stability category, were available. The wind direction records were used to decide the optimum alignment of the instrumented array of masts. The layout of the masts was assisted by predictions of dispersion behaviour. The wind speed and stability records provided guidance on the time of day and time of year at which particular trials conditions were most likely to be obtained.

4 Design of the Experiments

In each experiment, a fixed volume of 2000 m^3 of heavy gas was released from a twelve-sided container 14 m across and 13 m high fabricated from plastic sheeting. The container was fitted with a roof which was removed immediately prior to release. The sides of the container collapsed to ground level in a time of about 2 secs, leaving an upright cylinder of gas momentarily stationary. An example of the stationary cloud is shown in Fig 2.

The gas used was a mixture of Refrigerant-12 (dichlorodifluoromethane) and nitrogen, giving the capability of achieving initial densities relative to air from 1 to 4.2. Most of the experiments have been conducted at an initial relative density ratio of around 2.

The gas sensors employed a polarographic cell which measured the oxygen deficiency induced by the presence of the released gas. Two designs of sensor were available, with frequency response of 1 Hz and 10 Hz respectively. Both had a spatial resolution of about 2 cm. Typically, 170 of the 1 Hz sensors and 5 of the 10 Hz sensors were deployed on the instrumented masts.

A total of 10 tri-axial sonic anemometers was used to monitor turbulence structure in the approach flow, in the cloud and above the cloud. A meteorological mast was located 150 m upwind of the release position and monitored solar radiation and profiles of wind speed, temperature and humidity. The meteorological measurements permit the estimation of atmospheric stability according to any of the available stability classification schemes.

The gas in the container was marked with orange-coloured smoke prior to release. Cine-, video- and still-film records were obtained from cameras at ground level and on a helicopter hovering at 300 m.

The initial layout of the masts is shown in Fig 3 and the associated instrumentation is detailed in Table I. In addition to the 41 fixed masts, there were 4 trailer-mounted masts. The deployment of instruments was varied during the course of the trials to take account of the different types of trials being performed. Full details are included in the final report on the trials, McQuaid and Roebuck (1984).

The initial programme of unconfined releases over unobstructed terrain was sufficiently successful to encourage a second series of trials to study the effect of obstructions on the dispersion of heavy gas. The obstructions selected were as follows:

a) a 5 m high 2-dimensional impermeable fence, representing features such as a shore-line or a purpose-designed barrier. The trials were to investigate when complete blocking of the gas cloud occurred.

b) 10 m high 2-dimensional permeable screens, representing a belt of trees. The trials were to investigate any partial blocking or dispersion enhancement effects.

c) a 9 m cubed isolated building. The effects to be studied included the interaction between the density-stratified flow and the building and the interaction between the wake of the building and the flow, especially where the building was upwind of the spill point.

As a consequence of the sale of equipment belonging to the project, funding of a short series of supplementary trials was possible. The concensus view of sponsors was that some continuous release trials should be performed. These were achieved by pumping gas from the 2000 m^3 container to a point 50 m acrosswind. The gas, which had an initial density relative to air of 2, was released at a rate of about 250 m^3/min.

5 Results of the Trials

The trials began in July 1982 with a series of gas releases over unobstructed terrain and a total of 19 trials were conducted during this part (Phase I) of the programme. Of these, 3 were preliminary tests for system evaluation, 14 were at a density relative to air of approximately 2, 1 was at a density relative to air of 4.2 and 1 was at neutral density. A summary description of the Phase I trials is given in Table II.

The second part of the trials (Phase II) was concerned with the effect of obstructions on the dispersion of heavy gases. A total of 10 trials were carried out during Phase II, using the 3 different types of obstruction described in Section 4. Four trials were conducted with a 5 m high impermeable fence positioned on a semi-circular arc of 50 m radius from the spill point. Two trials were conducted using 10 m high permeable screens at the same position as the impermeable fence. Four trials were carried out using a 9 m cubed building positioned 50 m downwind of the source for 3 trials and 27 m upwind for 1 trial. A summary description of the Phase II trials is given in Table III.

Three continuous releases were conducted during June 1984. A summary description of the trials is given in Table IV.

6 Availability and Analysis of Data

Magnetic data tapes containing data collected during the trials and hard copy plots of the data are being prepared for all trials. Data tapes for Phase I and Phase II trials and hard copy plots for Phase I trials have already been made available to sponsors of the trials and those for remaining trials will be available before the end of 1984. The data will be made available to other organisations as soon as practicable.

Analysis of the large amount of data generated by the trials has begun but will necessarily be a long-term effort by the many organisations which have sponsored the trials. Some preliminary analysis is included in McQuaid and Roebuck (1984) and in the Proceedings of HSE's Symposium on the Trials, which will be published in 1985 as a special issue of the Journal of Hazardous Materials (Elsevier).

The results of preliminary analysis show that the trials have been successful in providing extensive and self-consistent data over a range of environmental conditions. The design of the experiments has been vindicated, in particular the numbers of gas sensors in the path of the gas clouds (as detailed in Tables II, III and IV) are remarkably high by the standards of field experiments.

7 Acknowledgements

Thanks are due to the organisations listed in the Annex for their financial support and to Shell UK Ltd and the Central Electricity Research Laboratories for the loan of equipment.

8 References

Hall D J, Hollis E J and Ishaq H (1982), A wind tunnel model of the Porton dense gas spill field trials. Report (R 294 (AP)), Warren Spring Laboratory, Stevenage, Herts.

McQuaid J (1982), Future Directions of Dense-Gas Dispersion Research, J Haz Matls, 6, 231-247.

McQuaid J and Roebuck B (1984), Heavy Gas Dispersion Trials at Thorney Island. Final Report on Contract Nos. SR.029.UK and SR.036.UK, Commission of the European Communities, Brussels.

Picknett, R G (1981), Dispersion of dense gas puffs released in the atmosphere at ground level. Atm. Env., 15, 509-525.

ANNEX

List of Sponsors of the Thorney Island Trials

HSE
British Gas Corporation
UK Department of Energy
Battelle Institute, Federal Republic of Germany
American Petroleum Institute
ICI plc
Insurance Technical Bureau
British Petroleum
Comitato Nazionale per l'Energie Nucleare, Italy
US Coast Guard
Commissariat a l'Energie Atomique, France
Institute for Air Research, Norway
Shell UK
Gaz de France
Electricite de France
Central Electricity Generating Board
Gas Research Institute, USA
Commission of the European Communities
Safety and Reliability Directorate, UKAEA
Amoco
Mobil
Texaco
UK Department of Trade
TNO, Netherlands
Ministry of Public Health and Environmental Hygiene, Netherlands
Union des Industries Chimiques, France
E I Dupont de Nemours and Co, USA
National Maritime Institute (now NMI Ltd)
Government of Sweden (Principal Dept's National Defence Research Institute, Fire Research Board, Civil Defence Administration)
Esso, UK
DSM, Netherlands
TransCanada Pipelines
Arabian American Oil Co, Saudi Arabia
Atmospheric Environment Service, Canada
Britoil plc
Total, Cie Francaise des Petroles
Det norske Veritas
Atkins Research and Development
British Nuclear Fuels Ltd

TABLE I: Instrumentation associated with the mast array shown in Fig 3

Mast Code	Instrumentation
A	5 cup anemometers 2 sonic anemometers 5 thermometers 1 solarimeter 2 relative humidity sensors 1 wind vane
V	1 wind vane
F	4 gas sensors (1 Hz)
M1	1 sonic anemometer 4 gas sensors (1 Hz) 2 gas sensors (10 Hz) 2 ultra-sonic gas sensors 1 smoke detector
M2 (a)	4 gas sensors (1 Hz) 2 sonic anemometers 2 gas sensors (10 Hz) 1 smoke detector
M2 (b)	4 gas sensors (1 Hz) 2 sonic anemometers 1 gas sensor (10 Hz) 1 smoke detector
M3	4 gas sensors (1 Hz) 3 sonic anemometers 3 gas sensors (10 Hz) 3 smoke detectors 3 ultra-sonic gas sensors
D	4 gas sensors (1 Hz) 1 wind speed 1 wind vane 1 relative humidity 1 thermometer

TABLE II: Summary Description of Heavy Gas Trials - Phase I

Trial Number	Wind speed* m/s	Pasquill Stability Catgory**	Initial Relative Density Ratio	Number of Gas Sensors Which Responded to Gas***	Remarks
004	3.8	B	0.97	22	
005	4.6	B	1.65	26	Gas container dropped in two stages
006	2.6	D/E	1.60	46	Container 2/3 full
007	3.2	E	1.75	57	
008	2.4	D	1.63	73	
009	1.7	F	1.60	62	
010	2.4	C	1.80	11	Wind changed direction at time of release
011	5.1	D	1.96	26	
012	2.6	E	2.37	65	
013	7.5	D	2.00	47	Wind speed rising during release and dispersion
014	6.8	C/D	1.76	50	
015	5.4	C/D	1.41	38	
016	4.8	D	1.68	45	
017	5.0	D/E	4.20	62	
018	7.4	D	1.87	60	
019	6.4	D/E	2.12	67	

* Wind speed at 10 m averaged over duration of experiment

** Pasquill Stability Category as assessed from observations, solar radiation, vertical temperature gradient, standard deviation of horizontal wind direction, Richardson number

*** Gas sensors of all types

TABLE III: Summary Description of Heavy Gas Trials - Phase II

Trial Number	Wind speed* m/s	Pasquill Stability Catgory**	Initial Relative Density Ratio	Number of Gas Sensors Which Responded to Gas***	Remarks
020	5.7	C/D	1.92	72	5 m fence
021	3.9	D/E	2.02	99	5 m fence
022	5.9	D/E	4.20	83	5 m fence
023	5.8	D/E	1.80	77	10 m screens
024	6.8	D	2.03	65	10 m screens
025	1.4	F	1.95	26	5 m fence
026	1.9	B	2.00	101	Building 50 m downwind
027	2.2	D/E	4.20	29	Building 50 m downwind
028	9.0	D	2.00	59	Building 27 m upwind
029	5.6	D	2.00	72	Building 50 m downwind

*   )
**) see Table II
***)

TABLE IV: Summary Description of Continuous Release Experiments

Trial Number	Wind speed* m/s	Pasquill Stability Category**	Initial Relative Density Ratio	Rate of Gas Release (m^3/min)	Number of Gas Sensors Which Responded to Gas***
045	2.3	E/F	2.00	260	86
046	3.4	D	2.00	260	24
047	1.5	F	2.05	250	79

*)
**) see Table II
***)

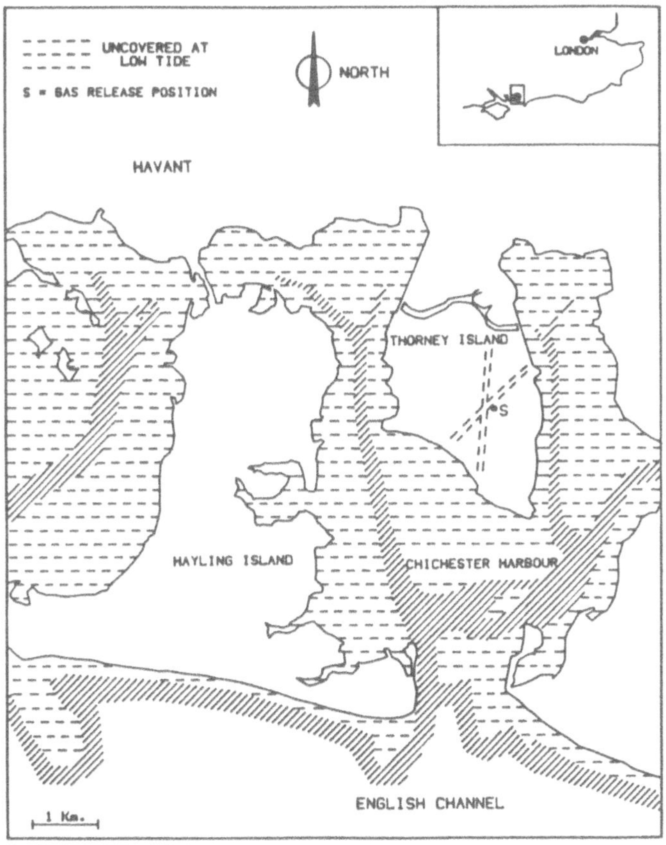

FIG 1 THE LOCATION OF THE THORNEY ISLAND SITE

Fig 2. The stationary gas cloud as the container reached ground level.

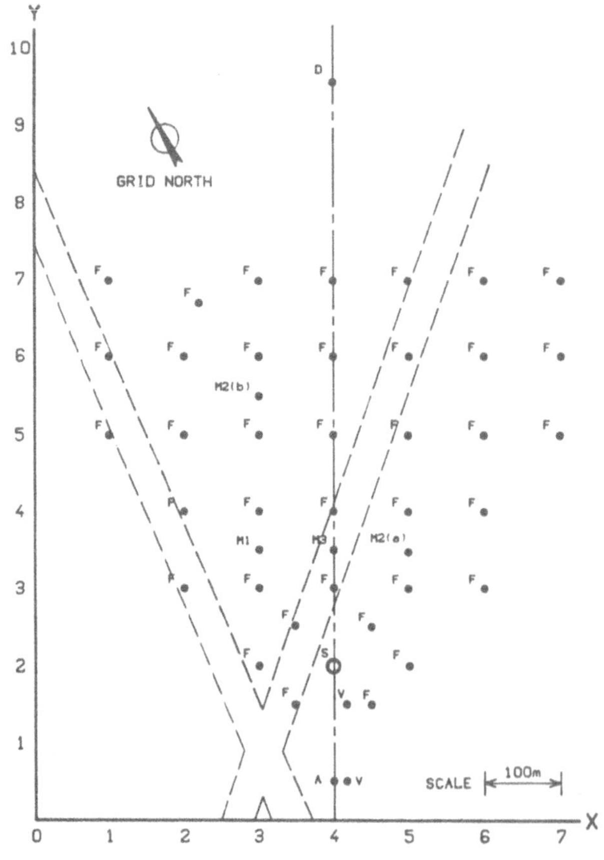

Fig 3 LAYOUT OF MASTS AT START OF PHASE I. FOR IDENTIFICATION OF MAST TYPES SEE TABLE I

Discussion

Comment: Riethmuller - VKI

Perhaps Jim McQuaid could say something about H.S.E. analysis?

Answer: J. McQuaid - HSE/London

HSE carried out a supplementary experiment during the Phase I Thorney Island trials. We set up an array of ground cameras on a circle of 300 m radius. The cameras pointed towards the centre and were synchronised to take still pictures at 0.5 second intervals. The experiment was done for three of the trials. The intention is that the pictures will be analysed to obtain the cloud height distribution during the period when total extinction of the light occurs. At later times when the cloud is more diffuse and only partial extinction occurs, it should be possible to obtain concentration distributions by tomographic techniques. This analysis of the pictures has not yet started.

ASPECTS OF THE DISPERSION OF DENSER-THAN-AIR VAPOURS RELEVANT TO GAS CLOUD EXPLOSIONS

C J WHEATLEY and D M WEBBER
Safety and Reliability Directorate
United Kingdom Atomic Energy Authority

Summary

The essential aim of the study presented here is to improve upon the understanding and prediction of the atmospheric dispersion of denser-than-air vapours, and thereby reduce the uncertainties in predicting hazards which might arise from the accidental release of a dense, flammable vapour cloud. In the first phase of the study, models for dispersion in the atmosphere of denser-than-air vapours are reviewed. It is found that a significant source of uncertainty in predictions of all models is the calculation of dilution caused by turbulence. This is due to spreading and stratification caused by the excess density of the cloud and to the interaction of the cloud motion with the ambient flow field. These effects lead to a complex field of turbulence. An additional, significant source of uncertainty is found to be present in '3D' models due to the use of coarse computational grids. A number of experimental tests are proposed which permit fundamental discrimination between the models with the object of reducing uncertainties. In the second phase of the study, a new 'box' model is proposed (A 'box' model is one in which only gross properties of the flow are predicted). All sources of turbulence are included in a way consistent with laboratory studies of entrainment in stratified flows. The prescribed concentration distribution models the initial 'mixed layer'/ 'gravity spreading' phase and the final 'passive' phase of dispersion with a smooth transition between the two. In the third phase of the study, implications of dispersion of denser-than-air flammable vapour clouds in open terrain for flame speeds following ignition by a weak source are assessed. It is concluded that flame speeds sufficient to cause significant overpressures cannot occur in unobstructed terrain.

1. INTRODUCTION

In the transport and storage of flammable materials, such as liquified natural gas≠ (LNG) and liquified petroleum gas≠≠ (LPG), it is remotely possible that an accident could occur resulting in the release of a large quantity of flammable vapour to the atmosphere. If this were to occur, then the vapour cloud would be transported by the wind. If it were ignited before diluted to a concentration below the lower flammable limit (LFL) it could very well cause loss of life as well as significant damage to buildings and equipment near the point of ignition. It is a potentially serious hazard, because of the relatively close proximity of storage sites and transport routes of such flammable materials to nuclear power plant (NPP) sites in general, and because of the severe damage that could be caused.

≠ Composed mainly of methane
≠≠ Composed of hydrocarbons with larger molecular weight than methane, such as propane and butane.

This study focusses on the dispersion of denser than air vapours (with emphasis on those aspects that are relevant to gas cloud explosions) because, whether the LNG or LPG is refrigerated or stored under pressure, any accidental release invariably leads to the formation of a denser-than-air vapour cloud. (For LNG this is because of the very low temperature of the vapour cloud formed.)

A crucial question which has to be addressed is: given that a release of a large quantity of flammable vapour occurs, how far will the cloud travel before it is diluted below the LFL? In many cases it is sufficient to establish that this distance is sufficiently small that the consequences of ignition prior to dilution below the LFL can be ignored. There is large uncertainty in estimating this distance. The reason for this is that field experiments on the dispersion of denser-than-air vapours which have been done to date have involved release of, at most, only a few tonnes of vapour. However, the mass of flammable vapour of concern in NPP safety assessments extends to the order of 10,000 tonnes. This demands considerable extrapolation which is done using mathematical models for the dispersion of such clouds. A comparison of these models shows a large variation of the predicted distance to LFL.

It is also necessary to consider the effects of possible ignition of the vapour cloud. Estimates with sufficient accuracy are consequently also required, both for the shape of the vapour cloud and for that mass of material within the cloud which is between the LFL and the UFL (the upper flammable limit). Large uncertainty also exists in extrapolations for these quantities. A further factor which might also influence the consequences of ignition is the influence on subsequent flame speeds of turbulence within the cloud prior to ignition.

The overall aim of the study reported here is therefore to reduce the uncertainties in those areas described above. The specific aims are as follows:

(i) To review the fundamental basis of existing models of heavy gas dispersion. In particular, to identify sources of uncertainty in the models and to devise tests which permit differences between the models to be fundamentally discriminated experimentally.

(ii) To develop a new box model which attempts to avoid these problems. (For reasons which will become apparent later we call this a 'generalized' box model.)

(iii) To assess the implications of turbulence within dispersing heavy gas clouds on flame speeds following ignition.

In this paper we summarise what was done and our main conclusions. The reader is referred to the final report of the study for details [1].

2. ASSESSMENT OF EXISTING MODELS OF HEAVY GAS DISPERSION

Dispersion of a denser-than-air vapour cloud is a complex, three-dimensional (3D), time-dependent process: the cloud spreads horizontally due to its excess density; it is deformed by the ambient flow field; dilution by vertical mixing is suppressed due to the stabilising density difference; and turbulence causing dilution can be generated in several different ways.

A large number of models have been proposed which make predictions for the dispersion of such clouds. Well over forty models are identified in this study. These range from the simpler so called 'box' models, in which only the overall behaviour of the dispersing cloud is predicted, to complex 3D models. In between these limits lie a variety of 'intermediate' models. Attention is paid to general scaling properties of the models and to sensitivity of predictions of the models to variations in the released mass of vapour, in the wind speed and in atmospheric stability. All types of model are considered. However, most attention is paid to box models, because of their inherent scrutability.

Analytic solutions are found for a wide class of box models. This facilitates quantitative comparison of a number of box models including several recent models. The following main results are obtained:

(i) In general, predictions of models vary widely.

(ii) Different models exhibit different scaling properties.

(iii) Different models exhibit different sensitivities of X_{LFL} (the distance from the point of release to the lower flammable limit) to changes in the wind speed, atmospheric stability, and mass of flammable material released.

The primary reason for the differences between the models lies in the treatment of turbulence and its effect on the rate of dilution of the cloud. A large number of mixing processes are identified. In no model are they comprehensively accounted for. There is also uncertainty over how any particular mixing process should be quantified. However, the results listed in (ii) and (iii) introduce the possibility of experimental discrimination between the models in a fundamental way. This removes one of the main difficulties in assessing box models: the presence of disposable parameters which can otherwise be adjusted to obtain agreement with experimental results.

An intermediate model is derived by using a form of shallow layer approximation to the Navier-Stokes equations. The model, in common with box models, requires that assumptions be made regarding mixing rates. This enables the basis of existing intermediate models to be studied. From this, the formulation of existing models and their relationship to box models are clarified. Because these models necessarily make similar assumptions concerning rates of mixing to those of box models, then uncertainties in this respect are similar to box models.

All the 3D models which we have encountered (ten in all) are comprehensively assessed.

In general these models consider approximations to the full equations in order to reduce computational difficulties. The following main results are found. The equations for conservation of contaminant cannot be decoupled from the equations which determine the flow field because of the importance of spreading of the cloud due to gravity. The hydrostatic approximation is invalid for heavy gas dispersion. Uncertainties exist over the form in which approximations should be made to eliminate sound waves and unwanted terms involving density fluctuations.

Numerical solution schemes use spatial grids which are coarse with respect to several aspects of the flow field. This in particular applies to the interface region between the top of the cloud and the atmosphere above. In no model have calculations been done to investigate convergence of the results at a function of grid size. The error incurred may be large. Time is also discretized. There is evidence to suggest, however, that the solution schemes provide numerically stable results providing suitably small time steps are used.

Turbulence closure assumptions are made in all models and in all cases they are shown to be deficient in a number of important respects. In particular, the gradient transport assumption is made in all models. This is shown in the assessment not to be valid for heavy gas dispersion because it cannot lead to proper modelling of entrainment across a density interface and cannot account for the spectrum of eddy sizes in the ambient flow field.

Published results are summarised and assessments are made from these of the sensitivity of X_{LFL} to changes in wind speed, atmospheric stability,and mass of flammable material released. From these results, X_{LFL} is predicted to increase with increasing wind speed. This is at variance with current understanding of the mechanism of dilution in a heavy gas cloud.

It is concluded that 3D models have serious deficiencies which cause predictions of these models to be unreliable. These deficiencies are essentially due to the use of coarse computational grids and inadequate models of turbulence. They are unlikely to be removed without using finer grids and sophisticated models of turbulence such as large eddy simulation. Such improvements are unlikely to be made in the foreseeable future.

In conclusion, there is wide variation of assumptions in all existing models, mainly with regard to quantifying the effects of turbulence on rates of dilution. An added uncertainty is present in 3D models due to the use of coarse computational grids. These problems in turn lead to wide variation in predictions for important quantities such as X_{LFL}.

3. REVIEW OF MIXING EXPERIMENTS AND DEVELOPMENT OF AN IMPROVED BOX MODEL

The aim here is to develop a box model with improvements in the following respects:

(i) Prediction of X_{LFL}

(ii) Prediction of the mass of flammable material in the cloud between the limits of flammability

(iii) Prediction of the cloud shape.

(ii) and (iii) are required as a starting point for hazard assessment involving ignition of the cloud.

Clearly, a crucial element of dilution of flammable vapour clouds is the modelling of entrainment rates. Since modelling mixing rates in flows with stable vertical density gradients is an important area of uncertainty, a basis for improvement in this area has to be established. This is done by considering the results of laboratory experiments on entrainment in stratified flows. These are reviewed and assessed with regard to implications for heavy gas dispersion.

It is found that, if the flow is stable enough, then a well mixed layer of heavy fluid will be formed with a sharp density interface at the top of the layer (we call this a 'mixed layer'). An explanation for the occurrence of such a layer is given.

Turbulence generated within the mixed layer, above the mixed layer and at the density interface can lead to dilution of the mixed layer. Dilution rates are quantified in terms of 'layer' Richardson numbers. Criteria for the occurrence of a mixed layer are formulated in terms of these Richardson numbers.

In conclusion, despite some remaining uncertainties, the review shows how to account for the many sources of turbulence including the effect of stratification due to the greater density of the flammable cloud.

A new box model is developed.

In developing such a model, apart from the need to reduce uncertainties with regard to turbulent entrainment, existing box models treat spatial variations of concentration unsatisfactorily. This last item is of importance when considering consequences of ignition - many current models unrealistically predict that all the initially released gas remains flammable until all the gas is diluted below the LFL. In practice, the mass of gas remaining flammable will steadily decrease to zero as dilution takes place. Because of this, the new model is based on a framework which comprehends concentration profiles more complex than those assumed in existing box models. We call the model 'a generalized box model'.

A particular concentration profile has been studied in detail. It consists of a uniform central region of radius R and height h and diffuse edge regions which are Gaussian in shape with thickness parameters σ_E and σ_T for radial and vertical diffuseness respectively. This profile enables the concentration distribution in both the gravity spreading/mixed layer phase and the passive phase of heavy gas dispersion to be described.

The equations for R and σ_E are straightforward generalizations of those in existing box models. As a result, the radial concentration profile is predicted to evolve smoothly from an initial rectangular distribution to a later Gaussian distribution. This smooth evolution in shape offers a significant advance over existing box models.

The equations for R and σ_E involve the quantity U_f, the spreading speed of a gravity current. Consideration is given to the formulation of an improved equation for U_f which, in particular, allows U_f to increase from zero.

The equations for h and σ_T were found to be more difficult to develop because of the complexities of entrainment in stratified flows. A process considered, which is not present in existing models but suggested by the review of laboratory experiments, is 'detrainment', whereby, relative to the density interface between the top of the cloud and the atmosphere above, material from the cloud is transported upwards across the interface into the atmosphere above and then rapidly diluted and carried away by the wind. This consequently leads to a reduction in mass of flammable material in the cloud below the density interface. This is in contrast to the counter process of entrainment whereby air is transported downwards across the density interface and dilutes the gas below.

Several alternative models are considered depending on whether detrainment is taken into account and on how it is taken into account. In all cases the vertical concentration profile is predicted to evolve smoothly from an initial rectangular distribution, corresponding to the mixed layer phase of dilution, to a later Gaussian distribution. Additionally, sources of turbulence leading to dilution are comprehensively accounted for in the equations. These features offer further significant advances over existing box models.

Subsidiary aspects of the equations for evolution of the concentration profiles are also considered. These include the determination of the length and velocity scales of turbulence in a heavy gas cloud and in the ambient flow field and the fundamental problem of how predictions from box models should be compared with experimental data.

The new model may lead to better predictions for gas cloud explosions in two main respects:

(i) Greater confidence may be had in predictions for maximum distances from the point of release of flammable material to the lower flammable limit.

(ii) More realistic estimates of source terms for combustion are provided. This includes the shape of the cloud, the mass of material between the lower and upper flammable limits, and the intensity of turbulence within the cloud.

Much information will be forthcoming from the recent field trials at Thorney Island [2], particularly with regard to concentration profiles, and so it is recommended that the next step in the work is to develop a computer program for the model in parallel with the analysis of the data from the Thorney Island experiments.

4. IMPLICATIONS OF HEAVY GAS DISPERSION FOR COMBUSTION

Following ignition of a flammable vapour cloud, a flame will propagate through the flammable region at a speed depending on the intensity of turbulence ahead of the flame. Significant overpressures caused by the flame will only occur if the flame accelerates to speeds of order 100 ms^{-1} or more. A study is made to determine whether the turbulence in the cloud prior to ignition could lead to such high flame speeds.

A necessary pre-requisite for such a study is knowledge of the effect of the turbulence ahead of a flame on the burning speed. This topic is therefore reviewed. Experimental results and theoretical models are considered.

There is uncertainty in quantifying the effect of turbulence when turbulent quenching of the flame is important, as will occur either in high intensity turbulence or in mixtures close to the flammable limit. However, apart from this, good evidence is found to support the hypothesis that the burning speed is proportional to the turbulence velocity scale. (This is at least a conservative parameterization when quenching is important).

Conditions necessary for the onset of flame acceleration are determined.

Enhancement of burning speeds due to both turbulence and partial blockage of the flow path are considered. It is found that both confinement and either obstacles (for the generation of turbulence) or partial blockage are required for the onset of flame acceleration. Quantitative criteria are formulated for this in terms of the fractional blockage ratio, the drag of obstacles in the flow path, and the adiabatic expansion ratio of the burning mixture.

It is concluded as a consequence of these findings that flame acceleration in unconfined vapour clouds of hydrocarbons such as LNG and LPG, following ignition by a weak source, cannot occur. In particular, it is concluded that turbulence arising from the effects of heavy gas dispersion has no bearing on flame acceleration. This finding is amply supported by both large and small scale experiments.

The main effect of the dispersion phase on flame speeds is due to the 'pancake' shape of the flammable region caused by gravity effects. Providing the cloud is unconfined, this leads to reduced flame speeds compared to an otherwise similar hemispherical cloud due to vertical venting of the combustion products. A simple model for flame speeds in unconfined vapour clouds based on this idea is proposed and the results are found to agree with experimental observations.

5. CONCLUSION

In conclusion, significant improvements in the understanding and modelling of the atmospheric dispersion of flammable vapour clouds have been made in this study. None the less, there remain significant uncertainties. These can only be reduced by thorough exploitation of experimental studies, on both large and small scale, in conjunction with the further development and testing of box and intermediate models.

Finally, although this study was done specifically with hazards to NPP arising from dispersion of flammable vapour clouds in mind, the study is of sufficient generality that it has ramifications for non-nuclear industry. It is relevant for atmospheric dispersion both of flammable vapours and of toxic substances.

REFERENCES

[1] WHEATLEY, C. J. and WEBBER, D. M., Aspects of the dispersion of denser-than-air vapours relevant to gas cloud explosions. To be published by the CEC (1984). (Numerous references are cited in this report, for which space is too limited here.)

[2] McQUAID, J., ROEBUCK, B. and WILDE, D. G., Large-scale field trials on dense vapour dispersion. To be published by the CEC (1984).

Discussion

Question: J. Thomas - British Gas

In his presentation the author noted that cloud turbulence was generally too low to promote significant flame acceleration. He also stated that there was evidence that in free clouds flame speed was proportional to wind speed. How general is this upward and onward trend.

Answer: C.J. Wheatley - SRD UKAEA

I have used the rsults of Raj who collated the available evidence from a number of experimental programmes, which clearly show that the flame speed in a flammable vapour cloud in open terrain of given roughness is proportional to the wind speed. For terrain of different roughness the constant of proportionality will be different. Of course, I do not expect the result to be generally true because a conservative relation between turbulence and flame speed has been used, which ignores quenching. Quenching will cause flame speeds to be less. On the other hand incomplete vertical venting of combustion products will cause flame speeds to be larger.

Question: P.M. Bradford - N.I.I.
You have described an "improved box model", but you state that uncertainties remain. Could you put this into context by suggesting to what degree of accuracy you believe it will now be possible to predict the distance to lower flammable limits for typical releases?

Answer: C.J. Wheatley - SRD UKAEA

I would not like to give a figure because I do not think that such a figure is obtainable from our study. There are many uncertainties which enter predictions and we have addressed just one of these, namely uncertainties associated with modelling the physical processes involved. We believe that we have made significant advances here which eventually will lead to greater confidence in applying models, especially when extrapolating to very large releases. However, there will remain uncertainties from other sources as I have mentioned.

Question: N.F. Scilly - HSE London

It is fairly unlikely that flammable clouds will ignite at wind speeds above 8 m/s (evidenced by the Shell trials at Maplin Sands).

Answer: C.J. Wheatley - SRD UKAEA

I do not accept that the evidence you cite shows in general that flammable vapour clouds cannot be ignited at large wind speeds. The main requirement for ignition is that a flammable region within a cloud should exist. Such flammable regions occur at both large and small wind speeds.

Question: N. Scilly - HSE London

Recent large scale work by British Gas suggests that unconfined clouds of hydrocarbon/air mixtures can accelerate to flame speeds of 60-80 m/s when obstacles are placed in their midst.

Answer: C.J. Wheatley - SRD UKAEA

I think that you have misundertood what I mean by confinement. By confinement, I mean that venting of the burnt gases does not occur. This can be achieved either by physical barriers or by configuration of the ignition point with respect to the cloud, for example, by ignition at the centre of a sphere of flammable mixture. The experiments by British Gas you refer to involved some degree of confinement in the sense I define above.

Question: G. Picciolo, ANIC (ENI) Italy

Use of actual models (3D and/or box models) to treat low concentration as occurs in risk assessment. Can the models used for flammable heavy gas clouds also be used for toxic clouds where the concentrations of interest are much smaller?

Answer: C.J. Wheatley, SRD-UKAEA

Toxic vapour clouds, like flammable vapour clouds, can also be denser-than-air and so in principle the same models can be used in either case. However, the concentrations of interest in either case are very different so that in flammable vapour clouds we are only interested in the early stages of dispersion where heavy gas effects usually dominate. This should be compared with toxic vapour clouds where, because we are interested in very low concentrations, it is found that passive dispersion effects can dominate. Box models can be used in either case providing they include all relevant dispersion effects. However, existing 3D models cannot be used to model dispersion of toxic substances because the cell size required to track the cloud to such low concentrations is nonsensically large.

Question: G. Schnatz, Batt. Inst.

Can you explain why you think transport within heavy gases is not gradient dependent? Do you think a spectral division of the different turbulence length scales and a separate calculation for each scale will help?

Answer: C.J. Wheatley, SRD UKAEA

The gradient transport assumption is not valid for modelling heavy gas dispersion in two respects, both essentially due to the large sizes of eddies in the flow field in relation to the size of the cloud. The first is the rate of entrainment across the strongly stratified region at the top surface of the cloud, which is not correctly calculated by gradient transport models. It is shown in our report that the entrainment rate calculated by 3D models is consequently dependent on the finite size of the grid used for numerical solution of the model.
The second is mixing due to turbulence in the ambient flow field where, as the cloud gets bigger, progressively larger eddies contribute to dilution. This leads to a stronger time-dependence of the growth in size of the cloud than predicted by gradient transport models.
In reply to the second part of your question I would say that a possibility is to use "large eddy simulation". This, however, is very expensive in computing time and size and would be difficult to justify in view of the other deficiencies of 3D models we have identified.

Question: J. McQuaid, HSE/Sheffield

In your description of the potential capability of 3D models, you say that only the VKI and HEAVYGAS models are suitable for transient flows. This conflicts with the views of the other modellers, who apply their models to transient flows, for example the results of the Thorney Island trials. What is your reason for saying that the other models are not suitable for transient flows?

Answer: C.J. Wheatley, SRD/UKAEA

I must emphasize that the assessement of turbulence models we have made is qualitative. What this means is that we do not know quantitatively what the importance of the various aspects of turbulence models I have listed is. In principle, all the aspects should be included but some may be more important than others. This applies to the question of modelling turbulence in transient flows where, in principle, time-dependent effects, such as those included in the VKI and HEAVYGAS models, should be included. This merely serves to highlight the uncertainties in modelling turbulence in heavy gas flows, which is my main point.

Parallel Sessions Research Area B

The protection of nuclear power plants against external gas cloud explosions

Session B2

Chairman: J.P. Zeeuwen (TNO/PML-Rijswijk)

Session secretary: J. Vilain (CEC/BXL)

VERIFICATION OF HEAVY GAS MODELS ON THE BASIS OF LARGE SCALE EXPERIMENTS

G. SCHNATZ and J. KIRSCH
Battelle-Institut e.V., Frankfurt

Summary

The main objective of the project has been the verification of a number of heavy gas dispersion models on the results of the Thorney Island large scale release tests. From the large variety of models published in the literature only such models have been selected which are frequently used and differ significantly in their basic assumptions.
The experimental data of 15 instantaneous heavy gas releases at Thorney Island have been processed. Comparisons have been performed between calculated results from the models and nine characteristic experimental data sets. For a number of models the results show good agreement for the downwind distance profile of the concentration at ground level. Similar results have been obtained for the crosswind spreading of the cloud. Large differences have been observed, however, for the concentration-time dependence for low wind speed conditions.
Some general model deficiencies have been identified. Recommendations for the applicability of the different models are given.

1. INTRODUCTION

The main objectives of this study are part of the objectives of the CEC research area B of the indirect action programme 'safety of thermal water reactors'. Aim of the research work performed in area B is the improvement of models predicting the impact of explosions of dense combustible vapor clouds on plant structures and systems. Realistic predictions of vapor cloud explosion consequences provide an essential input to the hazard assessment for nuclear power plants (and other technologies).

In the past a variety of different models for the prediction of the expansion the and dispersion of heavy gas clouds have been developed. These models can be divided into two categories:

- highly parametrized box models which consider the vapor cloud from an instantaneous release of heavy gas as a pancake-shaped radially expanding cloud with air entrainment through the top of the cloud as it drifts with the atmospheric wind field.
- models solving the differential equations numerically.

The work reported here focused on the models of the first category. The different physical approaches and models were verified and evaluated by the results of large scale experiments. In earlier studies purely theoretical comparisons between the results of various models have been made /1/, /2/. This work represents an improvement because for the first time the models have been compared systematically with uniform experimental data which were obtained by large scale spills.

2. MODELS INCLUDED IN THE COMPARISON

The comparison takes into account the following box models:

- DENS1 /3/
- DENZ /4/
- Eidsvik /5/
- Flothmann /6/
- Germeles and Drake /7/
- Picknett /8/
- van Ulden /9/

A detailed description of each model can be obtained from the quoted literature. The main characteristics of the models are the following:

- all models are so called box models which solve simplified differential equations for the expansion and dispersion of a heavy gas cloud; the equations contain various parameters which have to be adjusted in such a way that the results of the models match with results from experiments.
- apart from the model DENS1 by Meroney and Lohmeyer all models had been adjusted to field experiments, DENS1 had been adjusted to wind tunnel experiments.

3. IMPROVEMENTS OF THE PHYSICAL FOUNDATIONS OF THE MODELS

An analysis of the structure of the box models shows that the expression for the front velocity of the expanding heavy gas cloud is not exactly consistent with the conservation of energy principle.The equation for the front velocity which is presently used by all types of box models is given by

$$U_F = \sqrt{c \frac{\Delta\rho}{\rho} g H_0} \; \frac{R_0}{R} \qquad (1)$$

where c is a constant to be adjusted.

Keeping strictly to energy balance concept the following form results /10/:

$$U_F = \sqrt{2 \frac{\Delta\rho}{\rho} g H_0} \cdot \left(1 - \frac{R_0^2}{R^2}\right)^{\frac{1}{2}} \cdot \left(\frac{1}{1 + \frac{8}{3}\frac{R_0^4 H_0^2}{R^6}}\right)^{\frac{1}{2}} \qquad (2)$$

The main difference appears for the first phase of cloud expansion. The front velocity of equation (1) takes its maximum value for the time of release ($t=0$ and $R=R_o$) while as in equation (2) the front velocity starts from rest ($v=0$) and increases with time.

Equation (2) is based on balance of kinetic and potential energy and yields good agreement between experiment and theory for the first seconds of cloud expansion. To obtain good agreement for later phases of expansion and dispersion it is necessary to include the loss of energy due to friction effects. The complete energy balance consistent approach for the front velocity is discussed in /10/.

4. COMPARISON WITH HSE LARGE SCALE EXPERIMENTS

In 1982 and 1983 the British Health and Safety Executive (HSE) performed a series of large scale experiments in unobstructed terrain (Phase I). 2000 m^3 of freon-nitrogen mixture at ambient temperature have been released instantaneously. Gas concentrations and atmospheric parameters like wind speed and direction etc. have been measured at different elevations . The data are available on magnetic tapes and have been processed to suit the requirements of the comparison task. From 15 experiments with heavy gas in unobstructed terrain the results of 9 experiments were used in the comparison (table I).

Table I HSE-large scale trials used for the comparison

Trial	Wind speed	Atmospheric stability	Relative Density of Released Gas
7	3.2	E/F	1.75
8	2.4	D	1.63
9	1.7	F	1.60
10	2.4	D	1.80
11	5.1	D	1.96
12	2.6	E/D	2.37
13	7.5	E/D	2.0
14	6.8	E	1.76
17	5.0	C/D	4.2

Special attention has been paid to:

- trial 9 with low wind speed and stable atmospheric condition
- trial 13 with strong wind and near neutral stability
- trial 17 with high relative density of released gas (4.2 instead of about 2 for all other trials)

Before starting the discussion of the results a few remarks concerning the major characteristics of the measured concentration distribution of the the Thorney Island Trials will be given. Two categories could be identified which mainly differ in wind speed.

The low wind speed category results are characterized by:

- a strong vertical concentration gradient within the measured elevation range (0.4 m, 2.4 m, 4.4 m and 6.4 m)
- two subsequent maxima of concentration occure at elevations above 0.4 m, which are:
  - a first strong peak lasting only some ten seconds following the passage of the vapor cloud at gas sensor location (passage of the front vortex)
  - the second, extended maximum results from the increasing height of the cloud due to air entrainment through the top of the cloud
- measurable concentrations exist for a long time period before they are strongly diluted by dispersion (some hundreds of seconds)

In contrast the moderate and strong wind speed category

is characterized by:
- a weak vertical concentration gradient
- similar concentration time behaviour at all measured elevations without a second maximum
- short period of cloud existence

The main aspects of the model comparison can be reduced to three points:
- maximum downwind distance of flammable concentration (relevant for safety distances)
- crosswind spreading of the cloud (area covered with flammable concentrations)
- concentration-time behaviour of the cloud (cloud with concentrations between the flammability limits)

4.1 Results for the maximum downwind distance of flammable concentrations

The results are shown in figures 1 to 3 for trial 9 (low wind speed condition), trial 13 (strong wind speed) and trial 17 (high density experiment). The measured gas concentrations are 30 seconds moving time average values.
The peak concentrations of these time averaged functions are shown in figures 1 to 3 for all gas sensors which have detected any gas at 0.4 m elevation level. Three models show significant deviations from the envelope of the experimental results.

The model of Germeles and Drake overestimates the measured downwind concentrations for trial 9. Taking into account all other experiments this overestimation is supposed to be specific for all releases at stable atmospheric conditions. An analysis of the model structure shows that this deficiency is due to the unrealistically small dispersion parameters for stable atmospheric conditions.

The model DENS1 tends to underestimate the measured downwind concentrations for all compared experiments which is probably caused by the adjustment of the model parameters to wind tunnel experiments.

The model of Flothmann takes into account a vertical concentration gradient within the gas cloud. Because of the strong vertical concentration gradient especially for low wind speed conditions the calculated concentration decreases strongly with elevation. Consequently for the comparison of the maximum downwind distance of flammable concentrations the elevation of the measured concentration has to be considered (range of centimeters). This is illustrated in fig. 1 which shows both the calculated ground level concentration (0m) and the concentration for 0.4 m. The measured concentration for 0.4 m lies between these two curves predicted by the model of Flothmann.

Figure 4 shows a comparison for the elevation of 6.4 m for the strong wind experiment (Trial 13). Despite the fact that measured data are sparse for this elevation it can be seen that the models DENZ, Germeles and Drake and Eidsvik yield realistic concentrations even for distances up to 600 m downwind of the source. All other models tend to underestimate the concentrations at this elevation. The sharp decrease of cloud height within the first 200 m and the subsequent second maximum which is predicted by some of the models reflects the influence of two physical processes which are counteracting:

- the initial decrease of cloud height (slumping) caused by gravity and
- the subsequent increase due to air entrainement through the top of the vapour cloud.

4.2 Comparison of the prediction of crosswind spreading

The second major aspect of testing the model reliability is the crosswind spreading of the gas cloud. This effect is shown in figure 5 and figure 6 for trial 9 and trial 13 resp.. The figures show the lay out of the measuring instruments at Thorney Island with the release point at the coordinates 200/400. All ground level sensors (0.4 m elevation) which have seen gas are indicated in the figures. Those which have measured more than 1 % (time averaged) concentration are marked by black dots. The circels represent the calculated time development of the 1 % concentration contour line of each model. Good agreement results for nearly all models except the model of Germeles and Drake using the concept of a virtual source for the simulation of the drift.

A deficiency which is common to all models is the underestimation of the upwind spreading of the gas cloud for low windspeed conditions (except for the model of Germeles and Drake).

4.3 Comparison of the concentration-time predictions

Finally the time-concentration behaviour is shown in fig. 7 and fig. 8 for trial 9 and trial 13 resp.. Again the results for the low wind speed condition differ from those of the moderate and strong winds.

For low wind speeds all models are more or less significantly underestimating the period for which the gas cloud is within the flammability limits. In other words at low windspeeds the heavy gas cloud remains much longer inflammable than it is predicted by any of the models.

Fairly good agreement results, however, for moderate and strong wind speeds.

5. CONCLUSIONS AND RECOMMENDATIONS

The results can be summarized as follows:

- The comparison of results from predictions by box models with measured concentrations from large scale dispersion experiments shows that two models have to be ruled out because of significant differences between the predicted and the measured concentration values. In particular these are the model of Germeles and Drake (overestimation) and the model DENS1 (underestimation in all cases).
- For moderate and strong winds the agreement between model simulation and experimental values is much better than at low wind speed and stable atmospheric conditions.
- Upwind spreading of the cloud and time durations for the existence of concentrations within the flammability limits at low wind speed conditions are significantly underestimated by all models used within this study.

In general the comparison shows that by observing the specific limits of applicability some of the box models validated by this investigation seem to be useful tools for specific problems of consequence analysis.

REFERENCES

1. Battelle Institut: Critical appraisal of mathematical models of explosive gas cloud dispersion.
   Report to the Commission of the European Community, Contract No 462-78-7 ECI D (1981)
2. Havens, J.A.: Predictability of LNG vapour dispersion from catastrophic spills into water: an assessment.
   Report to U.S. Coast Guard (1977)
3. Meroney, R.N., Lohmeyer, A.: Gravity spreading and dispersion of dense gas clouds released suddenly into a turbulent boundary layer. Report for the Gas Research Institute, CER82 83RNM-AL-7 (1982)
4. Fryer, L.S., Kaiser, G.D.: DENZ - A computer programme for the calculation of the dispersion of heavy toxic or explosive gases in the atmosphere. SRD R152, UKAEA(1979)
5. Eidsvik, K.J.: A model for heavy gas dispersion in the atmosphere. Atmospheric Environment 14, 769-777 (1980)
6. Flothmann, D., Nikodem, H.J.: A heavy-gas dispersion model with continuous transition from gravity spreading to tracer diffusion. in : Heavy Gas and Risk Assessment, S. Hartwig (ed.), Reidel, (1980)
7. Germeles, A.E., Drake, E.M.: Gravity Spreading and Atmospheric Dispersion of LNG Vapor Clouds.
   Proc. 4th Int. Symp. Transport of Hazardous Cargoes by Sea and Inland Waterways,
   Report CG-D-24-76, 519-539 (1975)
8. Picknett, R.G.: Field experiments on the behaviour of dense clouds. Part 1. Main Report, Chemical Defense Establishment, Report Ptn/L 1154/78/1 (1978)
9. van Ulden, A.P.: On the spreading of a heavy gas released near the ground. in Loss Prevention I, 221 (1974)
10. Schnatz, G., Kirsch, J., Heudorfer, W.: Investigation of energy fluxes in heavy gas dispersion. in : Heavy Gas and Risk Assessment, S. Hartwig (ed.), Reidel, (1983)

Fig. 1: Comparison of calculated and measured concentration values as a function of source distance -experimental values are 30 seconds timeaverages of the maxima of each sensor, which has been within the gas cloud (results for trial 9, low wind speed)

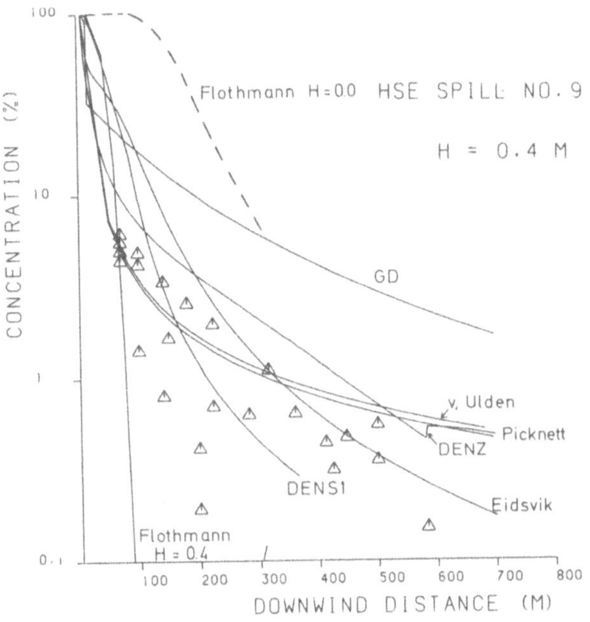

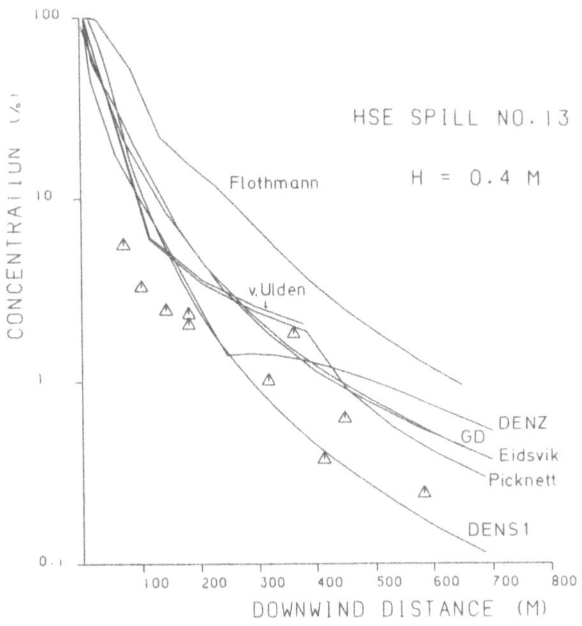

Fig. 2: Comparison of calculated and measured concentration values as a function of source distance (results for trial 13, strong wind speed)

Fig. 3: Comparison of calculated and measured concentration values as a function of source distance (results for trial 17, high density)

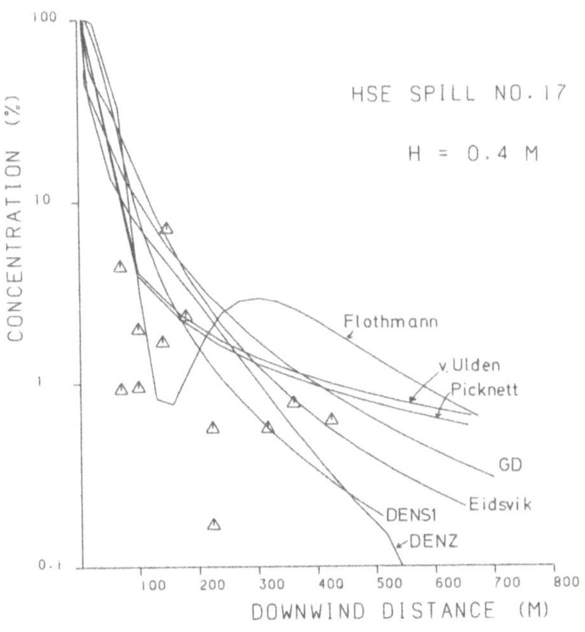

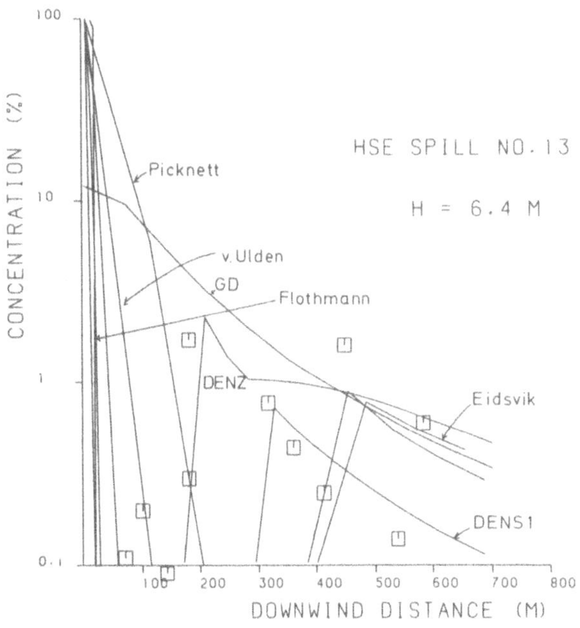

Fig. 4: Comparison of calculated and measured concentration values as a function of source distance (results for trial 13, strong wind and 6.4 m height above ground level)

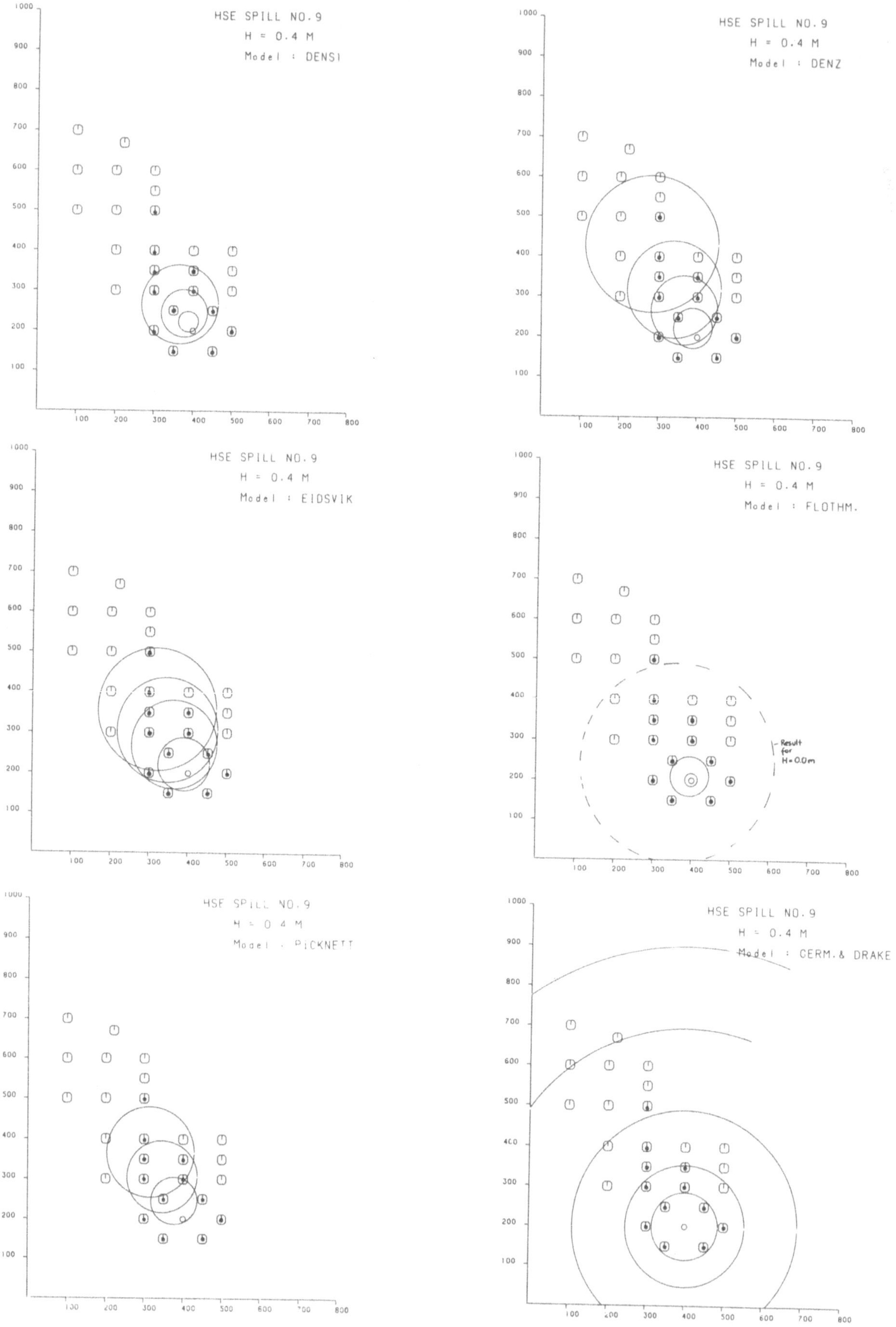

Fig. 5 a-f: Crosswind spreading of the cloud (results for trial 9, low wind speed)

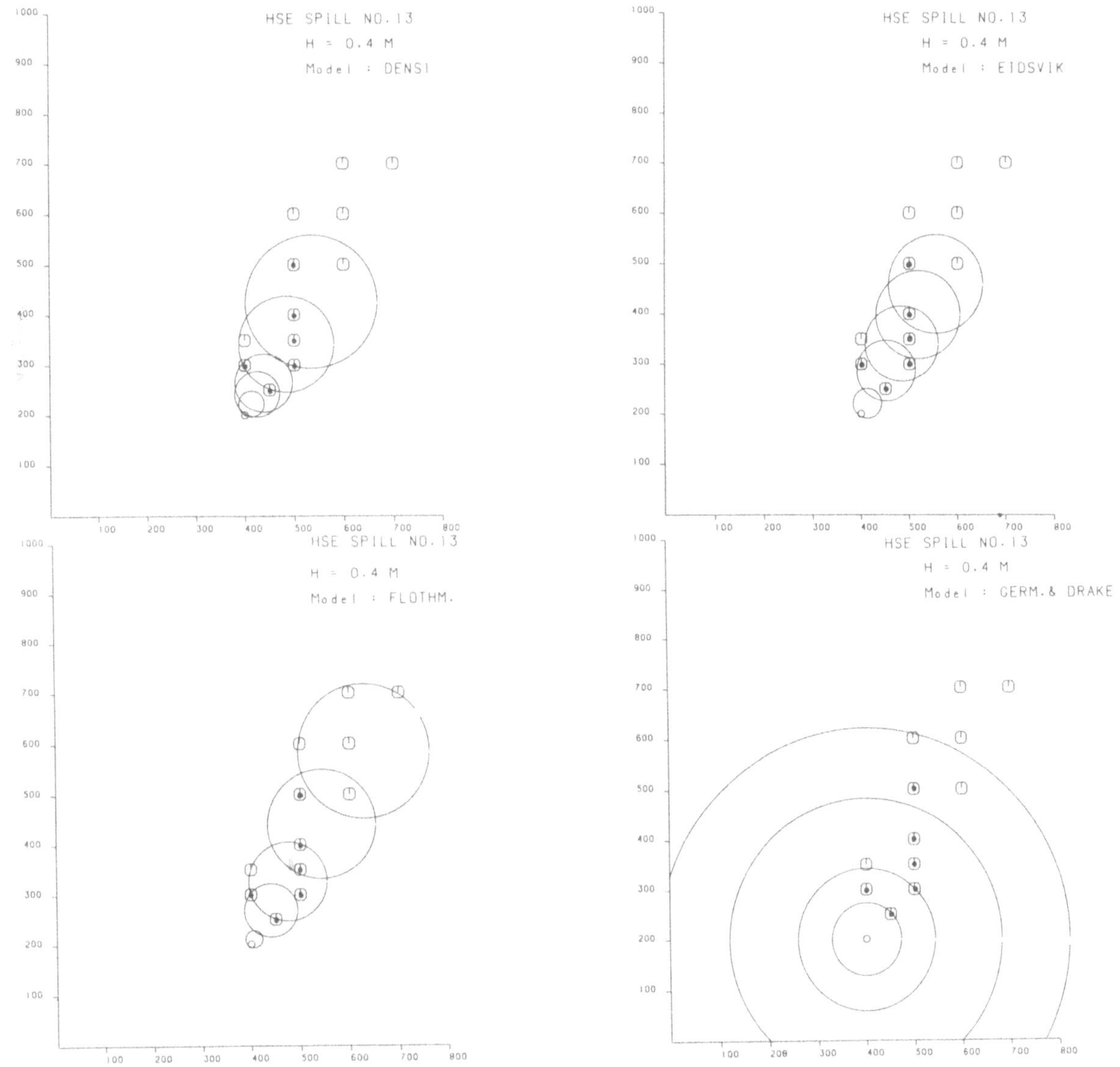

Fig. 6 a-f: Crosswind spreading of the cloud (results for trial 13, strong wind speed)

Fig. 7: Comparison for the concentration-time predictions for trial 9 (low wind speed)

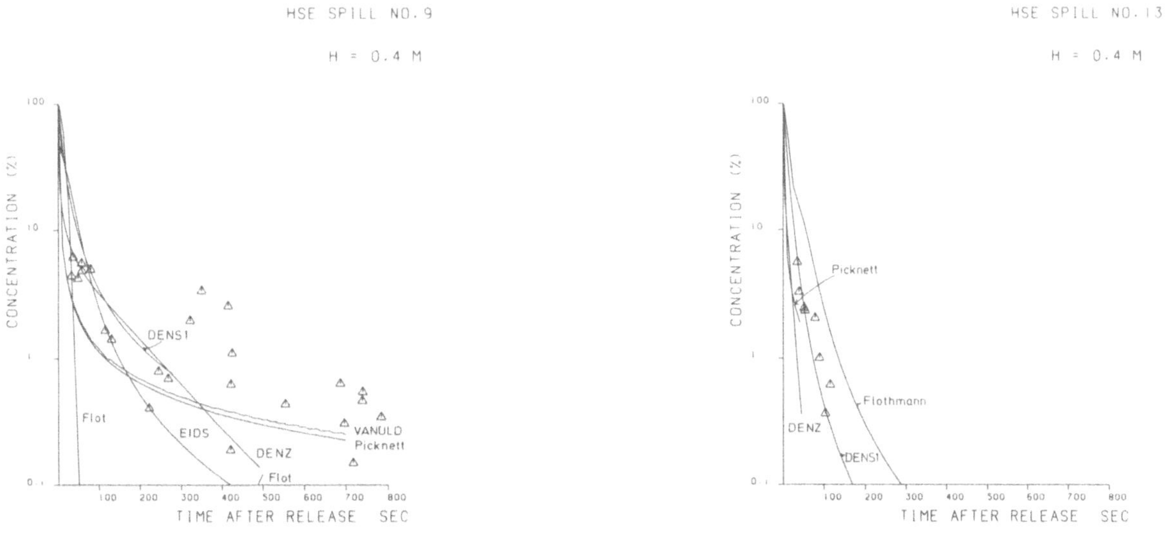

Fig. 8: Comparison for the concentration-time predictions for trial 13 (strong wind speed)

Discussion

Question: C.J. Wheatley - SRD UKAEA

Of the models for which you have found reasonable agreement with the data: to what extent do you think this is due to the values chosen for phenomenological constants and does your comparison test the models in a fundamental way?

Answer:G. Schnatz, Batt. Inst.

For the comparison no fitting of the free model parameters was performed. The agreement found for several models is based on the original values of the parameters, as they are proposed in the literature describing the models.

Question: J. Vilain, CEC Brussels

Is the second maximum of C(t) at low windspeed perhaps due to a multiple wave structure in the cloud? Is the time interval adequate?

Answer: G. Schnatz, Batt. Inst.

In contrast to the first maximum which is clearly generated by the front vortex and has a duration of only a few seconds, the second maximum is very broad and lasts around some hundreds of seconds. The second maximum also arrives later at the upper height levels. This indicates that it is not generated by a wave-type motion but by the growing vertical expansion of the cloud.

COMPUTER PROCESSING OF VISUAL RECORDS FROM THE THORNEY ISLAND LARGE SCALE GAS TRIALS

M.L. Riethmuller, J.M. Buchlin, S. Rickaert
von Karman Institute for Fluid Dynamics
Rhode Saint Genèse

Summary

Digital image processing of the airborne video record of trials Nos 007, 008 and 014 of the Thorney Island trials are performed. Comparisons are made with previous processing reported at this Symposium and a good agreement is observed. The processing is extended to the determination of the motion of the cloud edges.

1. Introduction

The present work consists in performing a computer processing of the visual records taken during the experiments organized by the HSE in the Thorney Island large-scale gas trials.

During the tests, concentration sensors are deployed over a large area and therefore, little information is known about the first seconds of release. However, for most of the tests, video and still recordings are taken, both from a helicopter and from ground level.

A processing of some of this visual material has already been carried out (1). The graphical technique described in (1) is straightforward but not suited to the processing of a large number of frames. Since this technique relies on the use of still frames, the interval between processings is governed by the growing rate of one frame every 1-2 seconds.

The digital processing of images from a video tape may produce data at a rate of 25 or 50 frames per second. This is particularly important if one wants to determine the initial accelerating motion of the cloud.

The present work aims at improving the knowledge of this phase of the release. The need for this information is emphasized by the fact that no gas sensor is in the cloud during the initital moment and therefore, data obtained should not only be a back up but also a complement to the sensor data.

2. Description of the digital image processing system

2.1 Hardware

The VKI Digital Image Processing System (DIP) is implemented on a PDP 11/34 computer. The general structure of this DIP is shown in figure 1.

The pictures to be processed may be on slides and will be imaged by a black and white video camera, or may be on video tapes. A frame grabber will convert each frame into digital form and store it as a 256x256 matrix. The content of this memory is read by a colour encoder which displays each frame on a colour monitor with 256 different colours encoding an identical number of levels of grey.

The frame grabber, the memories and encoder are connected to the DEC UNIBUS and there is software control on their operation. It is possible to freeze a certain frame and then dump the image memory in core memory or disk.

Once this is achieved, subsequent processing may be started. A processed image will be dumped from core memory to image memory and hence displayed on the monitor. It is also possible to use the plotter of the computer to obtain a graphical output.

Image processing most often involves the processing of large matrix and therefore may become a lenghty process. This is why the PDP 11/34 is linked to a VAX 11/780 which is far more powerful and able to handle extensive processing.

The operation of the whole system involves control over two computers and many peripherals. A specialized software has been written in order to fulfill all the necessary tasks.

2.2 Software

The existing software allows the user to perform the following operations:

- single frame grab;
- multiple frame grab;
- display frame with given number of arbitrary colours or levels (8, 16, 32 or 256). Also allows to select a range of colours or levels (and hence suppress a range);
- filter - spatially filters an image;
- display graphs of levels. A plot of the distributions of levels along a vertical or horizontal cursor is made;
- zoom - a part of the image is selected and enlarged to full screen size;
- normalization - an image is normalized with another reference one;
- determination of the histogram of densities;
- determination of contours;
- substraction of images;
- evaluation of areas (on well contrasted pictures).

This software package has been created following the requirements in multiple fields of interest and although it is capable of powerful functions, not all the functions required for the present objective exist. It is therefore necessary to develop a specialized software to complement the existing one.

2.3 Specialized software

Specialized functions have been added to the existing software in order to fulfil the present task. These are:

- Creation of a "window" in the image: in most cases, the cloud to be analyzed only covers part of the screen. The "window" is the minimum polygon which includes the cloud. This allows a

reduction in the number of data to be processed.
Alternatively, the window can be given the shape of a circle.

- Data compaction: the analysis of the contour of the cloud only requires binary data - cloud or no cloud. The image is transformed into binary form, thus reducing the number of data to be handled by a factor of up to 16.
- Determination of cloud contours.
- Fill overall area covered by the cloud with a single colour.
- Creation of a reference grid.
- Scaling on a reference length (runway most often).
- Scaling of image.
- Determination of:
  - location of center of cloud;
  - cloud contour coordinates;
  - area of cloud.
- Storage of results.

3. Results of the processing

3.1 Validation of the method

Trials Nos 007 and 008 have been the first ones to be processed in order to be able to verify the method. A good reference existed since a manual processing of the pictures taken during these tests had been performed in (1). In these references, cloud area and centroid location were plotted against time.

After processing with the DIP system, a comparison was made with those of (1) and good agreement was obtained, as shown in figures 2 and 3 for trial 007 and in figures 4 and 5 for test 008.

Because of the variable time delay between the release of the lid and the release of the bag, it was necessary to shift the results of (1) along the time axis. This was achieved by evaluating the time delay from the ground based video records.

For trial 007, this correction was unnecessary but it was needed for trial 008 in which a shift of 3 seconds was introduced and for trial 014 in which a shift of 4 seconds is used.

There still remains some doubts about exact bag release time since it seems that the clock used with the airborne camera is not always reset together with the ground based one. This became obvious from the latest video tapes. In all cases, the initial time was chosen as the time that the bag started to fall.

Advantage has been taken of the processing of trial of 014 to further validate the present method and it is evident from figures 8 and 9 that good agreement is obtained.

Smooth curves are drawn through the data without an automatic curve fitting procedure.

3.2 Results

Besides the duplication of the manual processing of SRD, further information has been obtained. Figure 6 shows the distance of upwind and downwind front from the source versus time. This distance is always counted from the source and is therefore positive for both.

On figure 7, the distance of these fronts from the centroid of the cloud is plotted against time.

The side fronts (transverse to the wind) have not been represented on this figure but they fall on the same common curve.

A similar exercise has been made for trial 014. Figure 10 shows the distance of fronts from the source while distances from fronts to centroid are plotted on figure 11. On the latter, side fronts have been represented.

On figure 12, the path of the centroid of the cloud has been plotted for trial 014. On the cloud path, time markers are indicated in seconds.

4. Discussion of the results

4.1 Assessment of the technique

It is apparent that valid results may be obtained even though the quality of video tapes is not perfect.

In some instances such as on figure 9, a relatively large scatter is obtained. This is due to the difficulty encountered in locating the source when the cloud is still hiding it. This problem has since been overcome by defining a set of auxiliary references by choosing prominent objects easily visible during this phase. The locations of these auxiliary references are determined using an image on which the source may be located.

4.2 Detailed results

The outcome of the processing of trial 008 is that, as see, on figure 7, the cloud is not affected by the wind during the first 25 s. The fronts move radially at equal velocity from the cloud centroid. This tends to indicate that during this period most of the process is dominated by gravity effect. This is to be expected since trial 008 corresponds to a relatively low wind speed.

On the contrary, trial 014 has been performed with wind velocities of the order of 5 to 6 m/s. Figure 11 shows upwind, downwind and side fronts displacement.

For the first 10 seconds, the cloud is seen to be expanding radially in a uniform way. After 10 seconds, the side fronts very rapidly stop expanding while the upwind and downwind ones move at almost constant velocity.

This would indicate that, in this particular trial, gravity was the dominant factor for the first 10 seconds of the release and atmospheric turbulence took over thereafter.

Finally, it is interesting to note that the cloud path of trial 014 shown on figure 12 is very well aligned with the average wind direction. This is not too surprising and confirms the coherence of the data.

5. Conclusions

At the present time, processing of the video tapes is still under way and a much more detailed analysis is needed to extract all useful information from these results. It is intended to use them to validate laboratory tests and numerical models.

Although incomplete, the present work demonstrated the high power of digital image processing in such an exercise and it is hoped that further useful information will be extracted in the near future.

Reference

1. A.J. Prince, D.M. Webber, P.W.M. Brighton: Thorney Island heavy gas dispersion trials - determination of paths and area of cloud from photographs.

Report SRD R318, Safety and Reliability Directorate, UKAEA, Culcheths, UK, 1984 (in press).

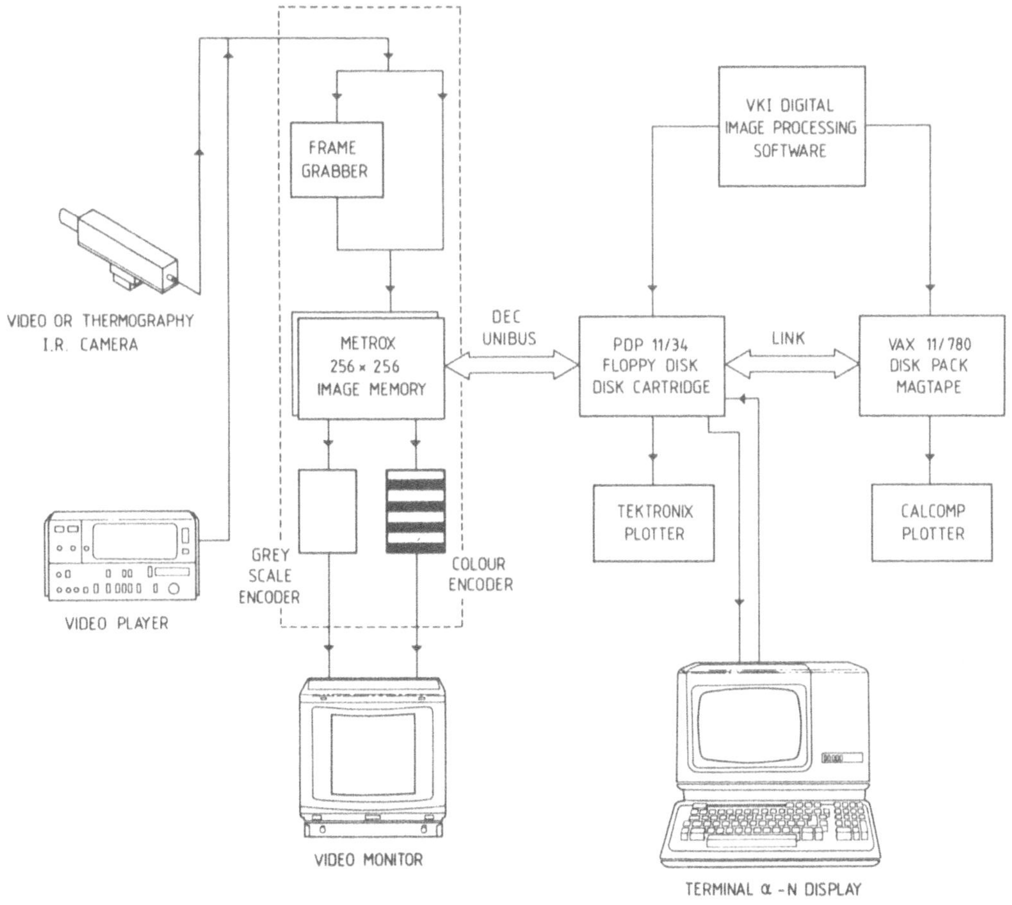

FIG. 1 - DIGITAL IMAGE PROCESSING SYSTEM

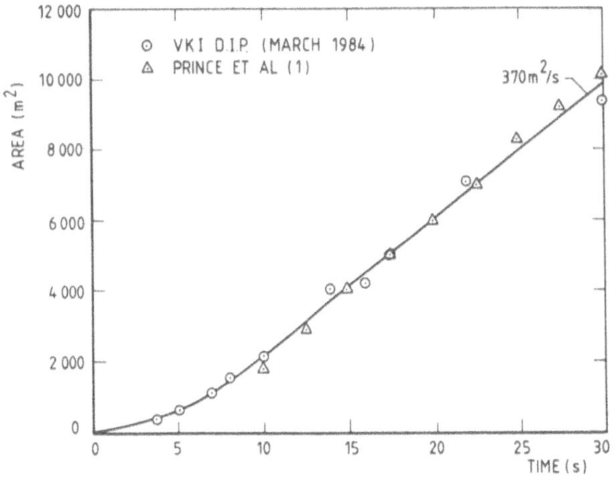

FIG. 2 - CLOUD AREA VS TIME
TRIAL 007 (WINDSPEED 3-4 m/s)

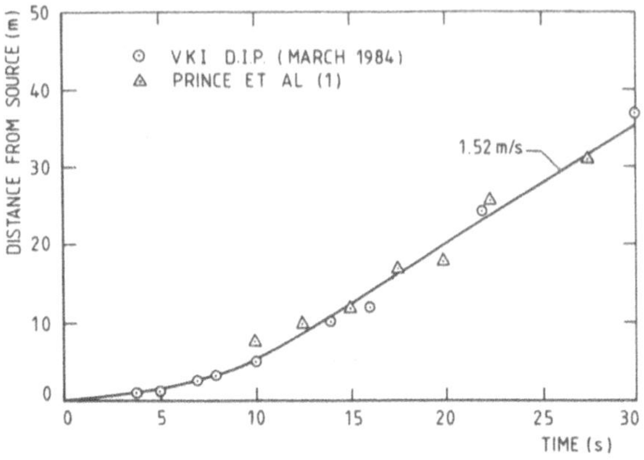

FIG. 3 - DISTANCE OF CENTROID FROM SOURCE VS TIME
TRIAL 007

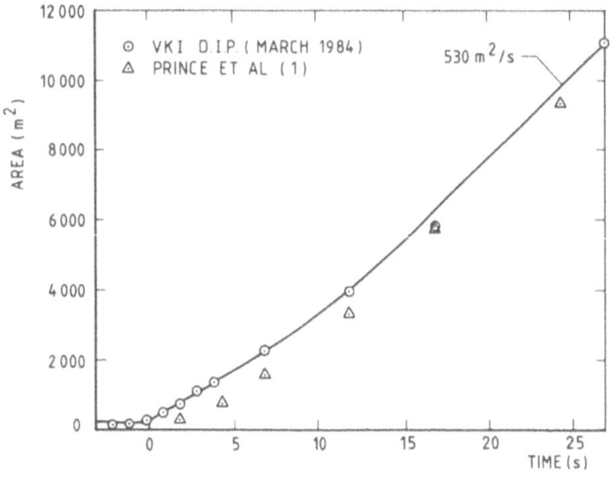

FIG. 4 - CLOUD AREA VS TIME
TRIAL 008 WINDSPEED 2-3 m/s

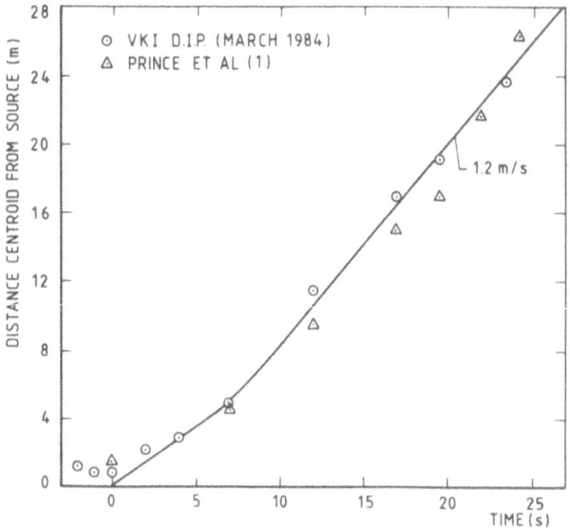

FIG. 5 - DISTANCE OF CLOUD CENTROID FROM SOURCE VS TIME
TRIAL 008

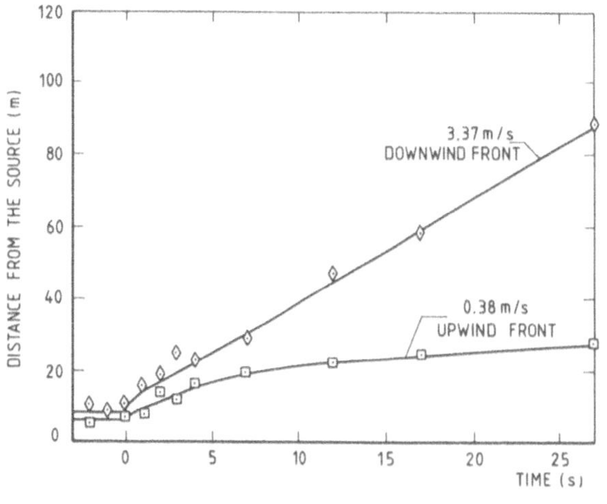

FIG. 6 - DISTANCE OF FRONTS FROM SOURCE VS TIME
TRIAL 008 VKI D.I.P.

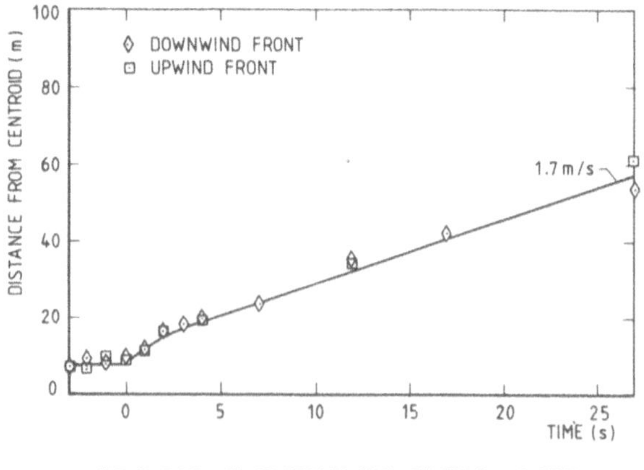

FIG. 7 - DISTANCE OF FRONTS FROM CENTROID VS TIME
TRIAL 008 VKI D.I.P.

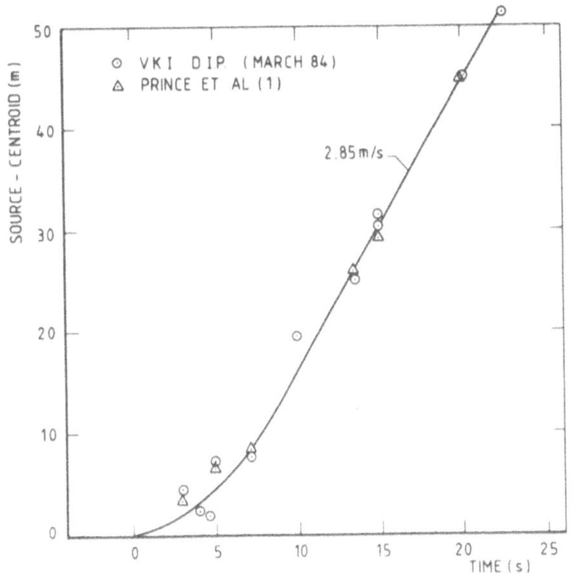

FIG. 9 - DISTANCE OF CLOUD CENTROID FROM SOURCE
VS TIME. TRIAL 014

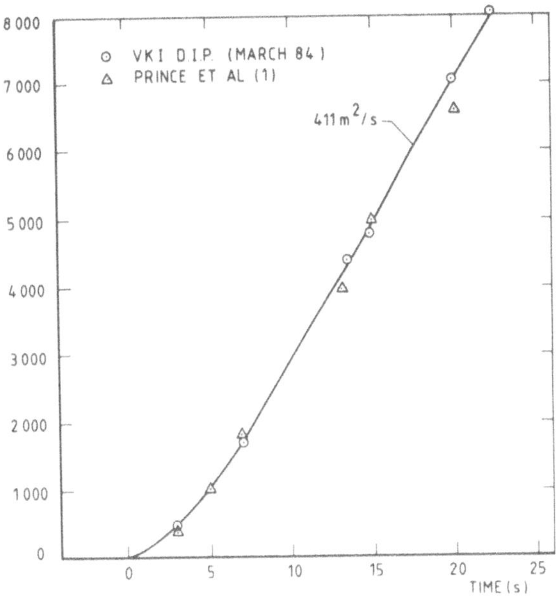

FIG. 8 - CLOUD AREA VS TIME
TRIAL 014 (WINDSPEED 5-6 m/s)

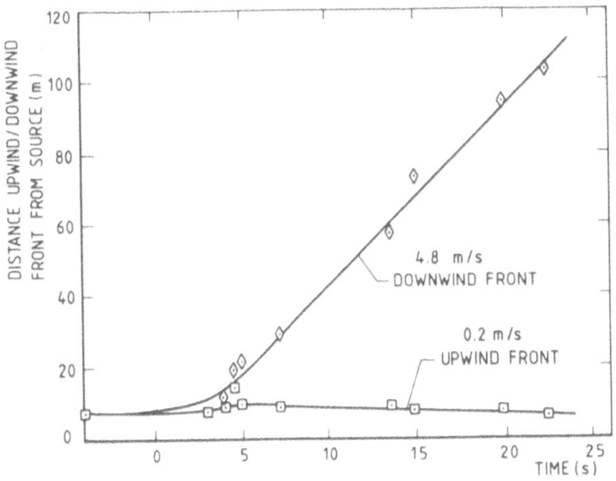

FIG. 10 - DISTANCE OF FRONTS FROM SOURCE VS TIME
TRIAL 014 VKI D.I.P. (MARCH 84)

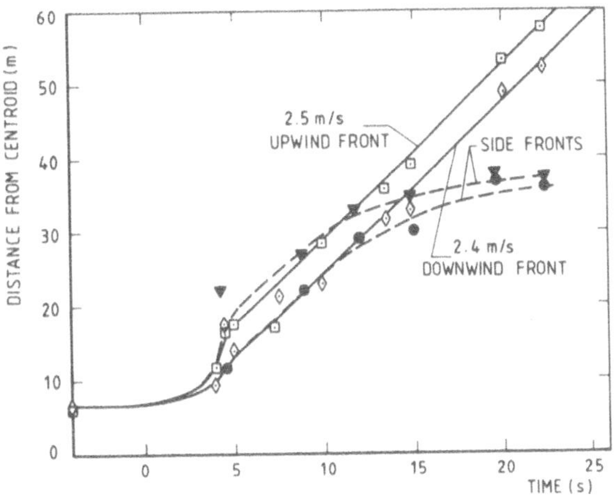

FIG. 11 - DISTANCE OF FRONTS FROM CENTROID VS TIME
TRIAL 014 VKI D.I.P.

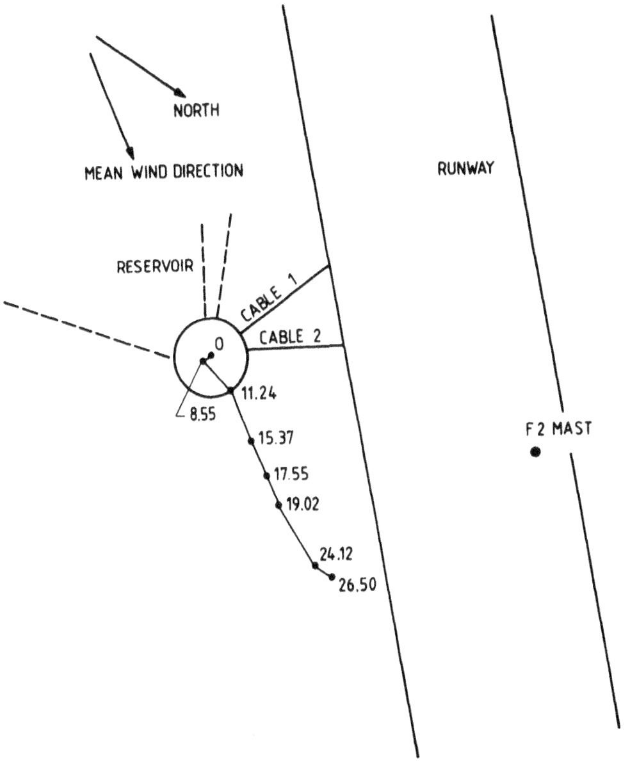

FIG. 12 - TRIAL 14 - CLOUD PATH

Discussion

Question: J.P. Schon - EC.CENTR./Lyon

What kind of smoke was used in the HSE experiments and what was the particle size?

Answer: J. McQuaid - HSE/Sheffield

The smoke was a military camouflage smoke. It was the same as used in the Porton Down trials by the Chemical Defence Establishment (as reported by Picknet, Atm. Env. 1980). The particle size was 0.5 micron.

Question: G. Picciolo ANIC

Is there a possibility for applying image processing to evaluate concentration?

Answer: L. Riethmuller - VKI/Rhode St. Genèse

It is true that optical density can be related to height of cloud and concentration. Using quantitatively this information could perhaps give some measure of a quantity depending upon both height of cloud and concentration.
However, we found that in the present case, we would not try to extract such information which would need far more work to be made available.
The HSE have taken the opportunity of the T.I. trials to experiment a multi-camera technique. Dr. McQuaid could perhaps say something about this technique.

Comment: G. Schnatz - Batt. Inst.

Remark to Chris Wheatley's point that the height of the cloud is determined from side photographs.

What will be seen from the side photographs of the heavy gas cloud is the height of the front vortex rather than the height of the heavy gas cloud.

Answer: L. Riethmuller - VKI/Rhode St. Genèse

Indeed, without a cross sectional view of the cloud, only the vortex height would be obtained.
Actually, we simulated such gas releases in the laboratory and used a laser sheet to visualize a cross section. The digtital image processing system was then used to obtain height and span of vortices.

Question: C.J. Wheatley SRD/UKAEA

I would just like to mention that SRD have extended their processing of the visual data to side-view photographs, from which it is also possible to obtain cloud area and position (assuming the cloud is circular) and, possibly more importantly, the cloud height which can then be compared with heights from the concentration data.
Can you envisage extending your digital image processing in the same way?

Answer: L. Riethmuller - VKI/Rhode St. Genèse

Digital processing of ground-based images has been considered, but we reckoned that the passage from a 2D processing to 3D would be too time consuming and would not, without great efforts, give better results than a "manual" processing.
In a larger research programme, it would be valuable to develop the software needed.

Transition from Slow Deflagration to Detonation

by W. Drenckhahn and C. Koch
Kraftwerk Union AG, Erlangen

1. Introduction

This contribution to the CEC-Programme is a literature study which encloses the literature to the beginning of 1983.

This report encludes also some new experiments done in the german safety programme of the PNP-project.

PNP means Prototype Nuclear Process Heat. This programme was sponsered by the Ministery of Research and Technology and Interatom.

My report contents the following points.

First I'll speak about the present licensing situation in the FR Germany to protect a nuclear power plant against external gas cloud explosions.
This for our company unsatisfying situation was the reason to participate at the CEC-Programme.

Then I will report about some recent work on damage analysis of real Vapour Cloud Explosions.

The Susceptibility to Detonation of different gases leads to a certain classification of gases concerning the explosivity of the substance used nearly a nuclear power plant.

The next point is concerning about experiments with explosions of a real vapour cloud. This contents the Maplin Sands and China Lake Experiments with LNG and LNP and the experiments with free Hydrogen-Air Clouds done by Pförtner in Karlsruhe.

The transition from deflagration to detonation will be discussed in several steps.

There is the well known phenomena of the transition in long tubes,the influence of obstacles on flame acceleration investigated in tubes and the influence of confinement on flame acceleration.

The last two points the transition from deflagration to detonation in partially confined hydrogen air clouds and the jet ignition of free gas clouds are large scale experiments done in the PNP project.

2. Present Licensing Situation in FRG

In 1976 in Germany was published a regulatory guideline "For the Safety of Nuclear Power Plants against Pressure Waves of Chemical Reactions".
At the time when the guideline has been worked out, there were some uncertainties about the mechanism and possible impacts of gas cloud explosions. The regulatory guideline contains two essential requirements (Fig. 1):

a) A pressure/time function for the layout of the safety related buildings. The pressure/time function shows the following design:

   The pressure increases in 0.1 s to 0.45 bar, decreases to 0.3 bar in another 0.1 s; then it has a quasistatic pressure. This pressure/time function is the conservative envelope for the deflagration of a gas cloud bordering on the reactor building.

b) The second requirement of the guideline is the observance of a mass-distance relationship. This relationship, determines the acceptable distance between a potential source of explosion and the safety-related items of a nuclear power plant. The relationship is:

$$R = 8 \cdot L^{1/3}.$$

This relationship assumes that the explosion mechanism is a detonation. The mass of the explosive material can be calculated with the method of the TNT-equivalent.

Both requirements of the guideline originate from physically different explosion mechanisms: the deflagration and the detonation of a gas cloud. Both explosion mechanisms show an essential difference in their pressure/time function, especially in the near field.

The second requirement, the assumption of a detonation, which in our opinion is the wrong instrument, often leads to difficulties in the licensing procedure of a nuclear power plant in Germany.
Therefore we are interested in changing this requirement and we believe that the knowledge about explosions mechanisms has increased in the last years so that the time is ripe to do something in this area.

3. **Damage Analysis of Vapour Cloud Explosions**

In last years much work has been done to analyze real explosion accidents. In the following, some recent papers will be discussed to show that the assumption of a detonation as the result of an accident is very unrealistic and a deflagration is the most probable explanation.

3.1 Giesbrecht

Giesbrecht (BASF) prepared an analysis of the accidents which occured in the chemical plants of BASF in 1943, 1948 and Flixborough 1974. He concluded that the explosions of vapour clouds were deflagrative and not detonative.

This can be explained by comparing the pressure/distance relationship, as gained from the damage analysis (full curve) with data calculated for a deflagration and measured for a detonation (Fig. 2). In this figure the maximum pressure is normalized to the border of the cloud r_G as a function of the distance from the explosion. One result is that in the vicinity of the undisturbed incoming waves pressures are lower by a factor of ten than those observed for a detonative blast-wave i.e. those gained by the calculation of Kogarko.

3.2 Wiekema

Wiekema, TNO, analyzed 165 vapour cloud accidents, which happened during the period 1920 - 1980. Clouds that were not ignited were excluded from the analysis.

The results of this study are the following.

- Nearly all the ignited vapour clouds were ignited within 1 kilometre from the location of the accident.

- Explosions occurred only in semi-confined situations, i.e. in the presence of obstacles. On the other hand, it is not true that if obstacles are present, an ignition will always result in an explosion.

- For delay times of more than 30 minutes, no explosion was recorded. Ignition following shortly after the release enhances the probability of an explosion.

- In nearly half of the accidents there were no fatalities.

- Some of the vapour clouds for which detailed information was available have been used to analyse the validity of a vapour cloud explosion model and the result was that a deflagration is more realistic for the accident damages than a detonation describtion.

3.3 Rischbieter

For the PNP-Project Rischbieter analysed some details from the Flixborough desaster.Beside the earlier damage analyses, with damages of buildings and structures, Rischbieter analysed the damages of slim structures, like utility poles. These damages were caused by the dynamic pressure. Different scenarios of deflagrative and detonative explosions were discussed. Because of the time I will not go into details and only present the main results. These are :

- a detonation of the whole vapour cloud can be excluded

- only in parts of the burning vapour cloud a higher pressure buildup could be observed

- the most probable explanation of the explosion damage is a fast deflagration.

At the end of this item we can summerize that all recent works on damage analysis of vapour cloud explosions came to the same result:
The assumption of a deflagration is more realistic to explain the explosion damages. This is in opposite to the knowledge 10 years ago.

4. Susceptibility to Detonation of Different Gases

In this section we will report about the present knowledge of the susceptibility of homogeneous gas-air mixtures to unconfined detonation for a series of commonly handled substances. This leads to a certain classification concerning the explosivity of the substance used.

For the layout of a safety structure, system or component it is often only necessary to know the one case leading to the strongest requirements. Therefore such a classification can be a useful tool to justify the danger of a special gas-air mixture.

In the review article of D.C. Bull, about the concentration limits for the initiation of unconfined detonation in fuel-air mixtures, it is shown that the detonability of mixtures of gas with air decreases in the order (Fig. 3):

acetylene, hydrogen > ethylene, propylene >
ethane > propane > butane > isobutane > methane

Methane in its most detonable concentration in air requires an initiation energy nearly forty thousand times higher than that for acetylene at its most detonable concentration.

5. Explosion of a Real Vapour Cloud

In this section, I will report on three examples of large scale experiments on free vapour cloud explosions. The first are the Maplin Sands Experiments done with LNG and liquid propane spills on sea, the second are the China Lake LNG-spill experiments with spill up to 40 m³.
The third example are experiments with free hydrogen-air clouds, simulated by hemispherical balloons, with nearly stoichiometric mixture and vloumes up to 2000 m³.

The results of the Maplin Sands Experiments are

- On the evidence of the trials performed at Maplin Sands the deflagration of truly unconfined flat clouds of natural gas or propane does not constitute a blast hazard.

- The maximum flamespeeds observed were 28 m/s for propane and 10 m/s for natural gas, average pre-mixed flamespeeds were 12 and 5 m/s respectively; all speeds are quoted relative to the ground.

- Flame generated overpressures were 0,8 mbar for propane and 0,4 mbar for natural gas, both measured in the immediate vicinity of the flame.

- Flame-generated pressure decayed as 1/R.

- No sustained flame acceleration, which might lead to faster flames developing in similar larger clouds, was observed.

- Expansion of the combustion products took place exclusively in the vertical direction.

5.1 China Lake Experiments 1980

During 1980 and 1981 the Lawrence Livermore National Laboratory did two series of field-scale, LNG spill experiments at China Lake.
The results of these experiments were very similar to the Maplin Sands Experiments.

5.2 Free Hydrogen-Air Clouds

In the safety research programme on gas cloud explosions of the (PNP)-project large scale experiments with free hydrogen-air clouds were carried out (Fig. 5) . The aim of these experiments was to measure the influence of cloud extension and of ignition energy on the flame speed and the pressure wave.

The free gas clouds were simulated by hemispherical balloons consisting of thin polyethylene foil. The balloons were filled with nearly stochiometric hydrogen-air mixtures.
Their volumes were 7,5 m^3, 50 m^3, 260 m^3 and 2100 m^3 (the corresponding radii: 1,53 m, 2,88 m, 5 m, 10 m).

The results of the tests are shown in Fig. 4 max. pressure about 60 mbar in the case of great ballon. A pressure decay of 1/r outside the ballon could be observed.

Pförtner showed in earlier experiments with methane-air clouds that the flame accelerates with path length because of the turbulence generated by the flame itself; the flame velocity increases with the 4th root of the cloud radius.

The main result of the present experiments with hydrogen is that this rule holds true for hydrogen, too. Fig. 5 shows the normalized flame front velocity f_T (or turbulence factor)

f_T is independent of the H_2-concentration and the temperature of the burnt mixture. Therefore f_T can be compared directly.

In order to discover a dependence on the ignition energy, the mixture was ignited in some cases by energies of 10 and of 1000 Joules of the exploding wire. The results did not differ significantly from each other.

So that we can summerize theses experiments:

- max pressure buildup 60 mbar
- Flame speed increases roughly with the 4th root of the radius
- no significant dependence on the ignition energy.

6. Transition of Slow Deflagration to Detonation

The phenomena of the transition from deflagration to detonation in narrow long tubes is well known in the literature, as well as that obstacles in the path of a propagating flame create turbulence resulting in the flame acceleration an high flame speed. Because of the time I will speak only about the influence of confinement on flame acceleration and the jet ignition of a gas cloud.

6.1 Flame Acceleration and Pressure Buildup in Partially Confined Hydrogen-Air Clouds

To study the influence of confinement on flame acceleration, large scale experiments with partially confined hydrogen-air clouds were carried out by Pförtner et al. The partial confinement is caused by two wooden walls which are located at a distance of 3 m. In some experiments one side of the lane was closed with an additional wall of 3 x 3 m^2 to increase the confinement (Fig. 6).

The lane colume of 90 m³ was filled with a mixture of about 40 Vol. % hydrogen, which was ignited in the centre of one end of the lane.

The highest measured flame speed was about 200 sec and the pressure was about 200 mbar (Fig. 7).
In order to increase the turbulence effects accelerating the flame, a fan was located near the ignition source at the closed end of the lane. The fan produced a turbulent flow field in the lane, which was maintained during the deflagration.

In two tests a transition from deflagration to detonation occured.
The Fig. 6 shows a plot of the plane speed and some typical pressure time curves of the second detonative trial.

To study the influence of the turbulence caused by the fan on the flame velocity the fan speed was varied. No further transition to detonation took place (fig. 7). Even 90 % of the maximum revolution rate was not able to produce the high turbulence field which is necessary for the transition from deflagration to detonation.

Let me summerize this item. In the case of the confined experiments, higher flame speeds were reached. Only under special conditions (confinement, high fan speed) a transition from deflagration to detonation occured. This is an important limitation for the construction of a nuclear power plant: to avoid arrangements which can produce similar effects.

6.2 Jet Ignition

The most probable type of scenario for a strong ignition source is that of the sudden rupture of the containment and a small local confined explosion giving rise to the ignition of the surrounding explosive mixture outside, by the turbulent jets of escaping hot product gases.

In closed vessels, Kogarko has shown that overall burning rates two orders of magnitude higher can be achieved with jet ignition, as compared to spark ignition in a closed vessel. Experiments in tubes were performed by Wagner et al and Eckhoff and during the last years by Geiger et al.

We will discuss only the last experiments.
Geiger et al simulated jet ignition of an unconfined gas cloud. Therefore experiments in model scale have been performed, to investigate pressure build up and flame propagation during jet ignition of a unconfined gas cloud. The free jet was produced by an explosion in a vented rectangular chamber (size 0.5 m x 0.5 m x 1 m) (Fig. 8).
In order to change the jet, orifices of different shape and size have been used. For large orifices a turbulence generator has been used in addition, for increasing the burning rate inside the chamber. The free gas cloud was simulated within a rectangular tent of thin polyethylene foil, with a size of 4 m x 1 m x 1 m.

In the experiments with ethylene-air mixtures done in the german program about safety of nuclear power plants the following results were obtained: In the case of the chamber without turbulence generator the largest value for the overpressure in the free cloud was 0,87 bar and it was produced by an approximately sonic free jet.

The sonic free jet was produced by a orifice ratio $F_O/F_G = 0{,}1$. For orifice ratious F_O/F_G larger than 0.2 the overpressure in the free gas cloud is significantly increased by a turbulence generator. The pressure maximum was 1,3 bar. In the most adverse case the overpressure is raised by a factor of five compared to the same chamber without turbulence generator.

In the experiments with ethylene-air mixtures a pressure build up was obtained This means, for a real explosion, that only local pressure build up can occur because of jet ignition.

For the PNP project Geiger et al performed, with the same experimental set up, jet ignition experiments with hydrogen-air mixtures.

Experimental details are shown in Fig. 9.

The tests showed that with jet-ignition of hydrogen-air mixtures a transition from deflagration to detonation can occur in the second gas cloud. But it is necessary that there are some structures in the outer cloud.

The structures were two walls, arranged parallel to the jet axis. The jet produced high rates of energy release and high turbulence values which lead to high burning velocities, so that the flame can produce very high turbulence in the wall layer itself and a transition from deflagration to detonation can occur.

In the test-arrangement used,the tranistion from the deflagration to the detonation only happened for stoichiometric mixture. Other compositions do not lead to a detonation.

7. Summary

a) Damage Analysis

- Detonation of the whole vapour cloud can be excluded
- only in parts of the cloud higher pressure build up could be observed
- the most probable explanation of the explosion damage is a fast deflagration

b) Free Vapour Cloud Experiments

- little flame speed acceleration
- little pressure build up

<1 mbar for Propane, LNG
60 mbar for hydrogen

c) Partially confined hydrogen-air clouds

- Flame speed acceleration up to 200 m/s
- pressure build up to 200 mbar
- with special conditions TDD is possible

d) Jet Ignition Experiments

- high flame speed acceleration
- high pressure build up (<2 bar, ethylene-air)
- with special conditions TDD happens in the case of rich hydrogen-air mixture

8. Conclusion

For the out of nuclear power plant our conclusion is:

- Even partial confinement of a hydrogen-air cloud leads to a pressure build up lower than 200 mbar

- Jet ignition of a gas cloud has to be avoided

- Detonative ignition has to be excluded

- Turbulence generating structures like grids, tubes, whirls after edges only produce a local pressure build up within the cloud and do not lead to a transition from deflagration to detonation.

Our starting point was that the assumption of a detonation of a free gas cloud for the lay out of a nuclear power plant is wrong. We hope that we could show in our work that our statement is true and that the time is ripe to change the german regulatory guide about chemical explosions.

Discussion

Note: No discussion has taken place on this paper.

1.)

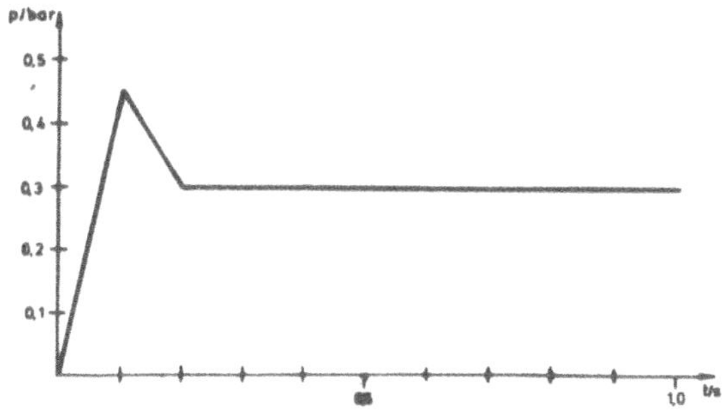

Pressure pattern on the nuclear power plant bilding

2.) $R = 8\,L^{1/3}$

Fig. 1

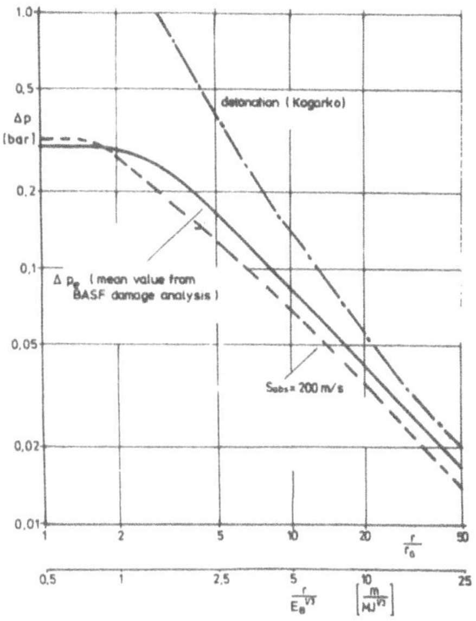

Fig.2 Equivalent TNT-mass for the BASF Damage Analysis and mere Detonation

ignition energy /g tetryl	gas mixture	concentration limit λ
1 g tetryl ≙ 4,27 kJ	hydrogen-air	0,95 - 1,23
	acetylen-air	1,10 - 2,85
10 g tetryl	hydrogen-air	0,64 - 2,1
	acetylene-air	0,65 -
	ethylene-air	1,05 - 1,65
100 g tetryl	hydrogen-air	0,48 - 2,88
	acetylene-air	0,46 -
	ethylene-air	0,75 - 2,35
	propane-air	0,96 - 1,35
	butane-air	0,96 - 1,35
	ethane-air	0,86 - 1,45
	methane-air no detonation	

λ = 1 is the stoicheiometric mixture

Fig. 3 Detonability of fuel-air mixtures as a function of concentration and ignition energy from D.C. Bull

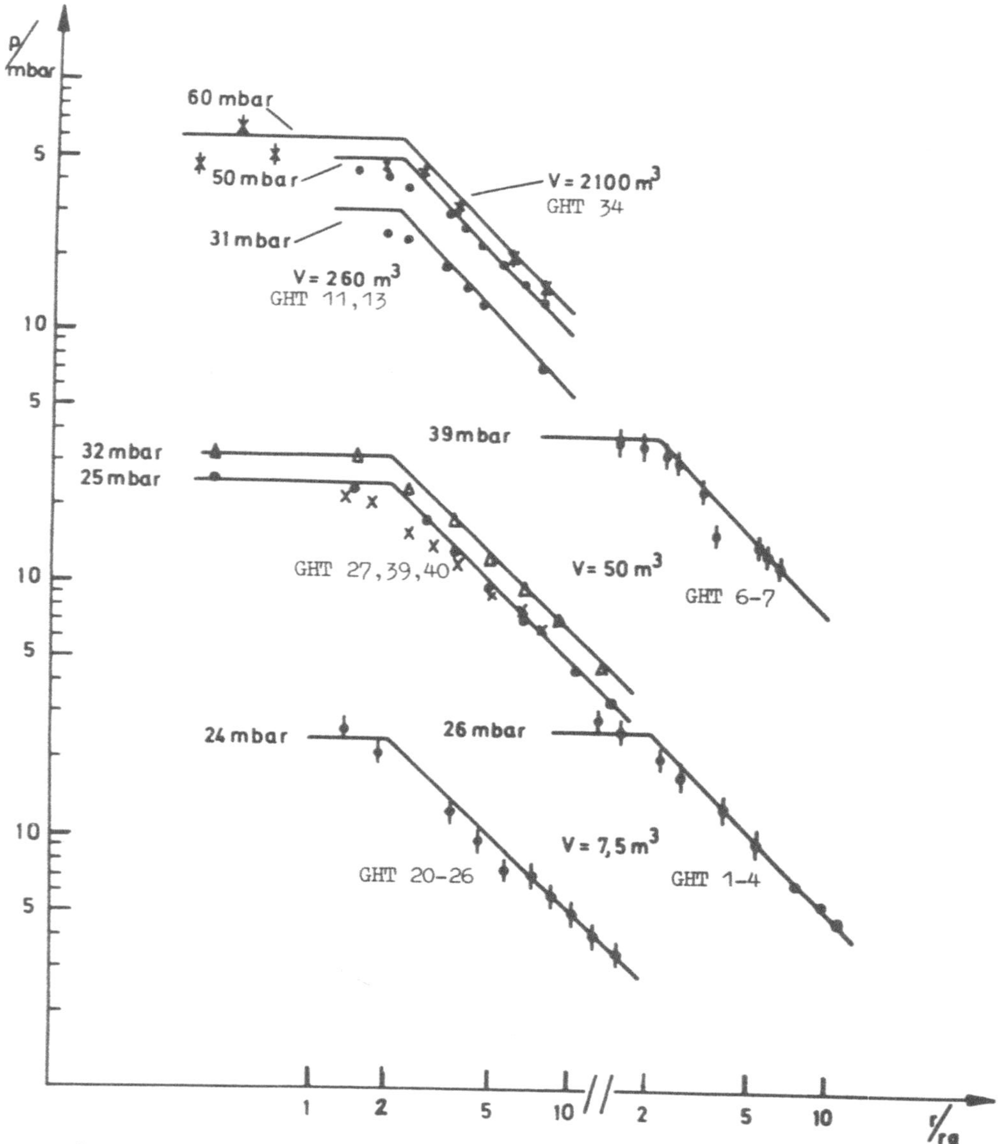

Fig. 4 : Measured pressure-distance relation of the free gas cloud tests

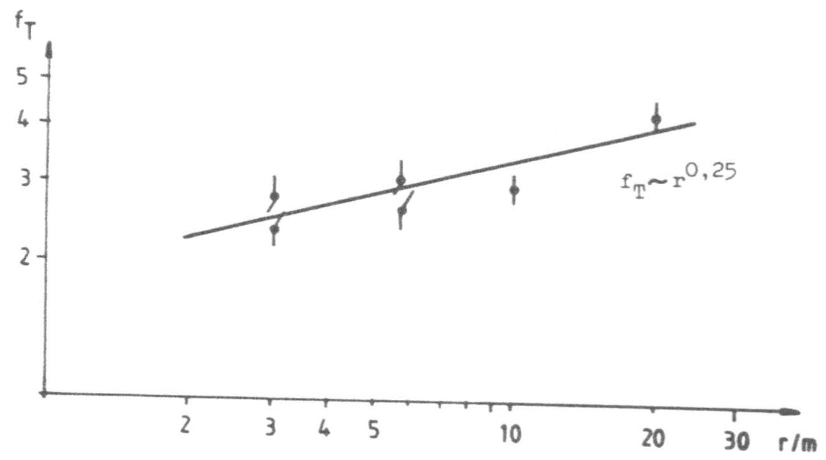

Fig. 5 : Flame front velocity as a function of the flame way.

f_T is normalized on the velocity of the laminar flame front.

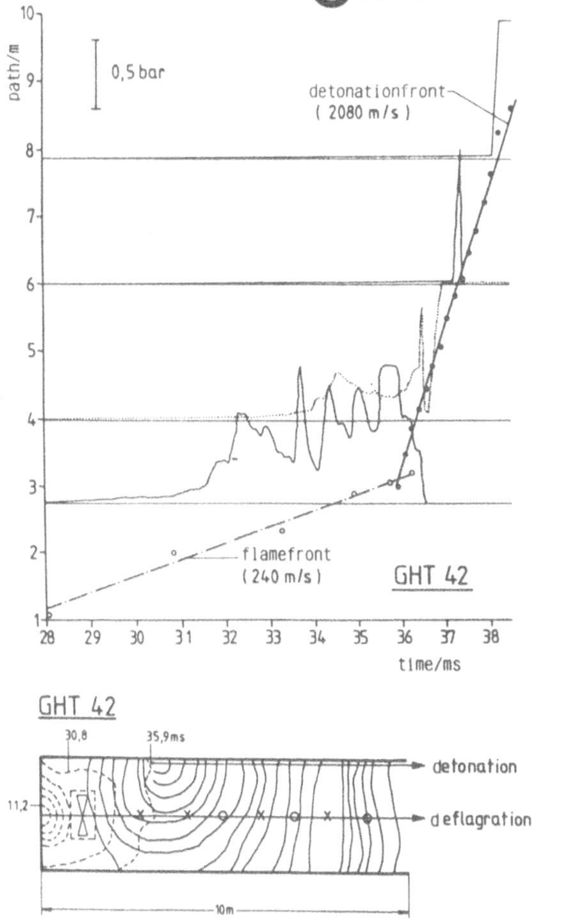

Fig. 6 :Top: Flame front velocity and typical pressure time curves at 3, 4, 6 and 8 m distance from the ignition point
Bottom: Flame speading; dashed lines: deflagration, full lines: detonation

Test Nr.	C_{H2} Vol %	Degree of Confinement	Fan r.p.m.	Temp. °C	S'_{max} (m/s)	Δp_{max} (mbar)	Remarks
GHT 35	37	0,65	-		164	120	
GHT 36	39	0,72	-		200	165	
GHT 37	39	0,72	-		196	200	
GHT 38	40	0,72	225 (1.0)	13	(600)	600	broken walls
GHT 41	38	0,72	225 (1,0)	5	2220	1.000	reinforced walls
GHT 42	36	0,72	225 (1,0)	12	2250	20.000	
GHT 43	41	0,72	113 (0,5)	18	250	300	
GHT 44	38	0,72	174 (0,75)	20	230	-	failure of Δp-measurements
GHT 45	41	0,72	174 (0,75)	19	280	450	
GHT 46*	41	0,72	203 (0,9)	15		480	
GHT 47*	37	0,72	203 (0,9)	16		420	

*) preliminary evaluation

Fig. 7: Lane trials (10 x 3 x 3 m^3), H_2-air mixture

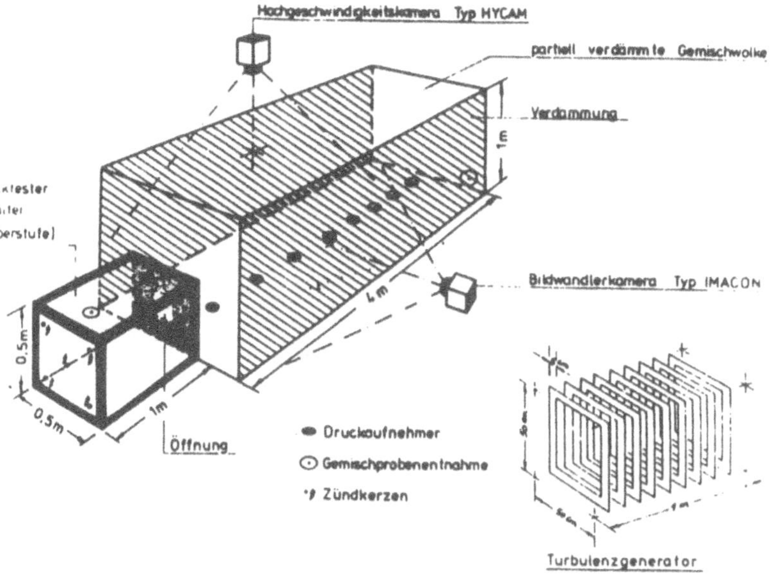

Fig. 8 : Testarrangement and -diagnostic

Test-Nr.	Orifices Ratio of F_0/F_G	Length of the orifice cm	Critical Diameter D_{KRIT} cm	turbulence generator yes	no	detonation yes	no	pressure P_{MAX} inside	outside	H_2 concentration Vol.% H_2 inside	outside
H 05	0,5	35,3	18	x		*		15	6	32	31
H 06	0,5	35,3	18	x			x	3	0,65	48,4	48,9
H 07	0,5	35,3	18	x				15	15	28	26,6
H 08	0,1	15,8	18		x	+			15	21.7 1)	
H 09	0,1	15,8	18		x		x		3	30,3	22
H 10	0,1	15,8	18		x		x	1,3	1	28.9	27.4
H,11	0,1	15,8	18		x	+		2.1	3*)	1)	1)

detonation in- and outside
* detonation inside
\+ transition from deflagration to detonation outside
1) probally stoeichiometric mixture
2) detonation at the end of the outer mixture cloud

Fig. 9 : Testresults of the second series

ON THE PROPAGATION OF THE PRESSURE PULSE DUE TO AN UNCONFINED GAS CLOUD EXPLOSION. A THEORETICAL ANALYSIS

J.A. ESSERS
Von Karman Institute and University of Liège, Belgium

Summary

A critical analysis of flow models used in computer codes for the simulation of the propagation in air of a pressure pulse due to a gas cloud explosion is presented. In particular, weaknesses of simple linear acoustic model are pointed out, and a more reliable non-linear isentropic model is proposed. A simple one-dimensional theory is used to evaluate as a function of the relative overpressure the speed of an incident normal shock-wave, as well as the strength and speed of the wave after reflection on a simplified rigid obstacle. Results obtained with the different models are compared to those obtained from the full Euler equations. A theoretical analysis of pulse deformation during its propagation is presented, and the ability of each model to correctly simulate that purely non-linear phenomenon is discussed. In particular, the formation of a sharp pressure pulse (shock-up phenomenon) is analyzed in detail.

From our analysis, the accuracy of the linear acoustic model for the evaluation of strength and speed of incident and reflected waves is found to be quite poor except for very weak overpressures. Additionally, such a model is completely unable to simulate pulse deformations. As a result, it should be expected to lead to important errors in the simulation of pulse interaction with non-rigid obstacles, even at very weak overpressures. As opposed to that very simple model, the proposed non-linear isentropic model is found to lead to an excellent accuracy in the prediction of all wave characteristics mentioned above and in the simulation of pulse deformation if overpressure is not too large.

1. INTRODUCTION

The protection of nuclear power plants against external explosions of heavy gas clouds is a relevant topic of nuclear safety studies. Several experiments have therefore been performed and computer codes been developed to study the propagation in air of the pressure pulse due to such explosions. The ultimate goal of those studies is to provide realistic inputs for the prediction of structure loadings and transient response. All flow models used in the various codes mentioned above can essentially be considered as belonging to one of the following two main categories. Some codes use the so-called Euler model. They numerically solve the general non-linear equations of inviscid perfect gas flows. The Euler model can lead to an excellent accuracy for a very wide range of relative overpressures ranging from weak deflagrations to strong detonations. However, because of the very complicated character of Euler equations, most of the codes have been developed for simple 1D situations corresponding to plane and spherical waves. A two-dimensional GASEX 2 code has however been constructed recently [1]. Several other codes use an approximate linear acoustic model leading to the well-known d'Alembert's equation. Because of the strongly simplified character of the latter, those codes could be applied to complex 2D/3D geometries e.g. corresponding a group of buildings in a powerplant. Let us e.g.mention the ZEPHIR [2] and ONEXSU [3] codes respectively using a panel method and a finite-difference technique to solve d'Alembert's equations, as well as a code developed to simulate the explosion of pancake shaped clouds [4]. Unfortunately, the relatively simple linear acoustic model can only lead to an acceptable accuracy for very weak overpressures.

In the frame of a detailed bibliographical study performed at Von Karman Institute [5,6], we presented a critical analysis of models used to simulate the propagation of pressure pulses in air. That analysis was based on a mathematical investigation strictly valid for 1D flows corresponding to plane or spherical waves, but the conclusions can be qualitatively generalized to more complex configurations. We also proposed a non-linear isentropic model leading to a potential flow equation. The latter is more complicated than the d'Alembert's equation, but is anyway much simpler than the general Euler system. That new formulation has been proved to lead to an excellent accuracy for weak and moderate overpressures, i.e. for most of the practical situations except strong detonations. The linear acoustic and non-linear isentropic models have been compared from the point of view of their ability to correctly simulate important phenomena such as : pulse deformation and the formation of a sharp compression front during pulse propagation (shock-up phenomenon), as well as shock wave propagation and reflection on a simplified plane or spherical rigid obstacle.

The main aim of the present paper is to summarize important conclusions obtained in the theoretical analysis mentioned above. A computer code has recently been constructed by the author at University of Liège [7] in order to verify the validity of those conclusions. With that code full Euler,non-linear isentropic and linear acoustic models can be used to compute 1D plane or spherical unsteady flows corresponding to the propagation and deformation of pressure pulses with any initial shape and overpressure level, and to analyze their interaction with a plane/spherical rigid or flexible wall. The main structural characteristics of the latter, i.e. elasticity, structural damping and inertia, can be simulated. That work, which is still under development is however not presented here. The results will be published in the near future.

2. GAS DYNAMICS EQUATIONS USED IN EULER, ISENTROPIC AND ACOUSTIC MODELS

To simplify the presentation, the equations of 1D inviscid compressible flows corresponding to plane/spherical waves propagation are only considered here. The Euler equations can be written in the following conservation form :

continuity : $$\frac{\partial}{\partial t}\rho + \frac{\partial}{\partial r}(\rho u) = -\frac{2\alpha}{r}\rho u \qquad (1.1)$$

momentum : $$\frac{\partial}{\partial t}(\rho u) + \frac{\partial}{\partial r}(p + \rho u^2) = -\frac{2\alpha}{r}\rho u^2 \qquad (1.2)$$

energy : $$\frac{\partial}{\partial t}[\rho(e + \frac{u^2}{2})] + \frac{\partial}{\partial r}[\rho u(e + \frac{u^2}{2} + \frac{p}{\rho})] = -\frac{2\alpha}{r}\rho u(e + \frac{u^2}{2} + \frac{p}{\rho}) \qquad (1.3)$$

where : - t and r are the time and space independent variables. For spherical waves, R is the radial coordinate measured from the centre of the cloud;

- α = 0 and 1 for plane and spherical waves respectively;
- p, ρ and u respectively denote the local values of static pressure, specific mass and gas flow velocity;

- for a perfect gas, internal energy per unit mass can be expressed as follows as a function of p, ρ and of the ratio γ of specific heats at constant pressure and volume (γ = 1.4 for air) :

$$e = \frac{p}{(\gamma - 1)\rho}$$

For reasons explained in section 5, the compression part of the pulse, initially due to flame acceleration in the cloud, tends to become sharper and sharper. A shock wave is finally formed that leads to an entropy increase. The latter can however be neglected with an excellent approximation if relative overpressure of the shock is not too high. The whole flow can then be considered as isentropic, and the following equation can therefore be written :

$$\rho = \rho_o \left(\frac{p}{p_o}\right)^{1/\gamma} \tag{2}$$

where subscript o indicates known constant quantities corresponding to the undisturbed atmosphere. Substituting (2) into (1.1) and (1.2), we obtain the equations of non-linear isentropic model :

$$\frac{\partial}{\partial t}\left(p^{1/\gamma}\right) + \frac{\partial}{\partial r}\left(u\, p^{1/\gamma}\right) = -\frac{2\alpha}{r} u\, p^{1/\gamma} \tag{3.1}$$

$$\frac{\partial}{\partial t}\left(u\, p^{1/\gamma}\right) + \frac{\partial}{\partial r}\left(u^2 p^{1/\gamma} + \frac{p_o^{1/\gamma}}{\rho_o} p\right) = -\frac{2\alpha}{r} u^2\, p^{1/\gamma} \tag{3.2}$$

Those equations can also be written in the following equivalent non-conservative form :

$$\frac{1}{\gamma p}\frac{\partial p}{\partial t} + \left(\frac{\partial u}{\partial r} + \frac{u}{p}\frac{\partial p}{\partial r}\right) = -\frac{2\alpha}{r} u \tag{4.1}$$

$$\left(\frac{\partial u}{\partial t} + \frac{u}{p}\frac{\partial p}{\partial t}\right) + \frac{1}{\rho_o}\left(\frac{p_o}{p}\right)^{1/\gamma}\frac{\partial p}{\partial r} + u\left(2\frac{\partial u}{\partial r} + \frac{u}{p}\frac{\partial p}{\partial r}\right) = -\frac{2\alpha}{r} u^2 \tag{4.2}$$

For a perfect gas, the local value of the speed of sound a can be expressed as a function of local temperature by the following relation :

$$a^2 = \gamma RT = \frac{\gamma p}{\rho} \tag{5}$$

From isentropic approximation, we can write :

$$\left(\frac{a}{a_o}\right)^2 = \frac{T}{T_o} = \left(\frac{p}{p_o}\right)^{\gamma-1/\gamma} \tag{6}$$

For very weak overpressures, additional approximations can then be performed. As induced gas velocity is then very weak, higher order terms in u^2 appearing in equation (4) are negligible. Convective derivatives $u/p\ \partial p/\partial r$ and $u/p\ \partial p/\partial t$ can also be neglected with respect to $\partial u/\partial r$ and $\partial u/\partial t$ respectively. As a result, air heating due compression in the pulse is also very weak. From equation (6), the local speed of sound a appears to be very close to sound speed a_o of the undisturbed atmosphere, and local static pressure p appearing in coefficients of equation (4) can be replaced by p_o. With those approximations, non-linear isentropic equations are transformed into those corresponding to the linear acoustic model :

$$\frac{1}{\gamma p_o}\frac{\partial p}{\partial t} + \frac{\partial u}{\partial r} = -\frac{2\alpha}{r}u \qquad (7.1)$$

$$\frac{\partial u}{\partial t} + \frac{a_o^2}{\gamma p_o}\frac{\partial p}{\partial r} = 0 \qquad (7.2)$$

Eliminating u between eqs(7.1) and (7.2), we obtain the following classical d'Alembert's equation, that clearly proves that all perturbations propagate with sound speed a_o of the undisturbed atmosphere :

$$\frac{\partial^2 p}{\partial t^2} = \alpha_o^2 \frac{1}{r^{2\alpha}}\frac{\partial}{\partial r}\left(r^{2\alpha}\frac{\partial p}{\partial r}\right) \qquad (8)$$

The approximations that consist in neglecting the effect of convection due to induced gas flow and atmosphere heating by the pulse on perturbations propagation speed are in fact rather severe. They will actually be proved to be much more restrictive than the isentropic approximation.

Equations (1), (3) (7) and (8) can of course be generalized to 2D/3D flows corresponding to more complex configurations. In particular, generalized form of system (3) corresponding to non-linear isentropic model leads to 3 and 4 coupled equations governing pressure and flow velocity components for 2D and 3D flows respectively. Thanks to the isentropic approximation, the induced gas flow due to a pulse propagation in a uniform atmosphere can be proved to be potential. The flow velocity vector can be expressed as the gradient of an unknown scalar potential function Φ. The latter is governed by a non-linear second order equation that can be obtained from a classical manipulation of non-linear isentropic equations. For 2D flows, that equation can be written as follows :

$$\frac{\partial^2\Phi}{\partial t^2} + 2u\frac{\partial^2\Phi}{\partial x\partial t} + 2v\frac{\partial^2\Phi}{\partial y\partial t} + (u^2 - a^2)\frac{\partial^2\Phi}{\partial x^2} + (v^2 - a^2)\frac{\partial^2\Phi}{\partial y^2} + 2uv\frac{\partial^2\Phi}{\partial x\partial y} = 0 \qquad (9)$$

where flow velocity components u and v in the x and y directions, and local sound speed a can be expressed as follows :

$$u = \frac{\partial\Phi}{\partial x} \quad ; \quad v = \frac{\partial\Phi}{\partial y} \qquad (10.1)$$

$$\left(\frac{a}{a_o}\right)^2 = 1 - \frac{\gamma-1}{2a_o^2}\left[2\frac{\partial\Phi}{\partial t} + \left(\frac{\partial\Phi}{\partial x}\right)^2 + \left(\frac{\partial\Phi}{\partial y}\right)^2\right] \qquad (10.2)$$

Equation (9) is in fact equivalent to the generalized form of system (3), but its solution is simpler.We therefore suggest to use it to obtain the potential field. The local values of velocity components u, v, of sound speed a, and of pressure p could then be easily obtain using eqs (10) and (6). To some extent, that second order equation can be considered as a generalized form of d'Alembert's equation. As far as we know, it has never been used to study the propagation of pressure pulses du to gas cloud explosions.

3. JUMP EQUATIONS THROUGH NORMAL SHOCK WAVES

Equations (1), (3) and (7) corresponding to the different models can be written in the following general conservation form :

$$\frac{\partial s_i}{\partial t} + \frac{\partial}{\partial r} f_i (s_j) = h_i (s_j, r) \qquad i = 1, \ldots, N \qquad (11)$$

The s_j are the fundamental unknown fields of the problem, and the f_i and b_i are functions of the s_j but not of their derivatives. The number N of those equations is equal to 2 or 3 depending on the model used. Those equations can be transformed using relative coordinates and flow velocity with respect to a referential moving with the shock. A classical manipulation of those transformed equations [5, 6], leads to the jump equations through plane or spherical normal shock waves. Those jump conditions are in fact algebraic equations relating the basic flow unknowns at point 2 located just behind the shock to similar quantities at point 1 just in front of the shock. The solution of those algebraic equations leads to the values of relative flow velocity w_2, pressure p_2 and specific mass ρ_2 as functions of w_1, p_1 and ρ_1 which are supposed to be known in region 1. For incident shocks, the latter e.g. correspond to undisturbed atmosphere.

From eq.(1) of the Euler model, the classical Rankine-Hugoniot jump equations are then obtained :

mass conservation : $\rho_2 w_2 = \rho_1 w_1$ (12.1)

momentum equation : $p_2 + \rho_2 (w_2)^2 = p_1 + \rho_1 (w_1)^2$ (12.2)

conservation of total enthalpy : $\frac{(w_2)^2}{2} + \gamma \frac{p_2}{\rho_2} = \frac{(w_1)^2}{2} + \gamma \frac{p_1}{\rho_1}$ (12.3)

From suitable combinations of those equations, it is possible to prove that $s_2 > s_1$, where s is the entropy per unit mass.

Equation (3) corresponding to the non-linear isentropic model also lead to jump equations (12.1) and (12.2). Equation (12.3) is however replaced by the following relation expressing the isentropic approximation :

$$\frac{p_2}{(\rho_2)^\gamma} = \frac{p_1}{(\rho_1)^\gamma} \qquad (13)$$

From eq.(7) of the linear acoustic model, we obtain :

$$w_1^2 = \gamma \frac{p_o}{\rho_o} = a_o^2 \qquad (14.1)$$

$$p_2 - p_1 = \rho_o a_o (w_2 - w_1) = \rho_o a_o (u_2 - u_1) \qquad (14.2)$$

From equation (14.1), we can conclude that in such a simplified model, the relative flow velocity is not arbitrary, but should always be equal to a_o. The normal shock waves therefore always propagate with a velocity equal to the sound speed of undisturbed atmosphere. The same conclusion is in fact valid for any other pressure perturbation. Equation (14.2) expresses that pressure variation throughout the shock is proportional to the variation of gas flow velocity.

4. NORMAL SHOCK-WAVE PROPAGATION AND ITS REFLECTION ON A SIMPLIFIED PLANE/SPHERICAL RIGID OBSTACLE

Consider a plane or spherical normal shock wave propagating with velocity V in a perfect gas at rest with pressure p_o, specific mass ρ_o, and sound speed a_o (undisturbed atmosphere). The shock wave leads to an increase of pressure and density. The gas behind the shock is then moved in the same direction as the latter. Let us denote respectively by $p_o + \Delta p$, $\rho_o + \Delta\rho$ and U the pressure, specific mass and induced gas velocity just behind the shock. The different jump conditions presented in section 3 can then be written with : $w_1 = -V$; $p_1 = p_o$; $\rho_1 = \rho_o$; $w_2 = U - V$; $p_2 = p_o + \Delta p$ and $\rho_2 = \rho_o + \Delta\rho$. Then solving the algebraic equations respectively corresponding to Euler and non-linear isentropic models, it is possible to express V/a_o, U/a_o and $\Delta\rho/\rho_o$ as functions of relative overpressure $\Delta p/p_0$. The non-dimensional entropy increase $\Delta s/c_v$ obtained from Euler model can also be expressed as a function of $\Delta p/p_o$. All of those analytical expressions and the corresponding graphs are presented in [5,6]. Note that for the linear acoustic model, we easily obtain from eq.(14) :

$$\frac{V}{a_o} = 1 \quad ; \quad \frac{U}{a_o} = \frac{1}{\gamma}\frac{\Delta p}{p_o}$$

Figure 1 shows graphs of non-dimensional shock velocity as a function of relative overpressure for the different flow models. The linear acoustic model clearly appears to strongly underestimate that velocity except for very weak shocks, whereas the non-linear isentropic model leads to an excellent accuracy even for rather high overpressures. The latter conclusion is not surprising if we consider (fig.2) the entropy variation obtained from the Euler model. That variation actually appears to be weak for weak and moderate overpressures.

We now consider the reflection of an incident plane (or spherical) normal shock wave on a plane (or spherical) rigid wall. The incident shock speed as well as the overpressure and flow velocity will now be denoted with subscript i. They are supposed to be known from the study presented above. The gas is obviously at rest behind the reflected shock. The speed V_r of the latter as well as overpressure Δp_r and the increase $\Delta\rho_r$ of specific mass behind the reflected shock are obviously unknown. All those unknown quantities can however be obtained by solving algebraic jump equations corresponding to Euler and isentropic models written with :
$w_1 = U_i + V_r$; $p_1 = p_o + \Delta p_i$; $\rho_1 = \rho_o + \Delta\rho_i$; $w_2 = V_r$; $p_2 = p_o + \Delta p_r$ and $\rho_2 = \rho_o + \Delta\rho_r$. For those two models, it is then possible to obtain the expressions and graphs of $\frac{V_r}{a_o}$, $\frac{\Delta p_r}{p_o}$, $\frac{\Delta\rho_r}{\rho_o}$ as functions of $\frac{\Delta p_i}{p_0}$. For the linear acoustic model, V_r is obviously again found equal to a_o, and $\Delta p_r = 2\Delta p_i$. Figures 3 and 4 show the graphs of V_r/V_i and of $\Delta p_r/2\Delta p_i$ versus $\Delta p_i/p_o$. The non-linear isentropic model again appears to lead to a good accuracy for weak and moderate overpressures, whereas the accuracy of the acoustic model is quite poor except for very weak shocks. It is also interesting to note that Δp_r is overestimated by the isentropic model thus leading to a safer wall design, whereas that overpressure can be strongly underestimated by the acoustic model.

Asymptotic expressions of V_i/a_o, U_i/a_o, V_r/a_o, $\Delta p_r/2\Delta p_i$ in terms of powers of $\Delta p_i/p_o$ are presented in table I. Those expressions which are valid for weak relative overpressures, can be used to compare the result obtained with the different models. From that table, the linear acoustic model appears to lead to relative errors of the order of $\Delta p_i/p_o$, whereas the error corresponding to the non-linear isentropic model are of the order of $(\Delta p_i/p_o)^2$

5. PULSE DEFORMATION DURING PROPAGATION

A pressure pulse measured relatively close to the exploding cloud does not exhibit any sharp shock, but rather a smooth compression part, characterized by a certain pressure rise time τ_o, and followed by an expansion part. Those two parts are respectively due to flame acceleration and deceleration. As shown on fig.5, the pulse can then be considered as a series of infinitesimal compression and expansion waves. Each such wave can be proved to propagate with a velocity W given by :

$$W = U + a \tag{15}$$

Pressure and gas flow velocity variations due to each infinitesimal wave are related to eachother by the equation :

$$dp = \rho \; a \; d \; U \tag{16}$$

In eqs.(15) and (16), ρ, a and U respectively denote the specific mass, the local sound speed and the gas flow velocity in front of the wave. A compression waves induces an infinitesimal gas acceleration behind it, and leads to gas heating and therefore to an infinitesimal increase of local sound speed. On the contrary, a gas deceleration and cooling, and therefore an infinitesimal decrease of sound speed are due to expansion waves. From eqs. (15), the following conclusions can easily be obtained. Because of gas acceleration and heating due to any infinitesimal wave in a compression waves packet, each compression wave propagate with a larger velocity than all waves propagating in front of it. As a result, all compression waves tend to bunch together and to form a shock waves. The opposite conclusion is valid for expansion waves packets, which tend to spread over larger and larger distances. The expansion part of the pulse therefore becomes wider and wider, whereas the compression part leads to a shock wave. The phenomena described above are due to purely non-linear effects. They can be suitably simulated by non-linear Euler and isentropic models.

We propose the following simple approximate theory to predict the main deformation characteristics of a plane pressure pulse, and namely to evaluate the time θ and distance λ required for complete shock wave formation. The first compression wave of the pulse propagate with a velocity equal to the sound speed a_o of the undisturbed atmosphere, whereas the last compression wave propagates with velocity $U_c + a_c$, where a_c and U_c respectively denote the sound speed and induced gas velocity at the end of the compression part of the pulse. The relative and average velocities W_r and W_a of those two waves are therefore given by :

$$W_r = U_c + a_c - a_o \qquad\qquad W_a = \frac{1}{2}(U_c + a_c + a_o) \tag{17}$$

Consider a pressure pulse with characteristic pressure rise time τ_0 measured at time t = 0. The initial thickness d_o of the compression part can be estimated as follows :

$$d_o = W_a \tau_o \tag{18}$$

The time θ and distance λ can then be evaluated as a function of relative velocity W_r. We obtain : $\theta = d_o/W_r$ and $\lambda = W_a \theta$. Then using (17) and (18), we can write :

$$\frac{\theta}{\tau_o} = \frac{1}{2} + \frac{1}{\frac{U_c}{a_o} + \frac{a_c}{a_o} - 1} \qquad (19.1)$$

$$\frac{\lambda}{a_o \tau_o} = \left[\frac{1}{2} + \frac{1}{\frac{U_c}{a_o} + \frac{a_c}{a_o} - 1}\right] \frac{\frac{U_c}{a_o} + \frac{a_c}{a_o} + 1}{2} \qquad (19.2)$$

For very weak relative overpressure, λ and θ are very large because $U_c/a_o \lll 1$, $a_c/a_o \simeq 1$, and relative velocity W_r is therefore very weak. For increasing values of $\Delta p/po$, θ and λ both decrease. For very large overpressures, θ finally tends to $1/2\ \tau_o$, whereas λ reaches a minimum value and finally increases to tend to infinity with $\Delta p/po$ because average velocity W_a then becomes very large. If the initial compression front is rather sharp, the values of U_c/a_o and a_c/a_o appearing in (19.1) and (19.2) can be approximately expressed as functions of $\Delta p/po$ using shock wave theory presented in section 4 for the Euler and the non-linear isentropic models. Graphs of θ/τ_o and of $\lambda/a_o\tau_o$ versus $\Delta p/p_o$ are shown on fig.6 and 7, and asymptoric expansions of those two quantities for weak values of $\Delta p/p_o$ are presented in table I. The agreement of the results obtained with the 2 models is remarkable.

Note that θ and λ can also be considered as good estimation of time and distance required for the width of the expansion part of the pulse to be multiplied by a factor 2. In a similar way, we can consider that during shock wave formation, the pressure rise time of the pulse and the thickness of the compression part will respectively become equal to $\tau < \tau_o$ and to $d < d_o$ (with $d/d_o = \tau/\tau_o$) after a time t and a distance l given by $t = \theta\frac{\tau}{\tau_o}$ and $l = \lambda\ d/d_o$. For plane waves, the maximum overpressure remains constant during pulse propagation. The simple theory presented above can therefore be used to simulate pulse deformation. That conclusion is however not valid for spherical waves, because overpressure then decreases as a function of radial distance. As a result, the values of θ and λ obtained from the initial values of $\Delta p/p_0$ using figs 6 and 7 are in general strongly underestimated. An iterative procedure using figs 6 and 7 to evaluate θ and λ for spherical waves accounting for the decrease of $\Delta p/p_o$ as a function of r has been proposed in [5,6]. In those papers, different laws that have been presented in the literature to express overpressure as a function of radial distance depending on flow regime, i.e. on pressure level, are also reviewed.

Even for plane waves, the theory presented above is very approximate. It is actually possible to prove that the values of θ and λ can strongly depend on the detailed shape of the compression part of the pulse, and in particular on the location of an inflection point in the initial profile of non-dimensional local sound speed a/a_o. That effect has been taken into account in a theory of characteristics presented in [5,6]. For the sake of simplicity, that theory has been applied to the equations of non-linear isentropic model for plane one-dimensional flows. Thanks to that simplification, Riemann invariants can be obtained along each characteristic line in the (r,t) plane. Pressure profiles can then easily be obtained at successive time levels, and shock wave formation can be simulated in detail. Additionally, the trace of the shock wave in the (r,t) plane can also be computed. The numerical calculations are very simple. They only require the iterative solution of a few non-linear algebraic equations using Newton's iterative technique, and the integration of an ordinary differential equation. The whole procedure can therefore be introduced in a programmable pocket calculator. For flows corresponding to spherical waves, no Riemann invariant can be found along characteristic lines. The technique mentioned above can then still be used, but the required computational effort becomes significantly heavier.

The theoretical analysis presented in this section clearly demonstrates an important drawback of the linear acoustic model. With such a model, all compression and expansion waves propagate with the same velocity a_o. As a result, pulse deformation cannot be simulated and a shock wave cannot appear if a shock is not present in the initial pulse. The linear acoustic model can therefore lead to important errors in the simulation of pulse interaction with a non-rigid obstacle even at very weak overpressure, because the dynamic response of such an obstacle and the overpressure of the reflected pulse do not only depend on the overpressure level but also on the shape of the incident pulse.

REFERENCES

1 MARTIN, D. : "GASEX-2 - a two-dimensional hydrodynamic code for use in gas cloud explosion calculations", SRD R 173, 1982.

2 AILLAUD, P.; PAROT, J.M. : "Calcul des surpressions exercées par une onde acoustique aérienne sur des structures réfléchissantes." Rapport DEMT/SMTS/LAMS/82-17A, 1982.

3 NGUYEN VAN CHI; WARLUZEL, A. : "Détermination des surpressions sur les bâtiments d'une centrale dues au passage d'une onde d'explosion survenue dans son voisinage.", EDF Report HE 041/8189, 1981.

4 AUTON, T.R.; PICKLES, J.M. : "The calculation of blast waves from the explosion of pancake shaped vapour clouds", Central Electricity Generating Board, RD/L/N2/0/78, April 1979.

5 ESSERS, J.A. : "Characterization of the pressure wave originating in the explosion of an extended heavy gas cloud. A detailed bibliographical study", Commission of the European Communities, Nuclear Science and Technology, Paper EUR 8482, 1983.

6 ESSERS, J.A. : "Characterization of the pressure wave originating in the explosion of a gas cloud", V.K.I. Lecture Series on unconfined gas explosions, February 1983.

7 ESSERS, J.A. : "Development and implementation of a pressure propagation code applicable in spherical geometry to Euler/isentropic/acoustic modelling. Comparative treatment of shock-up and reflection on simplified rigid or elastic obstacles." Progress Report-C.E.C. ECI-990-7221-83-B, July 1984.

Discussion

Note: No discussion has taken place on this paper.

PARAMETERS \ MODELS	EXACT SHOCK MODEL	ISENTROPIC MODELS	ACOUSTIC MODELS
NORMALIZED WAVE SPEED $\frac{V}{a_0}$	$1+\frac{\gamma+1}{4\gamma}\frac{\Delta p}{p_0}-\frac{(\gamma+1)^2}{32\gamma^2}\left(\frac{\Delta p}{p_0}\right)^2$	$1+\frac{\gamma+1}{4\gamma}\frac{\Delta p}{p_0}-\frac{(\gamma+1)(7\gamma-1)}{96\gamma^2}\left(\frac{\Delta p}{p_0}\right)^2$	1
INDUCED FLOW VELOCITY $\frac{U}{a_0}$	$\frac{1}{\gamma}\frac{\Delta p}{p_0}-\frac{\gamma+1}{4\gamma^2}\left(\frac{\Delta p}{p_0}\right)^2+\frac{3(\gamma+1)^2}{32\gamma^3}\left(\frac{\Delta p}{p_0}\right)^2$	$\frac{1}{\gamma}\frac{\Delta p}{p_0}-\frac{\gamma+1}{4\gamma}\left(\frac{\Delta p}{p_0}\right)^2+\frac{(\gamma+1)(13\gamma+5)}{96\gamma^3}\left(\frac{\Delta p}{p_0}\right)^3$	$\frac{1}{\gamma}\frac{\Delta p}{P_0}$
NORMALIZED REFLECTED WAVE SPEED $\frac{V_r}{a_0}$	$1-\frac{5-3\gamma}{4\gamma}\frac{\Delta p_i}{p_0}+\frac{(11-5\gamma)(\gamma-1)}{32\gamma^2}\left(\frac{\Delta p_i}{p_0}\right)^2$	$1-\frac{5-3\gamma}{4\gamma}\frac{\Delta p_i}{p_0}+\frac{23\gamma^2-50\gamma+39}{32\gamma^2}\left(\frac{\Delta p_i}{p_0}\right)^2$	1
NORMALIZED OVERPRESSURE BEHIND REFLECTED WAVE $\frac{\Delta p_r}{2\Delta p_i}$	$1+\frac{\gamma+1}{4\gamma}\frac{\Delta p_i}{p_0}-\frac{(\gamma-1)(\gamma+1)}{8\gamma^2}\left(\frac{\Delta p_i}{p_0}\right)^2$	$1+\frac{\gamma+1}{4\gamma}\frac{\Delta p_i}{p_0}-\frac{(\gamma-1)(13-8\gamma)}{24\gamma^2}\left(\frac{\Delta p_i}{p_0}\right)^2$	1
$\frac{\theta}{\tau_0}$	$\frac{2\gamma}{\gamma+1}\frac{p_0}{\Delta p}\left[1+\frac{\gamma+1}{2\gamma}\frac{\Delta p}{p_0}-\frac{\gamma+1}{16\gamma}\left(\frac{\Delta p}{p_0}\right)^2\right]$	$\frac{2\gamma}{\gamma+1}\frac{p_0}{\Delta p}\left[1+\frac{\gamma+1}{2\gamma}\frac{\Delta p}{p_0}-\frac{3\gamma-17}{48\gamma}\left(\frac{\Delta p}{p_0}\right)^2\right]$	∞
$\frac{\lambda}{a_0\tau_0}$	$\frac{2\gamma}{\gamma+1}\frac{p_0}{\Delta p}\left[1+\frac{3(\gamma+1)}{4\gamma}\frac{\Delta p}{p_0}+\frac{\gamma+1}{16\gamma^2}\left(\frac{\Delta p}{p_0}\right)^2\right]$	$\frac{2\gamma}{\gamma+1}\frac{p_0}{\Delta p}\left[1+\frac{3(\gamma+1)}{4\gamma}\frac{\Delta p}{p_0}+\frac{23\gamma+3}{48\gamma^2}\left(\frac{\Delta p}{p_0}\right)^2\right]$	∞

TABLE I - ASYMPTOTIC VALUES OF IMPORTANT PARAMETERS FOR WEAK OVERPRESSURES

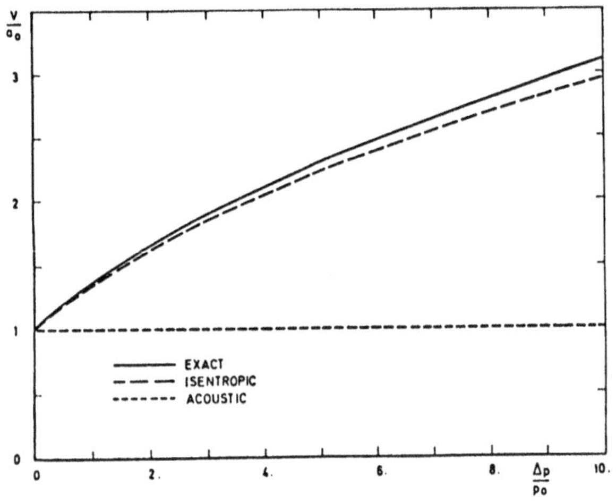

Fig.1 - Velocity of the wave front as a function of the pressure ratio $\gamma=1.4$

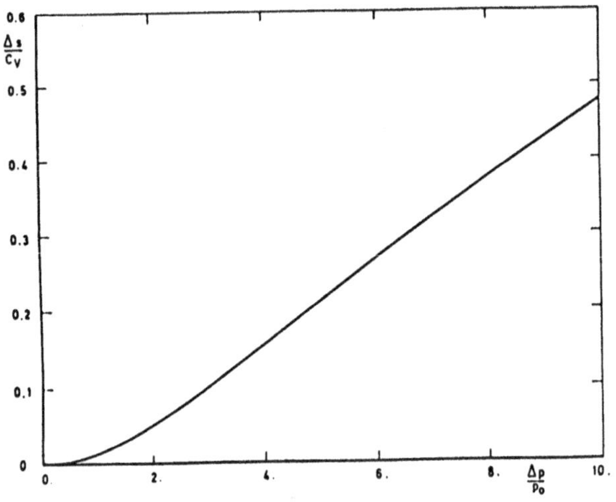

Fig.2 - Entropy variation through the pressure wave front

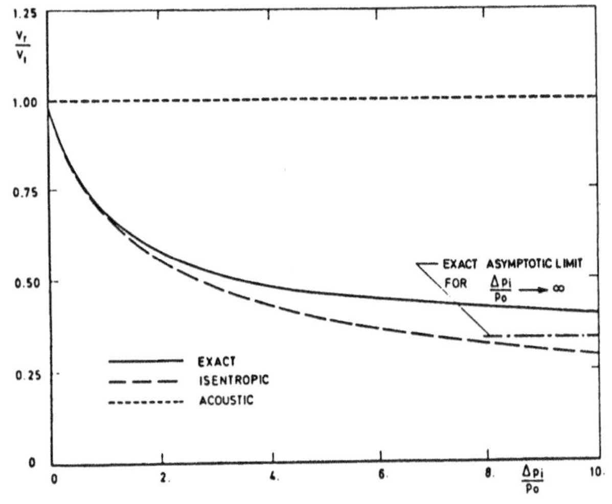

Fig.3 -Ratio of the reflected to the incident velocity of the blast wave $\gamma=1.4$

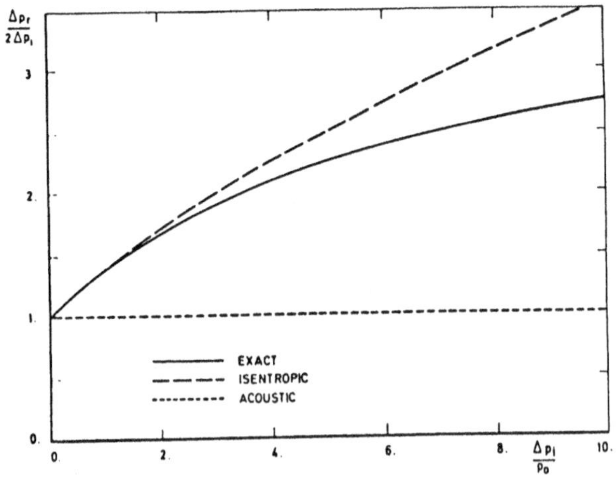

Fig.4-Overpressure due to the reflected wave $\gamma=1.4$

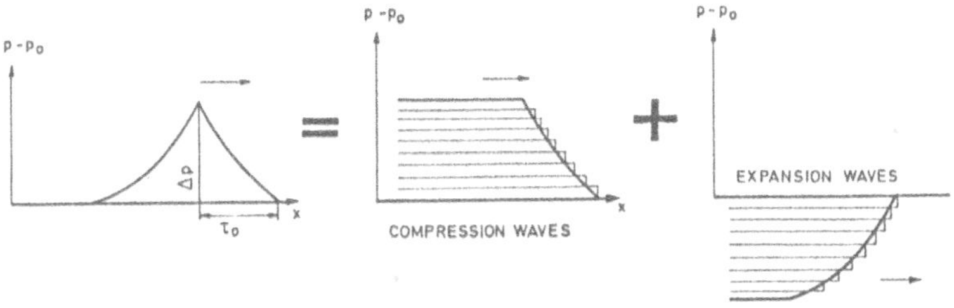

Fig.5-The pressure pulse as a superposition of compression and expansion waves

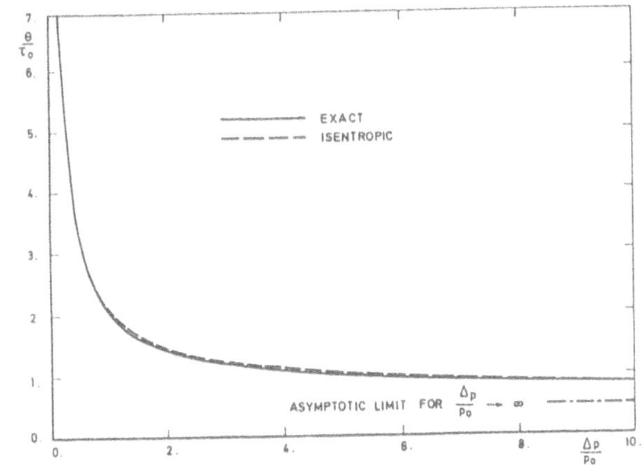

Fig.6-Time required to form a sharp wave front $\gamma=1.4$

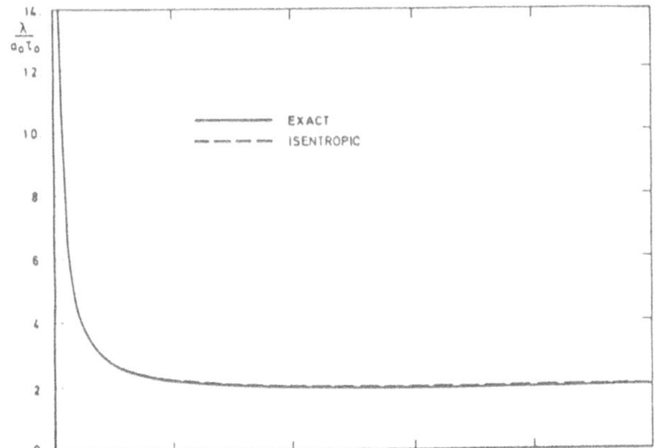

Fig.7-Time required to form a sharp wave front $\gamma=1.4$

Parallel Sessions Research Area B

The protection of nuclear power plants against external gas cloud explosions

Session B3

Chairman: M.L. Riethmuller (VKI/R.S.G.)

Session secretary: J. Vilain (CEC/BXL)

GAS CLOUD EXPLOSIONS AND THEIR EFFECT ON NUCLEAR POWER PLANT
BASIC DEVELOPMENT OF EXPLOSION CODES

S F Hall, D Martin and J MacKenzie
Safety and Reliability Directorate, UKAEA

Summary

The study of factors influencing the pressure and velocity fields produced by the burning of flammable substances has been in progress at SRD for some years. This paper describes an extension of these studies by using existing codes for a parametric survey, modifying and codes to produce more realistic representations of explosions and developing a two dimensional combustion code, FLARE.

The one dimensional combustion code, GASEX1, has been used to determine the pressure from a burning gas cloud for a number of different fuels, concentrations and burning velocities. The code was modified so that gas concentrations could be modelled. Results for concentration gradients showed the pressure depended on local conditions and the burning velocity.

The two dimensional code, GASEX2, was modified to model the interaction of pressure waves with structures. It was used, with results from GASEX1, to model the interaction of a pressure wave from the combustion of a gas cloud with a rigid structure representing a nuclear power plant.

The two dimensional code FLARE has been developed to model the interaction of flames and pressure waves with structures. The code incorporates a simple turbulence model with a turbulence dependent reaction rate. Validation calculations have been carried out for the code.

1. INTRODUCTION

The study of factors influencing the pressure and velocity fields produced by the burning of flammable substances has been in progress in SRD for some years. During this time several computer codes have been developed to aid in these studies. This present work is an extension of these studies, using existing codes for systematic parametric surveys, removing some of the limitations of the codes and producing a new code.

In general the combustion of gas clouds is poorly understood. The one dimensional combustion code GASEX1 has been used to conduct a parametric survey of the factors influencing the overpressures from a gas cloud explosion. The code was modified to allow concentration gradients to be modelled. It was then used to determine whether concentration gradients could cause flame acceleration.

The two dimensional hydrodynamic code GASEX2 has been modified so that the interaction of pressure waves with arbitrary rigid structures could be modelled. This allows more realistic modelling of the pressure wave from a gas cloud explosion interacting a rigid structure representing a nuclear power plant. After a number of validation calculations had been completed the code was used to model the interaction of pressure waves from a propane/air gas cloud explosion interacting with a nuclear power plant.

Although GASEX2 can model the interaction of pressure waves from a gas cloud explosion with a rigid structure it does not contain a combustion model so cannot model the interaction between a flame and a structure. The turbulence at a flame front caused by the presence of obstacles could be an important method of flame acceleration. The two dimensional combustion code FLARE was developed to overcome this limitation of GASEX2. FLARE, as well as containing a combustion model, contains a turbulence model and is able to model the complex interactions between flame and flow fields produced by complex geometrics. Validation calculations have been carried out with the code.

More details of this work can be found in reference 1.

1. GASEX1 Description

GASEX1 has been used to model the overpressures from the combustion of a gas cloud and the propagation of the blast wave into the surrounding atmosphere. The gas cloud and the surrounding atmosphere are modelled by one dimensional hydrodynamic equations which are solved by a finite difference scheme. The method is a Lagrangian scheme and is based on the method developed by Brode (2,3). The one dimensional scheme is constructed so as to allow the use of planar, cylindrical or spherical geometry. For gas cloud explosion calculations spherical geometry has been used. To ensure stability of the scheme when shocks are present in the solution artificial viscosity is used.

The rate of combustion is determined by a specified (input) burning velocity. Energy from combustion is deposited into a cell of the finite difference scheme. The amount of energy deposited depends on the burning velocity, the timestep used in the finite difference scheme and the unburnt gas density ahead of the combustion region. When all the fuel in the cell is burnt combustion continues in the adjacent cell. The energy released raises the temperature in the cell and hence the pressure which drives the gas flow ahead of the flame. A more detailed description of GASEX1 can be found in reference 4.

GASEX1 has been modified so that it can deal with concentration gradients of the combustible gas within the gas cloud. For spherical geometry the gas concentration varies with the radius. The change in gas concentration affects the density of the gas, assuming constant temperature and pressure within the gas cloud, and the heat release from combustion. For most common fuel/air systems the density change between the upper and lower flammability limits is relatively small, therefore for a concentration gradient within a gas cloud the effects resulting from the density differences are small and we have shown that they can be neglected. The major effect is from the change in the heat of combustion along the concentration gradient. Therefore the modifications to GASEX1 to allow concentration gradients assume that the density within the gas cloud is constant but the energy released by combustion varies with the gas composition. Details of the modifications made to GASEX1 to include concentration gradients are given in reference 4.

2. GASEX1 Calculations and Results

GASEX1 has been used to study the factors influencing the pressure fields produced by unconfined gas cloud explosions. A series of calculations have been carried out to investigate the effects of parameters such as fuel type, fuel concentration and burning velocity. The effect of concentration gradients has also been investigated. For comparison purposes the calculations as far as possible have used the same initial conditions. Spherical geometry was used for the gas cloud. The initial radius of the cloud was 12 metres and occupied 25 cells out of 125 cells used in the calculation. Initially the gas cloud and surrounding air were at a pressure of 1.013 bars and a temperature of 286°K. Ignition was at the centre of the gas cloud. Six fuels were used, acetylene, ethylene, propane, methane, ammonia and hydrogen. Two constant concentrations were used for each fuel, the lower flammability limit and the stoichiometric mixture, each for three burning velocities 30, 50 and 70 metres per second. A similar set of calculations were carried out using concentration gradients. Two situations were studied, one with the stiochiometric mixture at the centre of the gas cloud falling to the lower flammability limit mixture at the edge of the cloud, and one with the lower flammability mixture at the centre of the gas cloud rising to the stoichiometric mixture at the edge of the cloud. For these calculations the initial radius of the gas cloud was adjusted slightly so that for a given fuel both concentration gradients gave the same total energy release.

For the constant concentration calculations all the fuels produced pressure profiles with a similar shape. The pressure is fairly flat in the burnt gas then rises to a maximum at the flame front after which it decreases gradually to the level of that of the undisturbed gas. For the 30 metres per second burning velocity the overpressures are fairly low with only a small peak at the flame front. For the stoichiometric mixtures the maximum peak overpressure is 0.6 bars and for the lower flammability limit mixtures it is 0.2 bars. For burning velocities of 50 and 70 metres per second the overpressures are higher and the pressure peak at the flame front is sharper. For the stoichiometric mixtures the order of fuels for the highest overpressures, in decreasing order, is acetylene, ethylene, propane, methane, ammonia and hydrogen. It was found that this order was reflected in the order of energy release by combustion, in MJ/kg, multiplied by $\rho(\alpha-1)$ where ρ is the density and is the ratio of specific heats of the unburnt gas. The only exception was hydrogen which by this scheme is above ammonia. This suggests that with hydrogen additional factors are involved with its combustion. With lower flammability limit mixtures the highest overpressures are produced by ammonia followed by propane, methane, ethylene, acetylene and hydrogen. This order matches the order of the energy release by combustion.

The results showed that for determining overpressures the burning velocity is a more significant factor than the fuel type. Changing the concentration effects the pressure but the extent depends on the individual fuels.

The calculations with concentration gradients were carried out to determine whether gradients would cause flame acceleration and hence higher overpressures.

These calculations used the same gases as for the constant concentration calculations except for hydrogen. The results showed that the overpressures depended on the burning velocity and the local gas concentration. For the concentration gradient with the lower flammability limit mixture at the centre of the cloud and the stoichiometric mixture at the edge, the peak overpressure at the completion of combustion was almost the same as for the constant stiochiometric mixture but the peak was broader. The results of these calculations showed some flame acceleration or deceleration occured but not sufficient to produce flame velocities greater than the constant flame velocities of the stiochiometric mixtures.

Pressure profiles at the completion of combustion for the different fuels and concentrations are shown in Fig 1 for the 50 metre per second burning velocity and in Fig 2 for the 30 metre per second burning velocity. More details of the results are given in reference 1 and 5.

3. GASEX2 Description

GASEX2 (6) is a two-dimensional hydrodynamic code which uses the finite difference scheme known as the ICED-ALE technique (7,8). The code can use either a Lagrangian or Eulerian finite difference mesh. The latter was used for the interaction of shock waves from gas cloud explosion with structures because the Eulerian method does not produce a distorted mesh. The code can use either planar or axisymmetric geometry. To produce stable solutions when shocks are present an artificial viscosity term is used. Unlike GASEX1, GASEX2 uses an implicit scheme so that large time steps can be used in the calculations.

The code has been modified so that it can model the interaction of shocks with rigid structures. Two modifications have been carried out. The first allows structures to be modelled on the finite difference mesh and the second allows non-rectangular cells to be used when setting up the mesh. The first modification ensured that no fluid flow occurs into or out of cells designated as part of a structure. The second modification was incorporated into the code so as to enable a better representation of a curved structures. The method of producing non-rectangular cells is an iterative method based on that given in reference 9. The method, as developed for GASEX2 starts with a mesh consisting of equal sized rectangular cells. Designated cells are then moved to a given shape by an iterative method and the remaining cells are moved to fill in evenly the remaining space. The resulting shapes are constructed from straight lines and arcs of circles. An example of the type of structure produced by this method is given in Fig 5.

A further modification to the code allows different formulations of the artificial viscosity. Details of this and the modifications as well as details of the code can be found in reference 1 and 6.

A final modification made to this code was to allow data from a GASEX1 calculation to be used as initial data for a GASEX2 run. This would then be used to model the interaction of a shock wave from a gas cloud explosion with a rigid structure. A calculation using this procedure is described later.

4. GASEX2 Validation Calculations

A number of validation calculations were carried out on GASEX2. The first series of calculations modelled a one-dimensional shock tube. A high pressure region generates a shock wave with well defined properties which travels through the low pressure region. This problem can be solved theoretically and results can be compared. The calculations used values for the high pressure region ranging from 2.0 down to 1.2 bars. These calculations were undertaken to see how well GASEX2 modelled shocks. The results showed that, although GASEX2 gave good agreement with the theoretical model for the fluid properties behind the shock frong, the shock was spread over a number of cells. It was found that the lower the shock overpressure the greater the spreading of the shock. Reducing the parameter which determined the amount of artificial viscosity reduced the spreading of the shock to some extent but could lead to an unstable solution. Changing the artificial viscosity term was more successful for some cases. However, it was found that for the lower value of the high pressure region the spreading of the shock wave was not reduced by either of the above methods, suggesting that there is a limit below which shock spreading cannot be reduced. Details of these calculations are given in references 1 and 10.

The second series of validation calculations where for the interaction of a shock with a rigid structure. The first was a comparison with an experiment in which a shock, with a ratio of shock pressure to ambient pressure of 1.92 interacted with a square obstacle. The experiment is described in reference 11. The results from GASEX2 and the experiment were compared when the shock reached the end of the obstacle. Figure 3 compares the experimental isobars with those predicted by GASEX2. The agreement is fairly good but away from the obstacle for the highest value, the calculated isobar falls off rapidly with distance from the obstacle when compared with the experimental isobar. This effect is probably due to the resolution given by the size of cells used in the calculation.

The second calculation in this series was for the interaction of a plane shock wave with a cylindrical obstacle. This calculation showed the feasibility of using non-rectangular cells to construct an obstacle. Although the comparison was only qualitative the results from GASEX2 agreed reasonably well with the experimental results given in reference 12. More details concerning these comparisons are given in reference 10.

The final comparison calculation undertaken was between the codes GASEX1 and GASEX2. A GASEX1 combustion calculation for a stoichiometric methane/air gas cloud with a burning velocity of 50 metres per second was completed. The results when combusting was completed were stored and this data was used as initial data for a GASEX2 calculation. The results from the GASEX1 calculation subsequent to completion of combustion could then be compared with the results from the GASEX2 calculation. AS GASEX2 is a two-dimensional code and GASEX1 is a one-dimensional code a much coarser mesh was used for GASEX2 than for GASEX1. The comparison showed reasonable agreement but the details of the pressure profiles at various times did not show complete agreement due to the coarseness of the GASEX2 mesh. This comparison showed the feasibility of using data from a GASEX1 calculation an initial conditions for a GASEX2 calculation.

5. GASEX2 Results

Calculations with GASEX2 were undertaken to model the interaction of a shock wave from a gas cloud explosion with a rigid structure representing the layout of a nuclear power plant. GASEX1 was used to provide the initial data for GASEX2. At the completion of combustion the gas cloud had generated a burnt gas pressure of 2.0 bars and a peak pressure of 2.2 bars. Two main calculations were completed with GASEX2 with this data and with the same structure but with the initial centre of the burn gas cloud at a different position relative to the structure. The results from GASEX2 showed the buildup of pressure on the structure. Isobars from one of the calculations is shown in Fig 4. The structure is to the right of the figure. For this calculation the burnt gas cloud was initially opposite the circular structure which represents the reactor building. The finite difference mesh for the second calculation is shown in Fig 5. The circular region at the centre of the mesh is the initial burnt gas region. The pressure time profiles were compared for a number of points on the structure for the two calculations. An example is given in Fig 6. This shows the pressure buildup between the parts of the structure representing the reactor building and the fuel storage building. There is a slight difference between the peak pressures and the pressure rise occurs earlier for the second of the two calcualations. But for both calculations the confinement produces a pressure rise which is approximately 1 bar greater than the peak pressure in the initial burnt gas region.

6. FLARE Description

The computer code FLARE (1) has been developed to model the interaction of flames and pressure waves with structures. FLARE is a two-dimensional hydrodynamic code which includes both a turbulence model and a combustion model, with the rate of combustion determined by the turbulence generated by the flow. FLARE has been developed from the model presented by Hjertager (13). The equations describing the model are in conservative form and solved by the Flux Corrected Transport (FCT) technique (14-16). The fluid flow is described by four equations for the density, the momentum in two directions and the internal energy. The composition of the gas is described by two parameters, the mass fraction of fuel and the mass fraction of oxygen. The amount of combustion products and oxygen depleted air can be deduced from these. From these and the mass fractions of fuel and oxygen the density and internal energy, the temperature and pressure of the gas can be determined. The turbulence model used is a two parameter model and describes the turbulence by the turbulent kinetic energy (k) and the turbulent energy dissipation rate (ε). (17). The reaction rate assumes a one-step reaction with the 'reacting species' contained in separate eddies. The 'reacting species' are, for premixed gases, the fuel/oxygen mixture and the combustion products since it is assumed that the reaction is propagated as hot combustion products come into contact with the unburnt mixture through processes of turbulent mixing. The reaction rate is inversely proportional to the eddy lifetime of the turbulence, a quantity which can be determined from the turbulence parameters k and ε . The reaction rate is also proportional to the limiting species for the reaction which is either the mass fraction of fuel, the mass fraction of oxygen or the mass fraction of the combustion products.

The equations generated by this model are solved by a finite difference scheme incorporating the FCT method using Zalesak's method of flux limiting (18). The FCT method was chosen because of its superiority in dealing with shocks and discontinuities when compared with the artificial viscosity method, as it minimises the spread of shocks by limiting the diffusion which occurs in the method. The finite difference scheme allows a choice of boundary conditions and geometries including arbitrary shaped obstacles.

7. FLARE Validation Calculations

Many validation calculations were carried out on FLARE as the code was developed. At all stages comparisons were carried out with experiment and corresponding calculations using GASEX2. These comparisons gave us confidence that the numerical models and algorithrus were working as designed. For example, the validation calculation for GASEX2 described above for the interaction of a shock with an obstacle (see fig 3) was repeated for FLARE. The resulting isobars when the shock has reached the end of the obstacle are compared with the experimental isobars in Fig 7. The agreement between FLARE and the experiment is fairly good with the pressure contours from FLARE not falling off so rapidly towards the reflected shock as those given by GASEX2.

A final validation calculation was to compare the code with an experiment to test the combustion model.

The experiment (13) consisted of a pipe 10 metres long and 2.5 metres in diameter and having five equally spaced annular constrictions along the pipe, giving a blockage ratio of 0.3. The final annular constriction is at the open end of the pipe, and other end being closed. The pipe initially contained a stoichiometric methane/air mixture which was ignited at the closed end of the pipe. This was modelled by a mesh consisting of 30 cells along the pipe and 15 cells radially. It was assumed that, for ignition purposes that 10% of the mixture at the closed end of the pipe was burnt. The initial pressure was 1.01 bars, the initial density 1.237 kgm^{-3} and the initial temperature 271.6°K. Results for the pressure time profiles at the closed and open end of the pipe are shown in Fig 8. As FLARE does not model ignition processes the results are not expected to match the times of the experiment. Therefore the experiment result for the closed end was shifted along the time axis until the position of the peaks from the experiment and FLARE results agreed. Results from the experiment and FLARE show reasonable agreement for the overpressure at the closed end of the pipe, though the fall off in pressure is not matched by the FLARE calculation. The results for the open end of the pipe show less agreement; although the initial pressure rise is matched by FLARE the peak pressure is later and smaller. This poor agreement is due to the continuative flow boundary. The fact that the FLARE overpressures do not fall off as quickly as in the experiment is probably due to the fact that heat passing through the tube walls in not modelled. The lack of agreement between Hjtertager's computer results and those from FLARE is, perhaps, surprising but can be at least partly explained by noting that although the two codes solve the same equations the solution mthod is quite different.

8. Conclusions

The main conclusions of this paper are that:

i) In constant burning velocity models, variation of physical and chemical parameters makes some difference to the results but all of these differences are dominated by the choice of burning velocity, which is arbitrary.

ii) Following from this is the conclusion that constant burning velocity models are of limited usefulness in assessing the risk from gas explosions since they have little predictive capability.

iii) The computer modelling of the propagation of pressure waves (including shocks) and the interaction with structures in two dimensions is now fairly good. Comparison with analytic solutions and experiments gives us confidence in the use of GASEX2 and the Flux Corrected Transport method used in FLARE.

iv) Similarly, initial validation calculations indicate that the FLARE code will be very useful in modelling turbulent flame propagation, flame acceleration due to interaction with obstacles and the production and propagation of pressure waves.

References

1. HALL S F, MARTIN D and MACKENZIE J, 'Gas cloud explosions and their effect on nuclear power plant - phase 1: basic development of explosion codes' EUR 8955 en

2. BRODE H L, 'Numerical Solutions of a Spherical Blast Wave' Journal of Applied Physics vol 26 pp 766-775 (1955)

3. BRODE H L, 'Blast Wave from a Spherical Charge'. Physics of Fluids Vol 21 p217 (1959)

4. MACKENZIE J and MARTIN D, 'GASEX1-A General One-Dimensional Code for Gas Cloud Explosions'. SRD R 251 (1982)

5. MACKENZIE J, 'A computational survey of the factors influencing the pressure fields produced by unconfined vapour cloud explosions'. SRD R 262 (1983)

6. MARTIN D, 'GASEXII: A Two-Dimensional Hydrodynamic Code for use in gas cloud explosion calculations'. SRD R 173

7. HARLOW F H and AMSDEN A A, 'A Numerical Fluid Dynamic Calculational Method for all flow speeds'. Journal of Computational Physics Vol 8 pp 197-213 (1971)

8. HIRT C W, AMSDEN A A and COOK J L 'An arbitrary Lagrangian Eulerian computing Method for all flow speeds'. Journal of Computational Physics vol 14 pp 227-253 (1974)

9. AMSDEN A A and HIRT C W, 'A simple scheme for generating general curvilinear grids'. Journal of Computational Physics Vol 11 pp 348-359 (1973)

10. MARTIN D 'Validation calculations for the code GASEX2'. SRD R 252

11. BLEAKNEY W, WHITE D R and GRIFFITH W C 'Measurement of Diffraction of shock waves and resulting loading of structure'. Journal of Applied Mechanics pp439-445 (1950)

12. BRYSON A E and GROSS R W F 'Diffraction of strong shocks by cones, cylinders and spheres'. Journal of Fluid Mechanics vol 10 pp 1-16

13. HJERTAGER B H, 'Simulation of transient compressible turbulent

reactive flow'. Combustion Science and Technology vol 127 pp 159-170 (1980)

14. BORIS J P and BOOK D L, 'Flux Corrected transport. I SHASTA a fluid transport algorithm that works'. Journal of Computational Physics vol 11 p38 (1973)

15. BOOK D L, BORIS J P and HAIN K 'Flux corrected transport. II Generalisation of the method'. Journal of Computational Physics vol 18 p 248 (1975)

16. BORIS J P and D L 'Flux corrected transport. III Maximum error FCT algorithms'. Journal of Computational Physics vol 20 p 397 (1976)

17. LAUNDER B E and SPALDING D B 'Lectures in mathematical models of turbulence'. Academic Press (1972)

18. ZALESAK S T, 'Fully multidimensional flux-corrected transport algorithms for fluids'. Journal of Computational Physics vol 31 pp335-362 (1979).

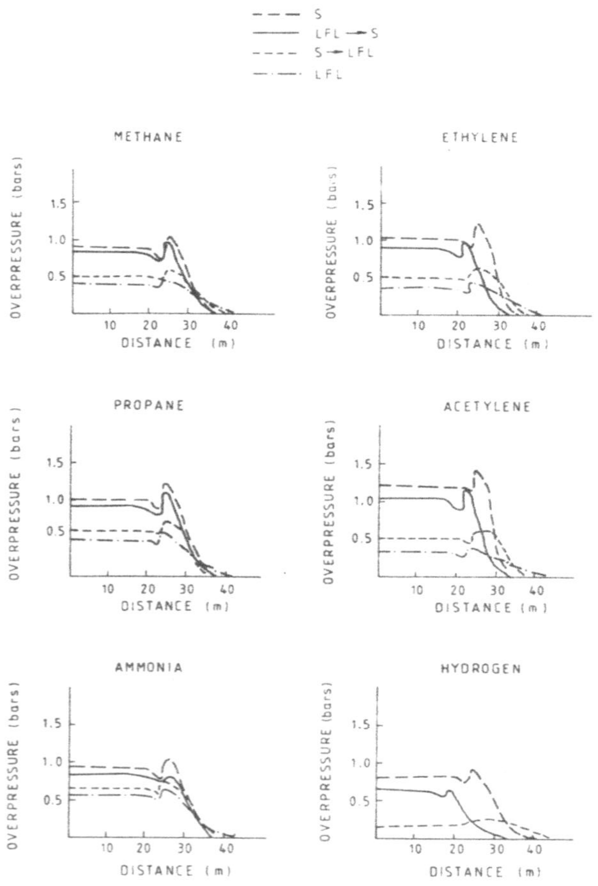

FIG. 1 OVERPRESSURE-DISTANCE PROFILES FOR DIFFERENT MIXTURES AT A BURNING VELOCITY OF 50 ms^{-1}

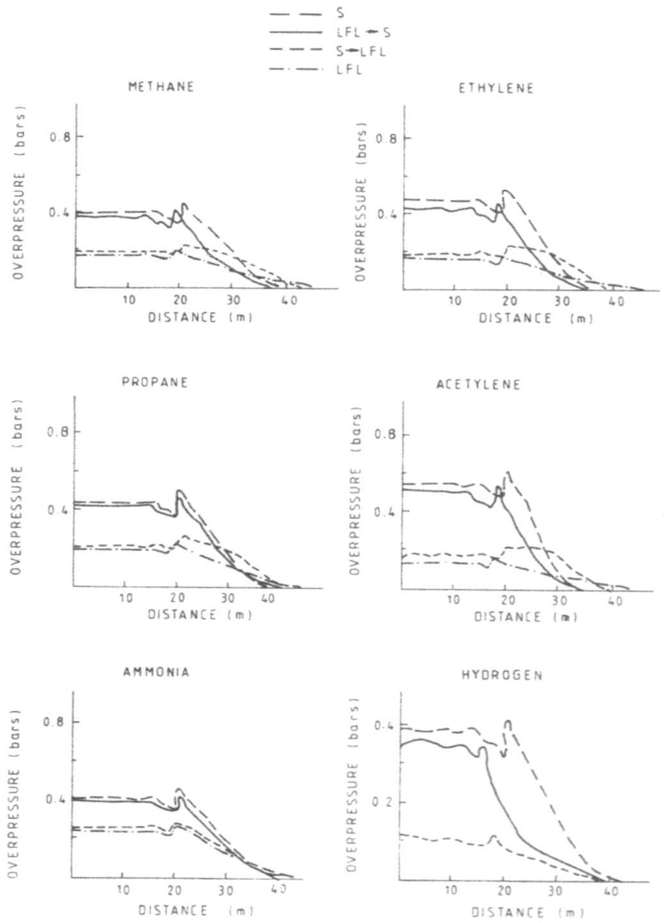

FIG. 2 OVERPRESSURE-DISTANCE PROFILES FOR DIFFERENT MIXTURES AT A BURNING VELOCITY OF 30 ms^{-1}

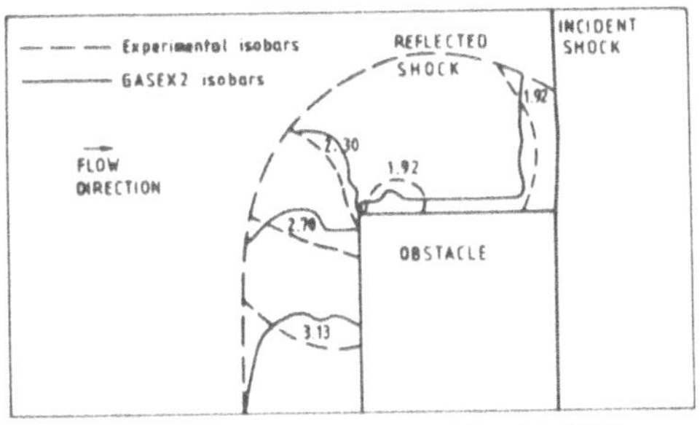

NUMBERS GIVE PRESSURE AS A RATIO TO AMBIENT PRESSURE.

FIG. 3 COMPARISON OF GASEX2 RESULTS WITH EXPERIMENTAL RESULTS FOR INTERACTION OF A SHOCK WITH A SQUARE OBSTACLE.

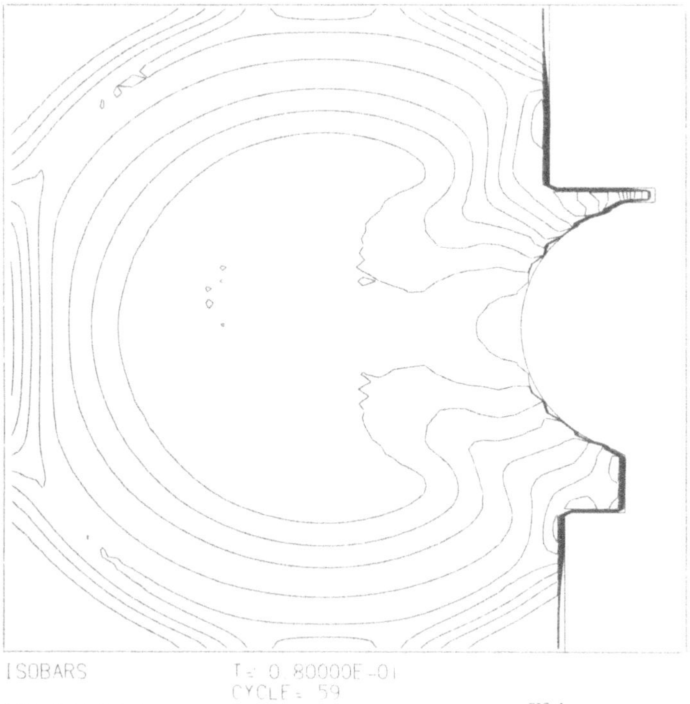

FIG 4

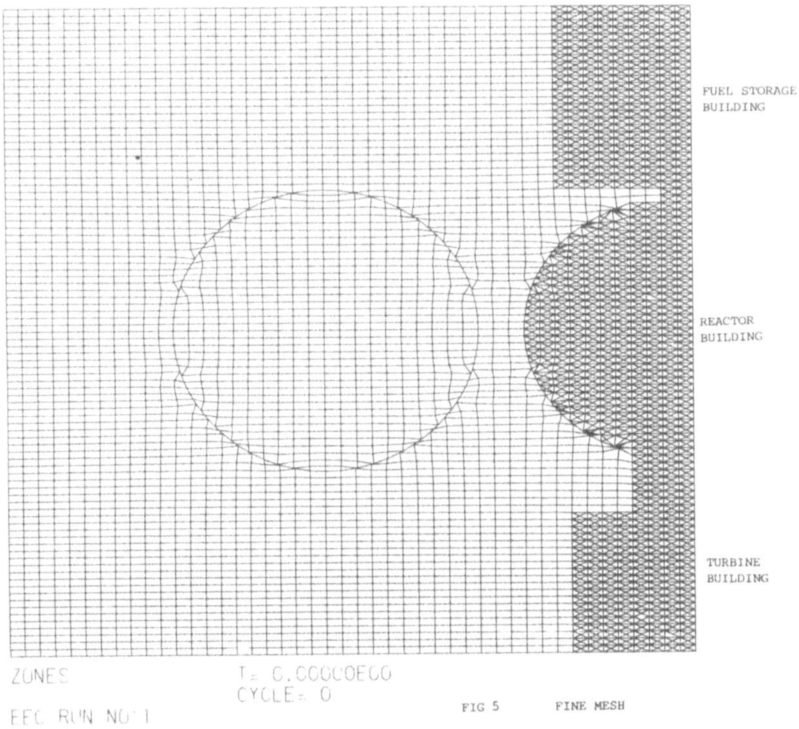

FIG 5 FINE MESH

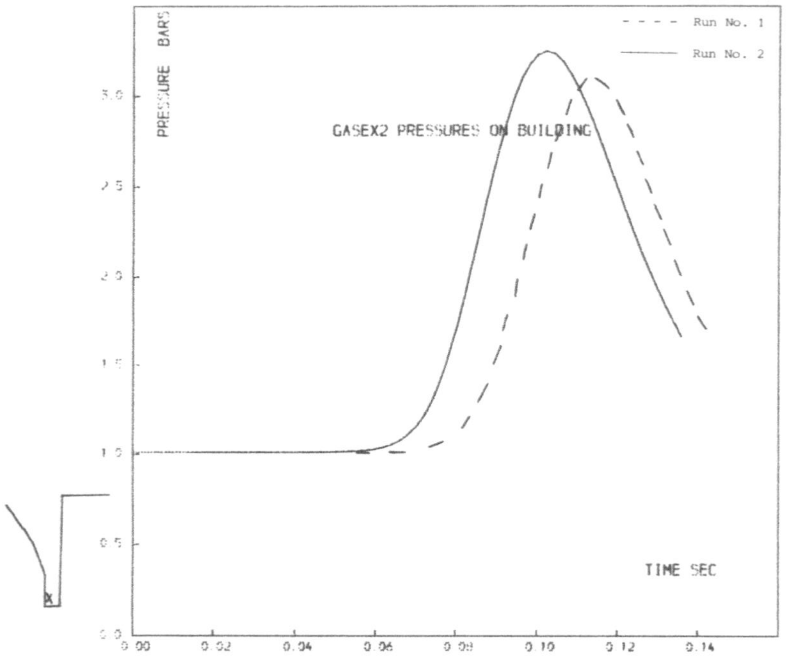

FIG 6

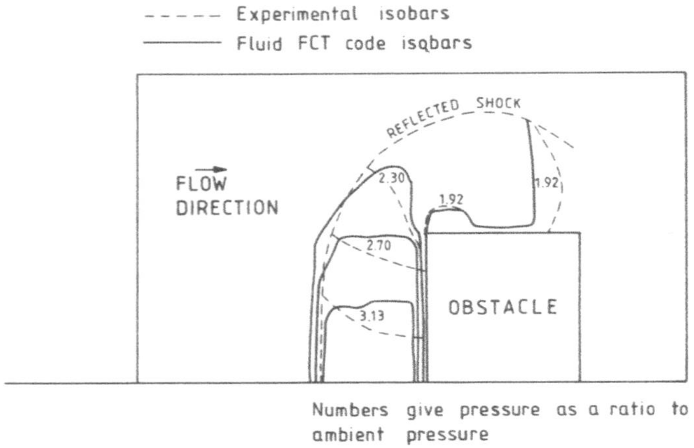

Fig 7 Comparison of FCT fluid code with experiment for interaction of shock on a block.

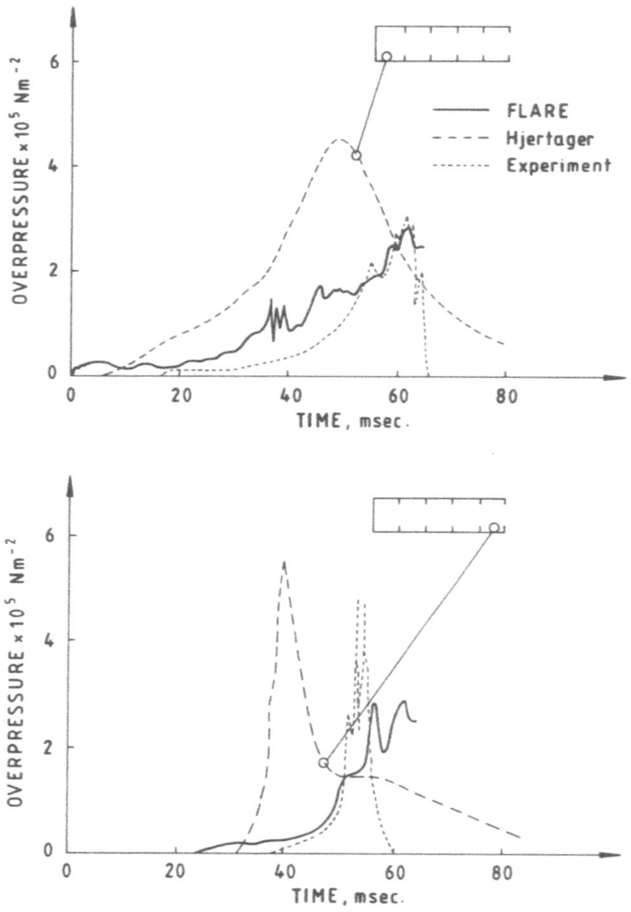

Fig. 8 Comparison of pressures in long pipe

Discussion

Question: J.A. Pappas - Norsk Hydro, Norway

In any computer code using a turbulence model for combustion, assumptions must be made on the initial values of the turbulence level or strength of ignition. In your calculations:

a) How strong an ignition did you chose?
b) How sensitive are the results to changes in the initial conditions?

Answer: S.F. Hall - UKAEA/SRD

In our calculations we started at a point where the flame filled the tube. This corresponds to a much stronger ignition than used in the experiments. The results do not appear to be very sensitive to initial conditions once the flame is established, i.e. once it has passed one or two obstacles and the flow-generated turbulence is controlling combustion.

Question: N.F. Scilly, HSE, London

Has the code been used to examine the interaction of blast waves on structures in stand-off positions?

Answer: S.F. Hall - UKAEA/SRD

Yes. Some results are presented in the final report.

Question: J. McQuaid, HSE

Has the GASEX code been compared with this same experiment? Despite the inapplicability of the constant burning velocity assumption in the GASEX code, would it nevertheless be possible to get agreement with an assumed value of burning velocity? How does this freedom to select the adjustable inputs (burning velocity or constants in K-ε model) compare between GASEX and FLARE?

Answer: S.F. Hall - UKAEA/SRD

GASEX has not been used to model the Hjertager experiment. By suitable choice of burning velocity it would be possible to get good agreement at a given point and at a given time, GasexI probably would not be able to reproduce either the pressure history at a point or the pressure profile at a given time.
In GasexI, the choice of burning velocity completely dominates all other parameters. In FLARE it seems that results are relatively insensitive to the choice of parameters in the K-ε model.

Question: A. Siebertz, CEN/SCK-Mol

How good is the eddy break-up model for modelling the flame phenomena? Do you see some improvements? Which ones?

Answer: S.F. Hall - UKAEA/SRD

At the time when the code was developed the eddy breakup model was really the only model available which had the potential to model turbulent flame propagation. Comparison with experiments shows that the eddy breakup model gives reasonable results, although there is obviously room for improvement. The two main areas of improvement necessary are in the modelling of the low turbulence/near-laminar flames and quenching. The modelling of quenching awaits some better understanding of the phenomena involved.

Question: J.R. Thomas - British Gas

1) Can you explain why the calculated results of Hertjager differ from your own where the method of approach seems similar. Was this due to the numerical scheme employed?

2) Professor Bradley at Leeds University has obtained experimental results of burning velocity as a function of turbulence which are available. Have you compared your predictions with these results?

Answer: S.F. Hall - UKAEA/SRD

1) I believe the differences are largely due to the difference numerical scheme employed. Other differences are the exact way in which ignition is modelled and some boundary conditions. These also may have some effect, but we have not performed a systematic study of the differences.

2) We have not compared our code with Prof. Bradley's results. It may, however, be valuable to do so.

VALIDATION OF BLAST SIMULATION CODES AND EXPERIMENTAL INVESTIGATION OF FLAME PROPAGATION IN THE PRESENCE OF OBSTACLES

J.P.ZEEUWEN, C.J.M.VAN WINGERDEN and A.C.VAN DEN BERG
Prins Maurits Laboratorium TNO, P.O.Box 45, NL-2280 AA Rijswijk

Summary

The research project performed by TNO can be divided into a theoretical and an experimental part. In the theoretical part three spherically symmetric blast simulation codes were used. The codes were sucessfully validated by simulating self-similar flow fields in front of constant velocity flames and small- and large-scale explosion experiments. One of the codes was adapted to simulate blast due to explosions of pancake-shaped clouds. Although the exercise was not unsuccessful, more experimental verification is needed to substantiate the conclusions drawn.
The experimental part was concerned with the study of flame propagation in the presence of repeated obstacles and partial confinement. The obstacle set-up consisted of parallel horizontal tubes, similar to a "pipe-rack". Partial confinement was introduced by mounting a horizontal plate over the obstacles. Many parameters were varied including the scale of the experiments. The main conclusion from the research was that although "pipe-racks" cause some flame acceleration really high flame speeds and therefore damaging overpressures can only be generated by introducing partial confinement.

1. INTRODUCTION

Since external gas cloud explosions may pose a threat to nuclear installations, the Commission of the European Communities decided to incorporate in the 1979-1983 indirect action programme "Safety of thermal water reactors" an area on the protection of npp against external gas cloud explosions.

Within this area TNO has performed a project consisting of two theoretical and one experimental parts:

- validation of 1-dimensional codes for computation of pressure waves caused by spherically symmetric unconfined vapour cloud explosions
- evaluation of the use of 1-dimensional codes as an approximation for computation of pressure waves caused by non-spherical unconfined vapour cloud explosions
- experimental determination of the behaviour of flames in the presence of obstacles.

A full account of the research is given in the final report, consisting of two parts: part 1 [1] reports on the theoretical research while part 2 [2] deals with the experimental research. In this paper a brief account of the research will be presented, emphasizing the outline of the project and the main results and conclusions.

2. VALIDATION OF 1-D COMPUTATIONAL METHODS

2.1. Outline of the project

All hitherto proposed methods to estimate the blast from an unconfined gas cloud explosion were based on a 1-D approximation. The 1-D codes were evaluated in the past, mainly by comparing the outcome of complex methods with those of more simple methods. A validation of 1-D codes using experimental results therefore seemed valuable.

Experimental results of spherically symmetric explosions were available from small-scale soap bubble explosion research performed by the University of Poitiers/ENSMA under CEC study contract [3] and were to be obtained from large-scale balloon experiments by CEA-EdF [4].

Three different computer codes were to be used for the validation:

- A simple piston model

The flow field generated by a gas cloud explosion is simulated by a piston movement and computed by the method of characteristics.
For low flame speeds a simple and very straightforward relation between flame path and piston path exists. The model is described in [5].

- The CLOUD-code

This code, initially developed by Oppenheim and his coworkers [6], brings about a solution of the Lagrangean conservation equations by a very straightforward finite difference scheme.
Sharp gradients and shock waves are described employing an artificial viscosity term in the equations. This artificial viscosity, however, is the cause of considerable numerical diffusion.

- The BLAST-code

This code, developed at the Prins Maurits Laboratory TNO [7], solves the Eulerian conservation equations using a Flux-Corrected-Transport-(FCT)-algorithm. A gas cloud explosion is simulated by energy addition within a Lagrangrean grid. The FCT solution is characterized by minimal numerical diffusion.
All three codes allow flame speed manipulation.

2.2. Summary of the results

To begin with flow fields in front of constant velocity flames were considered:

- The Piston model reproduces self-similar flow fields in front of constant velocity flames reasonably well for flame speeds lower than 100 ms^{-1}. At higher flame speeds self-similar pressure fields are less well approximated due to the a priori assumption of isobaric expansion in the model.
- Both CLOUD and BLAST accurately reproduce self-similar flow fields in front of constant velocity flames. In order to prevent strong fluctuations in the flow field parameters due to finite difference modelling of energy addition in the codes a minimal flame speed of 100 ms^{-1} is required.

Simulation of explosion experiments yields information on the dynamic properties of the models. Small-scale soap bubble experiments and large-scale balloon experiments have been considered. The models were only used in their flame speed range of validity as found in comparison with self-similar flow field.

- Broadly all models reproduce the measured pressure-time histories rather well. One example is given in Fig.1, where the pressure pulse measured for a deflagrating large-scale balloon filled with an homogeneous gas mixture is compared to the simulated pulse generated using the Piston model. The differences in shape, duration and time of arrival of the pressure waves must be imputed to the limited accuracy (due to flame wrinkling, buoyancy and asymmetry of the flame sphere in the experiments) of the input data.

- For the same reason it was inevitable that the measured maximum overpressures were systematically overestimated by the models. This is shown in Fig.2 where the maximum overpressures generated by BLAST are compared with those measured for soap bubble explosions. Maximum peak overpressures generated by the BLAST code systematically exceed those generated by the CLOUD code. Moreover BLAST solutions display more wave details than the CLOUD solutions. These differences are due to the more optimized use of numerical diffusion in BLAST.

- The blast decay described by the Piston model is in agreement with the acoustic law. In some of the simulated experiments blast decay was less strong, probably due to buoyancy and asymmetry.

The blast decay described by BLAST is in agreement with experiments and acoustic law. CLOUD exhibits a somewhat stronger decay due to strong numerical damping.

2.3. Discussion

Based on the extensive evaluation performed in this project it can be assumed that, within the range of validity, the codes can be used with confidence to model spherically symmetric gas cloud explosions with both steady and non-steady flames.

Whenever an experimentally derived flame speed is implemented in the code it must be borne in mind that the resulting pressure wave is very sensitive to details in the flame path and that therefore the accuracy of the computed pressure pulse is much less than the accuracy of the flame path measurement.

With the 1-D codes validated as a result of this project, the main emphasis should now be placed on the determination of the flame speed to be expected in accidental situations.

3. EVALUATION OF THE USE OF 1-D COMPUTATIONAL METHODS TO MODEL NON-SPHERICALLY SYMMETRIC EXPLOSIONS

3.1. Outline of the project

In reality a gas cloud explosion will not be spherically symmetric for a variety of reasons: cloud shape and gas concentration distribution will be very complex, ignition will not occur in the centre and often at more than one location and the topography will be irregular. To model such explosions it is necessary to have multi-dimensional codes.

On the other hand, far away from the explosion the flow field will always be sphere-symmetric which makes it possible in principle to use 1-D codes for some purpose: by manipulating the energy release rate the influence of cloud shape, ignition location etc. on the resulting pressure pulse in the far field can be studied.

In order to evaluate the validity of this approach, the second part of the project included a comparison of experimental results with predictions obtained in the above manner.

The computer codes to be used were those described in the previous chapter.

Experimental results were available from small-scale soap bubble explosions performed by the University of Poitiers/ENSMA [8]. Also results from medium-scale experiments performed in the present project (see next chapter) were used as far as possible.

From the experiments only results for low-flame-speed pancake-shaped explosions were available with sufficient detail to attempt simulation. For this reason only the Piston model could be evaluated.

3.2. Summary of the results

Evaluation of the Piston code was performed with respect to the capability of the simulation of blast due to explosions of pancake-shaped clouds. For this purpose the Piston model had to be modified.

- For constant low velocity flames Piston model results were compared to acoustic solutions, which showed that the modified Piston model could be regarded as an acoustic model with the possibilty of handling every imaginable flame path.
- Simulation of experiments showed that the shape and duration of the measured pressure waves were reproduced quite well.

The maximum overpressure of the cylindrical phase of flame propagation appeared to be in reasonable agreement with experimental results.

The initial phase of spherical flame propagation was simulated less well, probably for want of detailed flame path data and due to experimental conditions. Fig.3 shows a comparison of simulation and experiment: in this case the spherical phase is reproduced well because some experimental observations were used to modify the input data. The large peak at the end of the experimental pulse is caused by a flame acceleration caused by a support used in the experiments. The input data did not contain this acceleration.

- Blast decay described by the Piston model is less strong than in the experiments. The way in which blast decay is modelled is not realistic as a result of the asymmetry of the exploding pancake-shaped cloud.

3.3. Discussion

From the evaluation it emerges that there is a distinct possibility that the modified Piston code can be used for such limited applications as the study of the influence of source characteristics on the shape of the blast wave. Definite proof has not been found due to the lack of suitable data.

Since the Piston model does not model the blast decay close to the asymmetrical cloud in a realistic way, it cannot be used to model wave propagation outside the cloud in these cases.

Unfortunately it has not been possible to evaluate the two high-flame-speed codes CLOUD and BLAST in this respect.

4. EXPERIMENTAL STUDY OF FLAME PROPAGATION IN THE PRESENCE OF OBSTACLES

4.1. Outline of the project

One of the main topics of research in the gas cloud explosion field is and has been the determination of what causes flames to accelerate in gas cloud explosions. Of the many mechanisms proposed the presence of obstacles and confining structures in the gas cloud has received much attention.

In small-scale soap bubble experiments performed by the University of Poitiers/ENSMA [9] repeated horizontal obstacles were investigated. Due to the soap bubble technique only obstacles resting on the ground plate could be investigated and only a small number could be placed inside the soap bubble.

It was proposed as part of this project to complement these experiments by experiments inside a closed explosion vessel of 1-m^3 capacity, designed to withstand 100 bar overpressure. By placing a ground plate and obstacle array in the centre of the vessel, more freedom as to experimental conditions was obtained at the expense of blast wave measurements and some, possibly disturbing, reflections from the vessel wall.

Later during the project, an extension of this part was granted by CEC to perform medium-scale experiments in the open, mainly to study the influence of scaling.

In the experiments the obstacles were always round parallel tubes, of diameter 10, 20 or 30 mm (small-scale) and 100 mm (large-scale) parallel to the ground surface and supported at the same height from the ground ("pipe-rack" type configuration). Partial confinement was added in some of the tests by placing a (closed or perforated) plate over the obstacles.

It should be noted that these experiments are closely related to another programme performed partly in parallel to this research, in which another obstacle set-up was studied, viz. repeated vertical cylindrical obstacles on a small scale [10] and a large scale [11] and where scaling proved possible [12].

In the present programme a large number of parameters are varied, some of which are depicted in Fig.4.

4.2. Summary of results

A large number of tests were performed on two different scales. Using an asymmetrical obstacle array consisting of horizontal tubes a maximum flame speed of about 50 ms^{-1} and a maximum overpressure within the array of 2,5 kPa was established for ethylene-air mixtures. When partial confinement was introduced flame speeds of up to 310 ms^{-1} in the case of central ignition and 420 ms^{-1} in the case of edge ignition were found.

A maximum overpressure of 75 kPa (edge ignition) could be measured. In the case of central ignition the maximum overpressure amounted to 70 kPa. By varying the obstacle parameters the following main results were obtained:

With the introduction of obstacles (without partial confinement) various parameters could be studied.

By varying the obstacle separation (pitch P) an optimum region was found for P > 2,25 tube diameter (investigated up to 4,5 tube diameter).

By varying the gap between obstacles and ground plane an optimum was found when the obstacles were raised some distance (0,2 tube diameter) from the ground. With smaller and also with larger gaps flame speeds were lower (see Fig.5).

When obstacles were introduced between two plates flame speeds were lowest when the obstacles were fixed to one of the plates. When the obstacles were at a small distance from one of the plates flame speeds increased but slightly decreased again when they were mounted in the centre between the two plates (this effect was noticed on a large scale only).

By using different perforated top plates the degree of confinement could be varied. Maximum flame speeds were found when the top plate was closed. For all porosities (open areas) of the top plate exceeding 25% of the surface area flame speeds are alike.

If the height of the (closed) top plate is reduced flame speeds increase dramatically (see Fig.6).

In spite of side-relief terminal flame speeds resulting from edge ignition are exceeding those resulting from central ignition because the length of flame travel is doubled.

For tests with confined obstacles as well as for tests with unconfined obstacles non-dimensionalized flame speed results (using the laminar flame speed as a parameter for scaling reactivity) are alike for different types of gas.

Blast measurements performed on a large scale revealed a stronger than acoustic decay of blast with distance close to asymmetrical obstacle arrays.
In Fig.7 the shape of the blast waves generated is shown. This Figure also shows how the amplitude of the blast wave is changed by the introduction of partial confinement.

4.3. Discussion

In the tests a large number of parameters was varied. For the majority of the parameters considered in the present investigation effects on flame propagation were found to be similar in small- and large-scale experiments.

The overall conclusions from the investigation are in agreement with those of other research performed recently with obstacle arrays of different set-up [10-12].

a. Partial confinement plays an important role in vapour cloud explosions. In the set-up in this investigation the effect of confinement is dramatic:
   - maximum flame speeds in the presence of obstacles and partial confinement are about six times the flame speeds in the absence of confinement.
   - partial confinement alone (no obstacles) results in flame speeds that are twice as high as the maximum flame speeds in the presence of obstacles but without confinement.
b. Simple scaling of flame propagation in obstacle environments is possible as long as flame speeds are relatively low (i.e., when molecular transport processes are dominating). This is exemplified for one set-up in Fig.8, which shows that geometrical factors (obstacle diameter and initial flame speed) are sufficient for scaling.

In view of the fact that many configurations have been investigated at two different scales, it can be considered that for the time being sufficient experimental data are available for this particular type of obstacle set-up.

5. CONCLUSIONS

a. The main conclusions from the project are:
   a.1. One-dimensional computer codes for blast calculation have been validated using experimental results from spherically symmetric explosion tests over a range of conditions. Broadly speaking computations are in good agreement with the experiments.
   a.2. The use of one-dimensional computer codes to model non-spherical explosions is justified for certain purposes. Without further experimental verification care should be taken not to exceed the limits of applicability outlined in [1].

a.3. Flame propagation in the presence of repeated obstacles, especially when confinement is present, may generate blast of significant strength. Such structures must therefore be avoided in the vicinity of nuclear installations if the presence of flammable clouds cannot be excluded.

b. For future research the following recommendations follow immediately from this project:

b.1. Emphasis should be placed on the determination of the flame speed to be expected in case of accidents. Results of investigations such as the experimental part of this project are valuable input.

b.2. The evaluation of the use of 1-D codes to model multi-D-explosions should be extended as soon as detailed experimental data become available.

REFERENCES

[1] VAN WINGERDEN, C.J.M. and VAN DEN BERG, A.C., Prins Maurits Laboratorium TNO, report PML 1984-C65 (1984).

[2] VAN WINGERDEN, C.J.M., Prins Maurits Laboratorium TNO, report PML 1984-C67 (1984).

[3] CHAMP, M., MONTANELLI, T., DESHAIES, B., LEYER, J.C., Université de Poitiers, Laboratoire d'Energétique et de Détonique de l'ENSMA, Rapport final d'étude contrat CCE 463/78/9/ECI/F (1979).

[4] To be presented at this seminar.

[5] VAN DEN BERG, A.C., Prins Maurits Laboratorium TNO, report PML 1982-107 (1982).

[6] OPPENHEIM, A.K., Conference on Mechanisms of Explosions and Blast Waves, Yorktown, Virginia (1973). Proceedings paper no.I.

[7] VAN DEN BERG, A.C., Prins Maurits Laboratorium TNO, report PML 1980-162 (1980).

[8] LEYER, J.C., Combustion and Flame, Vol.48, pp.251-263 (1982).

[9] LEYER, J.C., Université de Poitiers, Laboratoire d'Energétique et de Détonique de l'ENSMA, Document Intermédiaire, Contrat d'Etude CCE no.572/79/9//ECI/F (1980).

[10] VAN WINGERDEN, C.J.M. and ZEEUWEN, J.P., J. of Haz. Mat. Vol.8, pp.139-156 (1983).

[11] ZEEUWEN, J.P., VAN WINGERDEN, C.J.M., DAUWE, R.M., 4th International Symposium on Loss Prevention and Safety Promotion in the Process Industries, Harrogate (1983). Proceedings (IChemE Symp. series no.80), pp. D20-D29.

[12] ZEEUWEN, J.P. and VAN WINGERDEN, C.J.M., Presented at the 9th International Colloquium on Dynamics of Explosions and Reactive Systems, Poitiers (1983).

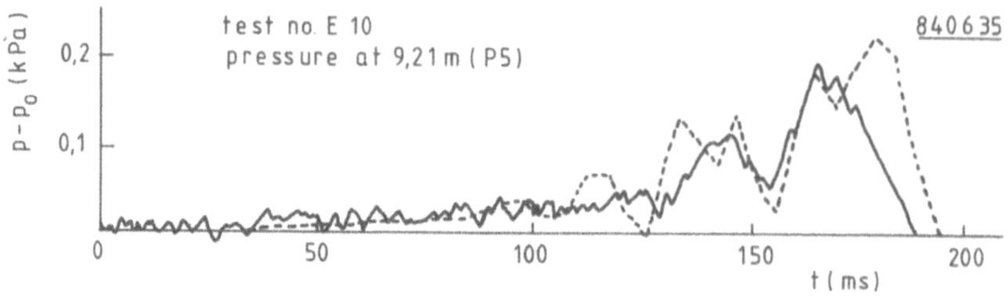

Figure 1. Simulation using the Piston model (dashed line) and measurement (full line) of pressure pulse (overpressure $p-p_o$ as a function of time t) due to deflagrating spherical cloud (large-scale balloon test, homogeneous gas mixture).

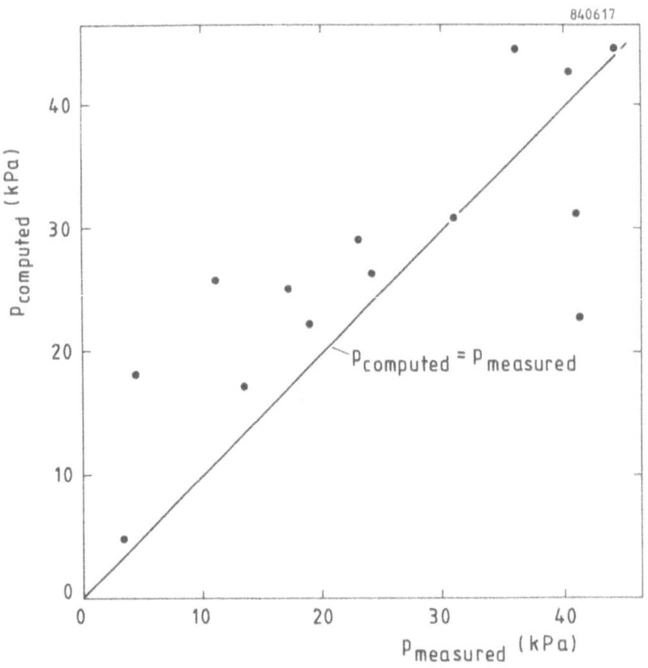

Figure 2. Maximum overpressure due to soap bubble explosions simulated by BLAST ($p_{computed}$) as a function of measured maximum overpressures ($p_{measured}$) at 0,34 m from the explosion centre.

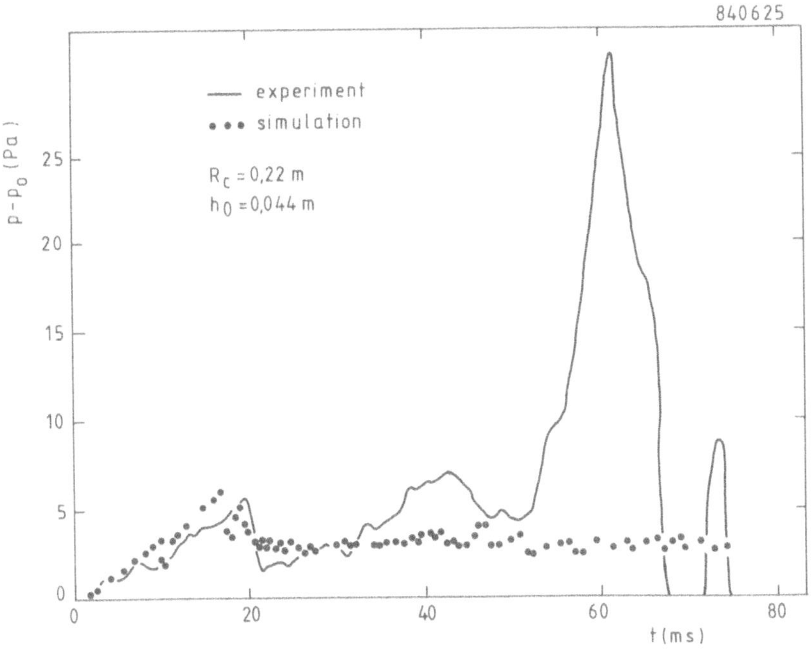

Figure 3. Simulation of a pressure wave due to an exploding small-scale pancake-shaped cloud.
(R_c = initial cloud radius, h_o = initial cloud height, measurement at 0,34 m from explosion centre).

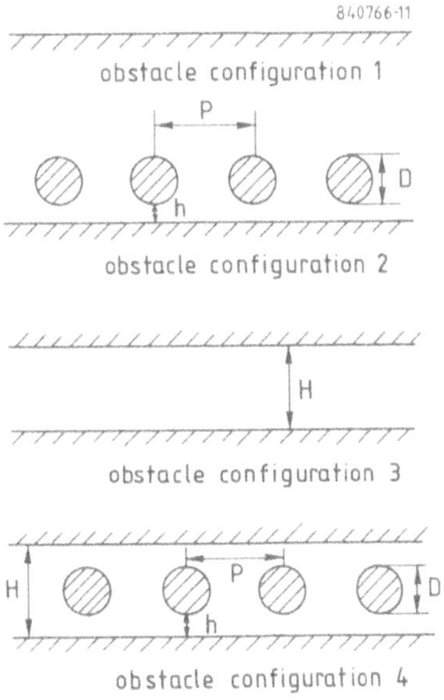

Figure 4. Overview of parameters varied in four obstacle configurations in the experimental investigation.
P = pitch (distance from centre to centre between obstacles),
D = obstacle diameter,
h = distance between obstacle and ground,
H = distance between top plate and ground.

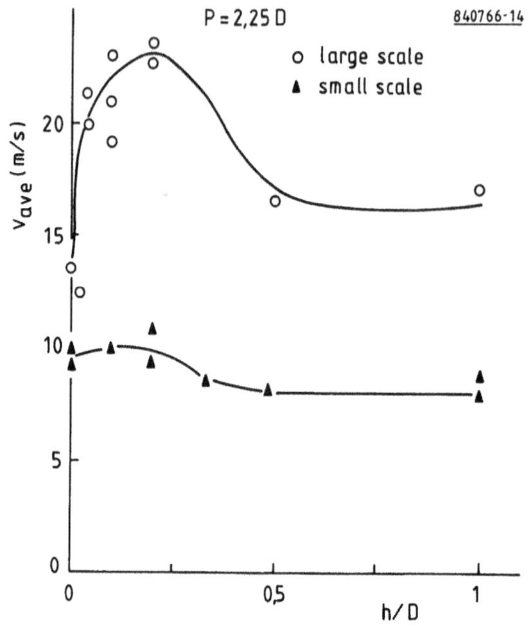

Figure 5. Effect of gap h between obstacles and ground plane on average flame speed v_{ave} (Pitch P = 2,25 D; no partial confinement; 7-8 vol. % ethylene-air).

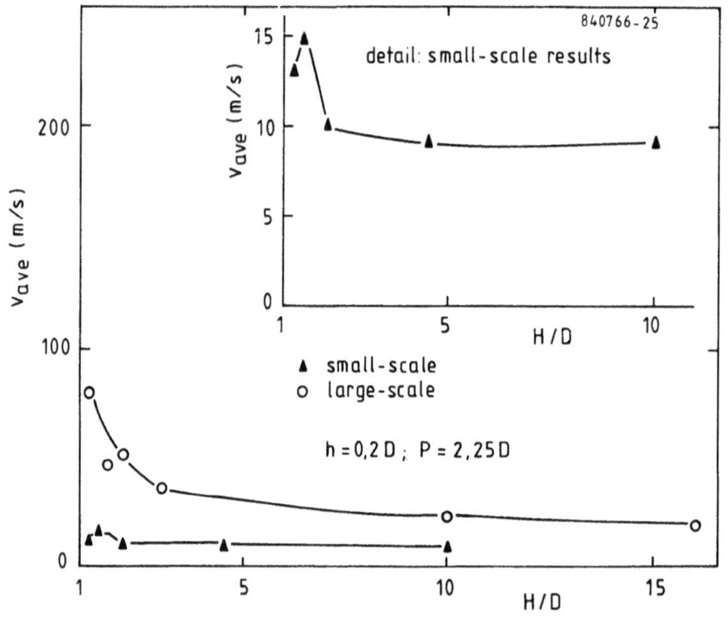

Figure 6. Influence of top-plate height on flame speed in a double plate configuration with obstacles (h = 0,2 D; P = 2,25 D; 7-8 % v/v ethylene-air).

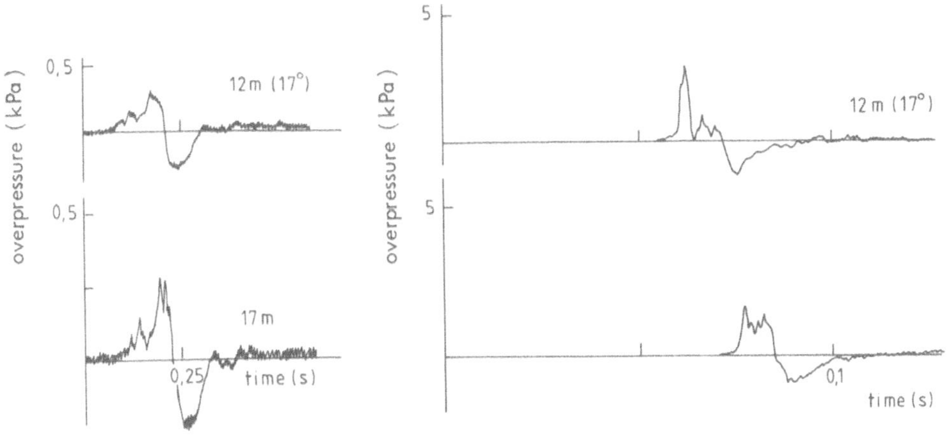

Figure 7. Blast overpressure at 12 m and 17 m resp. from the ignition source:
a: obstacle array over ground plate, no partial confinement (cloud height = 1/2 cloud radius),
b: obstacle array between two plates (H = 1,2 D),
(h = 0,2 D; P = 2,25 D; 7-7,5 % v/v ethylene-air).

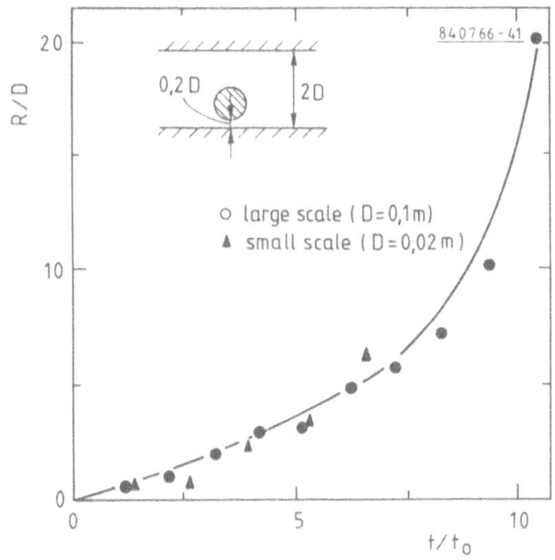

Figure 8. Simple scaling of vapour cloud explosion experiments. Partially confined obstacles (H = 2 D; h = 0,02 D; P = 2,25 D; 7-8 % v/v ethylene-air).
R = flame radius,
t_o = reference time scale = obstacle radius * inital flame speed.
The initial flame speed incorporates laminar flame speed (i.e. type of gas) and scale effects.

Discussion

Question: J.R. Thomas, British Gas

In your last slide you stated that simple scaling in combustion systems is possible provided the flame speed is low.
How low must this be?

Answer: J.P. Zeeuwen - TNO

The real criterion for the validity of simple scaling is not a value of the flame speed. However, as an order of magnitude a maximum flame speed of 100 m/s can be mentioned.

CHARACTERIZATION OF THE PRESSURE FIELD INDUCED BY THE EXPLOSIONS IN AIR OF A HYDROCARBON-AIR MIXTURE, WITH SLOW DEFLAGRATION OR FAST DEFLAGRATION

J. BROSSARD (1) - D. DESBORDES (2) - N. DI FABIO (3) - J.L GARNIER (4)
A. LANNOY (5) - J.C LEYER (2) - J. PERROT (6) - J.P SAINT-CLOUD (2)

(1) Université de Bourges-Orléans
(2) Université de Poitiers
(3) EDF-Direction de l'Equipement
(4) CEA-Département d'Analyse de Sûreté
(5) EDF-Direction des Etudes et Recherches
(6) CEA-Centre d'Etudes Scientifiques et Techniques d'Aquitaine.

Summary

The present research contract, carried out in the framework of a French working group CEA-EDF-ENSMA, was divided into four phases :

1) Phase 1 : Scaling effects on a pressure field generated by spontaneous accelerations of deflagrations in a homogeneous medium : these "CHARLES" tests performed on completely unconfined charges of ethylene-air and acetylene-air mixtures ($V \simeq 12\ m^3$) have demonstrated the high sensitivity of the pressure field to the flame acceleration, particularly at the end of the propagation.
2) Phase 2 : Effect of a sudden discontinuity in the concentration of the combustible gas on the deflagration speed in a heterogeneous medium : this discontinuity was obtained using two concentric latex balloons, filled with different hydrocarbon concentrations and also destroyed before firing.
3) Phase 3 : Study of pseudo-detonation modes for the explosion of spherical ethylene-air mixtures : these modes were sought by increasing an explosive plastic mass associated to the igniter.
4) Phase 4 : Study of the influence of cloud shape, ignition point location and simple obstacles in the flame path on the pressure field generated by the explosion.
   A retractable enclosure was designed, built and unfortunately destroyed by the second test.

1. INTRODUCTION

In the design and construction of nuclear power plants, it is necessary for safety reasons to take into account agressions from the outside, such as those of natural origin (earthquakes, floods...) and also agressions associated with human activities (falling aircraft ...). Among the latter group, dangers arising from the industrial environment of the power station or from the transport of dangerous materials in its vicinity must be considered. These take the form of a massive release of a dangerous product, generally a hydrocarbon, followed by the formation of a flammable air-hydrocarbon cloud. Depending on the weather conditions, such a cloud may drift, and if it is ignited, the pressure wave generated by the explosion may cause serious damages, even at considerable distances. In order to protect the safety functions of the plant, it has appeared necessary to assess the overloads that must be built into the structures.

About the year 1975, knowledge of the explosions of air-hydrocarbon mixtures in free atmosphere was not sufficient to develop a model representative enough for safety analysis. Consequently, a French working group, including people from the Commissariat à l'Energie Atomique (Safety Analysis Department, the Aquitaine Scientific and Technical Study Centre), Electricité de France (Equipment and Studies and Research Divisions), and the Universities of Bourges-Orléans and Poitiers, was set up to study and describe the phenomena of free-atmosphere explosions so as to provide the elements necessary for developing a safety rule applicable to the design of nuclear power stations. Although, in a real accident, the detonation of an air-hydrocarbon cloud is unlikely, the working group first undertook experimental studies on the detonation of gaseous mixtures, since this can be considered, at least in terms of the intensity of the pressure peak transmitted to the environment, as an intensification of the effects of deflagrations.

These tests lead to a methodology that could be used to predict the consequences of an accidental explosion considered as a detonation [1].

However, a detonation in free space is unlikely under actual conditions. Indeed, to obtain it, the experiments had to be carried out on mixtures for which detonation was ensured (sufficient ignition energy,reactive mixtures, optimum composition).

It was therefore decided, in the context of a collaboration agreement with the European Community Commission, to launch an experimental programme on variable-velocity deflagration in free atmosphere of an air-ethylene mixture.

The present paper describes the main results obtained in the framework of this collaboration (contract SRF 005-F.(S)), completing a previous communication [2].

2. PHASE 1 : deflagrations in a homogeneous medium : [3 and 4]

Phase 1 concerns explosions of spherical latex balloons ($V \simeq 12 m^3$, that is $R_o \simeq 1.4$ m), filled with an air-ethylene or air-acetylene mixture, with equivalence ratio in hydrocarbon ranging from 1 to 1.3. To obtain a strictly unconfined cloud, the latex balloon is destroyed with a shot-gun device, before firing the mixture at the centre of the cloud (fig.1).

At first the flame propagates with a constant speed, characteristic of the mixture (with acetylene $V_F \simeq 20$ m/s, with etylene $V_F \simeq 10$ m/s) out to a radius of about 1.6 m. Then the flame accelerates out to its terminal ($R_{F\,max} \simeq 2$ to 2.6 m) during the burning of the disturbed area at the border between the mixture and the environmental air, induced by the retraction of the latex wall. For example, for a homogeneous mixture of 9% ethylene in air, the constant speed is egal to 10 m/s and the maximal velocity $V_{F\,max}$ is 16 m/s, whereas the corresponding overpressures are 3 and 5 mbar (at $\frac{R}{R_o} = 2.3$).

A synthesis of results is given in table I. These results have been compared with those of an acoustical analysis previously carried out by ENSMA linking the overpressure $\Delta p(R,t)$ generated to the flame path history $R_F(t)$. R being the observation distance from the ignition point, if $R \gg R_o$ (R_o = initial radius of the explosive charge), the overpressure is [5] :

$$\Delta p(R,t) = \frac{\rho_o}{R} (1-\alpha) \left[2 R_F(\tau) V_F^2 + R_F^2(\tau) \frac{dV_F}{dt} \right]$$

with $\tau = t - \frac{R}{C_o}$

Where ρ_o and C_o are respectively the air density and sound velocity in the environmental air, α the ratio $\frac{\text{burnt gases density}}{\text{unburnt gases density}}$ (inverse of expansion ratio).

An example of this comparison is also given on fig. 2. The agreement between experience and computation is good.

Fig. 3 shows the field acceleration of the flame due to the presence of the disturbed zone induced by the withdrawal of the balloon.

Furthermore tests performed without preliminary destruction of the balloon evidenced that the velocity $V_{F\ max}$ and the pressure Δp_{max} are respectively 1.5 and 2.5 times the values above.

3. PHASE 2 : deflagration in a heterogeneous medium [6 and 7]

Phase 2 concerns explosions in a medium constituted by two mixtures with different hydrocarbon concentrations (ethylene only during this phase). Two spherical concentric latex balloons limit the two media. As above they are destroyed by the shot-gun device, before firing at the centre (fig. 1)

Four concentrations of air-ethylene were used : 5%, a lean mixture ; 6.6 near the stoichiometry (6.54%) ; 8.5% slighthy enriched ; 13%, a rich mixture.

The external balloon radius was always $R_o = 1.4$ m ($V_o \simeq 12$ m^3).

The internal balloon radius was either $R_i = 0.5$ m ($V_i \simeq 0.5$ m^3) or $R_i = 0.9$ m ($V_i \simeq 3$ m^3), to vary the values of flame front radius R_{FT} where the flame speed changes because of the change in concentration. After two tests in a homogeneous medium (air-ethylene mixtures at 5 and 13% in ethylene) to measure the flame velocity under these difficult conditions (V_F observed $\simeq 5$ to 6 m/s), a programme of eleven tests was carried out. The main results are given in table II.

This CHARLES 2 test campaign showed that the flame history R_F (t), because of the presence of a step in the combustible mixture concentration, was less reproducible and less predictable than that observed during CHARLES 1 tests in a homogeneous medium. In the CHARLES 2 tests, the acceleration is most often situated after the transition between the two compositions. Its location may be nevertheless influenced, in some cases, by the disturbance induced by the ignition and/or the internal envelope collapse. This is particulary true when the internal mixture is rich (13%) : the flame propagation is then fluctuating and several accelerations of greater or lesser intensity may occur. When the envelopes are not destroyed prior to ignition, the flame history R_F (t) is comparable to those observed in CHARLES 1 tests : the acceleration, particularly intense in this case, occurs at the end of the propagation. Here the maximum propagation velocities observed at the end of acceleration are higher than those in uniform air-ethylene mixture : 11 to 18 m/s in CHARLES 1 tests, 22 to 28 m/s in CHARLES 2 tests. It is to be noted however that these maximum values $V_{F\ max}$ are relatively independent of the composition couple chosen. Thus the presence of a concentration step seems globally to result in a multiplication factor of about 1.5 being applied to the maximum velocities for a uniform composition mixture with an equivalence ratio of about 1.2 (fig. 5) (16 m/s in CHARLES 1 tests instead of 25 m/s in CHARLES 2 tests).

It ensues that maximum overpressures are systematically higher than those corresponding to CHARLES 1 tests. The multiplication factor is approximately 2 to 3. The gap between the two series of values is due not only to the difference in maximum velocities, but also to the effects of the accelerations. The location of the acceleration plays a prominent part : the farther it occurs from the ignition point (R=0) the greater its contribution to the enhancement of the overpressure is. It should be added that the validity of the correlation model between the flame path R_F (t) and the induced pressure field $\Delta p(R,t)$ above, used in the interpretation of CHARLES 1 tests, was confirmed by the analysis of the CHARLES 2 campaign results. (fig. 4).

4 PHASE 3 : characterization of pseudo-detonation modes [8 and 9] :

The mixture tested is still an air-ethylene one, with an ethylene concentration of 8.5%.

The balloons (only one balloon in each test for phase 3) are spherical. Their volume is 12 m^3 ($\simeq$ 2.8 m diameter). The test balloon is destroyed before firing the mixture by the same shot-gun device that was used during phases 1 and 2. The ignition, is effected by a low energy igniter, boosted by a plastic charge. This charge, 5 g in the initial test, was increased until detonation was obtained, then decreased again to obtain deflagration. In a certain range (6 to 8.5 g) the regime induced is fairly aleatory (Table III and Fig. 6). The maximum flame velocity observed with the deflagration regime is about 32 m/s, 2.5 times the values observed in the CHARLES 1 tests. Table IV shows the overpressures obtained during some deflagration tests.

If one assumes that the flame front moves ahead with a constant velocity equal to the maximal value actually observed during the test, the use of the constant velocity model is generaly in good agreement with the experimental results. (fig. 7)

5 PHASE 4 : influence of cloud shape, ignition point location, obstacles :

A parallelepipedic-shaped (6.3 x 6.3 x 3.15 = 125 m^3) retractable canvass enclosure was designed and built (fig. 8). The walls of this enclosure are evacuated 300 ms before firing the explosive mixture (air-ethylene, $\simeq$ 8,5% in ethylene).

One test was performed on August 10th, 1984 with a low energy ignition and a deflagration regime was obtained. The flame velocity was about 10 m/s in the domain of constant speed, but reached 20 m/s at the end of the propagation, due to the same phenomena as those previously observed in phase 1 (turbulence at the limit of the hydrocarbon cloud). The recorded overpressure peak was $\simeq$ 1 mbar (at $\frac{R}{R_0}$ = 3). One other test was launched on August 28th, 1984 with the same mixture ignited at the centre of the lower face of the enclosure (centre of symmetry for a cubic volume of 6.3 x 6.3 x 6.3 m^3). Certainly because of an energy confinement due to the presence of the ground, a detonation regime was obtained .... and all the experimental facilities were destroyed.

6. CONCLUSION

The originality of this experimental work was that the mixtures tested were actually unconfined, exactly like a real cloud.

Flame accelerations, inducing overpressure peak increase, were shown to exist, both at the boundary of the cloud where the mixture meets the surrounding air through turbulence vortices and also at the interface between two layers with different concentrations.

A correlation between pressure and flame history using an acoustic model has been confirmed. Nevertheless, the prediction of the mechanical effects of a deflagration requires a complete and accurate knowledge of the flame history.

REFERENCES

1. BROSSARD, J et al. (1983) -Air blast from unconfined gaseous detonations IX ICODERS - Poitiers.
2. BROSSARD, J et al. (1983) - Experimental analysis of unconfined explosions of air - hydrocarbon mixtures - Characterization of the pressure field - 4th International Symposium on Loss Prevention - Harrogate.
3. PERROT, J (1981) - Essais CHARLES 1 - Déflagration de mélanges homogènes en milieu non confiné - Contrat CEEA/CEA - SRF/005 F - (S).
4. LEYER, J.C et al. (1981) - PHASE 1 - Rapport d'interprétation des essais "CHARLES" (déflagration lente) - Contrat CEEA/CEA - SRF/005 F - (S).
5. DESHAIES, B and LEYER, J.C (1981) - Flow field induced by unconfined spherical accelerating flames - Combustion and flame 40 (p. 141 to 153).
6. PERROT, J (1983) - Essais CHARLES 2 - Déflagraton de mélanges homogènes à concentration variable - Contrat CEEA/CEA - SRF/005 F - (S).
7. DESBORDES, D et al. (1983) - PHASE 2 - Rapport d'interprétation des essais CHARLES 2 : mélanges homogènes de concentrations variables.
8. PERROT, J (1984) - Essais CHARLES (Phase 3) - Compte rendu d'essais - Contrat CEEA/CEA - SRF/005 F - (S).
9. BROSSARD, J et al. (1984) - PHASE 3 - Rapport d'interprétation des essais CHARLES 3 : effets d'un amorçage fort sur la propagation de la déflagration - Contrat CEEA/CEA - SRF/005 F - (S).

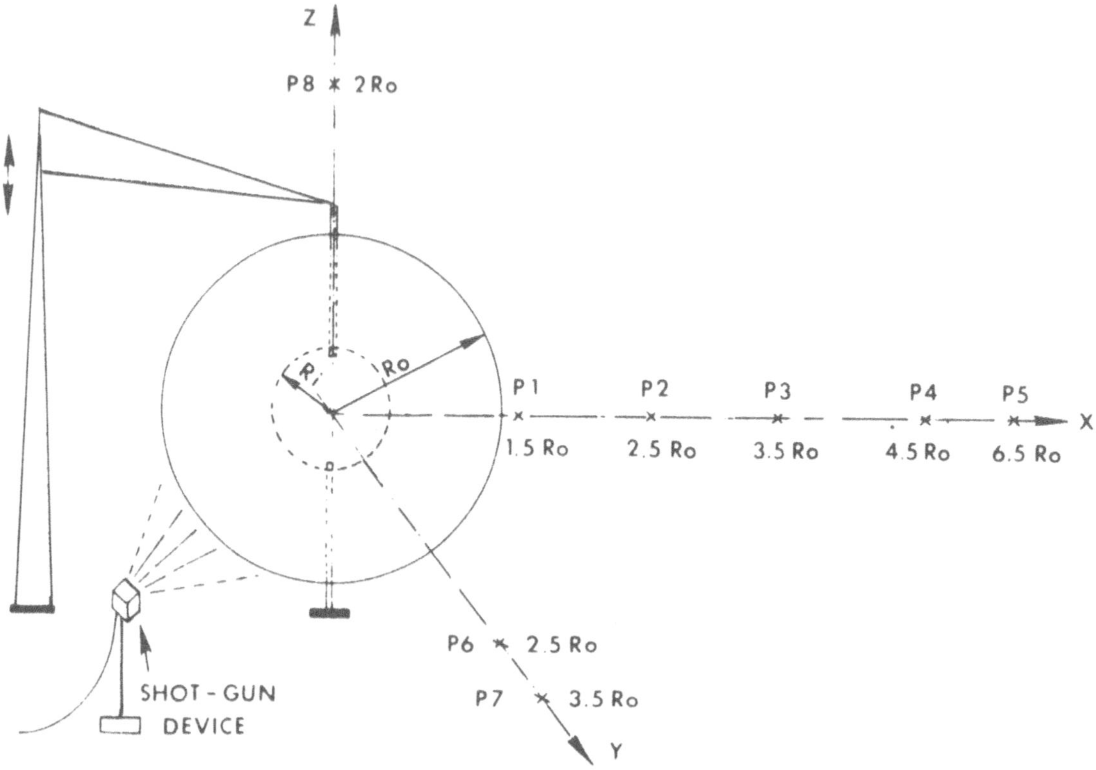

Fig. 1 – PHASES 1, 2 and 3 : SPHERICAL BALLOONS CARRIED BY AN ADJUSTABLE SUPPORT DEVICE (pressure transducer distances indicated are those of the phase 2 instrumentation)

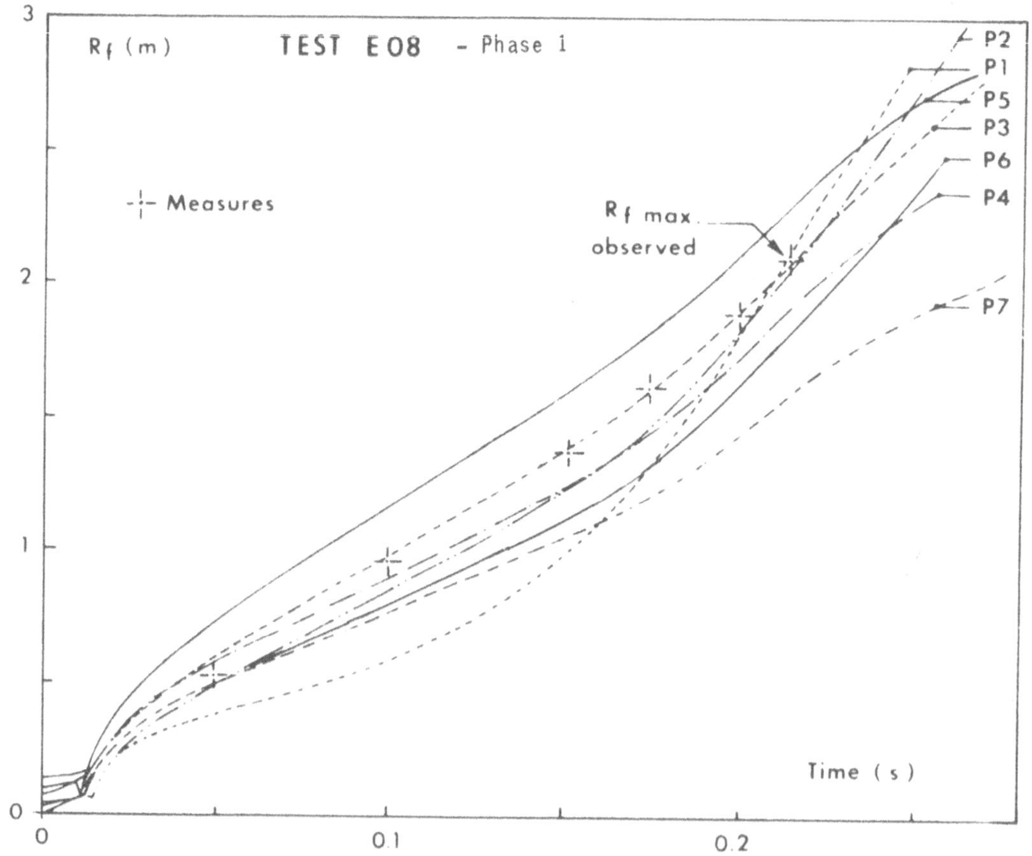

Fig. 2 – A COMPARISON BETWEEN OBSERVED TRAJECTORIES AND THOSE CALCULATED FROM PRESSURE RECORDS

Table I : PHASE 1. DEFLAGRATIONS OF HOMOGENEOUS MIXTURES - Synthesis of results - (E for ethylene - A for acetylene)

TEST N° Date	VOLUME m^3	HYDROC. %	V_F (m/s) Beginning of propagation	Max	ΔP (mbar) R (m) →	P_1 2.18	P_2 3.18	P_3 4.69	P_4 6.19	P_5 9.21	P_6 3.15	P_7 3.90	REMARKS
E 01 30 .7 .81	12	Air	-	-	-	-	-	-	-	-	-	-	Verification of shot-gun device noise and balloon removal noise.
E 02 15.09.81	-	-	-	-	-	-	-	-	-	-	-	-	Without balloon - Noise measurement with shot-gun device only (with a new pyrotechnical charge)
E 03 15.09.81	11.96	E 7.7	-	-		≠6.7 noise	9.4	6	4.5	3.8	8.8	5.5	Determination of the influence of partial confinement (burst without preliminary envelope destruction)
E4 29.07.81	12.06	E 8.45	11	15		≠4.6 noise	2.6?	1.6	1.1	1.1	2.2	1.5	
E5 16.09.81	12.14	E 9.06	8.4	16		No meas.	4.2	3.3	2.4	2.1	4.5	2.9	
E6 17.09.81	12.12	E 8.91	-	26		? sat.	14	9	6.8	6	15	9	No preliminary enveloppe destruction due to a spurious departure of the shot-gun device after firing the charge
E7 23.09.81	12.11	E 8.83	12	17.9		?	5.5	4.4	3.5	3	5.4	4.2	
E8 24.09.81	12.03	E 6.73	8.5	11		7.5	3	1.7	1.2	1.1	2.4	1.1	
E9 29.09.81	12.02	E 6.65	8.9	10.7		≠1.8 noise	1.5	0.9	0.7	0.6	1.5	0.6	
E10 30.09.81	12.10	E 8.76	12.7	17.6		≠1.1 noise	3.3	2.7	2	1.9	3.6	2.3	
E11 13.10.81	12.70	A 14.96	16.3	39		21.6	18.6	16.8	13	11.6	22.4	16.8	
E12 16.10.81	12	A 8	20	32		6.4	5.4	5.2	4	3.6	6	4.3	

Table II: PHASE 2. DEFLAGRATIONS IN A HETEROGENEOUS MEDIUM - Synthesis of results

(two concentrical spherical balloons)

TEST N° Date	Internal b. R_i m	Internal b. % C_2H_4	External b. R_0 m	External b. % C_2H_4	ΔP (mbar) R (m) → For E 16 →	P_1 2,1 2,8	P_2 3,5 id	P_3 4,9 id	P_4 6,3 id	P_5 9,1 id	P_6 3,5 id	P_7 4,9 id	P_8 2,7 id	V_F max m/s	Observations t_T is the time where the flame transits from internal to external composition.
E6 28.07.82	0,5	13	1.4	6.6		3.6	2.4	2.1	1.6	1.25	2.5	1.6	2.8	-	Slow down at t_T
E7 29.09.82	0.5	13	id	8.5		6.4	4	4	3.2	3	5.8	3.9	7	24	Internal balloon was not destroyed by the shot-gun device. Acceleration at t_T
E8 18.10.82	0.5	8.5	id	13		8.4	5.2	2 ?	4.5	1.2	2.4	1.5	6.4	22	Slight slow down at t_T stronger later Oscillations.
E4 bis 20.10.82	0.5	5	id	8.5		12	8.5	6	6	4	8	8	9	23	Distinct acceleration at t_T
E6 bis 03.11.82	0.5	13	id	6.6		3.8	3.6	2.2	2.4	1.2	2.2	2	4	27	Strong slow down at t_T
E10 30.11.82	0.9	5	id	8.35		3.3	2	0.9	1.5	0.5?	1	0.9	2.5	16	Oscillating flame. No significant speed variation.
E12 01.12.82	0.9	13	id	6.5		5.6	3.4	1.3	3	2.1	2	1.8	5.2	22.5	Oscillations in internal mixture Slow down at t_T
E13 02.12.82	0.9	13	id	8.5		5.6	6.4	3	4.8	3.6	5.6	4.2	7.2	23	Oscillations in the internal mixture. Distinct slow down at t_T
E14 11.01.83	0.9	8.5	id	13		4.4	8.4	6.4	10.2	6.4	7.5	6	12.5	27.5	Strong oscillations in the internal mixture No distinct acceleration at t_T. Doubtful test
E15 12.01.83	0.9	5	id	8.5		?	11.2	12.5	12.5	8.4	16.2	12	19	19.5	Intentionally no destruction of balloons before firing. Strong acceleration at t_T (V_F: 5m/s ≠ 20m/s)
E16 18.01.83	0.9	13	id	8.5		7.5	9.5	10.2	9.5	6.3	10.2	8.5	17	20.3 s	Intentionally no destruction of balloons before firing. Strong acceleration at t_T (V_F: 5m/s ≠ 20 m/s)

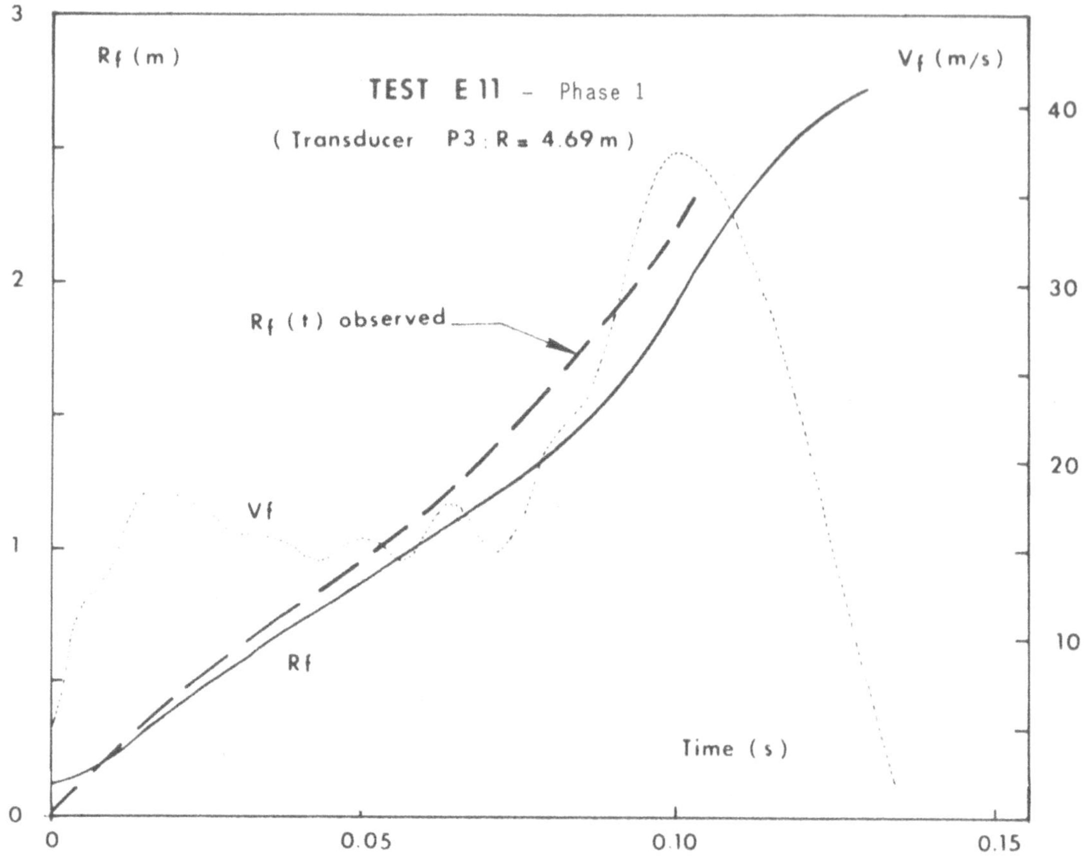

Fig. 3 – FINAL FLAME ACCELERATION (due to the withdrawal of the balloon)

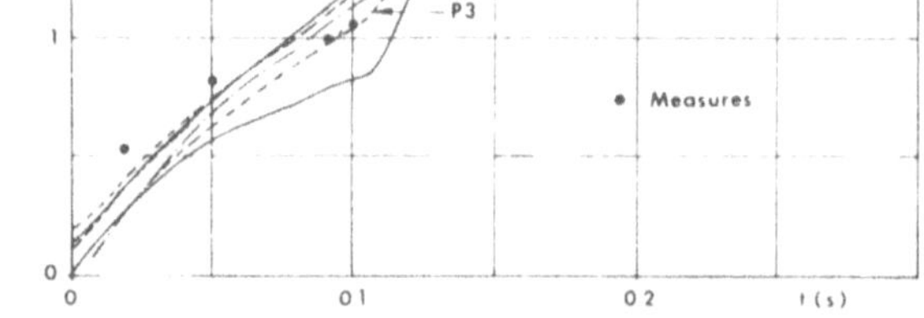

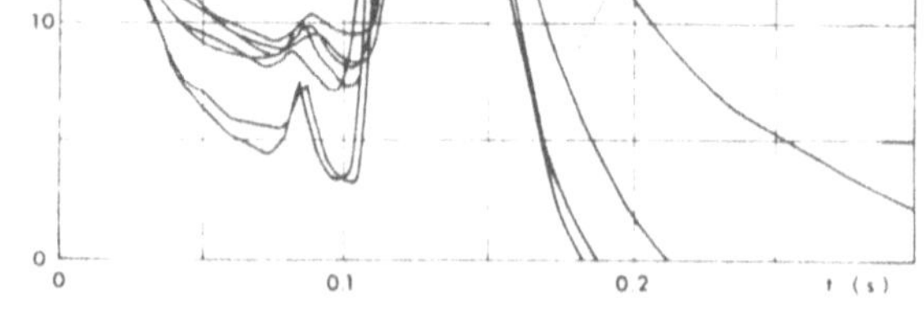

Phase 2

Fig. 4 – TEST 4 BIS – R_f (t)

Fig. 5 – TEST 4 BIS – V_f (t)

TRAJECTORY R_f (t) AND CELERITY V_f (t) CALCULATED FROM PRESSURE RECORDS (distances are given in table II)

TEST NUMBER	Plastic Mass (g)	REGIME
E1 21.06.83	5	Deflagration
E2 27.06.83	6	id
E2 bis 23.08.83	6	id
E3 30.08.83	7	id
E4 01.09.83	8	id
E5 08.09.83	10	Detonation
E6 13.09.83	9	id
E7 22.09.83	8,5	Deflagration
E8 25.10.83	8,6	Detonation
E9 03.11.83	8,5	id
E10 08.11.83	8	id
E11 09.11.83	6	Deflagration
E12 24.11.83	6	id
E13 10.01.84	7	Detonation
E14 18.01.84	6,5	Deflagration
E15 14.02.84	6,5	Detonation
E16 16.02.84	6	Deflagration
E17 23.02.84	6	Detonation
E18 01.03.84	5	Deflagration

No wall destruction before firing

Table III – Phase 3 – Régimes obtained

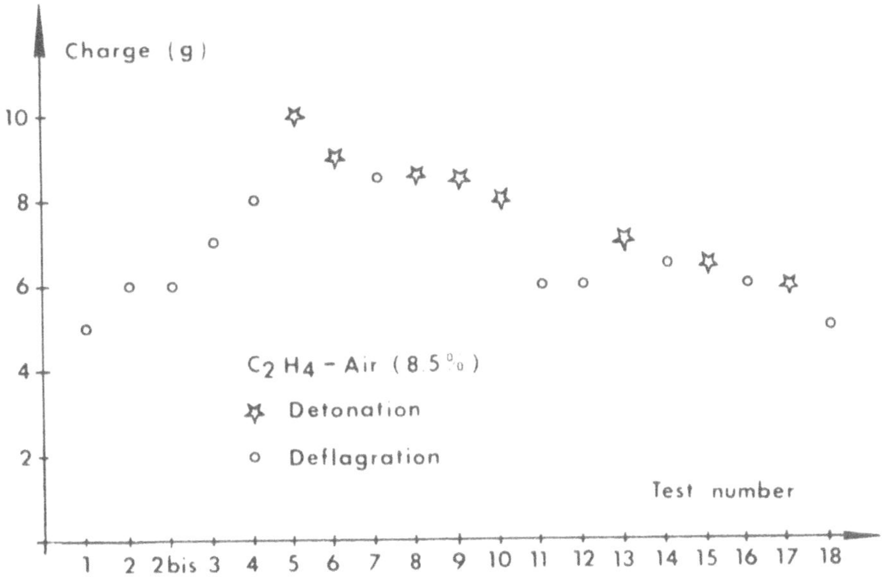

Fig. 6 – PROPAGATION REGIME VERSUS MASS OF PLASTIC – Phase 3

TESTS	Plastic mass (g)	Average velocity m/s	P_3 (4,9 m)	P_7 (4,9 m)	P_4 (6,3 m)	P_5 (9,1 m)	P_6 (28 m)
E_7	8,5	30 (?)	10	8	7,7	6	2,5
E_{11}	6	32	12	8	8	5,8	2,5
E_{12}	6	29	7	5	4	4	1,6
E_{14}	6,5	20	8	6,3	7	6	2
E_{16}*	6	—	8,5	7	5	6	2,6
E_{18}	5	26	2,4	2,2	1,4	1,4	0,4

* (balloon not destroyed)

Table IV - Average velocities and maximum overpressures obtained during some PHASE 3 deflagration tests

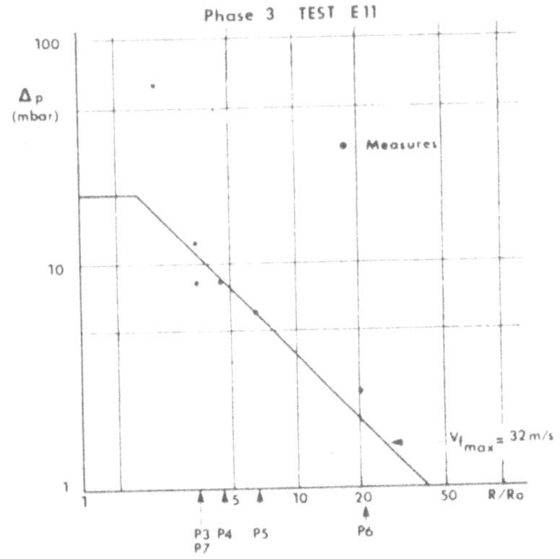

Fig. 7 – COMPARISON BETWEEN A CONSTANT SPEED MODEL AND EXPERIMENTAL VALUES

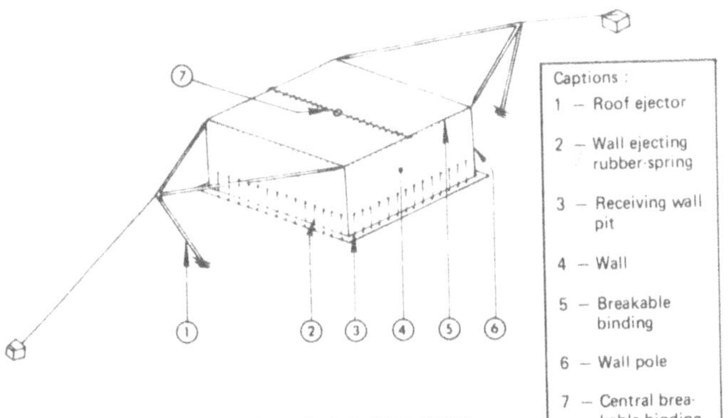

Fig. 8 – PHASE 4. RETRACTABLE ENCLOSURE

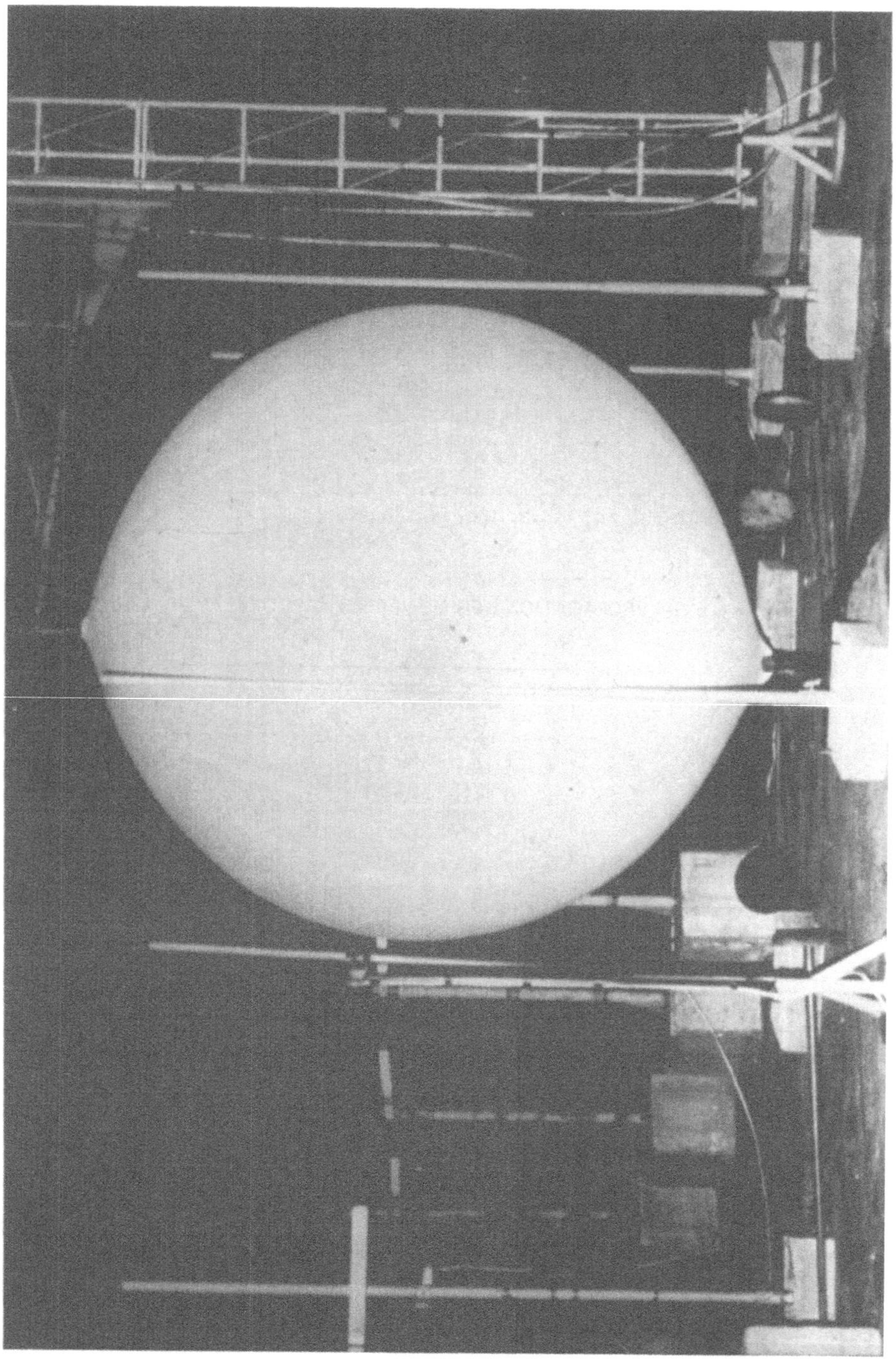

CHARLES 1 TO 3 TESTS : EXPERIMENTAL FACILITIES

Discussion

Question: C.J. Wheatley - SRD/UKAEA

You mentioned that you observed deceleration of the flame when passing from a lean or rich mixture to a stochiometric mixture. Are you sure that this is true?

Answer: J.L. Garnier - CEA/CEN-Far

These observations were made from high speed movie camera records. The degree of confidence is the one of camera records (600 frames per second).

Question: A. Siebertz - CEN/SCK

Do you know the necessary conditions for getting detonation or deflagration?

Answer: J.L. Garnier - CEA/CEN-Far

No, the boundary is fairly imprecise in spherical geometry (CHARLES Phase 3 tests). But in NPP safety analysis, which is the objective of these experimental studies, we must be very cautious to foresee the eventual energy concentration of the ignition source. The CHARLES 4.2 test, which led to the detonation of an air-ethylene mixture, is an example of the possibility of occurrence of such a regime with an igniter boosted by 5g of PETN plastic.

EXPERIMENTAL INVESTIGATION OF THE ACCELERATION OF DEFLAGRATION IN WAKE FLOW

H. WILLMS and G. ADOMEIT
Institut für Allgemeine Mechanik
RWTH Aachen, FRG

Summary

In real gas cloud explosions turbulence is produced by the flow field caused by the combustion process. But also turbulence can be produced by an initial flow field due to atmospheric wind or technical devices. This turbulence may lead to an increased rate of chemical reaction connected with high pressure levels. The reported laboratory-scale experiments are particularly designed in order to investigate the influence of a flow field present at the moment of ignition inside a partially confined hydrocarbon-air gas cloud. Experiments have been done using a flow channel capable to produce an unsteady flow field of combustible gas independent of the combustion process itself. The parameters which have been varied in these tests are the initial flow velocity, the gas-mixture composition and the geometry. The tests have been carried out mainly with stoichiometric ethylene-air and propane-air mixtures. The measured quantities are: pressure time history inside the test section, CH-radical radiation and also high speed photographs have been taken. The results show that the initial flow speed and the reactivity of the gas mixture investigated have a strong influence on the maximum overpressure and the duration of the positive pressure phase but also quenching effect may become importance for high flow velocities and mixtures of low reactivity.

1. INTRODUCTION

The topic of the project was the experimental investigation of the flame acceleration mechanism of the deflagrative combustion of partially confined hydrocarbon-air gas clouds in conditions of turbulence-producing geometry. The main point of the investigation was the study of the interaction between the flow field and the reaction zone and its influence on pressure build-up. An important characteristic of the experiments described is that the flow field has been produced by hydrodynamical means and not only by the expanding reaction products. Using a suitable experimental set up, it was possible to vary the flow velocity present at the moment of ignition in a wide range. The influence of an initial flow field must be taken into account, in connection with safety aspects of nuclear power plants or sites for chemical engineering, since in real gas cloud explosion such flow fields, which may have its origin in atmospheric wind, interaction with buildings, venting of buildings, etc. are usually present. Particularly, in the turbulent wake of structures, the flame surface can be folded and that may lead to an increasing burning rate and again to an increasing level of overpressure.

The tests have been done in a plane test section, consisting of two parallel plates, varying the geometry of obstacles inside the test section and the kind of gas mixture. The flow velocity has been changed as a parameter.

The measured quantities were

- pressure time history at different locations of the test section

- radiation intensity of chemical radicals (CH and C_2).

2. EXPERIMENTAL APPARATUS

Roughly, the experimental facility used can be divided in two parts: the part of the test section and the driving gas part. A sketch of the complete experimental set up is given in Fig. 1 and a sketch of the driving gas part is given in Fig. 2. The chief components of this experimental set up are the test section which contains the measuring volume, a short channel with a flame trap through which the fuel-air-mixture is pushed into the test section, an airbag-container with an airbag, an adjustable Laval-Nozzle to control the mass flux, a diaphragm support with a thin membrane to start the process, a rapid closing valve to stop the flow, and a driving gas vessel. Only the measuring volume of the test section and the airbag-container, as well as the channel between them, contain a combustible gas-mixture, all other devices contain air.

To prepare the experimental apparatus for a test, the open area of the test section is covered with a thin plastic foil, the test section and the airbag container is filled up with premixed gas. The pressure level of this part is the atmospheric pressure p_a. The fast closing valve is in open position and the driver vessel is filled up with compressed air. The process is started by cutting the diaphragm in the diaphragm support by means of a pneumatically operated knife. Then a shock wave moves through the laval nozzle into the airbag. The airbag is blown and pushes the mixture through the flame trap into the test section. A pressure transducer in the Laval nozzle is used to trigger an electronic delay unit, and this delay unit controls all other operations.

2.1 Description of the main components of the experimental set up

2.1.1 Test section

The measuring volume has a cylindrical shape and a size of 520 mm in diameter and 80 mm height. The test section used for the first series of experiments was made of two aluminium plates with windows of 280 mm diameter for the purpose of optical measurements. The later, major part of the investigation was performed with a window section of increased area. The front wall of the test volume was replaced by a new one made of plexiglass. This was possible because the tests have shown that the peak pressure inside the test section is usually below 200 mbar. The inner side of the two plates is prepared to build in obstacles of different kind. Figs. 3a to 3d show sketches of the different geometries which have been investigated. It should be pointed out that the test section is not a closed chamber but open between the two plates. This ring area is covered only with a thin plastic foil which is held by a rubber gasket during filling the test volume with the fuel-air mixture.

2.1.2 Airbag and airbag container

Safety aspects were the reason why we used an airbag inside an airbag container for separating the combustible mixture from the driving gas. The airbag operates as a nearly massless piston. The final design was only found after intensive experimental trials, which finally ensured that the airbag is blown uniformly. The airbag container is made of a plexiglass tube of 192 mm diameter and 500 mm length, so its volume is approximately 15 liters.

2.1.3 Laval nozzle

One important parameter which we have varied in our tests is the flow velocity through the inflow channel. This quantity has been controlled by an adjustable Laval nozzle with a quadratic profile. The lower bound of the nozzle is a circular segment of fixed shape while the upper one is flexible. The central portion can be raised or lowered by a spindle to any desired position. The ratio of the inlet and outlet pressure of the nozzle has been chosen as a critical one.

2.1.4 Diaphragm support and fast closing valve

The process is initiated by rupture of a Hostafan foil of 35 µm to 75 µm thickness (depending on the initial pressure difference). The diaphragm is clamped in a diaphragm support operated electromagnetically and is destroyed by a pneumatically driven knife.

To stop the flow from the driving gas vessel into the airbag a pneumatic piston-valve is used. The closing time is in the range of a few milliseconds.

3. HYDRODYNAMIC PROPERTIES OF THE EXPERIMENTAL SET UP

A combination of calculation and simultaneous measurement was chosen to determine the time dependence of the flow velocity. To achieve this, the pressure drop in the driving gas vessel and the pressure rise in front of the flame trap were measured. With this information the rate of mass-flow through the Laval nozzle into the airbag and further the unsteady flow velocity through the inflow channel can be calculated. The time dependent flow velocity is plotted in Fig. 4, using the height of the Laval nozzle as a parameter.

4. MEASURED QUANTITIES

4.1 Optical measurements

Measurements of chemiluminescence have been carried out in order to obtain information on the rate of chemical change within the test section. Information about the time-dependent rate of chemical change in a finite volume would be very useful for improving theoretical models describing unsteady turbulent combustion processes. But there are still a lot of open questions concerning data obtained with these measurements. Problems arisen in this connection are discussed in the final report of this project.

The arrangement which has finally been used for the measurements of chemiluminescence is shown in Fig. 4a. The desired wave band is selected by means of an interference filter. We used an interference filter Type Schott MA 3-0.3 with a half-width value of 2.7 nm and a transmission ratio of .53 . A partially reflecting mirror is used to split the beam into two parts. An additional orifice was inserted in the path of the first beam. In this way, only radiation from the inner area of the test section is measured at the first photomultiplier tube. The diameter of this inner area is approximately 280 mm. By means of the second photomultiplier tube radiation of the total test section has been measured, choosing either the 4316 A band of the CH-radical and also in some tests the C_2-band at 5165 A.

4.3 Pressure measurements

The pressure time history inside that test section has been measured by Kistler piezoresistive pressure transducers Type 4053 A 1 at different locations indicated in Figs. 3a to 3d. The transducers have been covered with a thin perforated brass plate to avoid thermal disturbances. So we are sure that the pressure measurements are not influenced by thermal effects. The pressure transducers are powered by piezoresistive amplifiers of Type Kistler 4601.

5. EXPERIMENTS PERFORMED

A total number of 302 tests have been carried out. Only tests nos. 171 to 302 have been utilized for final evaluation. Tests nos. 27 to 170 were done with different optical equipment and also the features of the driving gas part were different. Tests nos. 171 to 217 were done with stoichiometric ethylene-air mixture, varying the inflow velocity and the ignition time, and tests nos. 218 to 239 were done with stoichiometric propane-air mixture. The maximum flow speed through the inflow channel has been varied between 0 and 75 m/s.

Similar tests with an inflow channel cross section of 160 x 70 mm have been carried out with and without obstacles inside the test section, using stoichiometric ethylene-air and propane-air mixture. These tests are:

- tests nos. 240 to 259; propane-air; cylindrical obstacles Ø 40 mm, pitch 80 mm
- tests nos. 260 to 278; ethylene-air; cylindrical obstacles Ø 40 mm, pitch 80 mm
- tests nos. 279 to 290; ethylene-air; without obstacles
- tests nos. 291 to 302; ethylene-air; plate obstacles, 30 mm height, 2 mm thickness.

6. RADIATION CALIBRATION TESTS

The aim of the radiation intensity measurements was to determine the total rate of chemical change in a definite region. The idea of these measurements was that the oxidation of hydrocarbons produces several transient chemical radicals. If the total number of excited molecules of a certain species in the test volume investigated is proportional to the rate of chemical change, then the radiation intensity of a certain emission band of this radical should be proportional to the rate of chemical change, since the life time of these radicals is short compared with the duration of the combustion process itself. So, for a fixed optical set-up, the voltage taken at the photomultiplier tube should be equivalent to a definite volume of burnt gas per time unit. Various trials have been undertaken, in order to calibrate the radiation signal. These trials are based on different combustion chambers and steady state burners. Details are described in the corresponding final report.

Trying to associate the data of the calibration tests to the data of the main experiments does not give reasonable results. For example the fresh mixture which is burnt until the various peak values are reached is less than 1 % of the contents of the test volume. Dividing the maximum radiation intensity measured in the explosion tests by the specific radiation obtained in the calibration tests gives a value of 0.02 m^3/s burnt mixture. This value appears also too small. So at the moment it seems advisable to use the radiation signals with precaution as a measure of the rate of chemical change.

7. RESULTS

In all tests carried out there was always a steep rise in radiation intensity of the CH-band immediately after ignition. Simultaneously to this rise, a pressure wave is generated. The steepness of the initial rise of the radiation intensity signal and the level of the maximum overpressure depend strongly on the initial flow velocity, on the composition of the gas mixture and on the geometry of the test section.

Fig. 5 shows an example of radiation and pressure measurements in dependence upon time. The duration of the positive pressure phase is of the order of a few milliseconds, depending on the initial flow velocity and the gas mixture. Usually the duration of the radiation intensity signal is one order of magnitude larger than the duration of the pressure signal. This fact indicates that chiefly the first stages of combustion are responsible for the pressure wave generated.

In all tests which were performed, the peak pressure occurs during the increasing phase of the radiation signal and the pressure signal had nearly dropped to zero when the radiation signal has reached its maximum. The experiments have been evaluated with regard to the peak pressure, the time-integral over the positive pressure phase, the maximum radiation intensity, the maximum steepness of the initial rise of the radiation signal, and the time-delay between ignition and the different crest values. The scale on the ordinate of the diagrams described below concerning the radiation signal, represents only the voltage of the photomultiplier tubes because of the difficulties for calibrating these signals. The signal of the second photomultiplier tube has been adjusted to the signal of the first one by multiplication with the sensitivity ratio (6.3) of the two photomultipliers.

Fig. 6 shows the peak pressure in dependence upon flow velocity obtained in the tests using the inflow channel of 80 mm x 80 mm cross section and without any additional obstacles within the test section. For stoichiometric ethylene-air mixture, the pressure increases linearly with flow speed. In the case of stoichiometric propane-air mixture we obtained quite different results. The peak pressure is much less than the peak pressure in the ethylene experiments. At the beginning the peak pressure also increases linearly. For high flow speeds, however, the curve bends and for flow speeds greater than 50 m/s the pressure level decreases. At a certain value (here about 60 m/s) it is impossible to ignite the mixture by means of normal spark plugs. The maximum peak pressure reached in these tests is 85 mbar for ethylene and 31 mbar in the case of propane. Compared to these values the peak pressure in the experiments without initial flow speed is very moderate. Only few millibars are reached.

If the positive phase of the pressure signal is integrated over time, the data plotted in Fig. 7 are obtained. The behaviour of ethylene and propane is also quite different. For ethylene this quantity increases if we choose higher flow speeds, whereas for propane it decreases.

The evaluation of the measured radiation intensity curves yields qualitatively similar results (Fig. 8). The maximum radiation intensity in the case of ethylene-air mixture is slightly increasing for both photomultipliers. For propane-air mixture the radiation intensity increases with V_a, reaches a maximum and then it decreases until no ignition is possible with the spark plugs used. The shape of these curves is the same for the CH-radiation from the inner area as well as the radiation of the total area of the test section.

The maximum steepness of the first rise of the radiation intensity signal is plotted in Fig. 9. The behaviour of this quantity is quite similar to the behaviour of the maximum peak pressure. This is in very good agreement with results obtained by theoretical means /4/. For these calculations a gasdynamical model and a model of turbulent combustion have been assumed and there we found the same phenomena.

A possible explanation of the different behaviour of propane-air and ethylene-air mixture are quenching effects in regions with strong turbulence. Bradley /2/ found a similar effect for lean hydrogen-air mixtures. For ethylene-air mixtures this phenomenon has not been observed for the flow velocities investigated (up to 80 m/s). For ethylene quenching may occur at higher initial flow velocities than we can reach at the moment in our experimental apparatus. This question will be investigated in the future.

Fig. 10 shows the maximum peak pressure at the centre of the test section as a function of the initial flow speed for various obstacle configurations and an inflow channel cross section of 160 mm x 70 mm. The experimental set-up was capable to vary the inflow velocity from zero up to 43 m/s. It seems that the peak pressure is proportionally increasing with the flow speed for all kind of obstacles for both the ethylene and the propane mixture. The highest pressure levels, about 113 mbar, have been obtained with ethylene-air mixtures in the case of cylindrical obstacles. The peak pressure using obstacles is nearly 2.5 times as high as without. The peak pressure in the centre of the test volume using propane-air mixture, is always less than for ethylene-air mixture without obstacles. In most cases the peak pressure obtained in the tests with plates were below the peak pressure of the tests with cylinders.

Fig. 11 represents the peak pressure measured with the second transducer, which was located behind the last obstacles 120 mm from the centre. The distance to the border of the test section was 140 mm. This diagram looks similar to Fig. 10 but it is slightly different in detail. Particularly, the peak pressure in the case of ethylene-air without any additional obstacles was less than the pressure level of propane-air mixture with cylindrical obstacles. Furthermore, the ratio of peak pressures in the corresponding experiments was usually larger, whether with or without obstacles. This fact indicates that, if there are obstacles placed inside the path of the reaction zone, we can find high pressure levels in a larger region, indicating that the total mass reacted is larger in the latter case.

The difference between the two test gases propane and ethylene concerning the integral over the pressure signal, is comparatively small. The largest values, nearly 2.5 10^{-4} bar s, have been obtained for ethylene-air mixture with obstacles. A very interesting fact is that for ethylene this quantity is increasing with the flow speed at the centre of the test section. At 120 mm distance from the centre it seems that it remains on a constant value or that it is slightly decreasing, whereas for propane the behaviour is opposite. Since the pressure pulse for ethylene and propane are on the same level, the duration of the positive pressure phase is larger for propane than for ethylene because of the higher peak pressure of ethylene. It seems that the steepness of the first rise of the radiation signal is chiefly determined by the flow speed (Fig. 12). Generally, this quantity is increasing proportionally to the flow speed.

8. DISCUSSION

The experiments have shown that a flow field present inside a combustible gas cloud at the moment of ignition plays an important role with regard to flame acceleration processes and the pressure wave generated. For ethylene-air mixture and 75 m/s flow velocity within the inflow channel, the pressure level obtained is approximately 10 times as high as without any initial flow field. Also the strong influence of obstacles inside the burning gas cloud is evident. The turbulence produced in the vicinity of the obstacles leads to an increased burning rate. Obstacles can cause also an impediment for the outflow of the combustion products.

The area where higher pressure levels can be recorded is restricted to the vicinity of the obstacles and the surrounding of the sharp edges at the mouth of the inflow channel. Although the flame front propagation is maintained over a relatively long distance outside the test section, the peak pressure at the border of the test section is much less than the peak pressure at the centre of the test section.

For ethylene-air mixture, the peak pressure is proportionally increasing with the flow speed. So one can expect that this behaviour can be extrapolated to higher flow speeds than we can reach in our experimental facility. In the case of propane-air mixture it seems that for flow speeds higher than 45 m/s quenching effects become important. It should be pointed out that we have done only one test series using flow speeds higher than 45 m/s. So it remains to prove, if this value is independent of the geometry of the test section.

If quenching effects may occur even in stoichiometric ethylene-air mixture similarly to propane-air mixture is not known until now. For lean hydrogen-air mixtures, there exists already experimental and theoretical work concerning diminishing of flame velocity due to turbulence (Bradley /2/) respectively extinction of reaction zones due to heat loss (Peters, Smooke /1/). For the near future we are planning tests suitable to determine the flow velocity necessary for the occurence of quenching effects in dependence of the laminar flame velocity using non-stoichiometric mixtures of ethylene-air. Probably these results will be not transferable to other hydrocarbon-air mixtures of same laminar flame velocity, since quenching effects are dependent on details of chemical kinetics of combustion.

The radiation and pressure measurements have shown that the peak pressure chiefly depends of the time derivative dm_b/dt of the burnt mixture, respectively of the time derivative dV_b/dt of the burnt volume. A way to substantiate this fact is possible by simplifying the problem of cloud burning and wave propagation to a problem of source flow. For a source of constant strength the pressure field is only a function of space, but changes in strength of the source produce propagating pressure waves. Such changes in the burning rate can be caused by an enhanced flame velocity due to small scale turbulence or by an increase of the flame surface, as a result of flame folding in consequence of the flow field, and obviously by a combination of these two effects.

The absolute level of the maximum peak pressure obtained is moderate, especially if the relative strong degree of confinement of our experiments is taken into account. Schildknecht et al. /5/ (Battelle) found much higher pressure levels up to 3 bars for ethylene-air mixtures in a partially confined configuration, using a jet of combustion products for ignition. These experiments have been done at a much larger size. The contents of the test volume was 200 times larger than the volume of the test section used in our experiments. The experiments of Leyer /3/, who also used jet ignition in propane-air gas enclosed in soap bubbles, have shown pressure levels one order of magnitude smaller than in our propane experiments. Compared to the Battelle tests, they are three orders of magnitude smaller. In the tests of Leyer the total amount of gas mixture burnt is about 10 times smaller than in our tests. The large difference in peak pressure between the tests of Battelle and Leyer indicates that the geometrical size of the cloud is important and that is the reason why further tests at a larger scale are necessary for an estimation of the pressure levels which can be expected in real gas clouds.

REFERENCES

1. N. PETERS, M.D. SMOOKE: Fluiddynamic-chemical interactions at the lean flammability limit. Report, available at the Institut für Allgemeine Mechanik, RWTH Aachen. FRG.
2. K.J. AL-KHISHALI, D. BRADLEY, S.F. HALL: Turbulent combustion of near-limit hydrogen-air mixtures. To be published.
3. P. CHARUEL, J.C. LEYER: Effet de Pression engendre par l'Explosion de Charges Hydrocarbon-air non confinees amorcees par un jet de Gaz brules. E.N.S.M.A., Poitiers, France.
4. G. ADOMEIT, H. WILLMS: Maßstabseinfluß bei der Übertragung von Laborversuchen auf Explosionen in realer Größe. Abschlußbericht für die Gesellschaft für Reaktorsicherheit, Köln, FRG.
5. M. SCHILDKNECHT, W. HAAS, K. METZ and W. GEIGER: Versuche zur Freistrahlzündung von Wasserstoff-Luft-Gemischen im Hinblick auf den Übergang Deflagration - Detonation. Abschlußbericht für Fa. Interatom, Bergisch Gladbach, FRG.

Discussion

Note: No discussion has taken place on this paper.

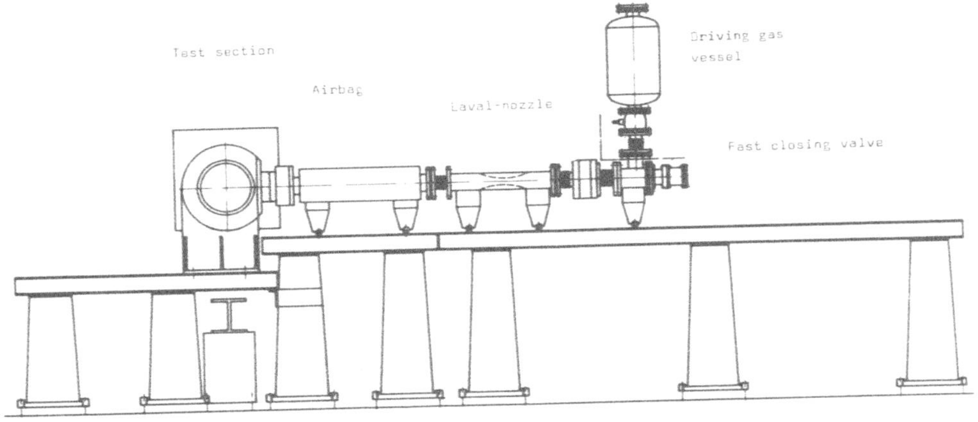

Fig. 1

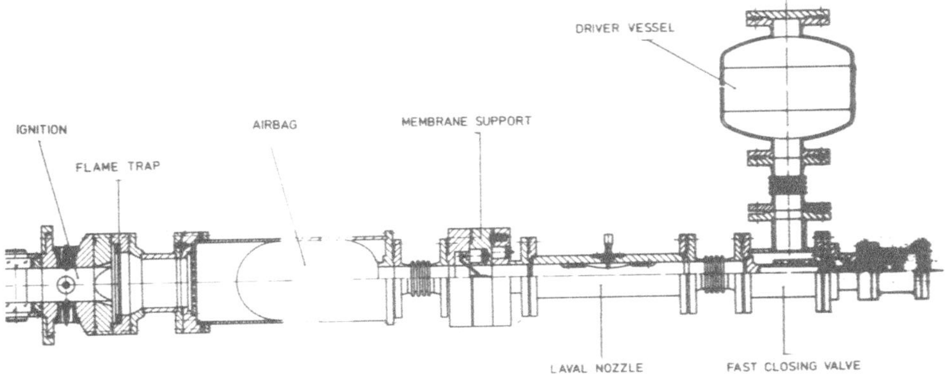

Fig.2 Outline of the driving gas part of the experimental set-up Constructional details are different from the experimental set-up in its final version, because improvements have not been drawn in this plot

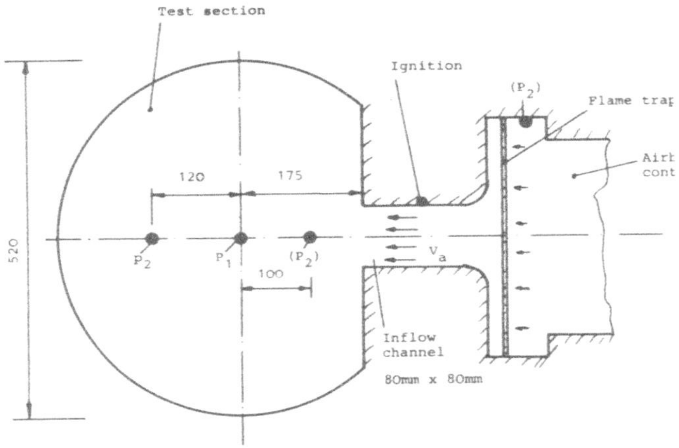

Fig. 3a

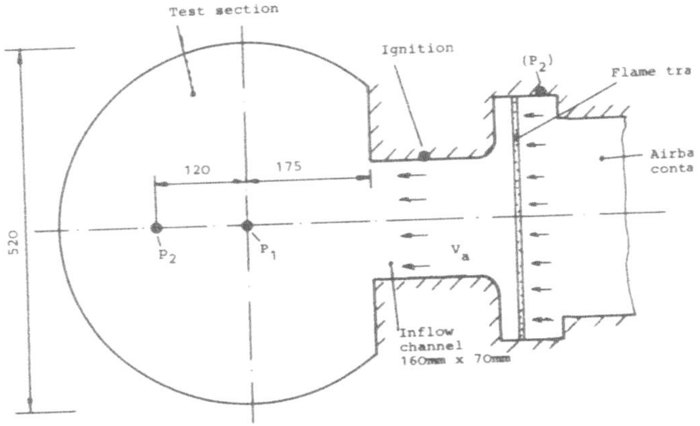

Fig. 3b

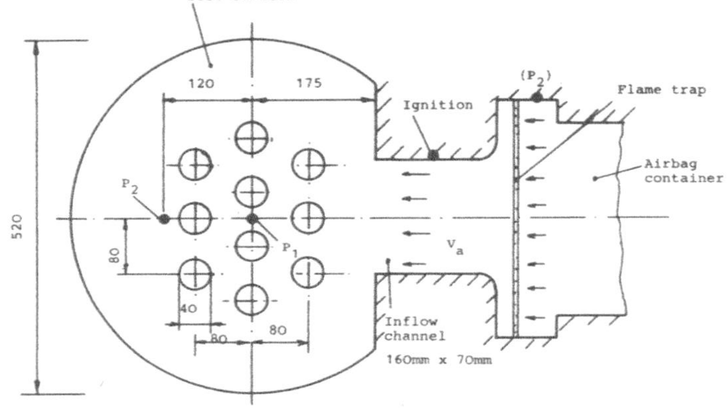

Fig. 3c

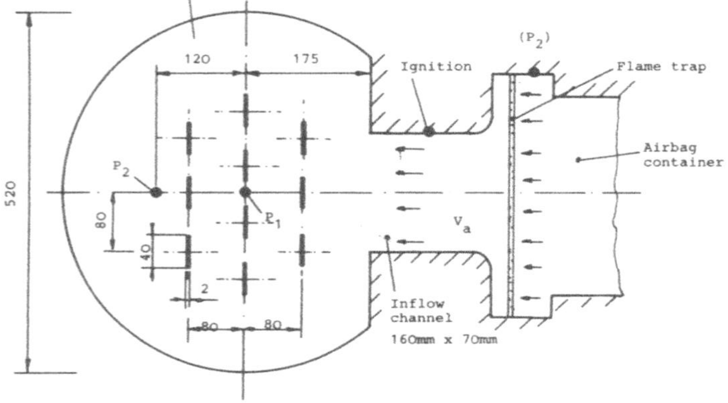

Fig. 3d

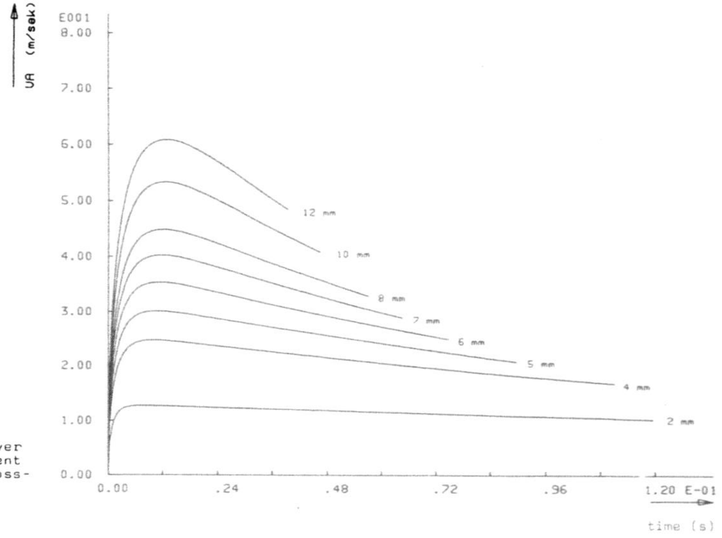

Fig.4
Flow velocity over time for different Laval-nozzle cross-sections

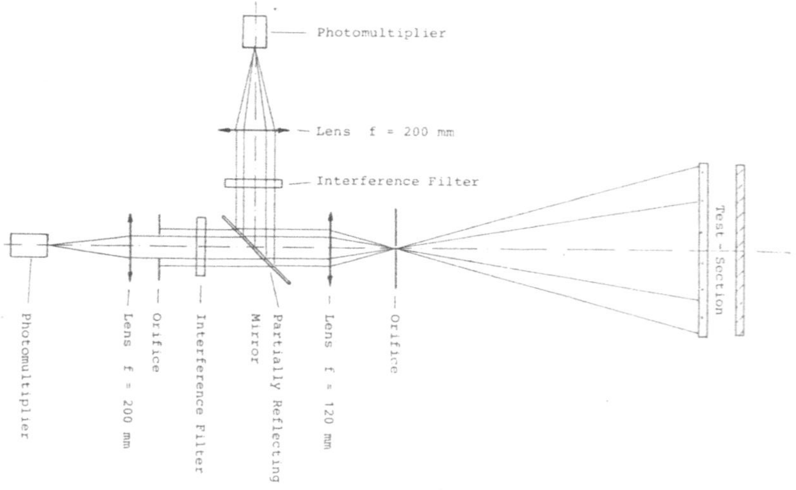

Fig. 4a Optical arrangement

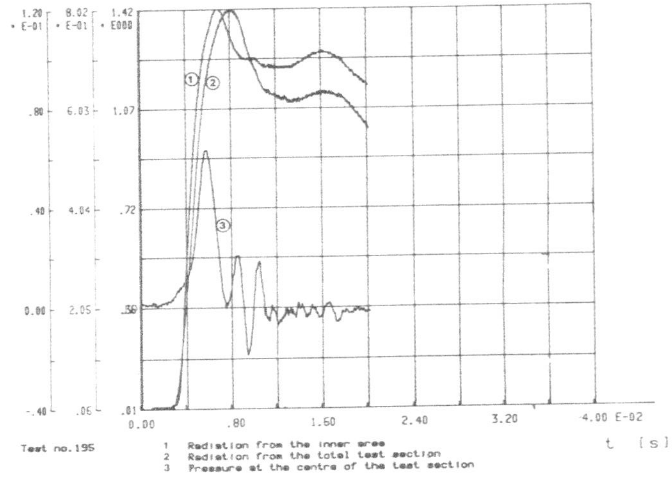

Fig. 5

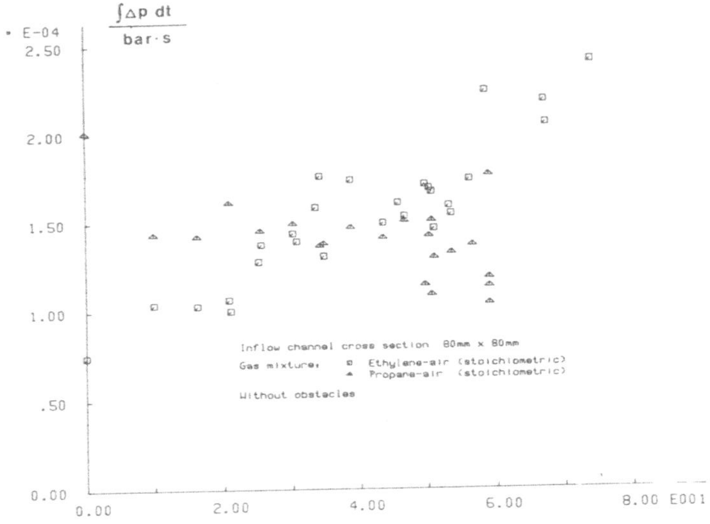

Pressure-pulse at the centre of the test section for Ethylene-air mixture and for Propane-air mixture

Fig. 7

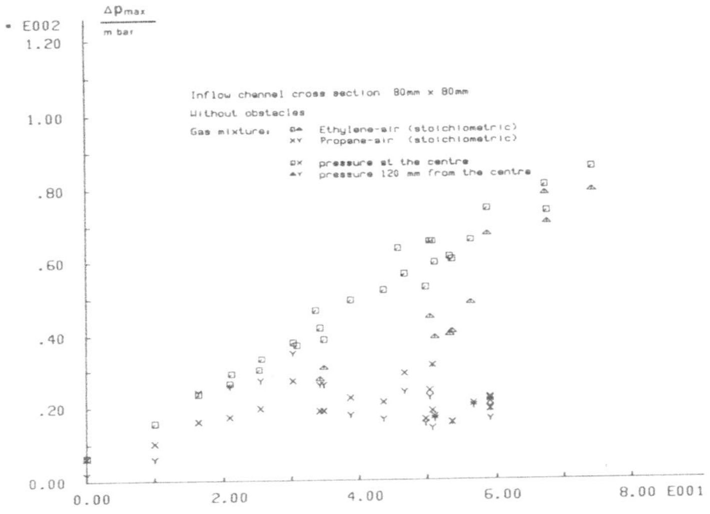

Comparison of the maximum overpressure for Ethylene-air and Propane-air mixture , Inflow channel cross section 80mm x 80mm

Fig. 6

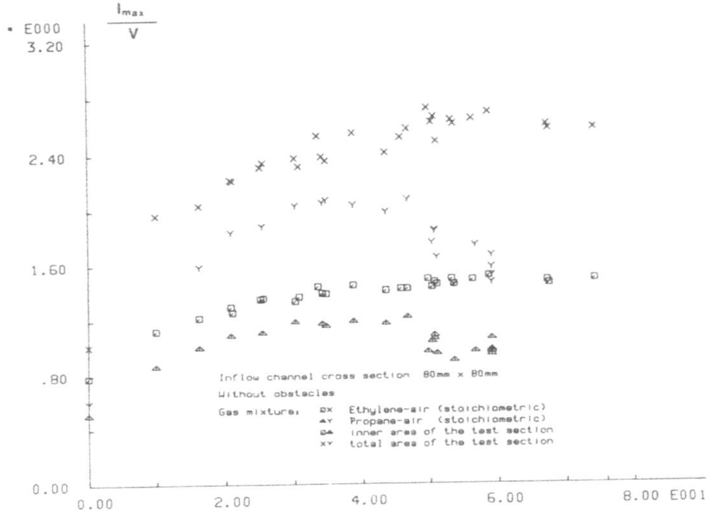

Maximum CH-radiation intensity from the total test section and maximum CH-radiation intensity from the inner area of the test section

Fig. 8

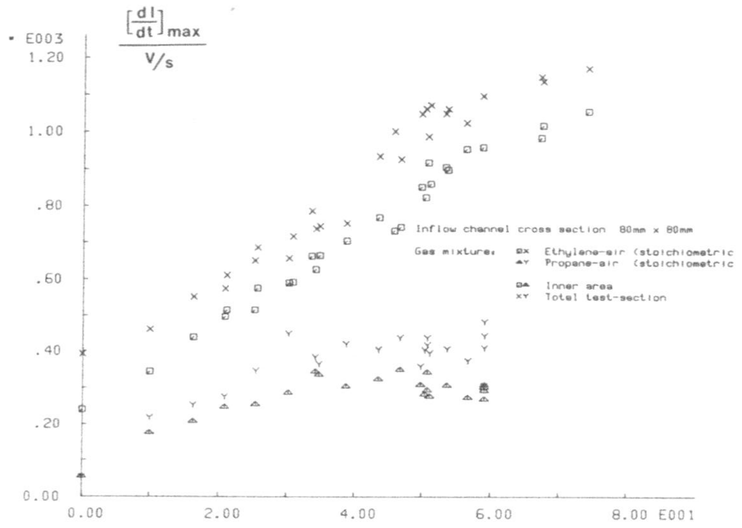

Maximum steepness of the first rise of the CH-radiation signal
from the total area of the test section and
from the inner area of the test section

Fig. 9

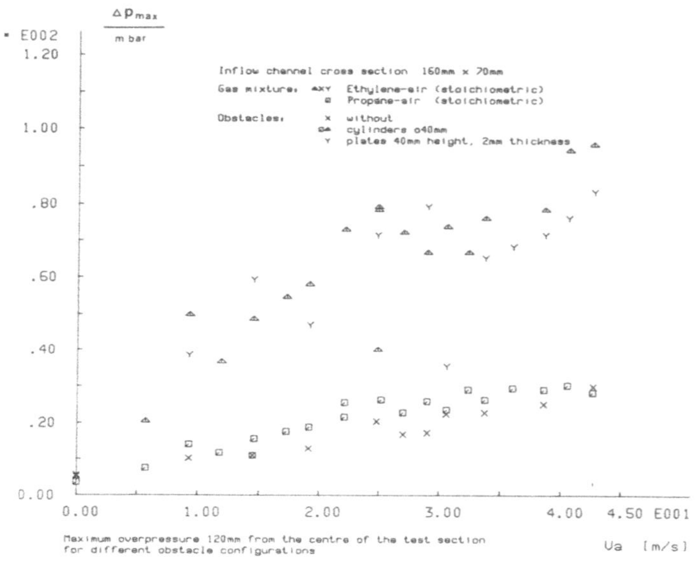

Maximum overpressure 120mm from the centre of the test section
for different obstacle configurations

Fig. 11

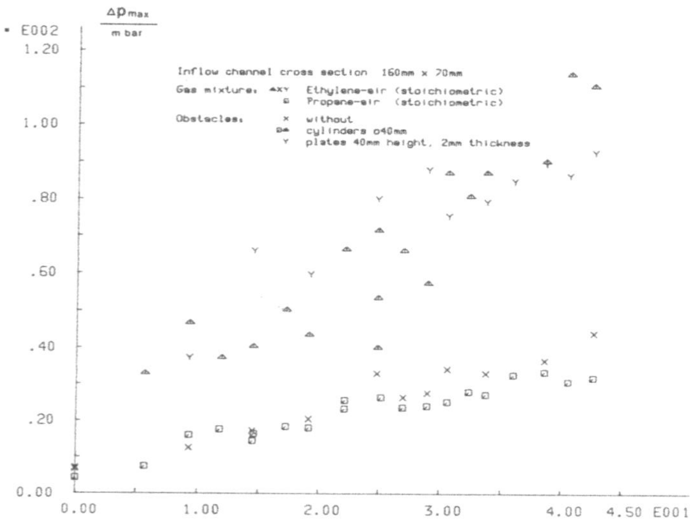

Maximum overpressure at the centre of the test section
for different obstacle configurations

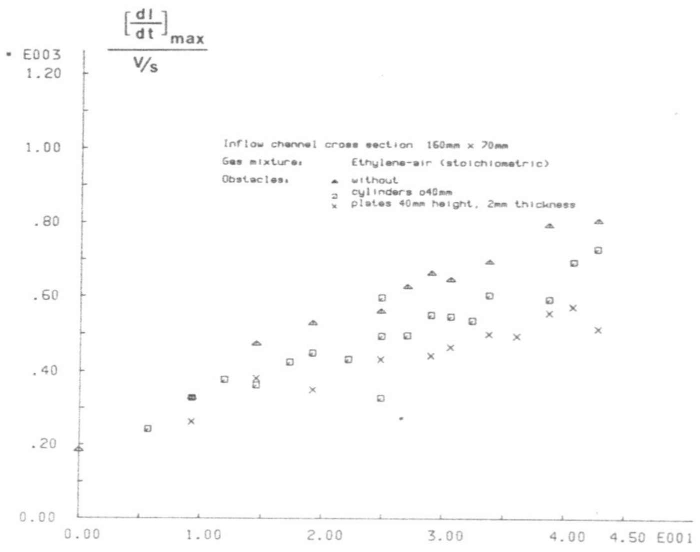

Maximum steepness of the first rise of the CH-radiation signal
from the inner area of the test section for different obstacle configurations

Fig. 12

Parallel Sessions Research Area B

The protection of nuclear power plants against external gas cloud explosions

Session B4

Chairman: A.R. Edwards (UKAEA/SRD-C)

Session secretary: J. Vilain (CEC/BXL)

CALCULATION OF OVERPRESSURES ON THE BUILDINGS OF A POWER PLANT DUE TO THE PASSAGE OF A SHOCK WAVE CAUSED BY AN ACCIDENTAL EXPLOSION

Contrat 13B/026 SRF

N. VAN CHI and A. WARLUZEL
Electricité de France
Direction des Etudes et Recherches (France)

PREFACE

The aim of this study is to determine overpressures on the external structures of a nuclear power plant submitted to a 0.2 bar accidental shock wave. These overpressures result from multiple reflections on various walls of the plant as well as focusing effects.

The problem has been studied by means of two methods : one theoretical, one experimental. The former is based on development and use of a three-dimensional numerical model. Integration of the acoustic propagation equation yields to the desired pressure field on the external structure of the power plant.

Because of the high number of mesh points needed for a good numerical accuracy, the application of this model is somewhat cumbersome.

The second method is experimental. It is based on the analogy between free surface hydraulic shock waves and shock waves in gases. It is a two-dimensional approach. The water level variations are mesured along obstacles representing the structures of the plant when these are submitted to an intumescence. The experimental results enabled us to test the numerical model in the case of reflections on simple two-dimensional geometries.

After being tested, the numerical model has been applied to the real geometry of a plant of two 900 MW units. The analysis of computational results shows that it is possible to combine two-dimensional results to obtain the first orders of magnitude of the overpressures.

I - THREE-DIMENSIONAL NUMERICAL MODEL

The amplitude of the shock wave (0.2 bar) allows the acoustic propagation approximation. Thus, the following equation is integrated in a three dimensional domain by a finite differences method :

$$\frac{\partial^2 p}{\partial t^2} - C^2 \Delta p = 0 \qquad (1)$$

1. Domain of computation and boundary conditions :

The geometrical domain is a parallelepiped in which stands the power plant constituted by two 900 MW groups.

The boundary conditions are :

$$\frac{\partial P}{\partial n} = 0 \text{ on the walls, the ground and the roofs.} \qquad (2)$$

$$\frac{\partial P}{\partial t} + c\,\frac{\partial P}{\partial n} = c^{te} \quad \text{] on the other boundaries of the domain} \qquad (3)$$

This condition of no reflexion is only valid for waves propagating perpendicularly to these boundaries. However distance between the plant and these boundaries is chosen sufficiently large so that parasit reflected waves by these boundaries reach the plant after that the interesting phenomena has occured. Therefore the domain dimensions are : 197 x 201 x 110 m (110 m height) (see Fig. 1). In order to obtain enough precise definition of the plant a variable mesh definition is used. The horizontal mesh is 69 x 66 points. There are 30 mesh planes in the vertical direction.

2. Resolution :

a) Spacial Laplacien (Δ) discretisation :

At each point of the orthogonal meshs we change the volume integral to a surface integral :

$$\iiint_V \Delta P \, dx\, dy\, dz = \iint_S \frac{\partial P}{\partial n} d \qquad (4)$$

For an elementary parallelepiped (Fig. 2) containing point q of spacial indice k,l,m and limited by the mesh indice k - 1, k + 1, l - 1, l + 1, m - 1, m + 1, we have :

$$\begin{aligned} V_{k,l,m} \; P_{k,l,m} = \; & a_{1\,k,l,m} (P_{k+1,l,m} - P_{k,l,m}) \\ + \; & a_{2\,k,l,m} (P_{k-1,l,m} - P_{k,l,m}) \\ + \; & a_{3\,k,l,m} (P_{k+1,l,m} - P_{k,l,m}) \\ + \; & a_{4\,k,l,m} (P_{k,l-1,m} - P_{k,l,m}) \\ + \; & a_{5\,k,l,m} (P_{k,l,m+1} - P_{k,l,m}) \\ + \; & a_{6\,k,l,m} (P_{k,l,m-1} - P_{k,l,m}) \end{aligned}$$

$a_{i\,k,l,m}$ being the surfaces of the elementary volume $V_{k,l,m}$.

When q is a boundary point and according to geometry $V_{k,l,m}$ and $a_{i,k,l,m}$ are modified. Such troncated volumes are shown on figure 2.

This enables representation of vertical walls, vertical troncated dihedrons and roofs. When Q is in plant we write $V_{k,l,m} = 1$ and $a_i = 0$.

b) Newmark scheme :

Integration of equation (1) on each cells gives the following system :

$$\frac{D\Delta^2 P}{\partial t^2} = C^2 AP$$

P vector of dimension K x L x M = N number of points

D square diagonal matrix (volumes of cells)

A square heptadiagonal symetric and negative matrix

Newmark scheme is then for equation (1) :

$$D(P^{n+1} - 2P^n + P^{n+1}) = C^2 \Delta t^2 \; A \; [\alpha P^{n+1} + (1-2\alpha)P^n + \alpha P^{n-1}]$$

$\alpha = 0$ explicit scheme

$\alpha = 0$ implicit scheme

This relation can be expressed as follows : BP = S, with

$$B = \frac{D}{C^2 \Delta t^2} \; I - \alpha A$$

I unity matrix

B symetric matrix, defined, with prevailling diagonal

The matrix B is inversed by a method of over-relaxation (vertical line by vertical line).

d) Computing time :

According to duration of incident wave (400 ms) and lenght of the plant the wave crosses the plant in about 1000 ms ; so integrating time must be the same. With an I.B.M. 3033 computer time of calculation is about half an hour for 200 000 mesh points.

c) Incident wave :

The incident wave with a triangular profil lasts 400 ms and its rise is 20 ms. Its speed is about 400 m/s. It is introduced on a vertical boundary plane of the domain by means of coefficient a_i.

3) <u>Results</u>

First the numerical model has been tested on simple geometries such a wall, a right dihedron and its results have been compared to experimental data (see chap. II). Figure 3 shows the focusing effect in a right dihedron. Then it has been applied to the real power plant.

Some results are given on fig. 4 to 6. Fig. 4 represents pressure infunction of time on a point of Diesel Yard (amplitude of incident wave = 1).

Fig. 5 represents iso pressure lignes at two instants of the propagation.

Fig. 6 shows focusing coefficients on the plant (ratio of local maximum pressure to incident maximum pressure).

II - EXPERIMENTAL STUDY

1) Installation

The tests and experiments were carried out in a flume. The shock wave is generated by mean of a shovel board moving at constant speed driving the fluid as a piston by a mechanical apparutus. The creation of a shock wave is a delicat problem ; so the speed and amplitude of the piston as well as the water depth (30 mm) have been chosen in order to obtain a shock wave with triangular profile, the front being steep and the wave lenght being of 2,10 m. So, according to the real wave length (160 m) the scale is 1/76.

2) Data

The structures (dihedrons representing the walls of the plant, etc.) are placed in the flume and the ratio between the raising of water level along the structures and the amplitude of incident wave gives the focusing coefficient. The raising of water level is measured by mean of resistivity probes.

Following figures show some experimental results and the comparison with the results of the two-dimensional version of the numerical model for simple geometries (schematic case).

Fig. 7 represents pressure version time on two points of a right dihedron.

Fig. 8 represents focussing coefficient in some structures (right dihedron and rectangular yard). We can see a good agrement between experimental and theorical results.

IV - RECOMBINATION OF THE SCHEMATICS RESULTS

The straight forward application of these two-dimensional results or their combination in order to solve three-dimensional problems enables us to give approximative values of overpressures on different sides of the plant. The comparison between these results and the results of the three dimensional code probates this approximative method.

V - CONCLUSION

This paper summarises the works of Direction des Etudes et Recherches of Electricité de France about indirect action program of the European Atomic Energy Community on the Safety of Thermal Water Reactors (area B).

First it sums up the elaboration and the results of a three-dimensional numerical model which computes pressure on all points of a power plant constituted by two 900 MW groups submitted to an accidental shock wave. Then it describes experimental data based on analogy between free surface hydraulic shock waves and shock waves in gases. The experimental results for shematic geometries are in good agreement with theorical results.

Owing to the heaviness of a three-dimensional numerical model, the application of these shematic results permits to obtain quickly some orders of magnitude and can be used during the definition of the plan of a power plant.

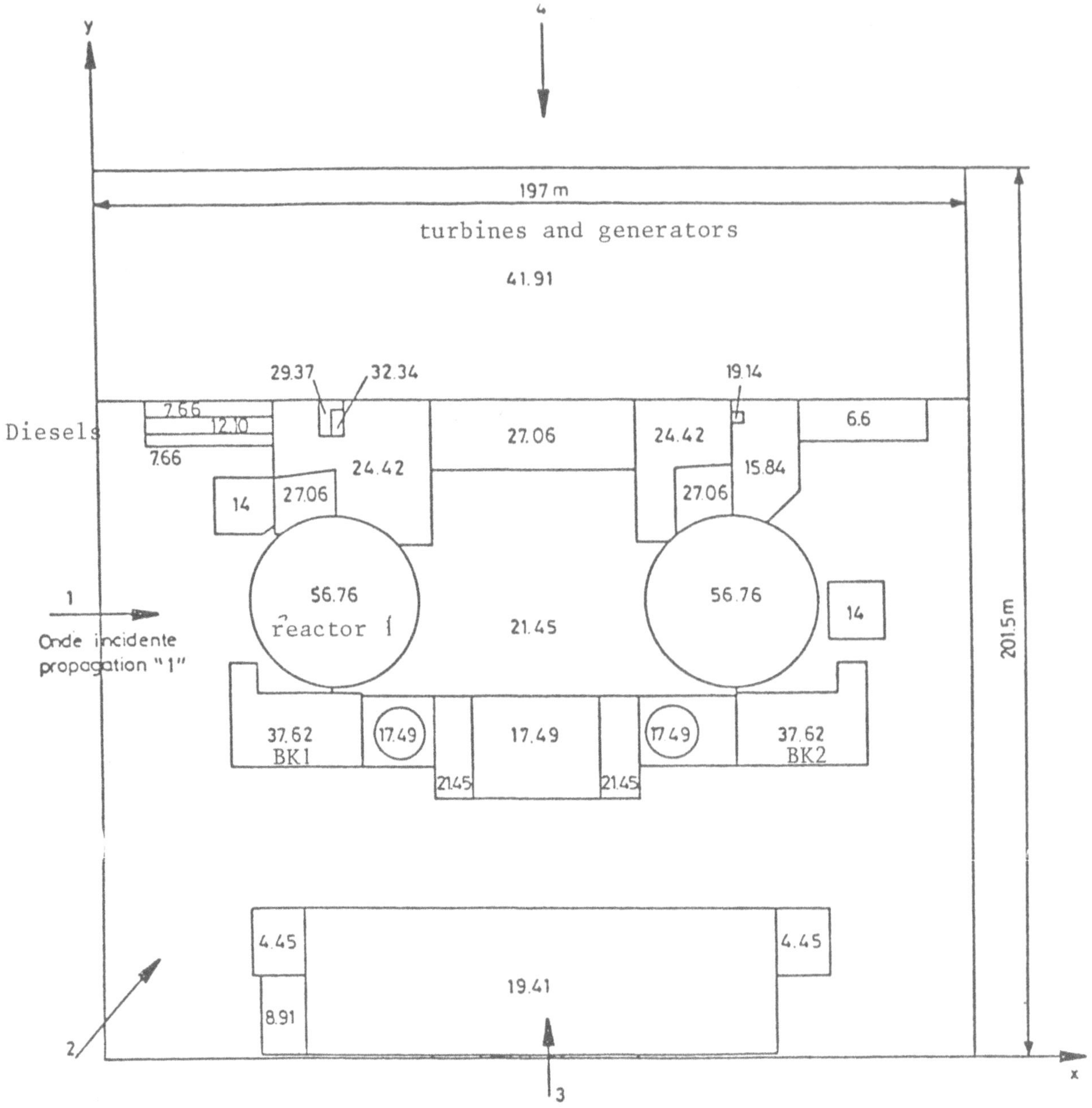

TOP VIEW OF THE PLANT
(ROOFS LEVELS IN METERS)

FIG. 1

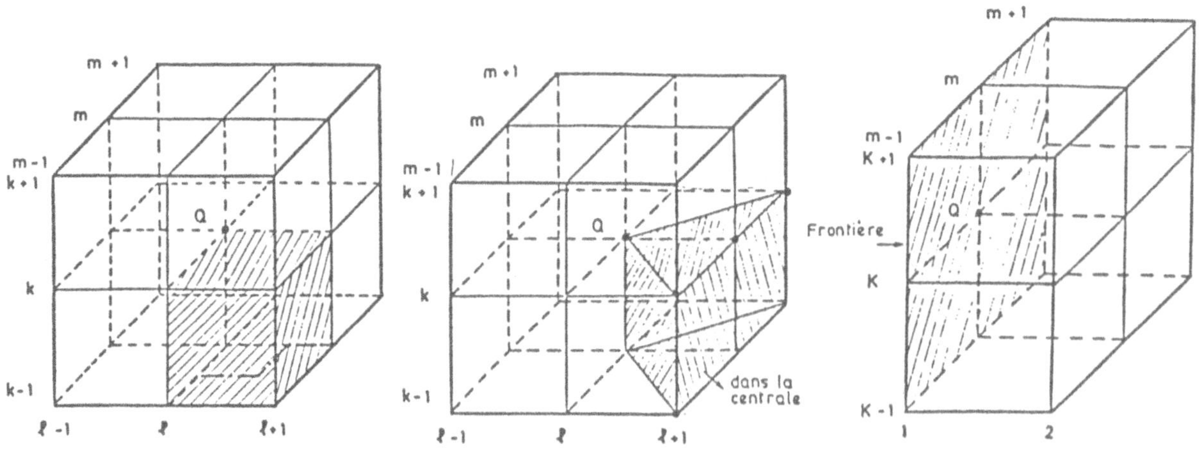

TRONCATED CELLS

FIG. 2

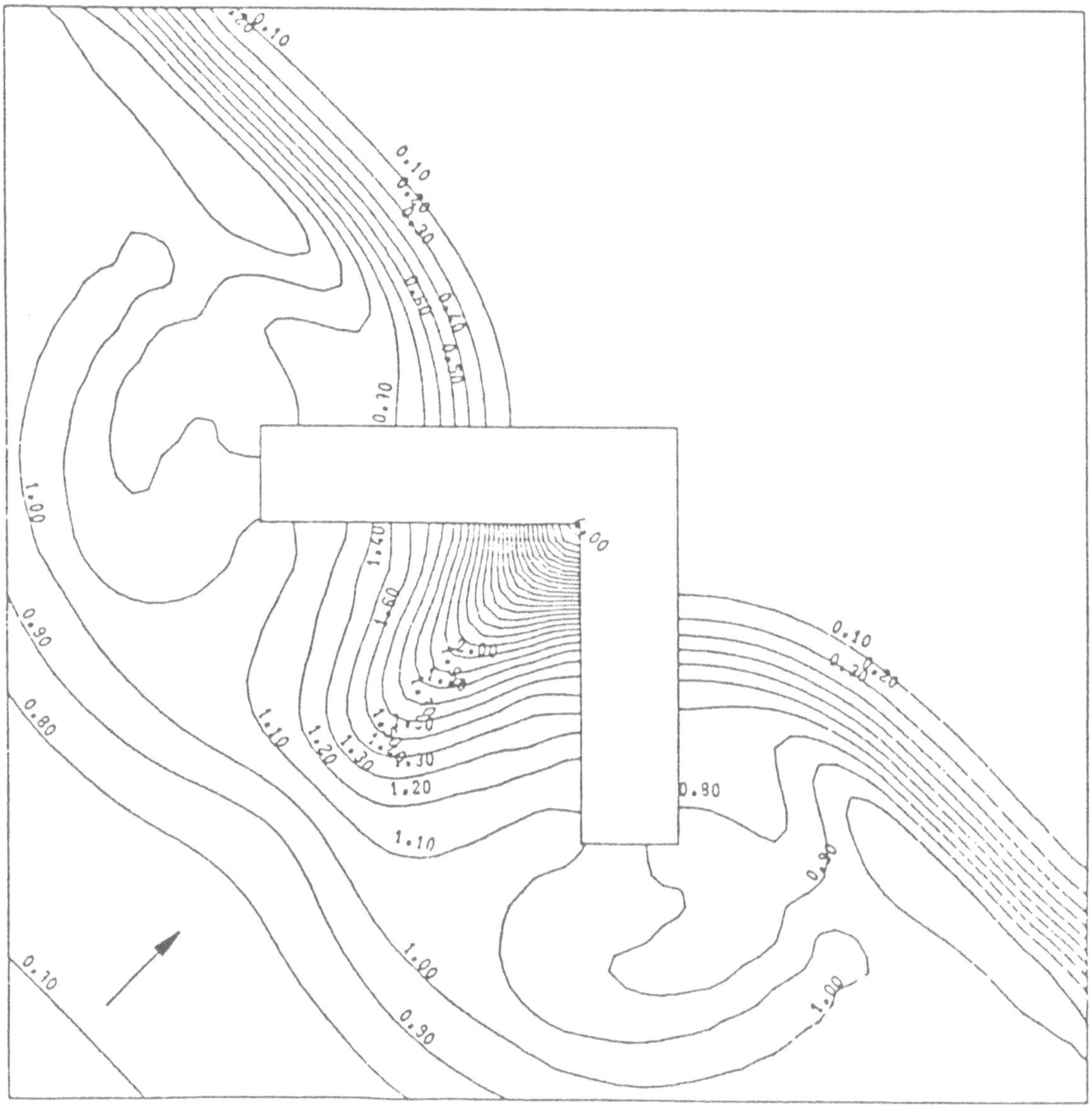

FOCUSING IN A RIGHT DEHIDRON
(WAVE AT INSTANT OF MAXIMUM PRESSURE IN THE TOP)

FIG. 3

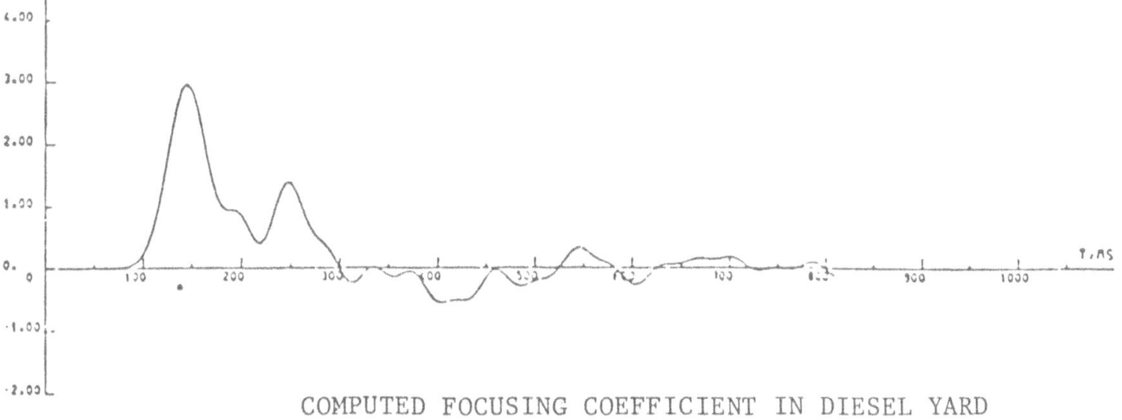

COMPUTED FOCUSING COEFFICIENT IN DIESEL YARD
(RATION BETWEEN LOCAL MAXIMUM PRESSURE TO INCIDENT
PRESSURE)

FIG. 4

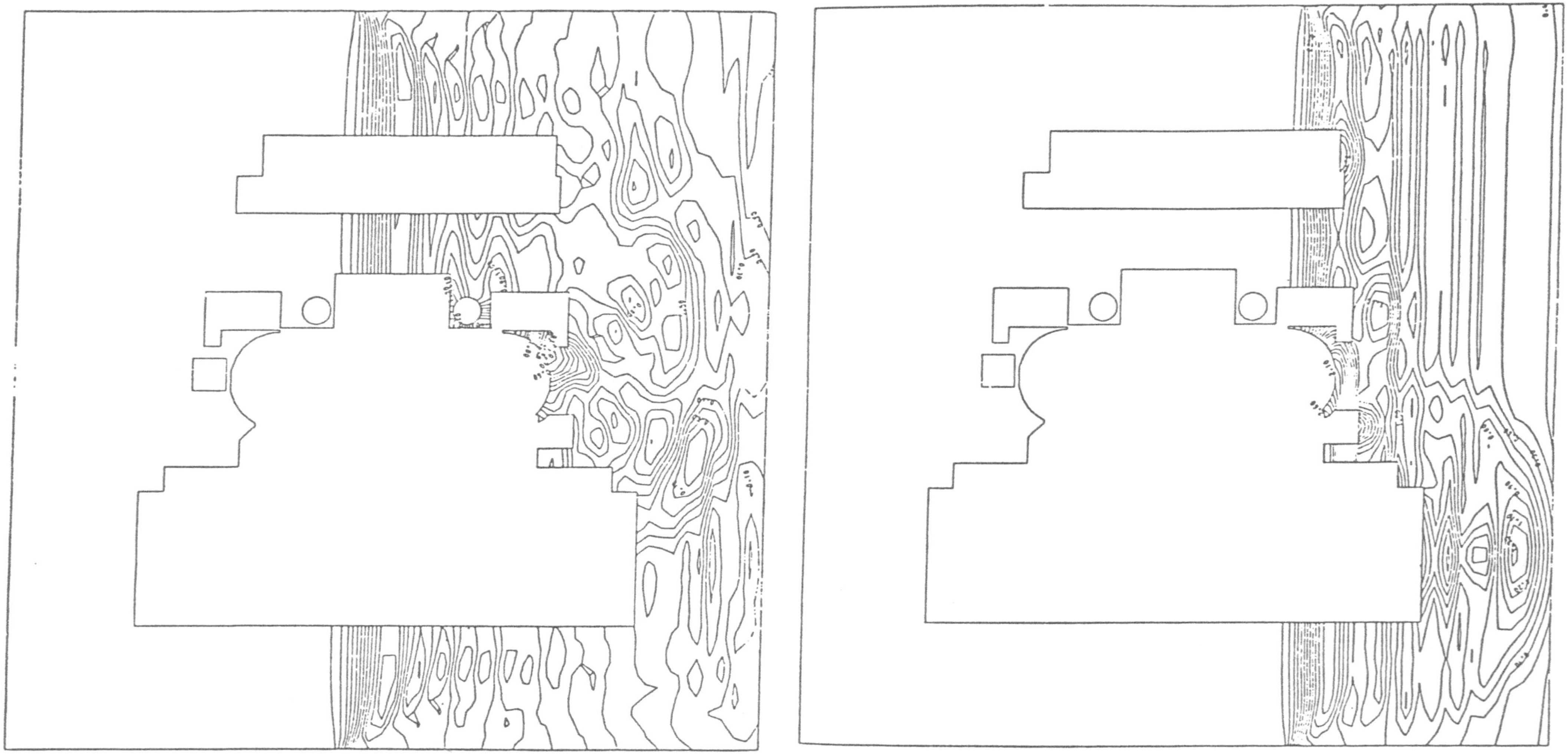

LIGNES ISO-PRESSURE AT TWO INSTANTS OF PROPAGATION AND AT LEVEL 2.5 m

FIG. 5

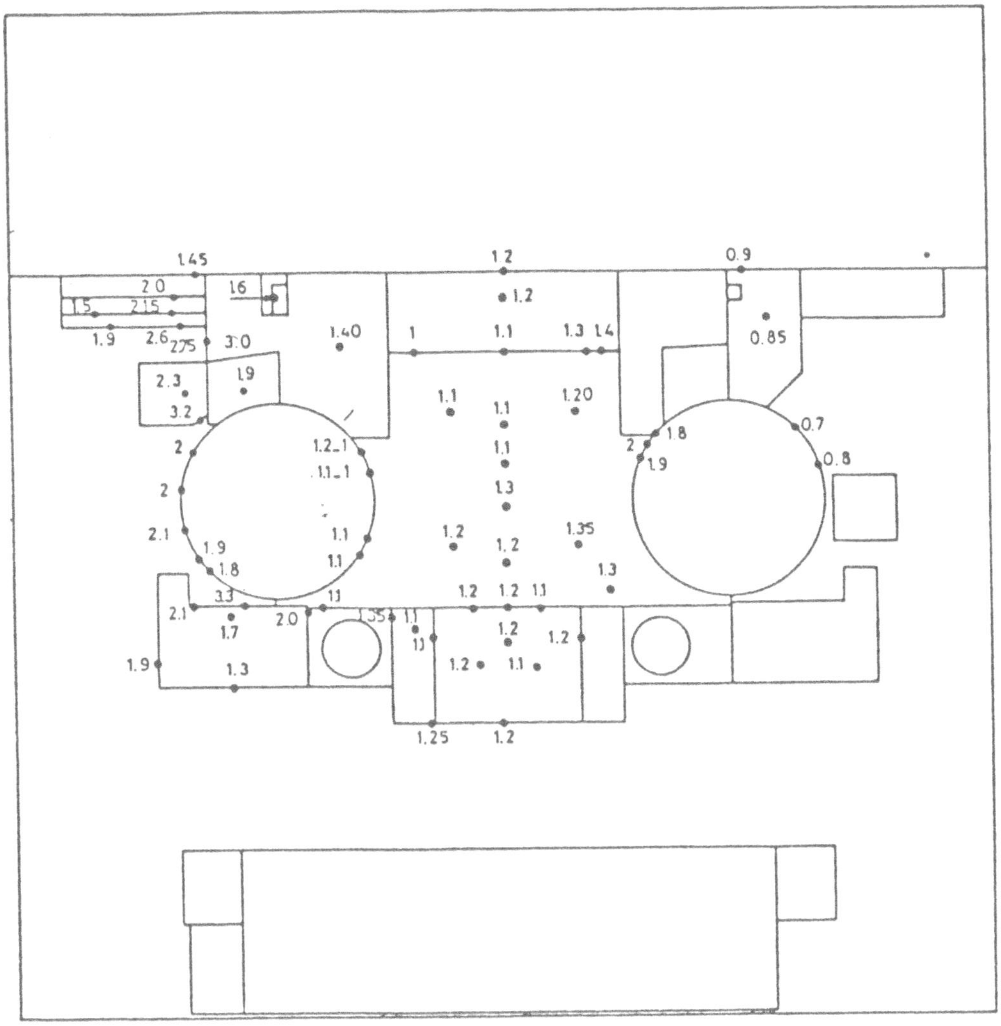

FOCUSING COEFFICIENTS (RATIO OF MAXIMUM LOCAL PRESSURE TO INCIDENT PRESSURE) FIG. 6

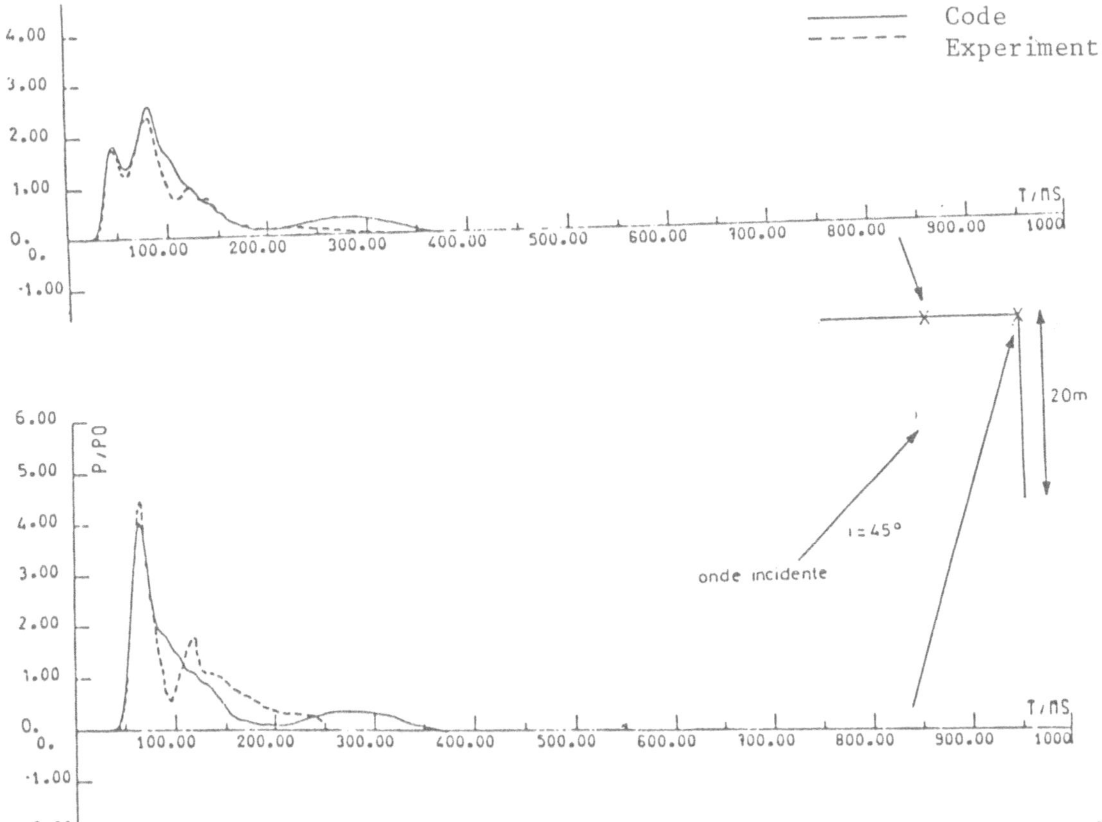

FOCUSING ON A RIGHT DEHIDRON FIG. 7

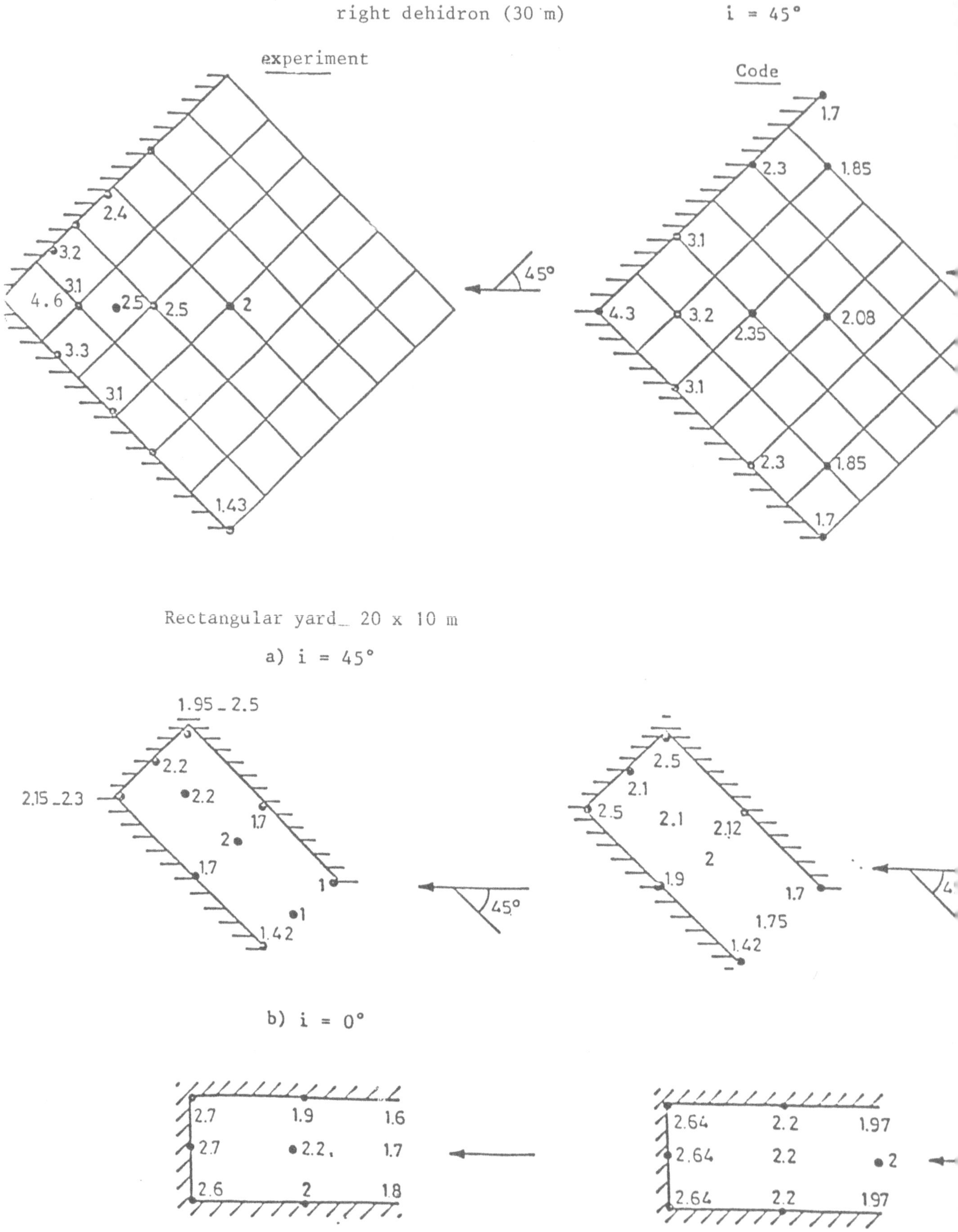

COMPARISON BETWEEN EXPERIMENTAL DATA AND COMPUTATION (2D VERSION) FIG.

Discussion

Question: A. Siebertz - CEN/SCK

What is the model which shows analogy between free surface hydraulic shock waves and shock waves in gases? What are the possible limitations of this model?
What model did you use for computing "your 2D hydraulic experiments" results?

Answer: A. Haughel - E.D.F./DER

The analogy between free surface hydraulic shock waves and shock waves in gases can be proven only for isentropic approximation in gases, that is for weak shocks. It was for us a mean to obtain easily experimental 2D results to calibrate the numerical model, but this one, as I mentioned, uses acoustic approximation.

BLAST WAVE PROPAGATION AND THE INFLUENCE OF OBSTACLES

S.I. Andersen, G.C. Larsen
Risø National Laboratory
Roskilde, Denmark

Summary

Pressure characteristics of blast waves simulating unconfined gasexplosions are investigated through a series of laboratory-scale experiments and larger scale field tests. Initially, the distance attenuation and the degree of rotational symmetry close to the source was investigated for the gas filled latex balloons used in these experiments, and for the field tests, the influence of different atmospheric conditions has primarily been established by using high-explosives detonators. The interaction between the blast wave and different boundaries and obstacles have been investigated in both laboratory-scale and field tests, and it is shown, that the influence is a very localized effect, leaving the blast wave unaffected at larger distances. The interaction with groups of building-like structures has been investigated in laboratory-scale tests, and it is shown that the pressures measured on the obstacle surfaces apparently are affected by both reflection and diffraction of the incoming blast wave. Pressure distribution and reflection factors for a single wall has been determined in the field tests, and relatively large reflected pressures have been measured. Correspondingly large pressures have been measured at certain locations at a group of buildings.

1. INTRODUCTION

The interest in hazards originating from the explosions of volatile solvents and combustible gases has increased considerable during the past decades, and both confined and unconfined explosions have been investigated in laboratory-scale experiments and in larger scale field experiments. Analysis of actual accidents has provided data on the damage experienced by accidental explosions, but the damage potential of unconfined explosions are still not well established.

Several factors may contribute to the uncertainty of damage caused by unconfined explosions, among which are different types of atmospheric conditions, different topographic configurations and different types of interaction with obstacles.

The present paper presents some of the results obtained in an experimental study (1), performed by Risø National Laboratory within the framework of the EEC "Indirect Action Program in Reactor Safety". The experiments comprise both laboratory-scale tests, in which different aspects of the source characteristics (attenuation, symmetry) were investigated as well as different obstacle configurations, and larger scale field tests, in which the influence of topography, obstacles and to some extent the atmospheric conditions were investigated.

2. EXPERIMENTAL PROCEDURE AND MEASURING SYSTEM

The deflagration-type blast wave were in most cases developed by burning of a mixture of methane, oxygen and nitrogen, contained in a latex balloon. The balloon were in all cases suspended in at least such a distance that the expansion of the balloon during the combustion process had finished without touching the ground in order to avoid influence from the ground on the combustion. The diameter of the approximately spherical balloon was 0,3 m in the laboratoryscale experiments, and ranged from 1.0 to 2.7 m in the field experiments. The mixture and approximate gasvolumes are given in table 1, where also typical pressures measured in 1 m respectively 5 m distances are shown.

In the laboratory-scale experiments, the mixture of the three constituents were made before the gas entered the balloon, and it was controlled by flow-meters. In the field experiments, the larger amount of gas were not premixed, but the balloon were always filled first with nitrogen and oxygen and finally with the methane. The amount of each constituents were measured by weighing the gas containers during the operation.

In both experiments the gas was ignited in the center of the balloon. A high-voltage spark was used for the small balloons, whereas the larger balloons were ignited by two small etectric detonators.

The blast wave pressure was measured with three different types of pressure transducers, table 2. The piezo-ceramic transducers were only usable for the low pressure (< 10 m bar) in the small scale experiments, whereas the hydrophones were only used in the field experiments. The B & K-microphones were used in both type of experiments at suitable locations, both as separate measuring points and for reference purposes. The signal from the transducers were recorded on a 21-channel analog tape recorder. The tape were later replayed and the signals digitized and stored on the central computer system for further analysis and reproduction.

3. SOURCE CHARACTERISTICS

As the intention of experiments was to study the relative importance from factors influencing the blast wave propagation, the ideal situation would have been a series of experiments with identical pressure-time curves close to the source, i.e. at a radius approximately corresponding to the largest fire ball radius. However, even for the laboratory scale experiments, this ideal situation could not be achieved and a series of measurements were therefore performed in both small-scale experiments and - to a minor extent - in the field experiments, where the expansion rate, the distance attenuation and variation in angular direction were studied. As the peak pressure varied from test to test, it was decided at a relatively early stage to normalize the results with the peak pressure, measured at a specific distance from the source. In some' cases, also the peak-to-peak pressure (positive and negative peaks) were used.

3.1 Expansion rate

As the latex balloons used were opaque, it was not possible to observe the flame front inside a balloon. However, as the latex skin acts as a membrane and presumably moves as the gas particles during the expansion period, high speed filming of the balloons yield some information on the expansion process.

In 14 tests, a Fastax 16 mm high speed camera were used with frame speeds ranging from 2000-6000 frames per second. Both the gas composition and the initial balloon diameter (i.e. charge size) were varied. In fig. 1 and 2, examples of the deformation of small, respectively large balloons are shown as function of time together with the corresponding measured pressure-time diagram at a specific distance. From the small balloons the time scale changed with the gas composition whereas the initial size influence both deformation and time. In the larger balloons, identical initial diameter and composition did not result in identical deformation patterns for the balloon expansion, and the blast waves were different. One of the reason for this is supposed to be that the ignition of the gas did not occur in the center, as this were very difficult to achieve especially in the field experiments.

3.2 Symmetry conditions

The pressure in a horizontal plane at different locations around the source were measured in a series of tests with the small balloons. As an example, the pressure in 1.0 and 2.2 m distance is shown in fig. 3 and in table 3 the reference value measured in 2.20 m distance and the dominant frequency in the blast wave is shown for these experiments. The peak pressure varies in the circumferential direction close to the source, whereas at larger distances rotational symmetry tends to be established. It should be noted, that there is a quite large difference between the frequency for the individual tests, but for the same test, the frequency are equal at the different locations. For the larger balloons, the pressure were measured in three different angular positions in 20 m distance, and the variation between the individual measurements corresponded well to the observations from the small balloons.

3.3 Distance attenuation

The peak-pressure in the blast wave from both the small and larger balloons applied in these tests showed reasonable agreement with the geometric attenuation at larger distances, i.e. the pressure is proportional to 1/R, where R is the distance from the source. However, close to the source, this is not true as shown in fig. 4, where the pressure-distance curve does not fit a 1/R-approximation for small values of R.

4. BLAST WAVE PROPAGATION

4.1 Influence of atmospheric conditions

The natural atmospheric variations as temperature, wind-speed and wind direction are normally supposed to be representative for the actual conditions when sampled over a 10 min. period. The blast wave propagation is very short compared to this period, and the influence from the different atmospheric conditions should then necessarily involve several tests during a 10 min. sample period.

For practical reasons, this has not been possible to perform using gas filled balloon explosions, and instead high explosive detonators or maroons have been used. The measurements were performed in one direction in up to 100 m distance from the source, and ten firings were done during a 10 minute period in a stable, neutral and unstable atmosphere.

The peak-pressures, normalized with the pressure measured in 5 m distance, is shown for the 10 minute period in fig. 5 for a neutral and an unstable atmoshpere. It is seen that for equal distances from the source, the peak pressure variations as function of the time are larger for the unstable atmosphere than for the neutral atmosphere.

For the three types of atmospheric conditions, the time average value and corresponding standard deviation are calculated for each measuring positions, fig. 6. These curves show, that the standard deviation is larger for larger distances for all types of atmospheres, and it is largest for the unstable atmosphere.

However, the difference between the individual atmospheres are less than the total deviations at distances larger than a few meters from the source, and it is seen that deviations amounts to 15-20% at distances up to 60 m for all three types.

These variations are now assumed to be valid also for the blast wave produced by the gas and have been used in the following experiments for the evaluation of the significance of different effects.

In most of the experiments, the shape of the blast wave remained unchanged within the measuring distance (100 m). However, in a few tests, a steepening of the positive phase were observed, fig. 7.

4.2. Influence from obstacles

Three different kind of "obstacles" has been investigated:

1) a bank of earth, 3 m high and 17.5 m long at the base and 8.5 m deep in the wave propagation direction.
2) a 3.3 m tall and 10 m long light-weight concrete wall.
3) a group of larch trees 4-6 m tall and growing on a plane area with a 1.5-2.0 m spacing.

The distance from the explosion center to the bank and to the wall was 30, respective 25 m, whereas the distance to the first trees were 20 m.

Only the bank influenced the pressure wave significantly as the peak pressure were reduced in a zone approximately 15 m behind the bank, whereas neither the wall nor the group of trees exhibited any influence of the overall behaviour of the pressure wave, fig. 8.

5. INTERACTION WITH OBSTACLES

5.1 Interaction effects

In a series of laboratory-scale experiments, rectangular boxes simulating a building were placed on a hard surface at 2.2 m distance from the source. The cross section of the boxes were square, 0.25 x 0.25 m and they had different length, 0.25, 0.50 and 0.80 m. The pressure was measured by embedded transducers in the box surface, and the position and mutual configuration of the boxes were varied. Both measurements with single boxes and two or more boxes at different distances were performed. The reference pressure were measured in 2.2 m distance opposite to the box position.

From the pressure measurements, reflection coefficients were established at different locations at the surface, fig. 9. The reflection coefficients were measured to 1.25-1.5 in most cases on the front side, and approximately 1.1 on the back.

Applying two boxes situated at equal distances from the source did not result in any interaction effect, whereas this was the case for two boxes placed one behind the other in the propagation direction, fig.9. The pressure on the backside surface of the first box is considerably larger at the center than for the single standing box, and is supposed to be due to reflection phenomenon between the two closely spaced boxes.

5.2 Reflection coefficients on walls

The pressures imposed to the light-weight concrete wall described in section 4.2 were measured by placing the hydrophone transducers close to the surface at various positions both at the frontside and the backside of the wall. Due to the limited number of measuring channels and transducers, several experiments had to be performed. The reflection coefficients are shown in fig. 10, where also the standard deviation are given at positions, where several measurements are available. The average over the front surface is 3.3, and the measured maximum value is as high as 5.5 for a single event, and 4.4 for a position where averaging is possible. The measured average value for the backside is 1.25, and it is surprising, that it is greater than unity.

Reflection coefficients were also measured at a group of buildings, a deserted farm house consisting of a two storey living house with two adjacent auxiliary buildings, all three forming a small courtyard, where one corner were closed as the two auxiliary buildings were joined together. Explosions were performed both outside and inside the courtyard, and pressures were measured at different external and internal locations at the buildings. The internal pressure, measured with all windows opened, corresponded to 70-80% of the corresponding free field pressure, and the pressure on the backside of the house to 50% of the same. Also, the pressure measured inside the courtyard augmented to only 75% of the corresponding free field pressure.

Measured reflection coefficients are given in table 4 in different locations at walls facing toward the blast wave as well as at the corner of the house (external corner) and in the corner between the two auxiliary buildings (internal corner).

The corresponding free field pressures, determined by the measurements in 5 m distance are also given. The reflection coefficients varies from 2.0 to 3.8 for the central part of the wall, and the coefficients measured at ground level are in agreement with the measurements on the free-standing wall and at higher pressure levels to the coefficients in the internal corner. The two low values for the internal corner for 3 and 4 mbar are measured for explosions performed outside the courtyard, in which case living house may still have some shielding effect at the opposite side of the courtyard.

6. CONCLUSIONS

The main conclusions from these experiments are the following.

- The influence on blast wave propagation from different atmospheric conditions (stable, neutral, unstable) are negligible compared to the overall scatter introduced by natural turbulence in the atmosphere. This influence is quite large and may give more than 25% variations in peak-pressures, emanating from the same event.

- In most cases, the influence on wave propagation from different obstacles with characteristic dimensions corresponding to buildings are negligible. However, some local shielding effects have been observed.

- When buildings are closely spaced, reflection between the buildings can result in slightly higher pressure on faces pointing away from the source.

- Reflection coefficients as high as 4-5 has been observed at certain locations on building structures, and the coefficients may vary locally from test to test.

Reference

(1) "Gas Explosion Characterization, wave propagation", Final Report, CEC-Contract No. 025SRDK, Risø-R-499, Risø National Laboratory, 1984.

Discussion

Note: No discussion has taken place on this paper.

Table 1.

Balloon dimensions, gas composition and examples of measured positive phase peak pressures.

Balloon diameter	Volume (m^3)	Composition $CH_4/O_2/N_2$	Peak Pressure (mbar)	at distance (m)
0.30	0.014	16/32/52	5	1.0
1.26	1.0	25/50/25	9	5.0
1.68	2.5	-	69	-
2.16	5	-	130	-
2.72	10	-	140	-

Table 2.

Transducer type used for pressure measurements.

Type	Diameter (mm)	Frequency Range (±2 DB) (Hz)
B & K 4147 Microphone	12.7	0.01-18.000
" 4136 "	6.4	3.9 -70.000
" 8100 Hydrophone	21.0	0.1 -80.000
" 8103 "	9.5	0.1 -125.000
Risø Piezoceramic	16.0	43-1500

Table 3.

Average values of frequency measured in 8 positions at equal distance from balloon.

Measuring radius (m)	1.0	1.0	1.6	1.6	2.2	2.2
Peak ref. pressure (mbar)	2.9	1.2	3.3	2.5	2.8	3.8
Average Frequency (Hz)	155	210	170	159	198	172
Standard Deviation (%)	2	6	1	1	2	1

Table 4.

Measured reflection coefficients on building.

Peak Pressure at measuring position (mbar)	Central part af building at level 0.5 m	2-3 m	5 m	External Corner	Internal Corner
17	3.4	2.0	1.4	2.6	
5	3.4	2.2	2.4	2.6	
15				3.5-4.2	
24	4.2	3.2,3.8	2.2		
4					1.7
3					1.5
4-5		1.7*		1.8	
30		2.6			5.1
21		2.0			4.5
24		2.5			5.4
28		2.5			
~2		3.6,2.2		1.9	
~2		2.8,1.5*		1.6,1.1	

* Measured on wall not perpendicular to propagation direction.

Fig.1. Expansion curve and corresponding pressure curve, 0.25 m balloon.

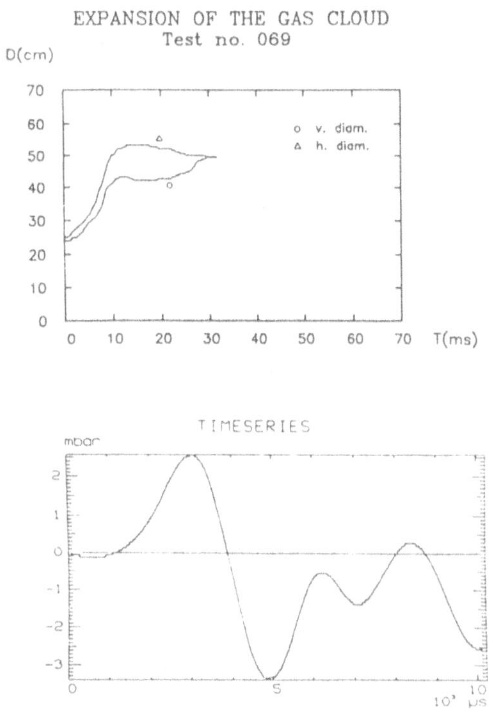

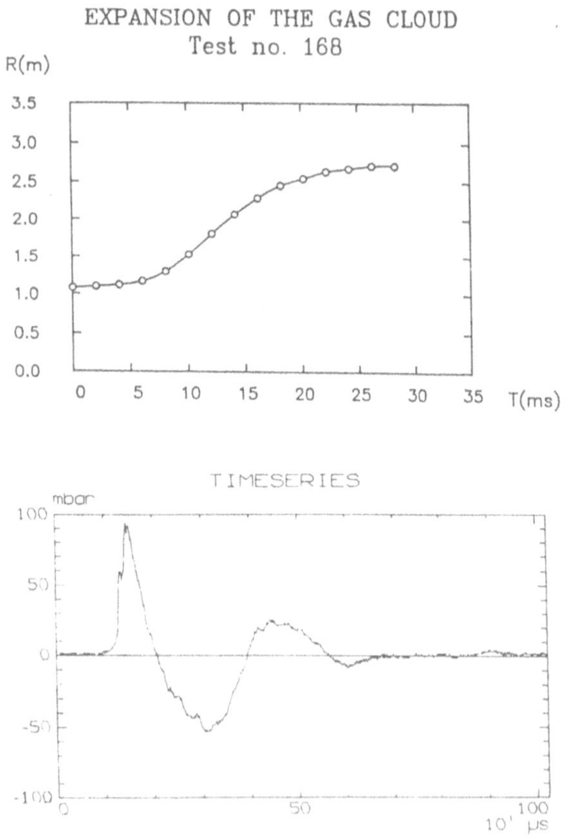

Fig.2. Expansion curve and corresponding pressure curve, 1.0 m balloon.

Fig.3. Pressure distribution in circumferential direction in 1.0 m and 2.2 m from 0.30 m balloon.

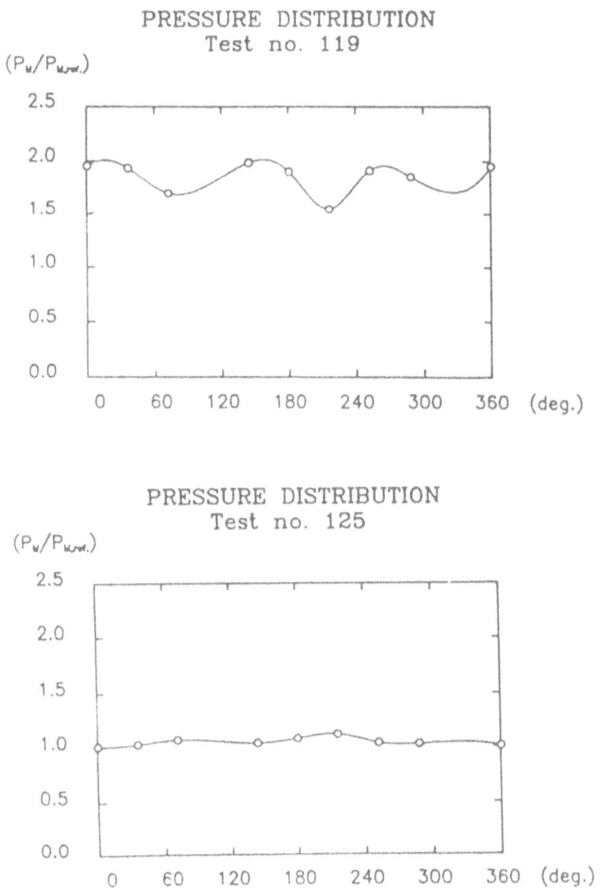

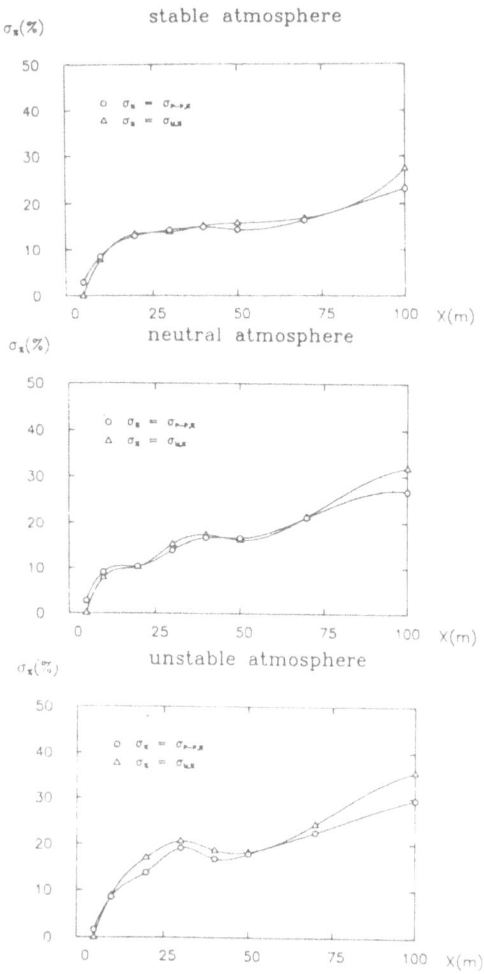

Fig.6. Standard deviation of measured peak pressure as function of distance from the source.

Fig.5. Normalized peak pressure from 10 maroon tests.

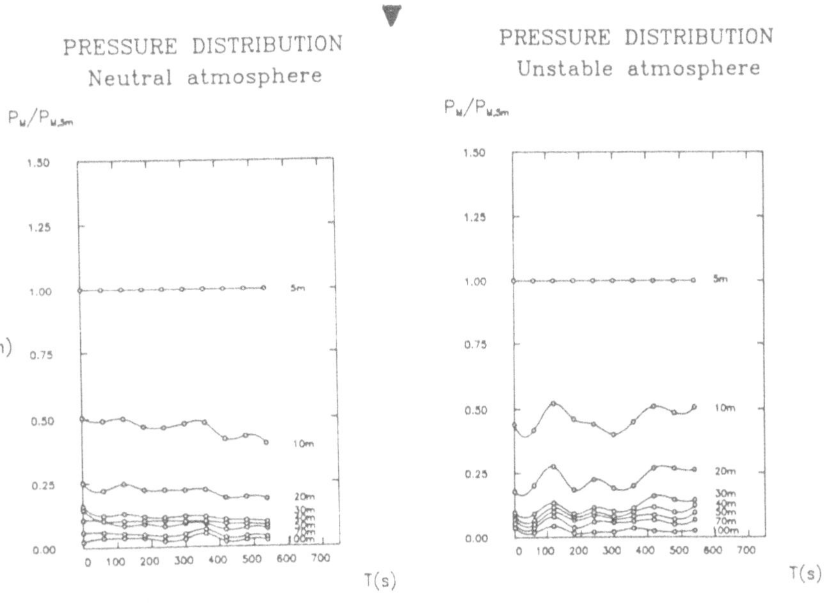

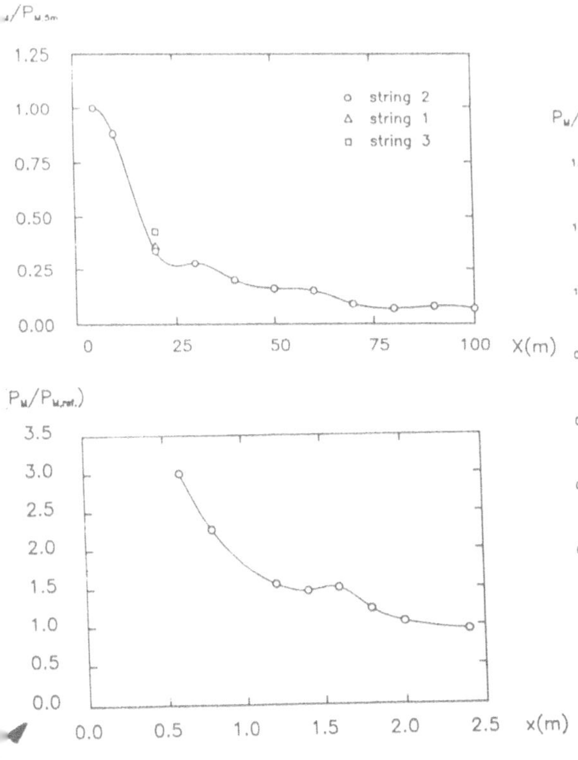

ig.4. Normalized peak pressures measured within 2.5 m distance for 0.30 m diameter balloon and 100 m for 2.7 m diameter balloon.

Fig.7. Steepening of wavefront, 2.7 m balloon.

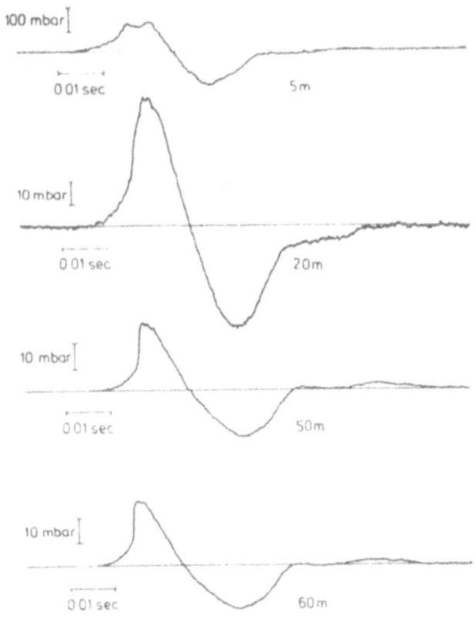

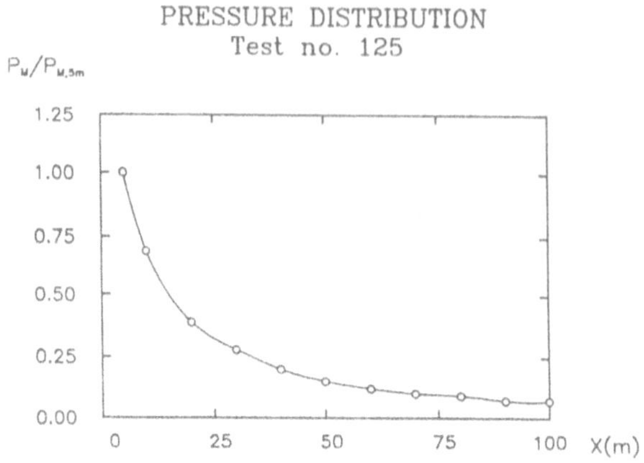

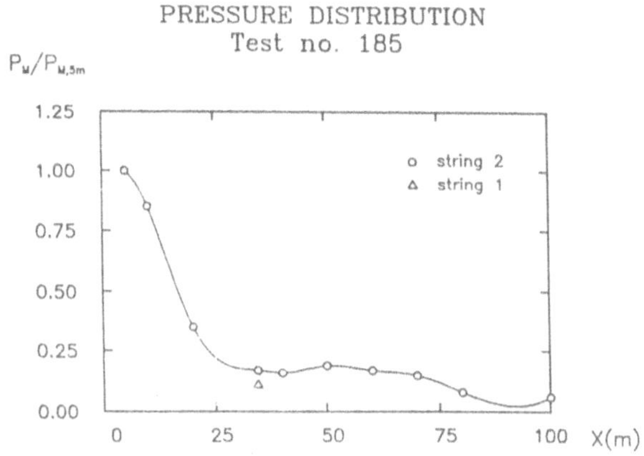

Fig.8. Distance attenuation for 3 m high bank of earth (test no. 185) and 3.3 m tall wall.

Fig.9. Normalized peak pressures measured on front and back of box shaped obstacles.

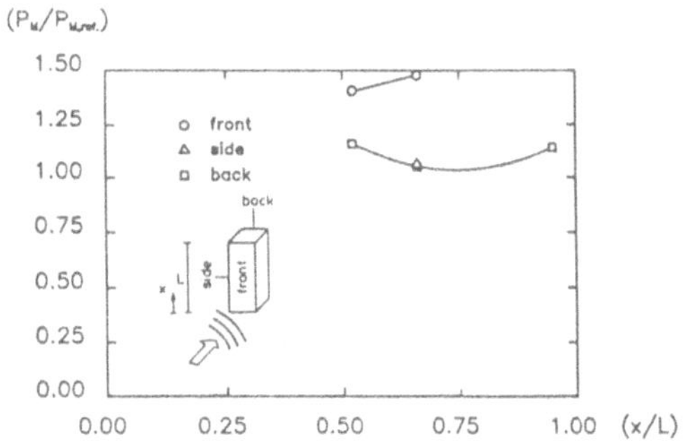

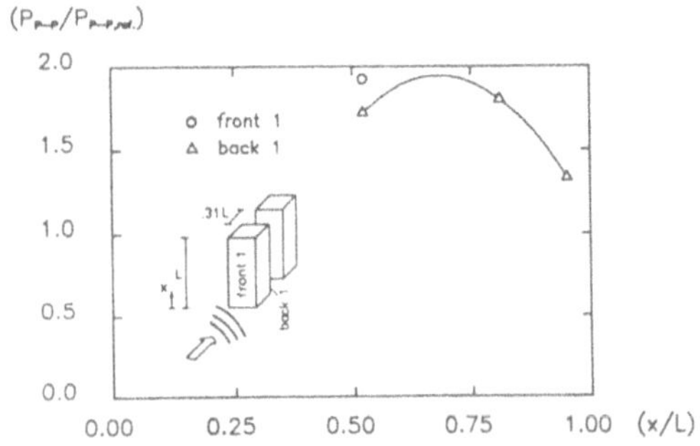

Fig.10.Reflection coefficients measured on front and back of 3.3 m wall.

CALCULATION OF THE WALL PRESSURE FIELD GENERATED ON A GROUP OF BLUIDINGS BY AN EXTERNAL EXPLOSION

C. BERRIAUD and P. VERPEAUX

Abstract

The aim is to work out a tri-dimensional code calculating the overpressure distribution on any structure when an air shock wave arrives specially in the case of geometry providing multiples waves reflection. The computer code ZEPHIR was written with the assumption that the phenomena agrees with the acoustic laws. An integral formulation leads infinite tridimensional problems to a bidimensional one on a surface S limiting the real D. The surface S is discretized with triangular elements and replaced by the polyedre constituted by the mesh. The pressure is calculated at the gravity center of the triangular elements with interpolation between several time steps.

The ZEPHIR codė has been validated at first against analytic solutions obtained for simple obstacle geometries.

Tests were performed with models of PWR buildings at a scale of 1/40, a cylinder for the reactor building and a parallelepiped for the fuel building, used adajaced or separated by a gap. Plastic explosive permits to obtain the pressure time history of an hydrocarbure explosion at the scale. The results are in good agreement with the computation.

The code ZEPHIR is characterized by a relatively cheap utilization.

1. INTRODUCTION

The calculation of the external explosion effects on building must be conducted through three independent steps as following :

1 - Determination of the waves characteristics in free field from the peculiar conditions of the accident.

2 - Overpressure distribution calculation on the building from wave characteristics.

3 - Calculation of the building response to the overpressure distribution loading.

There are already methods to perform the first step. The last step is actually the most known and codes are available to calculate any structure. But the middle step was not easy. The complexity of tridimensional equation leads generally to an expensive computation and most of the engineers are using bidimensional codes.

The aim of this work was to work out a cheaper tridimensional code to obtain results good enough to be used in the most part of civil problems. So the useful configurations are characterized by a low surpressure and a particle displacement little in relation with the building dimension : the phenomenon can be considered as linear and one can suppose that the acoustic equation is valid in these cases. The explosions bringing on strong wind as nuclear ones or chemical ones closed to small building are excluded.

2. SYMBOLS

A	: solid angle at M_o
c	: sound velocity
D	: diameter of the cylinder
d	: distance MM_o
$d\sigma$	: surface element at point M
i	: parameter to be adjusted to avoid computation instabilities
M	: point of the surface Σ
$\vec{n}$	: external normal vector to Σ at point M
P	: maximum static overpressure in free field
Pr	: theoretical reflected overpressure in free field
p(M,t)	: overpressure at point M and time t
R	: ratio of time necessary to cross the element over the time step
r	: distance from M_o to the source of pressure
S(t)	: source value at time t
t	: time
t_c	: time coefficient used to modify the time step
$\Delta \ell$	: mean dimension of an element
Δt	: computation time step
Δx	: distance between gravity centers of two elements
Σ	: surface of the structure
Σ'	: surface of the structure excluding the proximity of point M_i
τ	: reduced time = $\frac{2c}{D}$ t

3. ZEPHIR CODE

3.1 Hypothesis

Alembert's equation describes the physical problem under some more assumptions :

- the overpressure amplitude ($\simeq$ 0.1 bar) is small in front of atmospheric pressure and effects of flow are neglectable. It is acoustic waves.
- particulate displacement is small in front of structure dimensions. Fluid-solid interaction is "acoustic".
- a complete reflexion on boundaries is also assumed because they are perfectly rigid and motionless.

Many workers tried to resolve this equation but rapidly difficulties to resolve the problem in an infinite acoustic space arise because of actual means of computation. ZEPHIR code is formulated with boundaries integral equations solved with a technique of time retarded potentials, then infinite 3D problem switchs in a 2D problem over the surface of the structure.

3.2 Numeric formulation

The code ZEPHIR computes pressure over the surface of a 3D structure under any incident waves functions of time S(t) coming up from a pin point source situed at finite distance. Under acoustics assumption the equation giving at time t_o pressure at point M_o of the surface is ($\vec{\text{grad}}\, P.\vec{n} = 0$) :

$$(4\pi - A)\, p(M_o t_o) = \iint_\Sigma \left[p(M, t_o - \frac{d}{c})\ \text{grad}\ \frac{1}{d} - \frac{1}{cd}\ \dot{p}\ (M, t_o - \frac{d}{c})\ \text{grad}\ d \right] n\ d\sigma$$

$$+ 4\pi\ \frac{S(t_o - \frac{r}{c})}{r} \qquad (1)$$

The real structure is replaced by a set of flat triangular elements.

The use to flat element leads to a great simplification of the initial equation. In fact, it is possible to set at the gravity center M_i of each element i

$$2\pi\, p(M_i, t_o) = \iint_{\Sigma'} \left[p(M, t_o - \frac{d}{c}) \text{ grad } \frac{1}{d} - \frac{1}{cd} \dot{p}\,(M, t_o - \frac{d}{c}) \text{ grad } d \right] n\, d\sigma + 4\, \frac{S(t_o - \frac{r}{c})}{r} \qquad (2)$$

because : - solid angle A around M_i is equal to 2π

- $\overrightarrow{\text{grad}}\, \frac{1}{d} \cdot \vec{n}$ and grad $d.\vec{n}$ are equal to zero over the i element because $\overrightarrow{\text{grad}}\, d$ is included in the plane of the element.

The second part of equation (2) is decomposed over the j elements ($j \neq i$).

$$2\pi\, p(M_j, t_o) = \sum_{\substack{j \in \Sigma \\ j \neq i}} \iint_j \left[p(M, t_o - \frac{d}{c}) \text{ grad } \frac{1}{d} - \frac{1}{td} \dot{p}\,(M, t_o - \frac{d}{c}) \right] n\, d\sigma + 4\pi\, \frac{S(t_o - \frac{r}{c})}{r} \qquad (3)$$

for both parts of the integral of the above equation we assume the pressure is constant over the element during an interval of time.

$$\left. \begin{aligned} A_{ij} &= \iint_j \text{grad} \frac{1}{d}\, n\, d\sigma \\ B_{ij} &= \iint_j -\frac{1}{cd} \text{grad}\, d.n.d\sigma \end{aligned} \right\} \iint_j \left[p(M, t_o - \frac{d}{c}) \text{ grad} \frac{1}{d} - \frac{1}{cd} \dot{p}(M, t_o - \frac{d}{c}) \right].n.d\sigma = p\, A_{ij} + \dot{p}\, B_{ij}$$

A_{ij} : solid angle issued from the gravity center of i and containing element j. These quantities are computed at the beginning of the run. At that time we have chosen to do an exact computation of these quantities using spheric trigonometric formulas (Huillier's formula). It is justified to perform such computation because it is important to check :

$$\sum A_{ij} = 2\pi$$

Which is necessary in order to verify the balance of energy.

B_{ij} term is more difficult to compute, we choose a numerical integration method. Using :

$$\iint_j \frac{\text{grad } d}{d} \cdot n\, d\sigma = \iint_j - d \text{ grad } \frac{1}{d}\, n\, d\sigma$$

Let us assume $B_{ij} = - d\, A_{ij}$.

This last substitution is done on a set of triangles which are integration elements.

Having a closer look on the equation (3), we can see that $p(M_o t_o)$ is only dependent on :

- pressure over the structure until the time $t_o - \Delta t$. Δt is the time necessary to the wave to advance a characteristic length of the element,
- the derivate of the pressure up to $t_o - \Delta t$.

Assuming known t the problem is now to discretize $p(M, t)$ for $0 < t < t_o - \Delta t$ and for $M \in \Sigma$.

For the spatial problem we assume that the value is constant and equal to the pressure at the gravity center.

Let us assume that the pressure is known at any Δt up to $t_o - \Delta t$.

The equation (3) can be set to (4) :

$$2\pi\, p_i(t_o) = \sum_{j \neq i} \left(A_{ij}\, p_j\, (t_j) + B_{ij}\, \dot{p}_j\, (t_j)\right) + 4\pi\, \frac{S(t_o - \frac{r}{c})}{r_i} \tag{4}$$

with $t_j < t_o - \Delta t_i$

and where the only problem is the interpolation of :

- $p_j\ (t_j)$ at the different times
- $\dot{p}_j\ (t_j)$.

In the figure 1 the pressure to take in account is the one between C and D limits of j element.

A natural hypothesis is that $p_j(t_j)$ is only dependent of j which leads p to be on line AB.

In order to have $p_j(t_j)$ we first do a linear interpolation between the closest time of t_j : t_{j1}, t_{j2}

then, $\dot{p}$ is : $\dot{p}_j = \dfrac{p_j(t_{j2}) - p_j(t_{j1})}{t_{j2} - t_{j1}}$

This method called method one leads to time instability.

The instability makes not sure of the precision of results. So a considerable effort has been made in order to obtain a stable method.

It seems better to have a mean value of p during the time step over j. So we did an interpolation between less close times up to A - B,

- the closest : option 0 in the code which is in fact the method 1,
- the ith closest : (Figure 2) option i in the code (method 2).

All results presented here were obtained with this last version.

3.3 Practice and limits of use

The structure must be meshed taking into account on one hand the smallest size and on the other hand the smallest number of elements. Distance between two gravity centers of element must be as constant as possible over the whole mesh. Function of speed wave propagation and the size of element we have a time step value Δt. To this time step a frequency $f_t = \frac{1}{\Delta t}$ is a cutting frequency. For phenomena existing at a higher frequency than f_t, it is necessary to short the time step and to stabilize the numeric computation.

The code calculates the distance between the gravity center of the neighbouring meshes and the time step is taken from the lowest found distance Δx_{min} by $\Delta t = \frac{\Delta_{x\ min}}{C}$. The user can modify Δt by introduction of a coefficient t_c. So Δt is then equal to $\frac{\Delta_{x\ min}}{C} \times t_c$.

As indicated in 3.2, the i parameter is introduced to prevent numerical instability.

A practical rule is to take $i \cong \frac{R}{2} = \frac{\Delta \ell}{2C\ \Delta t}$ where $\Delta \ell$ is the mean dimension of an element. Since $\Delta t = \frac{\Delta_{x\ min}}{C}$, $i \cong \frac{\Delta \ell}{2\Delta_{x\ min}}\ \frac{1}{t_c}$.

Generally $\Delta \ell$ is about the double of Δx_{min} and $i \cong \frac{1}{t_c}$ but with some meshing the optimum can be different (i is an integer and t_c a decimal number). It is recommended to take i equal to the integer immediately above the obtaind value.

In the case of a parallelepiped (fig. 3) 5 calculations were made, resumed in the following table. The source is a pressure pulse equal to 1 with a width equal to 3 time steps of the first calculation.

t_c	1	1	0.7	0.7	0.7
i	0	1	1	2	3
$\frac{1}{t_c}$	1	1	1.4	1.4	1.4
recommended i value	1	1	2	2	2
Example figure n°	4	5	6	7	8

One sees :

- instabilities when i is lower (fig. 6) or higher (fig. 8) than the recommended value. They are very strong for i = 0 (fig. 4),
- no instabilities for the recommended values :
  $t_c = 1$ $i = 1$ (fig. 5)
  $t_c = 0.7$ $i = 2$ (fig. 7)
- the same level of the peak but a more precise curve with $t_c = 0.7$ (fig.7) than with $t_c = 1$ (fig. 5), specially in regard of the rectangular pulse shape.

As any explicit computation in time ZEPHIR is under condition $\frac{c\Delta t}{\Delta x} \leq 1$, computation cost increases with the ratio structural dimension versus wave lengths. In fact, this ratio must be smaller than 2 to 5.

An idea of the computation cost is given by the following example : for a group of a cylindrical building and a parallelepipedic one with 143 nodes and 260 elements the computation cost for 54 time steps is between 2000 and 3000 francs on IBM 3081. The memory filling is 204 K octets.

4. EXPERIMENTAL DATA

In order to validate the ZEPHIR Code experimental data were searched. Some results of test made for "Departement de Sûreté Nucléaire" of "Commissariat à l'Energie Atomique", were available [1] and a complementary test program was performed. The characteristics on the whole are resumed in table I.

Models were made of welded steel plates or prestressed microconcrete. In all cases the cylinder was 1,4 m height and 1 m in diameter and the parallelepiped 0,9 m height, 0,75 m long and 0,5 m wide. These dimensions are corresponding about to a french PWR nuclear plant at the scale 1/40. Material of explosion and distances to the models were chosen to obtain an overpressure time history on the models at the scale (The reference is a triangular shape with 0,2 b overpressure and 0,4 s duration).

5. ZEPHIR VALIDATION

Two series of calculations were performed :

In the first one comparisons were got with analytical solution computation (Infinite cylinder with a pressure step as source and with an exponential source - sphere under a pressure step - dihedron under a pressure step).

In the second one they were got with experimental results (cylinder under a detonation - cylinder and parallelepiped joined under a deformation - cylinder and parallelepiped with a gap between them under a detonation - cylinder and parallelepiped joined under a deflagration).

For instance in fig. 10 and 11 comparisons are made between results of ZEPHIR computation on the cylinder meshing of fig. 9 and analytical solution computation on an infinite cylinder. In the same way fig. 13 shows a comparison between MIMOSA test and calculation on the meshing of fig. 12.

All these comparisons lead to the following conclusions :

- There is a good agreement between ZEPHIR calculations and analytical (fig. 10) or experimental results (fig. 13).
- But a parasitic peak (fig. 11) appears on the rear part of the buildings at a time corresponding to a direct way through the front part. The true wave passing round the building arrives later. The parasitic peak is then easily identifiable and can be cancelled in interpretation of results.
- Theorically the acoustic assumption is not true for overpressure higher than 0.2 bar. Nevertheless tests performed up to 0.9 bar (fig. 14) give results a little higher than the ZEPHIR computations but still in good agreement with them.

6. CONCLUSION

The ZEPHIR Code is now available to computation of the wall pressure field generated on a group of buildings by an external explosion. The choice solution permits calculations on three dimensional structures at reasonable price in acoustic field, but also with higher pressure ; in the last case, the pressure peak obtained can be a little too low.

A simple rule is given to choice coefficients concerning the time step to avoid instability computation.

A parasitic peak can appear at the rear face of the structure. It is corresponding to a direct wave way through the structure and appears at the time given by the sound velocity. So it is easy to detect and to cancel it.

REFERENCE

1. B. BARBE and al.
Behaviour of a PWR containment submitted to an external explosion.
Paper J 10/9 6th SMIRT, Paris, 17-21 August 1981.

TABLE I

Tests Characteristics

Test	Structure	Incident overpressure				Source		
		Type	Static pressure in Pa	Duration in ms	Direction: Angle from parallelepiped axis	Material	Quantity	Distance in m
AMEDEE 2/6	Cylinder	Detonation	7 000	18		Air-Ethylene 8%	215 m^3	110
MIMOSA 79-1	Cylinder and joint parallelepiped	"	22 000	7	0	Plastic	17 kg	22
2	"	"	23 000	7	- 45°	"	"	"
3	"	"	22 000	7	- 90°	"	"	"
4	"	"	24 000	6	0	"	"	"
5	"	"	22 000	7	- 45°	"	"	"
6	"	"	21 000	8	- 90°	"	"	"
MIMOSA 80 [1]								
$S_o/2$	Cylinder and joint parallelepided	"	10 200	9.1	- 20°	Plastic	12 kg	35
S_o	"	"	20 500	9	"	"	22 kg	27
$2S_o$	"	"	35 000	9	"	"	41 kg	22
$3S_o$	"	"	57 000	9.4	"	"	64 kg	20
$4S_o$	"	"	75 000	10	"	"	94 kg	19.5
$5S_o$	"	"	109 000	9.5	"	"	160 kg	20
SESAME 1	Cylinder and joint parallelepiped	"	20 000	7.5	- 22°5	Plastic		22
2	Cylinder and distant "	"	15 000	7	"	"		21.9
3	"	"	22 000	9	"	"		23.8
4	Cylinder and joint parallelepiped	Deflagration	6 800	18	"	{ air, O_2, C_3H_8	1500 l, 2000 l, 700 l	3.5

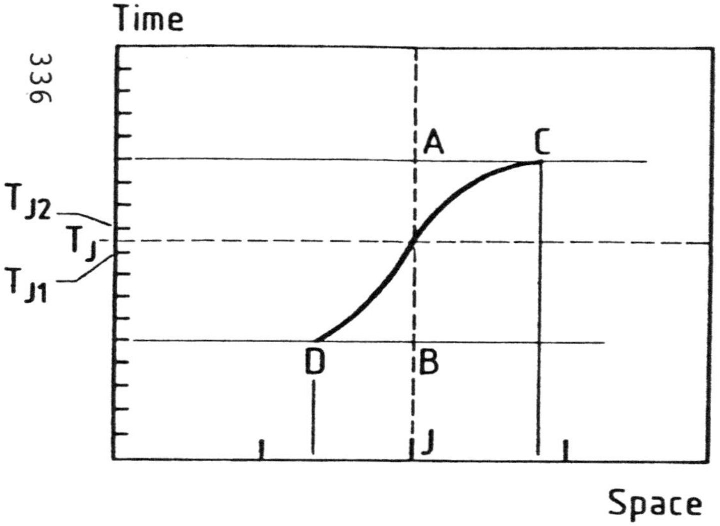

Fig. 1

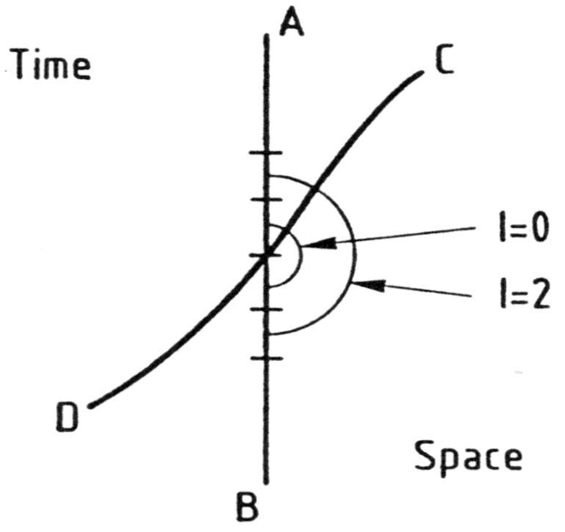

Fig. 2

Mesh length 15 height 9 width 15

Number of nodes 48
Number of elements 82
step $\Delta x = \sqrt{2}$

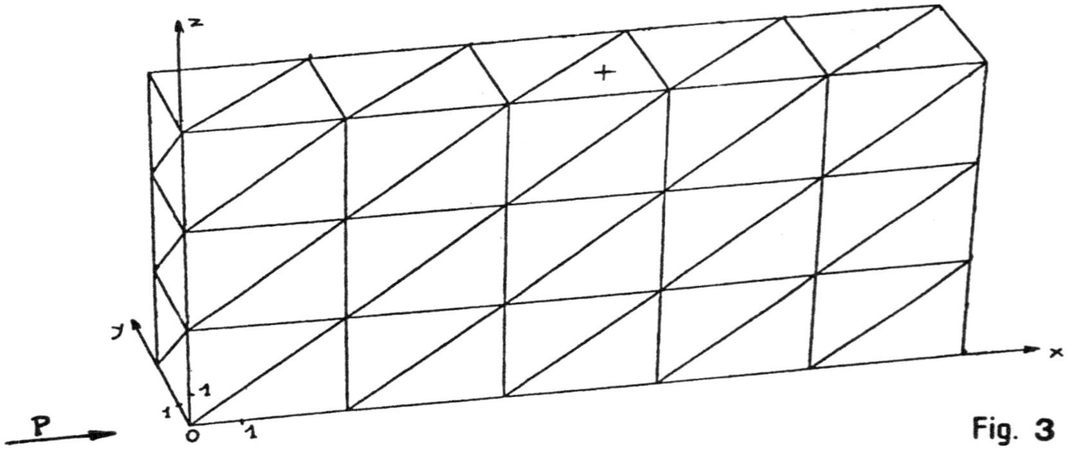

Fig. 3

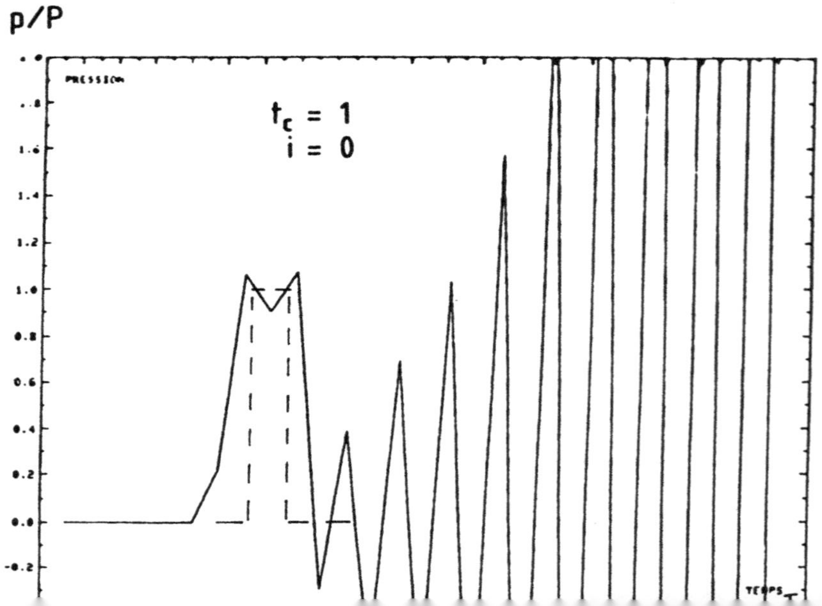

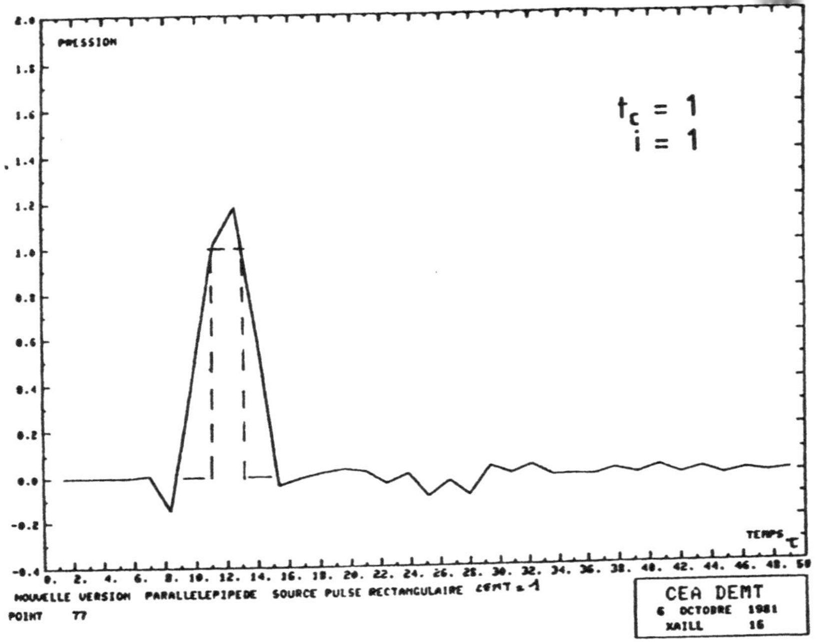

Fig. 5

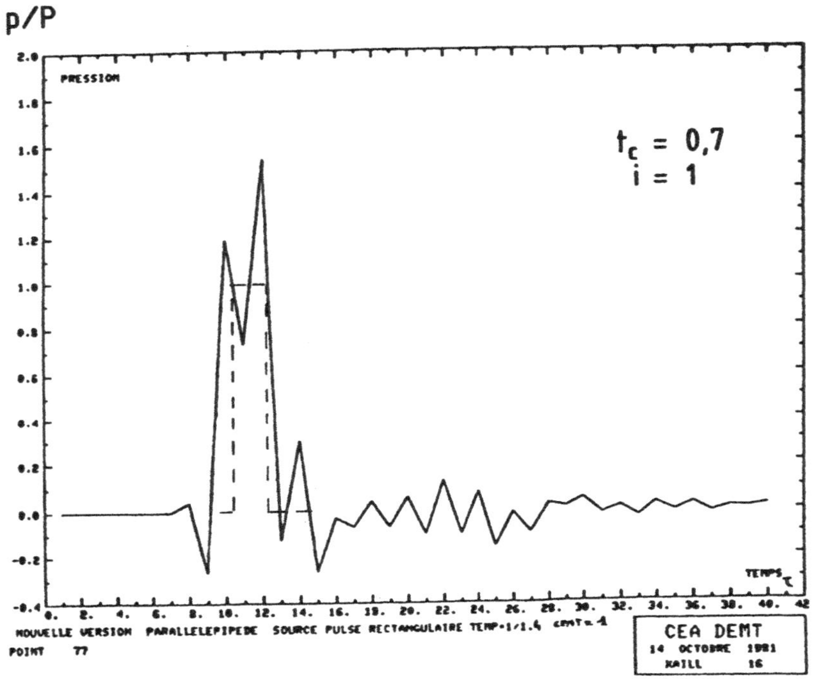

Fig. 6

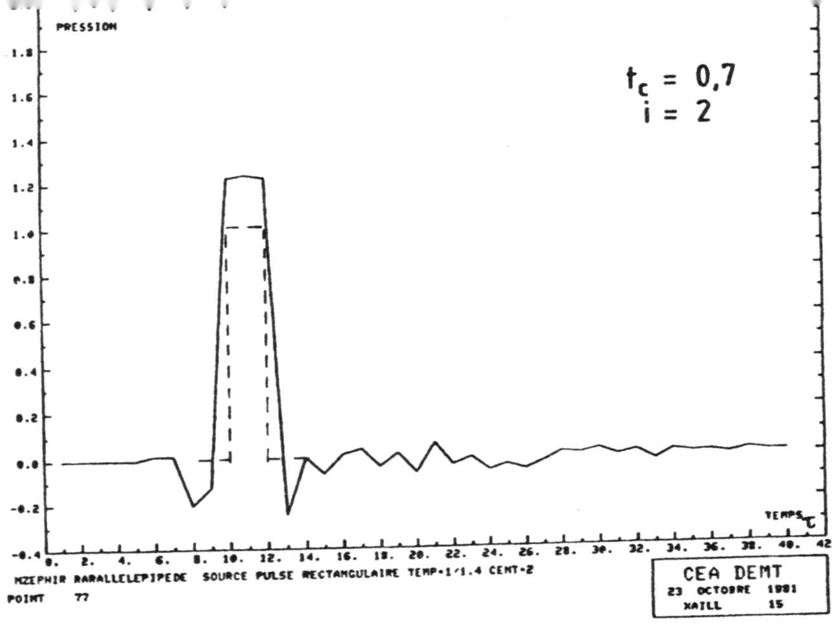

Fig. 7

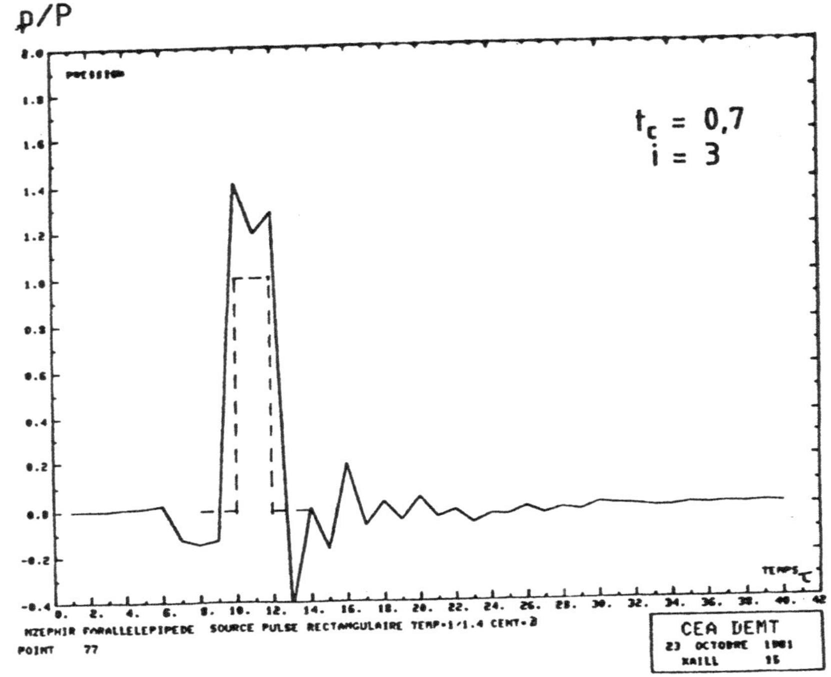

Fig. 8

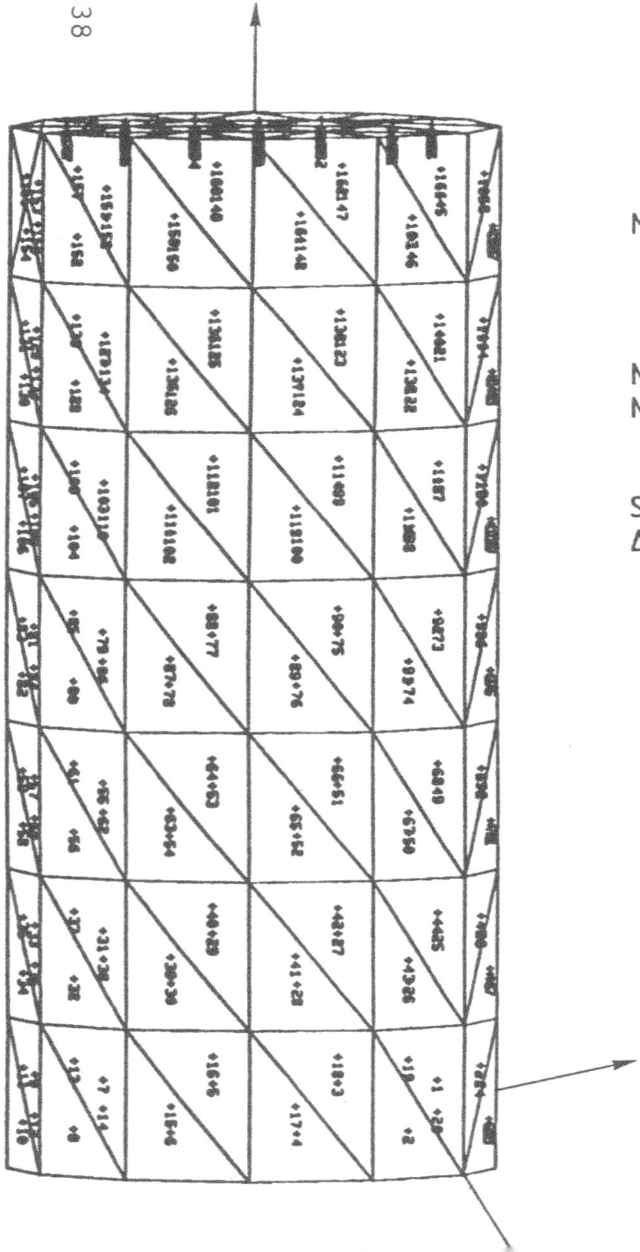

Mesh radius 1
height 3.62

Number of nodes 103
Number of elements 192

Step
$\Delta x = 0.165$

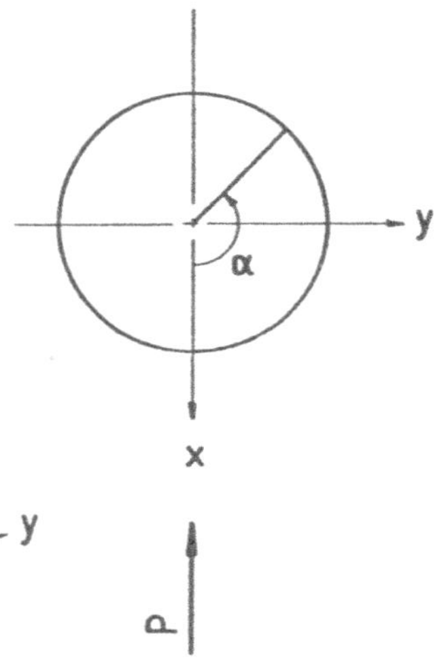

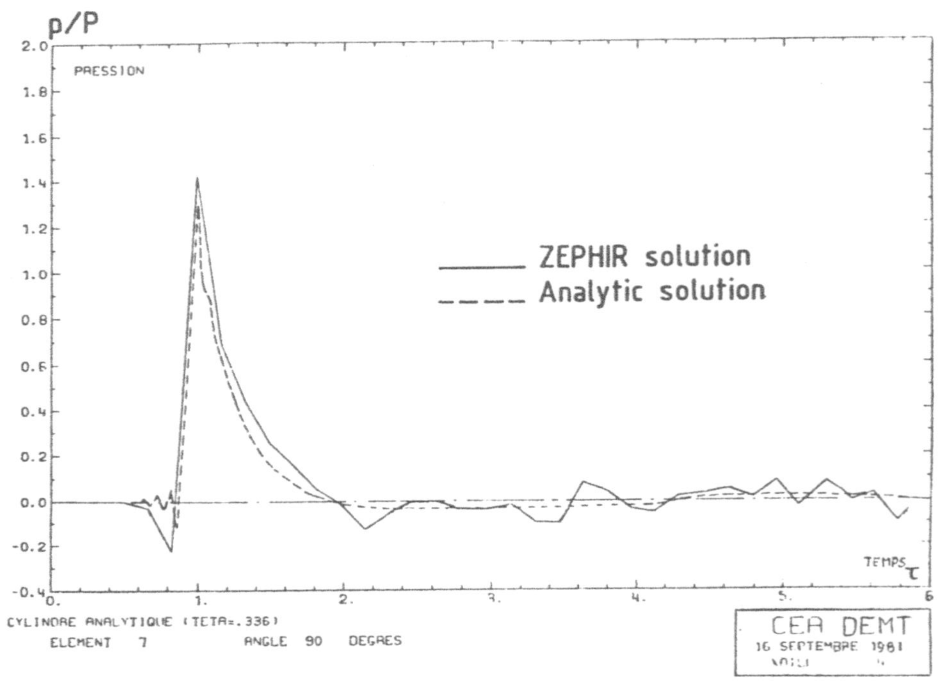

Fig. 10

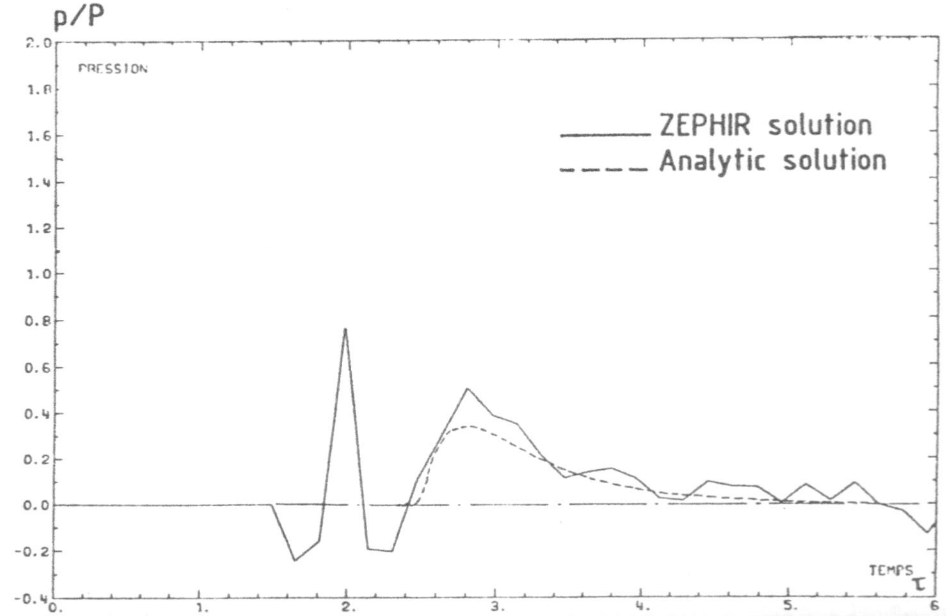

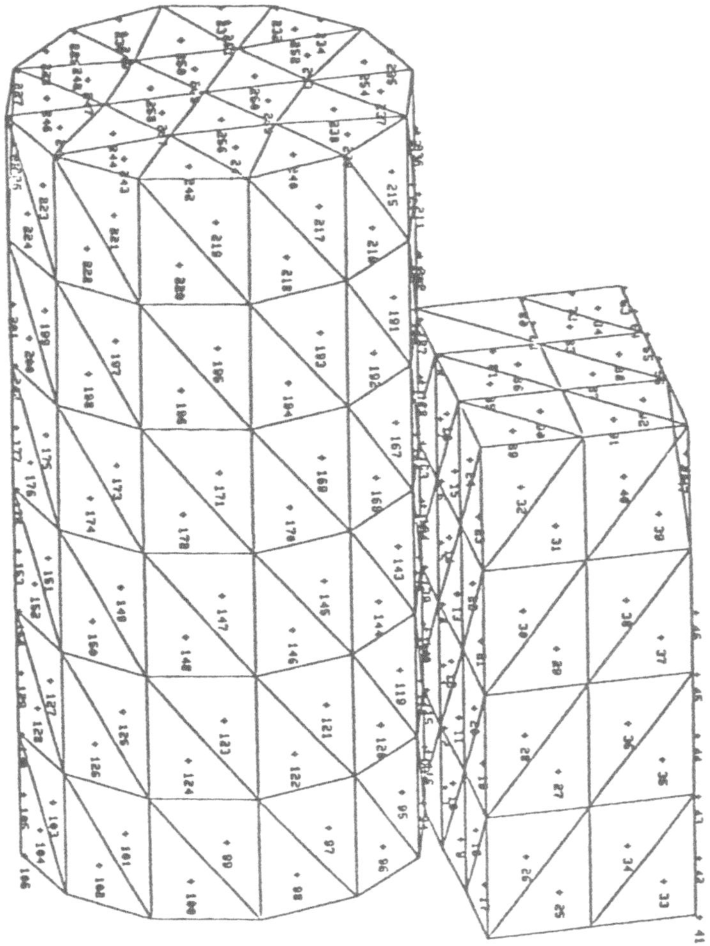

MIMOSA mesh

Fig. 12

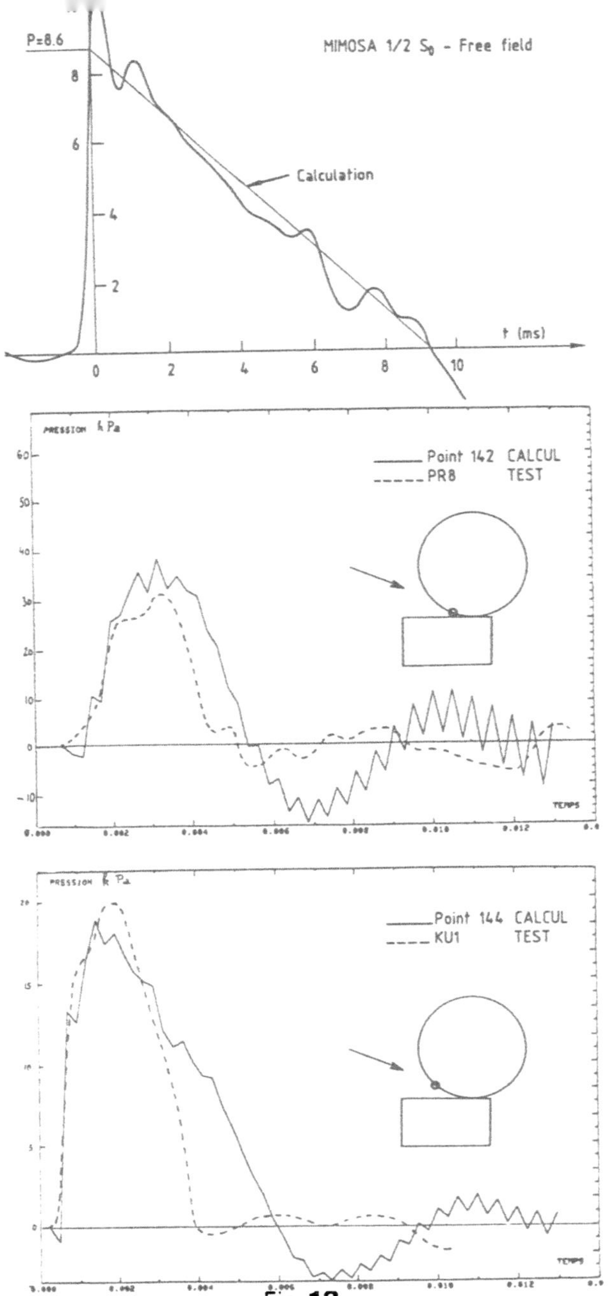

Fig. 13

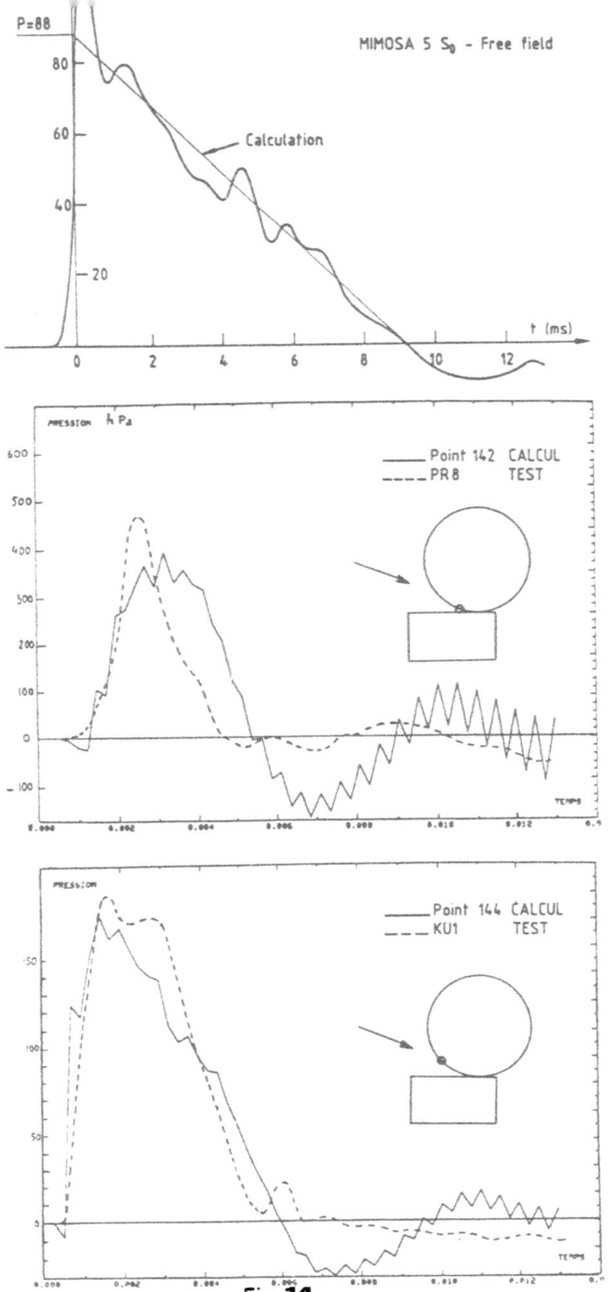

Fig. 14

Discussion

Question: A. Hauguel - EDF/DER

In the MIMOSA test case, why have you changed the incident wave field? This seems to make more difficult the comparisons of pressure field on the building.

Answer: C. Berriaud - CEA/CEN-Sac

The consequence of the ZEPHIR code calculation is a certain smoothing of time history. In a first comparison one can simplify the source time-history, but I agree that a more accurate comparison should be interessant.

RECENT RESULTS OF EXPERIMENTAL STUDIES ON THE TRANSITION OF A DEFLAGRATION INTO A DETONATION AND SOME ASPECTS FOR THE TRANSPOSITION OF THESE RESULTS TO CONFINED EXPLOSIONS

H. Pförtner

Fraunhofer-Institut für Treib- und Explosivstoffe (ICT)
7507 Pfinztal 1, Germany

ABSTRACT

For the safety of a nuclear power plant it is of substantial interest to know realistic values of the overpressure build-up in a chemical explosion. To obtain a further insight into the complex pattern of events in this field experimental investigations were carried out in free and partially confined homogenous clouds with hydrogen as a fuel. The results of the experiments in an unconfined hemispherical configuration show a distinct dependence of the flame velocity on the cloud size which does not exceed, however, an upper limiting value. The experiments under partially confined conditions without additional turbulence indicate that the flame velocity and consequently the pressure field is governed by the hemispherical flame front initially formed; i.e. the flame velocity is not affected by a partial confinement. To increase the turbulence effects further experiments were carried out with a fan and with jet ignition. In the case of fan induced turbulence the flame velocity was strongly influenced by the fan speed. At a certain fan capacity the transition of a deflagration into a detonation was observed. With jet ignition produced by venting an explosion chamber into the partially confined cloud the transition occured even with hydrogen concentrations in air as low as 22 Vol%. The results which are discussed on the basis of turbulence models and with respect to their transposition to confined explosions show that high damaging overpressures or the transition from a deflagration into a detonation are only generated by strong turbulences as a consequence of flow and/or obstacles.

INTRODUCTION

Extensive studies and the accidents which occur despite intensified primary explosion protection have shown that chemical explosions cannot be excluded. Accordingly, additional protective measures must be taken, irrespective of the probability of occurence of such an event. To estimate these measures further information about the generation of chemical explosions and their effects on buildings and plants must be compiled. Especially for the safety of a nuclear power plant it is of substantial interest to know realistic values of the overpressure build-up produced by chemical explosions.

Despite the fact that a lot of experiments carried out in this field and most of the accidents have shown that a deflagration is by far the most probable mode of combustion in an accidental situation, a detonation with considerably higher overpressures is discussed, too.

Therefore, the most important question is to know under which conditions the transition from a deflagration into a detonation will occur.

EXPLOSIVE EVENTS IN A FUEL AIR MIXTURE

If an explosible fuel air mixture is ignited, there will always be, initially, a combustion wave whose rate of propagation is given by the laminar burning velocity S_o and the ratio of the densities ahead of and behind the flame front E. Now, as a result of turbulence effects this velocity could increase as the distance covered by the flame increases, which means that the velocity would have to be expanded by a turbulence factor f_t to get the actual flame velocity S_t

$$S_t = S_o \cdot E \cdot f_t \qquad (1)$$

The formation of a shock wave as a precondition to the change from a deflagration into a detonation is therefore determined by the nature of the process of flame acceleration which is closely linked with the rate of energy release. Flame propagation at normal flame velocities, well below the sound velocity of the unburnt gas mixture, will certainly not result in a detonation. On the other hand, there must be a detonation if the energy density, as a function of time, is high enough to initiate a shock of sufficient intensity. In a free cloud this high power density necessary to initiate a detonation is normally only available from high explosives.

This statement is confirmed by many experiments which have been carried out in the past. When igniting a free cloud deflagratively with flames, exploding wires or pyrotechnical fuses no events resembling a detonation were observed. The overpressures were less than 100 mbars. Even if a gas detonation was transmitted into a free spherical cloud a fast deflagration was merely triggered whose velocity decreased with increasing flame path. When the detonation was transferred into a plastic tube filled with stoichiometrically premixed ethylene/air mixture, however, it was propagated over the whole length; i.e. the propagation of a detonation obviously depends on the geometry of the cloud, too.

On the other hand, 8 g, 30 g and 80 g of high explosive were sufficient to initiate a gas detonation in free stoichiometric ethylene/air, propylene/air and propane/air mixtures, respectively. In methane/air mixtures it was not possible to initiate a detonation even with 2500 g of high explosive. In a real cloud of propylene/air with its concentration gradients 300 g of high explosive did not suffice to initiate a detonation compared with 30 g in a stoichiometric mixture.

In summary it can be said, indeed, that free gas clouds can be made to detonate only by sufficiently strong detonative initiation, if at all.

If a fuel air mixture is ignited deflagratively at the closed end of a long pipe, however, there is always the transition from a deflagration into a detonation after a certain induction period or induction distance.

This is due to the fact that the compression waves preceeding the flame overtake themselves producing a shock wave. These shock waves raise the pressure and cause a flow of unburnt gas which becomes more and more turbulent due to friction at the wall. The flame front is accelerated continously, until the transition from deflagration to detonation occurs. This transition is characterized by the induction distance X_i which depends on the pipe diameter d according to

$$X_i = c \cdot d^n$$

where c and n are constants for a given gas mixture and a given state [1,2].

Other investigations [3] have shown that the induction distance in pipes also depends inversely on the burning velocity so that in turbulent mixtures with their higher burning velocity this distance is much shorter.

Under these aspects in a confined space like a room the transition from deflagration to detonation is very unlikely, too, because of its geometric dimensions.

Nevertheless, there still exist the idea that the acceleration of the flame due to turbulence effects could become high enough to cause a detonation.

RESULTS OF RECENT EXPERIMENTAL INVESTIGATIONS

To get a further insight into the flame acceleration mechanisms within the PNP safety research program experimental investigations were carried out in free and partially confined clouds the aim of which was to measure the flame velocity and the overpressure depending on the cloud size and the partial confinement. The experiments under partially confined conditions were extended to fan induced turbulences and jet ignition to increase the turbulence effects. Because of the high burning velocity and the importance in the PNP-project hydrogen was used as a fuel. These experiments were performed under a contract of Interatom.

The free gas clouds were simulated by hemispherical balloons consisting of thin polyethylene foil. The balloons were filled with nearly stoichiometric hydrogen/air mixtures with volumes between 7.5 and 2100 m^3 corresponding to balloon diameters between 3.0 and 20.0 m. The gas mixtures were ignited by exploding wires in the center of the balloon. In this way a good spherical symmetry of the pressure field and the flame propagation without any reflections of the pressure wave was obtained. The overpressures were measured by piezo-resistive gages installed on the floor. The flame propagation was filmed by high speed cameras. The maximum overpressures measured were between 22 mbar and 63 mbar corresponding to flame velocities between 39 and 84 m/s (Table I). These results show that there is a evident dependence of these values on the cloud size due to flame induced turbulence increasing with the volume of the fuel/air mixture. The recalculated turbulence factors, i.e. the ratio of turbulent and laminar burning velocity, seem to indicate that there exists a limiting value which is reached asymptocally so that the flame front velocity cannot increase arbitrarily with increasing flame path.

This result which is in agreement with statements of Strehlow and Wagner [4,5] is logical insofar as the fluctuation velocity u' cannot exceed the flow velocity

$$u = (E - 1) \cdot S_t \qquad (2)$$

were S_t is the turbulent burning velocity and E the expansion ratio, i.e. the ratio of the densities ahead of and behind the flame front, respectively. So, several turbulent flame models lead to a maximum turbulence factor between (E-1) and E.

Karlovitz [6] correlates the flame induced turbulent velocity component with the difference in the kinetic energies of the expansion and the normal velocity component. In connection with this consideration the experimental results lead to the relation

$$f_t = 1 + (E-1) \left[1 - (1 + 0{,}02\ r_0)^{-2}\right]^{1/2} \qquad (3)$$

Abdel-Gayed et al. [7] have shown that the eddy diffusivity is proportional to $[1-\exp(-t/t_L)]$, where t is the elapsed time for cloud growth and t_L a scaling parameter. The time t is linked with the cloud radius r_0. As the turbulence factor is proportional to the square root of the diffusivity, it can be described by the relation

$$f_t = 1 + (E-1) \left[1 - \exp(-0{,}038\ r_0)\right]^{1/2} \qquad (4)$$

The checkup of these equations for other fuels but hydrogen like methane and propane confirms their general validity.

In Fig. 1 Eq.(3) and (4) are compared with the experimental values for a stoichiometric hydrogen/air mixture with E = 6,93.

Very similar results were obtained under partially confined conditions. The partial confinement consisted of two parallel walls, 3 m high and 10 m long in a distance of 3 m closed at one end. The ignition took place in the center of the closed side. The results of these experiments which were carried out with rich hydogen/air mixtures show that in absence of an additional turbulence the flame velocity is not influenced by the confinement and it is determined by the initial hemispherical propagation of the flame front (Table II).

Only when the flame leaving the confinement a stronger acceleration due to the jet-like expansion of the gas mixture ahead of the flame front was observed. By this jet additional turbulence effects were induced so that the flame front velocity reached values in the unconfined space up to 200 m/s corresponding to overpressures of more than 200 mbar.

In order to increase the turbulence effects accelerating the flame in a more defined way a fan was located near the ignition source at the closed end of the lane. The fan had a diameter of 1.25 m and a maximum capacity of 24,000 m^3/h. It produced a turbulent flow field in the lane which was maintained during the combustion process.

Different experiments with 50, 75, 90 and 100 % of the full fan capacity were carried out. While in all cases the flame front velocity between the ignition source and the fan was about the same as in the experiments without additional turbulence it increased strongly after the flame had passed the fan (Table III). So, with lower fan capacities the flame velocities reached values between 170 and 190 m/s which, in a certain distance from the fan dropped to a velocity of about 110 m/s. With the full capacity, however, a flame front velocity of 222 m/s was measured. After a flame path of about 3.5 m from the ignition point the flame was still accelerated and the transition from deflagration to detonation occured near the wall. As the flame velocity is well below the sound velocity of the unburned mixture it is assumed that the high flow velocity induced by the high flame velocity interacted with the boundary layer near the wall producing still higher turbulences which initiated the detonation. This was the first time that such a transition was observed in a more or less free cloud only confined by two parallel walls.

Figure 2 shows the flame contours and the path/time dependence of the flame front which results in a detonation velocity of about 2080 m/s. The overpressures measured within the lane were as high as 25 bars.

The flame acceleration produced by the fan can be explained in a very similar way as in the case of flame induced turbulences only. In connection with the theory of Karlovitz [6] Kozachenko and Kuznetsov [8] consider the wrinkling of the flame surface increased by the turbulent fluctuation velocity u' of the flow and a linear dependence of the turbulent burning velocity, i.e. the turbulence factor on u'. This consideration leads to the relation

$$f_t^* = 1 + (E-1) \left[1 - (1+0.02\ r_0 + u')^{-2}\right]^{1/2} + 0.36\ (E-1) \cdot u' \quad (5)$$

Fig. 3 shows the comparison between the calculated and the experimental values which are in good agreement for hydrogen and some hydrocarbons as well.

The aim of the tests with jet ignition was to check the possibility whether the transition from deflagration into detonation could be scaled, after corresponding investigations carried out by Battelle-Institute [9] have shown that there is also such a transition in a stoichiometric hydrogen/air mixture under specific conditions. Fig. 4 shows the flame contours and the flame path/time diagram of such a test in a mixture of 22 Vol% hydrogen with its distinct transition from deflagration into detonation. In this case the venting ratio, i.e. the ratio of the venting area to the entire front area, was 0,10. In contrast to this result in a mixture of about 20 Vol% hydrogen no transition occured. With a venting ratio of 0,30 the same results were obtained in spite of a coarser turbulence structure.

Lee [10] has shown that each fuel mixed with air occupies a certain cell size when reacting detonatively which depends on its concentration. For a stoichiometric hydrogen/air mixture a detonation cell width of about 1.5 cm is obtained. Compared with the tests carried out by Battelle-Institute all geometric dimensions were expanded by a factor of 3. If there exists a scaling law with the cell size, a transition must occur up to a cell width of about 4.5 cm in the larger setup which corresponds to a hydrogen concentration of 20,8 Vol%; that means that below this concentration there is no transition. This assumption is fully confirmed by the experimental investigations. At least in this special setup the transition from deflagration into detonation can be scaled by the cell size, indeed.

Despite the fact that all these investigations just described were performed in free and partially confined clouds only, it is to be thought that these results could be transposed to confined spaces, too. In all cases, in a pipe and a partial confinement as well, the generation of a sufficiently high turbulence plays the dominant role in triggering a detonation.

In a large building like a containment there are several compartments connected with a floor. If the compartment and the floor are filled with a hydrogen/air mixture, the compartment will act like a explosion chamber vented by the passage so that within the floor the transition from deflagration into detonation can happen. Or, within a large compartment there are many setups which form a partial confinement; due to the ventilation a flow is always present so that a similar event could occur, especially in connection with obstacles. In this case, also a jet ignition could cause the transition.

In summary it can be said that as a consequence of these results such a transition is not to be excluded, when independent on the degree of confinement, partially or completely confined, a sufficiently high turbulence is produced. The premise is, however, the presence of a hydrogen concentration within a certain range.

REFERENCES

[1] Bartknecht, W. — Moderne Unfallverhütung 11 (1967)

[2] Griesche, G — Depotber. 21/70-Inst. f. Bergbausicherheit, Leipzig

[3] Ginsburgh, J.; Bulkley, W.L. — Chem.Engng. Progress 59 (1963) 82

[4] Strehlow, R.A. — 13th Loss Prevention Symp. (1980)

[5] Wagner, H.Gg. — PTB-Mitteilungen 91 (1981) 248

[6] Karlovitz, B. — Selected Comb. Problems - Butterworth Sci.Publ., London (1954) 248

[7] Abdel-Gayed, R.G.; Bradley, D.; Lwakabamba, S.B. — First Specialist Meeting (Int.) of the Comb. Inst. (1981)

[8] Kozachenko, L.S.; Kuznetsow, J.L. — Fizika Goreniya i Vzryva 1 (1965)

[9] Schildknecht, M. — "Versuche zur Freistrahlzündung von Wasserstoff-Luft-Gemischen im Hinblick auf den Übergang Deflagration-Detonation" - Bericht des Battelle-Instituts, Frankfurt/M. (1983)

[10] Lee, J.; Knystautas, R.; Guirao, C.; Benedick, W.B.; Shepherd, J.E. — Second Int. Workshop on the Impact of Hydrogen on Water Reactor Safety (1982)

r_0 (m)	c (Vol%)	M_0 (g/mol)	M_F (g/mol)	T_0 (K)	T_F (K)	S_0^{293} (m/s)	$S_0^{T_0}$ (m/s)	E (-)	S_F (m/s)	f_t (-)	$\bar{f}_t$ (-)
1,53	30,1	20.8	24,1	283	2378	2,59	2,44	7,25	-	-	
1,53	28,6	21,2	24,5	279	2341	2,41	2,22	7,26	46	2,85	
1,53	29,2	21,0	24,4	282	2374	2,48	2,32	7,25	46	2,73	
1,53	29,1	21,0	24,4	282	2369	2,47	2,31	7,23	42	2,51	2,58
1,53	29,1	21,0	24,4	282	2369	2,47	2,31	7,23	41	2,45	
1,53	29,5	20,9	24,3	281	2378	2,52	2,35	7,28	45	2,63	
1,53	29,2	21,0	24,4	281	2373	2,48	2,32	7,29	39	2,31	
2,88	27,2	21,5	24,8	286	2273	2,23	2,14	6,89	43	2,92	
2,88	29,4	20,9	24,3	279	2374	2,51	2,31	7,32	50,5	2,99	2,96
2,88	29,5	20,9	24,3	279	2376	2,52	2,32	7,32	54	(3,18)	
									korr.	2,96	
5,00	31,0	20,5	23,9	281	2368	2,68	2,50	7,23	59	3,26	3,46
5,00	25,9	21,9	25,1	283	2195	2,06	1,94	6,77	48	3,65	
10,00	29,7	20,9	24,2	283	2378	2,54	2,39	7,26	83,6	(4,82)	4,48
									korr.	4,48	

Table I: Results of the balloon tests

r_0 (m)	c (Vol%)	M_0 (g/mol)	M_F (g/mol)	T_0 (K)	T_F (K)	S_0^{293} (m/s)	$S_0^{T_0}$ (m/s)	E (-)	S_F (m/s)	f_t (-)	$\bar{f}_t$ (-)
1,50	35	19,5	22,5	281	2318	3,11	2,90	7,15	49	2,36	
1,50	39,1	18,4	21,0	272	2209	3,40	3,00	7,11	54	2,53	2,39
1,50	39,0	18,4	21,0	276	2213	3,40	3,07	7,02	49	2,27	

Table II: Results of the lane tests without additional turbulence

r_0 (m)	c (Vol%)	M_0 (g/mol)	M_F (g/mol)	T_0 (K)	T_F (K)	S_0^{293} (m/s)	$S_0^{T_0}$ (m/s)	E (-)	Q (m^3/h)	S_F (m/s)	$f_t^{\bullet}$
1,50	41	17,9	20,3	291	2178	3,46	3,42	6,60	12 000	172	7,62
1,50	38	18,7	21,4	293	2260	3,35	3,35	6,74	18 000	176	7,79
1,50	41	17,9	20,3	292	2179	3,46	3,44	6,58	18 000	179	7,91
1,50	41	17,9	20,3	288	2175	3,46	3,36	6,66	21 600	190	8,49
1,50	37	18,9	21,8	298	2290	3,28	3,38	6,66	21 600	186	8,26
1,50	36	19,2	22,1	285	23,02	3,20	3,05	7,02	24 000	222	10,37

Table III: Results of the lane tests with fan induced turbulence

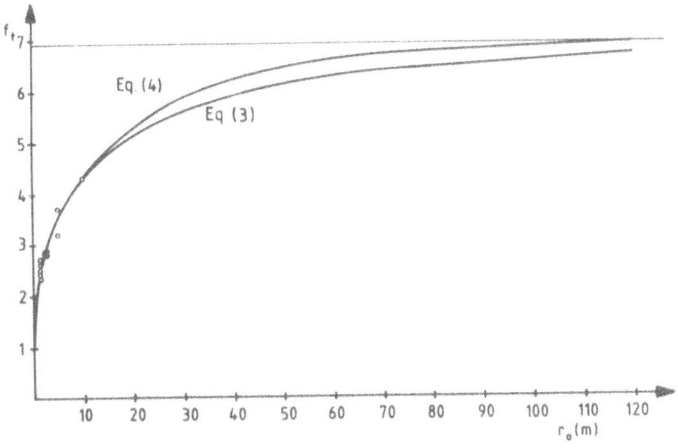

Fig. 1 Turbulence factor f_t as a function of the cloud radius r_o for a stoichiometric hydrogen/air mixture (E=6.93)

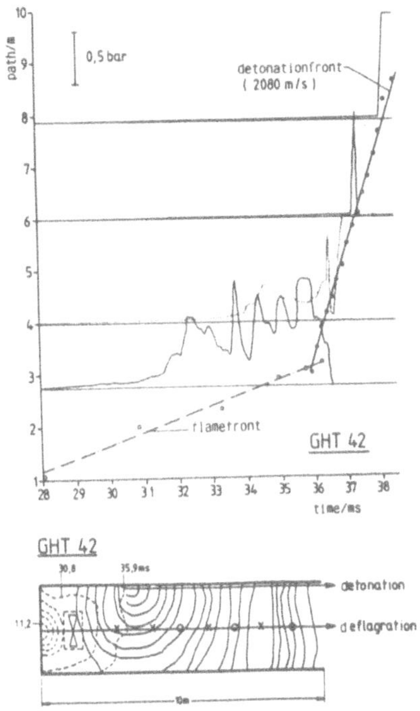

Fig. 2 Top: Flame path/time diagramm and typical pressure/time curves
Bottom: Flame propagation with time. Dashed lines: deflagration, full lines: detonation

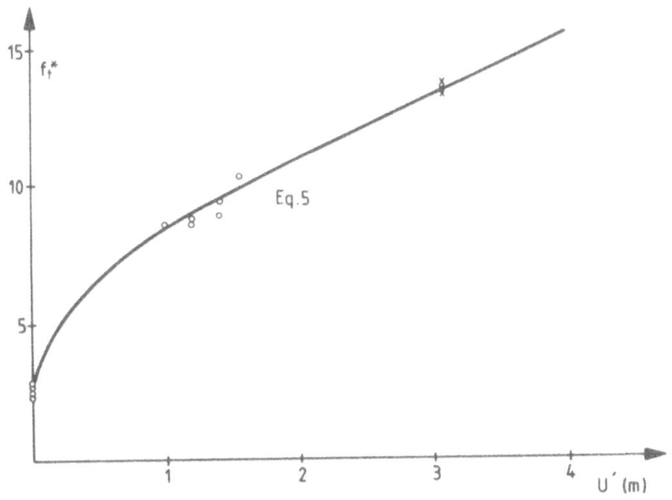

Fig.3 Turbulence factor f_t^* with additional turbulence as a function of the turbulent fluctuation velocity u' for a stoichiometric hydrogen/air mixture (E=6.93,r_o=1,5m)

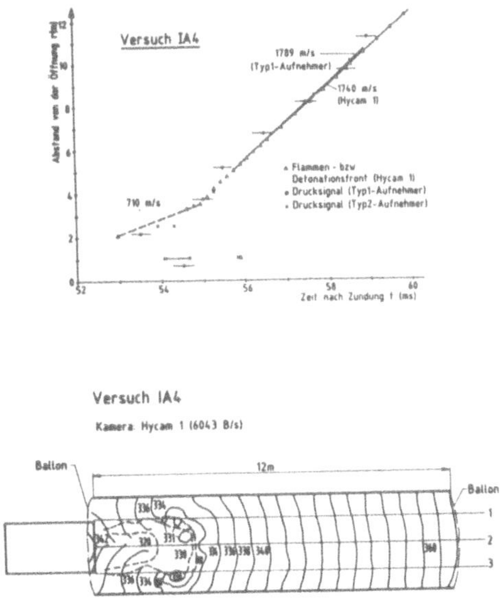

Fig.4 Flame path/time diagram and flame contours with jet ignition in a hydrogen/air mixture with 22 Vol% H_2

Discussion

Question: J. Thomas - British Gas

In your presentation you noted that the turbulent flame-speed factor was directly proportional to the intensity of the external turbulent flow. Is there not a possibility that, at high turbulence levels, combustion will begin to quench for many hydrocarbon fuels?

Answer: H. Pförtner - Fraunhofer Inst./Pfinztal

That's right. The turbulence factor will not increase arbitrarily, also in this case. As I have mentioned before, a turbulence factor greater than 25 requires very high turbulent fluctuation velocities, which can only be produced under extreme conditions. On the other hand, according to Bradley and coworkers, there seems to be a quenching of the flame at high turbulence intensities, indeed.

Parallel Sessions Research Area C

The release and distribution of radioactive fission products in the atmosphere following a reactor accident

Session C1

Chairman: J. Kretzschmar (Janssen Pharm./Beerse)

Session secretary: B. Tolley (CEC/BXL)

REVIEW OF SPECIFIC EFFECTS IN ATMOSPHERIC DISPERSION CALCULATIONS

B Y UNDERWOOD, P J COOPER, N J HOLLOWAY, G D KAISER+ and W NIXON

Safety and Reliability Directorate, UKAEA

Summary

This work consists of a series of ten individual review Chapters - written between 1980 and 1983 - together with a summary document linking and overviewing the work. The topics covered are as follows: 'Plume Rise in Nuclear Safety Studies'; 'Dry Deposition'; 'Wet Deposition'; 'Atmospheric Dispersion in Urban Environments'; 'Topographical Effects in Nuclear Safety Studies'; 'Coastal Effects and Transport over Water'; 'Time-Varying Meteorology in Consequence Assessment'; 'Building Effects in Nuclear Safety Studies'; 'Effect of Turning of the Wind with Height on Lateral Dispersion'. Although the reviews are, on the whole, general in approach, emphasis has been given where appropriate to the impact of various phenomena on the assessment of reactor accident consequences. In general the work focusses on the 0-100 km range of distance downwind of the source. The reviews fulfil several functions: they serve as introductions to the subject areas; they outline theoretical and experimental developments; they act as reference documents providing a copious source of references for more detailed investigation of particular points; they raise unresolved technical issues and attempt to indicate principal uncertainties; they point to areas requiring further development.

1 INTRODUCTION

This contract involved the writing of ten review chapter listed in Table 1. A principal motivation for reviewing these topics was the recognition that the 'first generation' computer codes written to assess the consequences of major reactor accidents account for the phenomena under consideration at only a very rudimentary level, if at all. Many of the subject areas have undergone considerable development in the last few years and the need for up-to-date reviews was apparent. Although some of the technical areas have been generally reviewed in the literature, the emphasis in this work has been directed towards their impact on the calculation of reactor-accident consequences and this has dictated to some extent the selection of material included.

In general the reviews focus on the 0-100km range of downwind distance. In practice, for a zero-threshold, linear dose-risk relationship, a large contribution to the collective-dose integral used to calculate latent cancers may arise from distances $>$ 100km, albeit from a large number of very small individual doses. However, since the integration tends to attenuate sensitivity to many details of the modelling, a common approach in consequence assessment has been to make rather simplistic assumptions about the $>$ 100km range. The attention given to the longer range depends on which consequence parameters are considered most important, as discussed in section 4.2.

+ Present address: NUS Corporation, 4 Research Place, Rockville, USA - MARYLAND 20850

The reviews have several uses: they serve as introductions to the subject areas; they outline theoretical and experimental developments, particularly recent ones; they act as reference documents, providing a copious source of references for more detailed investigation of particular points; they raise unresolved technical issues and attempt to indicate principal uncertainties; they point to areas requiring further development.

Each of the reviews contains a summary and a detailed set of conclusions which will not be duplicated here. However, in the following section a very brief outline of the scope and principal conclusions of each review will be given to set the scene for an overview of the study as a whole.

2 SCOPE AND PRINCIPAL CONCLUSIONS OF INDIVIDUAL REVIEWS

2.1 "Plume Rise in Nuclear Safety Studies" (G D Kaiser)

By agreement with the CEC this review was not as comprehensive as the subject might appear to warrant; the reason for this was the existence of a prior contract between the CEC and SRD specifically dealing with plume rise (albeit in a fast-reactor content) which discussed many of the salient points. However, the current review by Kaiser fills a gap in the literature on plume rise in an accident-consequence context by raising important issues besides those connected with the calculation of mean plume trajectory: ie what are the conditions under which a buoyant plume can lift itself out of a turbulent building wake; what is the ground-level concentration underneath a rising plume; what are the factors determining the final height to which a buoyant plume may rise? It is concluded that more attention has to be directed to these questions if a reliable quantification of the impact of plume rise is to be made. To this end SRD commissioned wind-tunnel investigations of 'lift-off' and of the concentrations under a rising plume; preliminary results are described in the report, as are the results of large-scale field trials (the METEOTRON project) in France.

The discussion in general reinforces the conclusions of the earlier contract report, namely that

1. plume rise can in some circumstances bring about a considerable reduction in the predicted number of early deaths and morbidities, and
2. the predicted reduction in the number of latent effects is much less marked.

2.2 "Dry Deposition" (B Y Underwood)

Previous assessments of reactor-accident consequences have shown significant sensitivity of some consequence parameters to the magnitude of the dry deposition velocity. The review first explains physical concepts associated with the theory of dry deposition, including the resistance-chain analogy, then goes on to review separately gas and particle deposition velocities since the issue of what is the physico-chemical form of the principal radionuclides released is not prejudged. Different surface types - smooth, rough, water, urban - are treated separately, with the emphasis on vegetative canopies. For gases the applicability of standard mass-transfer correlations is assessed, and the importance of the surface-resistance component highlighted. The extensive empirical data on iodine deposition to grass are reviewed. For particles, theories of transfer to smooth surfaces are described in some detail since this gives insight into the physical mechanisms involved; however, it is also demonstrated that the micro-roughness characteristics of natural surfaces have a strong effect on deposition. It is shown how the net dependence on particle size results from a number of different physical processes dominating in different size regimes, leading to a characteristic minimum in the 0.1 - 1 μm size range.

The review concludes that deposition velocity depends strongly on characteristics of the released material, for example whether particle or gas; particle size especially has a strong influence. This sensitivity implies a need to better define source-term parameters. Secondly, deposition velocity also depends strongly on surface characteristics, indicating perhaps that details of land usage may need to be incorporated into consequence assessments. Not much is published on deposition in urban environments. Recent filtration-type models for deposition to vegetative canopies are promising and warrant further development. Finally, deposition velocity depends on atmospheric conditions, although the dependence is not usually dominant.

The report also reviews techniques for incorporating dry deposition into dispersion models based on both Gaussian and K-theory approaches; some useful analytic solutions in the latter context are described. The limitations of the 'source depletion' assumption are viewed from numerous angles. However, it is concluded that errors attributable to the manner in which dry deposition is incorporated into a dispersion model can readily be reduced to a level which is insignificant compared to the effect of uncertainties in the magnitude of the dry deposition velocity.

2.3 "Wet Deposition" (B Y Underwood)

As with dry deposition, some consequence parameters are strongly sensitive to wet deposition rates. The review discusses the many components in the quantification of these rates, pointing to areas of the subject requiring further development. It treats separately below-cloud and in-cloud scavenging; separate sections are devoted to particle scavenging and gas scavenging (principally iodine).

For the below-cloud scavenging of particles, the scavenging coefficient concept is introduced; the coefficient depends principally on rainfall parameters and collection efficiencies of drops for particles. Many processes contribute to the collection efficiency, eg Brownian diffusion, inertial impaction, phoresis and electrical attraction; theoretical developments and experimental information are reviewed. It is found that the properties of the released particulate have a significant effect on scavenging, especially size distribution, indicating the need for a more detailed source term specification than hitherto made available. The dependence of scavenging coefficient on rainfall characteristics is usually parameterised simply in terms of rainfall intensity but the limitations of this approach are discussed. For gases the reversibility of uptake by water drops introduces important differences in comparision to particulate scavenging and these additional features are discussed with particular reference to iodine releases.

The review goes on to outline the complex problems which arise in the quantification of the in-cloud component of (particle) scavenging, in particular those related to the rate of uptake of released material by cloud systems. The theoretical formulations of the many processes contributing to in-cloud scavenging - eg nucleation, turbulent diffusion and electrical attraction - are explained, and the data available from field studies is reviewed. The rate coefficient for in-cloud scavenging is strongly dependent on characteristics of the released aerosol which are in general imprecisely known and on certain cloud parameters which may well not be known in a consequence-assessment context. The sensitivity of results to these factors needs to be assessed. Typically, in-cloud scavenging rate coefficients turn out to be higher than equivalent below-cloud coefficients particularly for intermediate particle sizes.

A separate section of the review is devoted to the implications of spatio-temporal variations in rainfall intensity, in particular the distortion of consequence results due to the use of time-invariant, source meteorology. Monte Carlo simulation and statistical modelling approaches to time variation are described. Three problems of the 'source meteorology' assumption are discussed: the use of Eulerian statistics in a Lagrangian context; the spatial structure of rainstorms; regional-scale climatologicalvariations. Some attempts to account for the latter two are described and the conclusion drawn that further work is required in these areas.

2.4 "Atmospheric Dispersion in Urban Environments" (N J Holloway)

This review addresses the particular situation of a source upwind of an urban centre - of relevance to a reactor-accident release - and does not deal with the wider topic of air pollution arising from sources within the complex itself. The report first discusses the factors which distinguish dispersion over the city from dispersion in flat, rural terrain, principally a significant increase in surface roughness and a change in thermal characteristics, the latter arising from not only 'self-heating' (due to industrial and domestic activity) but also, for example, from differences in evaporation characteristics and surface albedo. The review then briefly surveys techniques for modelling atmospheric dispersion in 'mixed' boundary layers in general, describing recent developments which go beyond the conventional Pasquill-Gifford approach; this serves as background to a discussion of special features observed at a rural-urban transition, in particular the growth of an internal boundary layer or, at low windspeeds, circulation patterns. The emphasis is on simple models for handling the phenomena, especially 'box' models (based on conservation-of-material equations) in situations of upward surface heat flux; however, a heirarchy of techniques is described including complex computer codes incorporating turbulence models which are seen as providing a standard of comparison for simple models. An important new input in this review is the working through of particular examples which illustrate how a box model can be 'patched' into the standard Gaussian dispersion scheme.

The review concludes that urban effects have a significant impact on computed concentrations: the increased level of turbulence tends to lower ground-level concentrations (except in fumigation situations). The interaction of cross-wind shear with increased vertical dispersion is highlighted as an area for further investigation.

2.5 "Topographical Effects in Nuclear Safety Studies - a Review" (P J Cooper)

Most basic models used to describe the atmospheric dispersion of releases from reactors are appropriate to flat or gently-rolling terrain. This review discusses the additional factors to be taken into account in complex terrain, consisting of hills, ridges, valleys etc. First, the characteristics of the flow over isolated features - such as a single hill or ridge - are outlined and then dispersion of passive pollutants in such flows is described, with emphasis on factors tending to increase or decrease dispersion (with resulting impact on ground-level concentrations). These descriptions draw on information from both wind-tunnel experiments and field studies. Factors tending to increase dispersion rate are, for example, increased levels of mechanical turbulence due to flow separation, recirculation etc, and the deflection of plumes around obstacles; factors tending to decrease dispersion rate are, for example, convergence of streamlines and physical barriers (such as valley walls). Details of the flow and dispersion are noted to be stability dependent. The discussion of isolated features serves as a background to a section on more realistic complex topography: the discussion concludes that, in general, horizontal dispersion is increased whereas vertical dispersion tends not to be influenced so much.

The review continues with a discussion of physical modelling, outlining the requirements for good wind-tunnel (or water-channel) simulation. Information on some experiments performed to date is presented in tabular form and the conclusion is drawn that there is scope for further work in this area. The report also describes various types of mathematical approaches to modelling diffusion in complex terrain, ranging from simple extensions of the Gaussian formulation to advanced turbulence-closure schemes; the emphasis is on the simpler models since it is feasible to include these in consequence-assessment codes. Information on many of the models developed to date is presented in tabular form. Besides diffusion modelling, codes handling topography require trajectory-tracking algorithms plus windfield modelling, and various appropriate techniques are described. It is pointed out that further work is needed to model windfields in non-neutral stabilities.

The review advocates that as a first step towards assessing the impact of complex terrain on reactor-accident consequence calculations, the terrain surrounding actual European reactors should be surveyed from this viewpoint.

2.6 <u>"Coastal Effects and Transport over Water"</u> (W Nixon)

First the review discusses dispersion over water in comparison to dispersion overland, pointing out the implications of the generally smoother surface (with consequent lower generation of mechanical turbulence) and of thermal differences. The report then goes on to discuss flow characteristics at a sea-land interface - for example the growth of an internal boundary layer (IBL) -and their impact on dispersion in a coastal environment, in particular fumigation and trapping. Theoretical and experimental work on the IBL is discussed; some of the simple models and experiments purporting to validate them are critically assessed. A separate section is devoted to the phenomenon of sea-breeze circulation; the conditions under which it develops, its characteristics and its implications for dispersion are discussed.

The review then considers the heirarchy of models developed to describe dispersion at coastal sites, ranging from extensions of the simple Gaussian model to complex turbulence models. Gaussian 'fix-ups' to handle 'trapping' and 'fumigation' situations which results from the presence of an IBL are described in some detail. The turbulence-modelling codes, such as the code APT (<u>A</u>tmospheric <u>P</u>ollution <u>T</u>ransport) which applies second-order closure of the turbulent flow equations, are seen as 'benchmark' facilities for testing and tuning the simpler models. Further exploitation of the APT code for this purpose is advocated. Attempts of varying complexity to model pollutant dispersion within a sea-breeze circulation are described; the review highlights the need for more field data to test these models.

The review concludes in general that evaluation of the pollution hazards at coastal sites requires consideration of the potential impact of the special features of coastal meteorology and dispersion behaviour. It points out the scope for further work in the following areas: categorization of over-water dispersion conditions; determination of the impact of plume buoyancy on 'trapping' scenarios; development of relatively simple models of IBL growth of wide applicability; characterization of sea/land transitions for a wider range of conditions than hitherto investigated.

2.7 "Time-Varying Meteorology in Consequence Assessment" (B Y Underwood)

This review first discusses qualitatively the inaccuracies due to using time-invariant, source meteorology in consequence assessment. Although it is recognised that spatial and temporal variations are interlinked, they are discussed separately for convenience. Three problems are identified in connection with the use of source meteorology: Eulerian statistics in a Lagrangian context; spatial structures; regional-scale climatological differences. Since these are discussed to a certain extent in other review chapters no further discussion is given in this one which instead focusses on the impact of changes in weather conditions (at source) as the plume travels downwind. Simple models which maintain conditions at time of release throughout plume travel results in distortion of frequency-consequence distributions and examples are provided. The point is emphasised by an examination of some statistics on the persistence of weather conditions.

Following this, techniques for handling time-variation in consequence assessment are described with emphasis on Monte Carlo simulation using recorded meteorological data: the codes CRAC and MARC (which allow hourly changes in windspeed, stability and rainfall intensity) are described. A separate section examines the need for tracking trajectory curvature and describes briefly some techniques for accomplishing this, concluding that there is as yet no quantitative evidence that this level of modelling is required in consequence assessment purely from the standpoint of time-varying meteorology. Published comparisons of calculations with/without time-varying meteorology are reviewed; these show differences of up to a factor of 10 in (annual) average concentrations for continuous releases (at distances $\lesssim$ 100 km). New comparison calculations were performed for this study using the code CRACUK in order to illustrate the impact on frequency-consequence distributions. These reveal large differences in the 'peak' number of early fatalities but less impact on delayed effects, demonstrating the necessity of taking into account time-varying meteorology for some consequence parameters.

2.8 "Building Effects in Nuclear Safety Studies - A Review" (P J Cooper)

Flow distortion in the vicinity of a building has an important influence on the near-field (first few km) dispersion of material released from the building. Clearly in the reactor-accident context this is of direct relevance to the safety of site personnel and of members of the public, if any, living in close proximity to the site. Although such close-in behaviour can be expected to have a relatively small impact on traditional risk indicators (based on numbers of acute or delayed fatalities, for examples) - except, perhaps, where it may interfere with the "lift-off" of a buoyant plume - it may still be an important factor when all aspects of accident impact are considered.

The review draws on the results of numerous wind-tunnel investigations to provide an up-to-date description of the phenomenology of flow around an isolated building and of its consequent influence on dispersion; a useful compendium of details of the experimental conditions in the various investigations is provided in tabular form in a subsequent section. The dependence of the dispersion on, inter alia, building details (shape, orientation) and on release position is described comprehensively. Generally for downwind distances of between 15 and 50 building heights, effects due to the building are diminishing, becoming insignificant beyond about 50 building heights.

Following this the review discusses numerous simple models for including the effects of an isolated building within dispersion calculations; the majority of the models are prescriptions for 'patching up' the basic Gaussian formulation in the wake region although one line of development is directed towards calculating concentrations in the 'cavity' or region of recirculating flow. The modified Gaussian models are generally tuned to predict well ground-level axial concentrations and cannot be expected to reproduce details such as concentrations profiles. Comparisons between experimental results and the simple models lead to be conclusion that many of the latter are broadly comparable in accuracy. Although the simple models have some significant limitations, they give considerably better concentration estimates then those obtained from a Gaussian model appropriate to a flat surface, and their use is thus recommended. Techniques for handling groups of buildings are still at a rudimentary stage of development.

A brief description of more complex theoretical and numerical models is provided. In general, those capable of handling realistic situations are still at a developmental stage and there is scope for further work in this area.

2.9 "Effect of Variations in Mixing Height on Atmospheric Dispersion Calculations (B Y Underwood)

This topic was expected to constitute only a minor part of the review exercise in view of an earlier American study demonstrating little sensitivity of the results of consequence assessments to deficiencies in mixing-height parameterisations [1]. However, the latter study considers only one generic type of modelling for mixing height, and the current review is more general.

Common parameterisations for the mixing height in consequence codes covering the 0-100 km range ignore some aspects of its variability by, for example, associating a unique mixing height with a given stability category or by utilising seasonal mean values. Calculations using a basic Gaussian model including dry deposition (but not wet deposition) are used in the review to check the impact on ground-level concentrations of changes in mixing height: the potential range of the mixing height parameter is deduced from experimental data. Generalised sensitivity indicators are also devised to approximately assess the impact on health effects. The main conclusion of these calculations is that inadequacies in mixing-height modelling are unlikely to cause errors in consequences greater than a factor of 3, although in practice the magnitude of the effect will depend on details of accident release, population and weather, and in extreme circumstances could be higher.

The impact of mixing-height parameterisations on long-range (>100km) calculations of concentration and collective dose is also discussed: the diurnal variation in mixing height is an important factor here. An illustrative example shows that the dependence of the pattern of ground-level air concentration on the timing of release within the diurnal cycle of mixing height may be substantial. However, this variability tends to be attenuated by the integration involved in calculating collective dose.

2.10 "The Effect of Turning of the Wind with Height on Lateral Dispersion" (B Y Underwood)

This review assesses the impact of including in the prescription for σ_y (plume cross-wind standard deviation) a contribution from turning of the wind with height. This contribution is expected to become significant a few tens of km from the source. It considers the theoretical and experimental information available, concluding that the effect can have an important impact on σ_y values beyond about 20km downwind; approximate parameterisations introduced to account for it are described. Although the effect can have a large influence on ground-level concentrations, the impact on numbers of health effects is expected to be less: acute effects are confined usually to within a few 10's of km of the release point and delayed effects involve an integral which washes out much of the sensitivity to σ_y values. Generalised sensitivity indicators are devised to approximately quantify the effect on numbers of acute and delayed fatalities : these point to typically less than a factor of 3 impact.

The review also contains an appendix on σ_y parameterisations in general, serving as background material to the main report.

3 SYNTHESIS

Although each review contains its own conclusions and recommendations, a number of common themes can be identified. (Reference to particular review chapters is denoted by underling in the following).

3.1 Simple models

In the context of calculating consequence for reactor accidents it is important to develop simple models, allowing rapid computation, to represent the additional phenomena under consideration here. This is necessary because of the large number of runs of the code likely to be required for a probabilistic risk assessment and supporting sensitivity studies. Of particular use are extensions or 'fix-ups' of the Gaussian model since this is an explicit formulation still widely used in the dispersion modules of consequence-assessment codes. Some such extensions are already standard, viz the technique of 'multiple reflections' to account for an impervious 'lid' to the mixing layer, the use of the 'source depletion' assumption to incorporate dry deposition into the Gaussian framework and the 'exponential plume depletion' approximation to account for the effect of wet deposition on the amount of material airborne.

The reviews produced for this study describe other examples of the use of simple models and Gaussian 'fix-ups'. Techniques have been developed to account for the trapping and fumigation effects at an internal boundary layer within a Gaussian framework as described in the reviews of coastal effects and urban effects. The review of topographical effects describes techniques to account for increased turbulence and streamline distortion in complex terrain within Gaussian formulation. Similarly the effect of turning of the wind with height of lateral dispersion has been allowed for via a simple prescription for increasing the Gaussian σ_y at downwind distances >20km. Numerous Gaussian extensions to allow concentrations in building wakes to be predicted are described in the review of building effects. The review of time-varying meteorology describes the 'virtual source' method of accounting for changes in atmospheric stability whilst retaining a Gaussian framework. Segmented plume techniques enable the effects of wind-direction changes to be taken into account both for time-varying meteorology and for topographical applications.

The development of simple 'fix-ups' should go hand-in-hand with the development of more realistic - and usually more complex - models for the phenomena concerned. These serve as 'benchmarks' against which the simple models can be checked and 'tuned'. Examples of this function are provided by the discussions of the APT computer code (using second-order turbulence closure) in the reviews of urban effects and of coastal effects. Similarly, more realistic numerical models of turbulent, recirculatory flow are mentioned in the reviews of building effects and topographical effects. The review of dry depositon contains numerous examples of how the 'source depletion' approximation has been tested against more realistic techniques for incorporating dry deposition into dispersion modelling.

3.2 Increased data requirements

Another important 'theme' emerging from many of the reviews is that increased realism in modelling often leads to a requirement for substantial amounts of extra data, much of which is not readily available in an appropriate form.

To account for the dependence of dry deposition on surface characteristics requires data on land usage, with the additional complication of seasonal variations. A proper assessment of the contribution of in-cloud scavenging to wet deposition requires specification of raincloud characteristics (besides rainfall intensity) for particular rain episodes. The quantification of both dry and wet deposition involves a knowledge of source term characteristics at a level of detail not readily available to date.

The inclusion of modelling for the influence of site-specific topography on dispersion implies that detailed wind fields must be constructed, either by interpolation from multi-station data (usually only available on a coarse spatial mesh) or by modelling the flow over/around specific features identified in a survey of the surrounding terrain. The quantification of coastal effects requires, for example, that air-sea temperature differences (or equivalent data) be known in addition to the usual dispersion parameters, and also whether or not a sea-breeze circulation exists. To account for the overall influence of urban effects on the dispersion of material emitted upwind of a city requires characterisation of the effective urban roughness and surface heat flux (which depends on a number of variables). An assessment of the pattern of contamination within the urban 'canopy' would require a more detailed representation of the spatial structure within the city.

Even proper quantification of mixing height variations requires more meteorological data than currently utilised in accident-consequence assessments, for example in terms of time-integrated heat fluxes or time-varying temperature profiles. Similarly, any attempt to do better than the simple 'fix-up' for including the effect of turning of the wind with height on laterial dispersion would require meteorological data on actual horizontal wind shear conditions to be specified.

In conclusion, it is important to bear in mind that a move towards greater realism in modelling often brings with it a requirement for additional data and that the collection, processing and incorporation of such data may present a challenge as great as - if not greater than - that of developing a more realistic mathematical model.

4 RECOMMENDATIONS FOR FUTURE WORK

4.1 Broad requirements

Most of the review chapters contain specific recommendations for future work and these will not be restated here. However, it is worth reiterating the underlying basic requirements for progress in the various areas. First, research at a fundamental level (both theoretical and experimental) into the topics of turbulent atmospheric flows and diffusion provides a vital background to the development of appropriate models for practical applications. Thus there is a continuing need to examine separate phenomena in idealised geometries etc to improve basic understanding, for example dispersion near isolated hills or ridges in the context of topographical effects, dispersion in the wake of cuboid structures in the context of building effects, dispersion at an interface of surfaces of different roughness or surface heat flux and single-raindrop collision efficiencies. Physical modelling (wind tunnel, water flume) has an important role to play at this level. Secondly, the capability to handle more realistic - and usually more complex - situations has to be developed, for example dispersion in complex terrain, dispersion in the vicinity of groups of buildings or plume rise from a building wake. Both numerical and physical modelling are important in this respect. A practical requirement is that the experience gained from modelling these complex situations should be translated (as far as possible) into simple, improvised model extensions capable to being incorporated into consequence assessment codes.

Thirdly, it is essential that field trials be performed to provide verification and validation of models under realistic conditions.

4.2 Priorities

It is natural to expect that this exercise of producing 10 reviews should lead to some insight into priorities for further development. However, a basic difficulty with this is that any ranking of subject areas according to their impact is dependent on a specification of which objectives or end-points of the calculations are perceived to be most important. For example, if the frequency-consequence (f-N) distribution of acute fatalities resulting from major accidents is deemed to be the dominant factor, then modelling assumptions for the distance range beyond about 50km have no impact. If, however, the f-N distribution for latent cancers is reckoned to be the most important factor then some aspects of modelling at ranges $>$ 100km may have a significant influence.

In practice the decision as to which end-points are important has to be even more specific than this : if, for example, the high frequency/low consequence 'tail' of the acute-fatality f-N distribution is judged to be of paramount importance, priorities for improvement might be quite different from those obtained by assuming some moment (for example the mean) of the f- N distribution is the primary risk indicator; there is as yet no consensus as to the relative weight to be placed on results in the tail end of consequence distributions. On the other hand, the number of health effects may take second place to the areas of land contaminated, in which case deposition processes would assume an even greater importance than otherwise. In the following it will be assumed that the conventional end-points - numbers of acute and delayed fatalities, areas of land contaminated and number of people moved - are of concern, without being more precise about relative weights.

To guide a decision on priorities it is necessary to perform sensitivity studies using consequence-assessment codes which include modelling extensions to handle the phenomena under consideration. Fairly rudimentary models - for example the Gaussian 'fix-ups' discussed in section 3.1 - would be adequate to test gross sensitivity. It has to be borne in mind that checking the sensitivity of, say, ground-level concentrations to the various effects does not necessarily provide an accurate indication of how sensitive the conventional end-points will turn out to be in actual consequence assessment : for instance, the integration involved in calculating latent cancer fatalities washes out the sensitivity to many factors. Thus it will be necessary usually to carry the calculation through to numbers of effects or, at least, to devise indicators which are more closely related to the conventional end-points than is ground-level concentration (nb the calculations performed for the chapters on the effects of mixing height variations and on the effect of turning of the wind with height on lateral dispersion). In the former case it has to be remembered that the sensitivity found may by a function of, inter alia, the precise source terms and population distributions used and not necessarily generalizable to other situations.

In the last few years a number of sensitivity studies of the above type have been performed, and in some cases have been used to draw conclusions about the impact of certain of the phenomena under consideration here. The studies of Ritchie et al [2,3] concerning rainstorm-structure modelling have been thoroughly discussed in the chapter on wet deposition; the indication is that failure to account for rainstorm structure could introduce significant distortion into the high consequence/low frequency tails of certain consequence distributions. The studies of Sprung and Church [1,4] have been reviewed in the chapters on the effect of mixing height variations and on the effect of turning of the wind with height on lateral dispersion respectively. Although limited in scope they lend support to other evidence that these two effects have fairly low impact. Aldrich [5] also examined the horizontal windshear effect and reached a similar conclusion. Several sensitivity investigations were performed in the context of the German Risk Study [6]: deposition rates (wet in particular and 'dry' to a lesser extent) and the height of plume rise were found to be parameters of particular importance. The Probabilistic Risk Assessment Procedures Guide [7] contains extensive tables specifying qualitatively the sensitivity of reactor accident consequences to models and parameters used in all phases of the consequence-assessment process. Many of the indications are based on judgement rather than on evidence drawn from quantitative sensitivity studies. However, in addition to the studies mentioned above the tables draw extensively on the results of the CSNI- sponsored benchmark exercise [8] in which standard problems were performed by numerous organisations using a wide variety of consequence-assessment codes, some of which included modelling for certain of the specific effects under consideration here. Although it is difficult to assess the sensitivity to isolated effects or parameters from the results since often a number of changes contributed simultaneously to differences, the following conclusions were drawn : the results confirmed the above-noted sensitivity to dry and wet deposition parameters and to plume rise modelling; they also indicated a moderate sensitivity to modelling for trajectory curvature (noted as a 'major' sensitivity in the PRA Procedures Guide) and low sensitivity to mixing heights.

In the course of preparing the current set of reviews, Underwood [9] performed several sensitivity investigations : the chapter on the effect of turning of the wind with height on lateral dispersion contains the results of calculations which, although limited in scope, indicate a fairly low impact on consequence parameters. Similar calculations were performed for the study of the effect of mixing height variations which, again, indicate only modest impact. The chapter on time-varying meteorology contains some illustrative calculations which indicate the importance of modelling temporal variations (particularly in rainfall intensity), at least for the high-consequence tail of acute-fatality f-N distributions. The reviews of wet and dry deposition contain recommendations for sensitivity studies on the effect of source-term parameters, particularly particle size distribution.

In the context of assessing the safety implications of locating a PWR at Sizewell in the UK, the National Radiological Protection Board (UK) has recently performed some sensitivity studies [10] to supplement the consequence assessment [11] performed for degraded-core accident sequences. Relevant to the present discussion they find only modest sensitivity to dry deposition velocity and to the energy content of the plume (determining plume rise).

Summing up, then, from the particular viewpoint of the conventional consequence end-points and subject to caveats noted above, a consensus view emerges that improvements in the modelling of wet deposition should appear high on any list of priorities, and the review of the topic produced for this study is correspondingly extensive. On the other hand, further work in the areas of cross-wind shear, mixing height variations and building wake effects (for passive plumes) should appear fairly low on the list. The reported evidence on dry deposition and plume rise is equivocal, reflecting in part the dependence on particular conditions and end-points, but on balance these topics are likely to appear moderately high on a list of priorities. It is evident that future consequence assessment codes should account for time-varying meteorology but, anyway, this type of modelling is already becoming standard. As for the other topics reviewed - urban effects, coastal effects and topographical effects - there appears to be insufficient quantitative evidence so far to determine what priority should be afforded them, and future studies should seek to clarify the position, with due consideration of the site-specific nature of the effects involved.

It should be borne in mind that the perception of priorities will continually evolve as understanding increases and as attitudes towards risk develop. It should also be remembered that the level of precision sought after in the atmospheric dispersion/deposition phase of the risk assessment process should be consistent with the level obtainable in other phases, such as characterisation of source terms and health physics modelling.

REFERENCES

[1] SPRUNG J L and CHURCH H W, Sandia National Laboratories Report SAND76-0618 (1977).

[2] RITCHIE L T, BROWN W D and WAYLAND J R, Sandia National Laboratories Report SAND79-0376 (1976).

[3] RITCHIE L T, BROWN W D and WAYLAND J R, Sandia National Laboratories Report SAND76-0429 (1980).

[4] SPRUNG J L and CHURCH H W, Sandia National Laboratories Report SAND76-0619 (1977).

[5] ALDRICH D C, Sandia National Laboratories Report SAND79-2081 (1979).

[6] VERLAG T U V RHEINLAND, German Nuclear Power Station Risk Study, Technical Appendix 8 (1981).

[7] USNRC, United States Nuclear Regulatory Commission Report NUREG/CR-2300 (1982).

[8] ALDRICH D C et al, International ANS/ENS Topical Mtg on Probabilistic Risk Assessment, Port Chester, NY (1981).

[9] UNDERWOOD B Y, EURATOM Report EUR8935 EN (1983).

[10] KELLY G N et al, National Radiological Protection Board (UK) Report NRPB-R142 (1982).

[11] KELLY G N and CLARKE R H, National Radiological Protection Board (UK) Report NRPB-R137 (1982).

TABLE 1

LIST OF REVIEWS

		Author
	Summary Document	B Y Underwood
Ch 1	Plume Rise in Nuclear Safety Studies	G D Kaiser
Ch 2	Dry Deposition	B Y Underwood
Ch 3	Wet Deposition	B Y Underwood
Ch 4	Atmospheric Dispersion in Urban Environments	N J Holloway
Ch 5	Topographical Effects in Nuclear Safety Studies - A Review	P J Cooper
Ch 6	Coastal Effects and Transport over Water	W Nixon
Ch 7	Time-Varying Meteorology in Consequence Assessment	B Y Underwood
Ch 8	Building Effects in Nuclear Safety Studies - A Review	P J Cooper
Ch 9	Effect of Variations in Mixing Height on Atmospheric Dispersion	B Y Underwood
Ch 10	The Effect of Turning of the Wind with Height on Lateral Dispersion	B Y Underwood

Discussion

Question: G. Fraser - CEC/Lux

Bearing in mind the impracticability of fully validating dispersion and deposition models because of the wide range of possible values for many of the parameters involved and the multiplicity of possibile combinations of these values, what degree of confidence might we hope to have ultimately in the results produced by the models?

Answer: B. Underwood - UKAEA/SRD

I cannot give a quantitative answer to the question of overall uncetainty in the consequence assessment process. However, it is worth pointing out that currently there is considerable activity concerning the question of uncertainty and a number of groups worldwide are attempting to give estimates. The outcome of this exercise impinges on the use to which PRA calculations can be put.

Question: B. Tolley - CEC/BXL

What is the progress and content of the extension work to this contract? Particularly the significance of F.P. source term characteristics on dispersion/consequence modelling.
Does it appear that the current big effort on FPST definition is broadly sufficient. Or does a lot more need to be done?

Answer: B. Underwood - UKAEA/SRD

Task I of the extension broadly examined the question of source term characteristics from the viewpoint of consequence assessment i.e. what information would be needed to properly quantify dry and wet deposition after release. It also examined a number of processes occuring post release which would alter these characteristics, identifying an area of modelling which has conventionally not been addressed.
It does appear that recent source studies are contributing towards providing information of the kind required by consequence assessors.

Question: B. Vanderborght - SCK/CEN

Did I understand properly that in the ranking of specific effects building wake, urban, coastal and topographical effects get a low priority?
I noticed that just those four topics have been studied in area C of the programme.
I suppose that the problem of "experimental validation" is a high priority for every topic in your ranking order.

Answer: B. Underwood - UKAEA/SRD

The slide showing tentative rankings was perhaps misleading. It was not intended to show coastal effects, topographical effects and wake effects as having low priority but rather having uncertain priority at present due to lack of quantitative sensitivity information. I must stress that the rankings provided were very tentative and also referred to specified calculation end points.
Yes, experimental validation is vitally important, as is brought out in

the individual review chapters.

Question: Goddard - Imp. Col./London

Could I ask for the reasons for a particular ranking of plume rise as a significant factor? Is this in respect of a plume rising or not, or is it in respect of current uncertainties in the trajectory, say, of a plume rising from an isolated stack?

Answer: B. Underwood

The ranking is based on the potential impact of plume rise on conventional consequence-assessment end-points. In terms of uncertainties, the question of detailed plume trajectory has received much more attention than other points of concern in consequence assessment, such as a criterion for lift-off from a building wake.

THE EFFECTS OF BUILDINGS ON LOW-LEVEL ATMOSPHERIC DISCHARGES

P.M. Foster and A.G. Robins
Central Electricity Generating Board
Central Electricity Research Laboratories
Leatherhead, Surrey
England

Summary

This study of building influenced dispersion provides a coherent assessment of both analytical and physical modelling techniques from comparisons with full scale data obtained at an operational nuclear power station. Plume sections showing the elevated structure and dilution of full scale plumes have been obtained using scanning lidar equipment and dual smoke/SF_6 tracers. Complementary wind tunnel studies have been made and validation studies between full and model scales have been carried out. The limitations of some popular simple dispersion models have been assessed from comparisons with the measurements.

1. INTRODUCTION

The effects of buildings on the dispersion of low level releases of radioactivity are most important in determining the peak individual doses and ground contamination in the immediate vicinity of the plant and in determining the source geometry for the longer range dispersal of the released material. This is particularly the case for the more probable, but less severe accidents where activity may be released from a low stack at roof level or via other openings in the containment building. It is also the case for low probability, large scale releases in which the hot emission plume does not rise 'cleanly' from the ground and/or building. In either case the effects are very complicated and have not been directly amenable to analytical methods of investigation. However, as a matter of necessity various simple analytical models have been proposed or adapted by the licensing authorities and utilities throughout the world which purport to deal with some of the more important aspects of building influenced dispersion, principally the enhanced dispersion which is produced by the building wake and the influence of source position. In the main these models are simple generalisations of the results derived from physical models; relatively little work has been done in the full scale which can adequately test many of the assumptions made, particularly concerning the elevated structure of the emission plume.

The building wake dispersion study undertaken by the CEGB was aimed at providing a coherent assessment of both analytical and physical modelling techniques from comparisons with full scale data obtained at an operational nuclear power station. The project was divided into three parts:

(i) to develop the necessary full scale measuring techniques,

(ii) to conduct a range of plume dispersion and mapping experiments at a suitable nuclear site at both full and model scales,

(iii) to assess the results of (ii) and compare them with analytical modelling predictions.

This paper describes the work which has been carried out to date with emphasis on parts (ii) and (iii).

2. FULL SCALE MEASUREMENTS

2.1 Choice of Site and Emission Source Positions

The full-scale experiments were conducted at Oldbury Power Station (Fig. 1). This is a magnox station with concrete pressure vessels and is situated by the river Severn about 20 km north of Bristol. This site was chosen both because of the flatness of the surrounding terrain and because the building effects were likely to be dominated by the single, large reactor building (approximate dimensions 150 m × 50 m × 54 m high).

The emission source positions investigated are shown in Fig. 2. Most full scale releases were made from one position, an outside balcony at the 24.6 m level on the north side of the reactor building (position 2), so that data consistency and variation with range could be assessed. However, some releases were also made from the building roof (position 3) and at ground level (positions 5, 6 and 7).

2.2 Experimental Technique and Instrument Deployment

A primary requirement of the project was the development of a technique enabling quantitative mapping of a building influenced plume at full scale. This was achieved by releasing simultaneously both smoke and SF_6 gas tracers, the smoke being monitored by lidar equipment and the SF_6 being monitored using bag sampler arrays. Scanning with the former provided vertical cross sections of the plume showing the relative distribution of elevated material in the plume whilst the latter provided ground level profiles showing the absolute distribution of material. Comparing the low level lidar results with those of the bag sampler array then enabled the lidar plume section results to be calibrated absolutely as well as providing a check on the general quality of the data.

The lidar was a pulsed ruby type with a pulse energy of 0.1J and a maximum fire rate of 1 Hz. It has an operational range of up to about 10 km, although for the present application most measurements were made within 0.5 km to 1.5 km range from the lidar. The backscatter data was stored on floppy diskettes, the signal digitisation being made with 8 bit resolution at a 0.05 μs sampling interval (equivalent to a spacial range resolution of 7.5 m along the line of sight. The smoke source was of the zinc oxide type and was produced by pyrotechnic grenades which burnt for about 16 minutes each.

The SF_6 bag samplers were of a multiple port design which enabled upto four consecutive 15 minute samples to be taken at one site. The samples were analysed by gas chromatography. From calibration comparisons made between two independent CEGB laboratories and the British Aluminium Research Centre, an absolute accuracy of within 20% was indicated. The samplers were deployed at sites marked at 10° intervals on the site boundary fence at about 125 m range from the centre of the reactor building, and at 5° intervals on two outer arcs at approximately 400 m and 700 m range. The arcs subtended a total angle of 135°, ranging from N to SE (Fig. 4). During each release between 10 and 20 samplers were usually deployed along the arc covered by the lidar.

Meteorological conditions were monitored in a variety of ways but for purposes of analysing the plume data the wind speed and direction measurements on a 50 m mast were used. Only category D stability conditions were investigated, this being appropriate to the complementary wind tunnel study.

2.3 Results and Data Processing

Over a period of about 15 months measurements were attempted on about 20 separate occasions of which approximately 25% were aborted due to adverse wind conditions. Twenty four 15 minute releases were made in total of which 20 provided useful data. The latter are summarized in Table I.

TABLE I. SUMMARY OF SUCCESSFUL TRACER EXPERIMENTS

lease mber	Source Position	50 m Wind		Approx. Distance to Sampling Arc (m)	Tracers Used	
		Direction θ	Speed (m/s)		Smoke	SF_6
3	2	17°	11	125	√	√
4 (?)	2	7°	10	125	√	√
1 (?)	2	32°	11.5	400	√	√
2 (?)	2	30°	11	400	√	√
3	6	32°	6.5	125/400	-	√
4	6	7°	6	125/400	-	√
0	2	32°	13	125	√	√
1	2	27°	14	125	-	√
2	2	37°	13	125	√	√
3	2	37°	15	125	√	√
4	2	27°	6	400	√	√
5	2	27°	5	400	√	√
6	2	27°	5	400	-	√
7	2	32°-57°	5	400	√	√
8 (?)	7	37°	3.5	125	√	√
2	5	27°	6	125	√	√
3	5	27°	6.5	400	√	-
4	3	37°	11	125	√	-
5	3	27°	11	125	√	√
6	3	27°	11	400	√	√

For consecutive releases in which the wind conditions held steady the data was analysed as a single 30 minute release, e.g. releases 22/23 and 24/25. A question mark has been placed on those experiments with suspect results. For release 4 the SF_6 concentrations were unexpectedly high (e.g. factor ×3) and for releases 11, 12 and 28 either bag sampler failures or changes in wind directon produced poor definition of the plume profile at ground level. However, although the concentration data for these four experiments may be questionable, the general plume dimensions and relative distribution of plume material is still available from the lidar results.

Data validation and processing presented a major task due to the large quantities of data involved. During an individual 15 minute tracer release period the lidar will scan through the plume about 30 times at 0.5° intervals of elevation, accumulating between 0.2 to 0.3 million backscatter records. These are subsequently validated by inspection and then averaged at each elevation angle (Fig. 3). Using a digitized site plan (Fig. 4) the bag sampler data is projected along the wind direction onto the lidar line of sight in order that the ground level SF_6 and low level smoke plume profiles may be compared and the lidar backscatter results calibrated (Fig. 5). The resultant data may then be used to display the concentration structure of the plume either in contour form (Fig. 6) or as cross-wind or vertically integrated profiles (Figs. 7 and 8). In these last three figures the concentration data have been reduced to the non-dimensional form: $CU_h h^2/Q$, where U_h is the wind speed at building height h and Q is the source strength.

The example which has been chosen to illustrate data processing, i.e. release 35, is one of the better examples. In some cases faults with bag samplers or wind recordings, poor alignment between the lidar sight line and the sampler array, or a shift in wind direction during the release could all produce a loss in data quality. However, except for release 27 in which a severe change in wind direction occurred, the quality in all cases was sufficient to produce acceptable plume sections. The crucial factor is to ensure that a sufficient number of plume scans is achieved during the release. In a previous study utilizing lidar to investigate building influenced plumes at the Duane Arnold Energy Centre (1) only 15 or so scans were made and the resultant plume sectons were of relatively poor quality. In the present study usually 25 or more were achieved.

3. WIND TUNNEL MEASUREMENTS

3.1 Experimental Techniques

The wind tunnel investigation was conducted in the large CEGB wind tunnel at the Marchwood Engineering Laboratories. This facilty has a 23 m long working section of 9 × 2.7 m cross section, which enabled the work to be carried out at a scale of 1/300, the region within 700 m of the reactor building being modelled. For all wind directions the approach flow characteristics were typical of a rural, neutrally stable, atmospheric boundary layer, with a height of 360 m and a surface roughness length of 0.1 m. The velocity profile exponent was about 0.17, and the relationships between the velocities at the boundary layer top, U_e, the level of the reactor building roof, U_h, and at the equivalent of 10 m above ground, U_{10}, were:

$$U_h/U_e = 0.72 \qquad U_{10}/U_e = 0.54 \quad .$$

Disperson characteristics within this flow had previously been found to be typical of the neutrally stable atmosphere for ranges out to 1 km and sampling times up to about 30 minutes. Details of the characteristics of the boundary layer are to be found in (2). For short range, building affected dispersion there is in neutral conditions little direct effect of boundary layer height, and hence the 1.2 m height used was entirely suitable.

Dispersion studies were carried out using hydrocarbon trace gas techniques (3). Flow visualisation and other qualitative observations were also made. For the quantitative work, the wind tunnel was operated at a free-stream speed of 2 ms^{-1} and neutrally buoyant helium-propane mixtures were emitted from 9 mm diameter sources at speeds of about 0.7 ms^{-1}. In the experimental work, averaging times were chosen to ensure good repeatability, generally to 10%, of measured quantities. An extensive range of source locations and wind directions were investigated, the resulting data set being summarized in Table II.

TABLE II. SUMMARY OF AVAILABLE MEAN CALIBRATION DATA

Sources	Range (m)	Wind directions, θ°
Centre-line concentrations, C(x,C.L.,0) (*, not source 1)		
1,2,3,4,5	60- 700	0,22.5,45,67.5,90,112.5*,135*,180,337.5
2,3,4,5	700-1350	ditto
A,B	160- 680	0,90
Contours of ground level concentration, C(x,y,o)		
1	60- 800	0,22.5,45,67.5,90,180
2,3,5,	ditto	0,22.5,45,67.5,90,112.5,135,180,337.5
4	ditto	45,90,112.5,135,337.5
Vertical cross-sections, C(y,z)		
1,2,3,5	120- 680	0,22.5,45,67.5,90,180
2	150	5,10,15,20,25,30,35,40
4	150- 680	45,90
A,B	120- 680	0

Ground level concentration contours were obtained from 5 lateral profile measurements in the range from 60 to 700 m, each profile being drawn from data at 16 points. The centre line distributions were determined from these data, together with additional profiles at 1050 and 1350 m. The vertical cross sections were each evaluated from 5 to 7 vertical profile measurements, at 7 points per profile. The tabulated output from the wind tunnel computer system presented concentrations in the non-dimensionalized form: $CU_e h^2/Q$, since U_e was routinely measured by the tunnel reference pressure difference. Further data analysis and plotting was performed manually.

3.2 Results

As is clear from the data set summary, considerable information concerning the effects of source position and wind direction is afforded by the results. A good general impression of these effects is given by Figure 9, which shows the non-dimensional ground level concentration on the plume centre-line at 150 m downwind, for sources 1 to 5. Sources 1, 4 and 5 were at ground level and the results indicate up to a factor of 4 variation either with wind direction for a given source, or with source location for a fixed wind direction. Results for sources 2 and 3 were always within a a factor of 2 of one another, frequently much closer. For most wind directions there was upflow at the position of source 2, carrying emissions towards roof level and producing a general behaviour like that for source 3. Both sources produced peak concentration levels for the wind direction of 67.5°, since this was the orientation for strongest aerodynamic downwash. Since 3 was at the centre of the roof, a similar peak was found at 112.5°, which of course was not to be expected for source 2.

Figure 10 shows some centre-line concentration distributions for sources 2 and 3. In both cases factor of 2.5 variations were observed at the equivalent of 1 km downwind. The 0° and 90° results apply to the broadside and end-on building orientations, and the 45 and 135° results to what would be equivalent orientations for a simple block building. Even for the symmetrically placed roof source, pronounced differences were found within the first 200 m, though further downwind concentration levels were very close to each other. For 45° and 135°, the plumes from sources 2 and 3 behaved rather similarly, in an overall sense, but not at 0° and 90°.

Typical of the ground level concentration contours are the results for source 3 and wind directions 0° and 45° shown in Figure 11. The 0° behaviour was indicative of an elevated emisson, albeit strongly perturbed by the building wake, whereas in the other case aerodynamic downwash and related effects were strong enough to mask the fact that the emission was from roof level. However, the vertical cross-section measured at 150 m, and shown in Figure 12, did display an elevated maximum, which had disappeared at 300 m downwind. It is clear from the figure that wake effects greatly enhanced the downward spread of the plume, the ratio of the peak to the surface maximum concentration being 1.3. In contrast, for a wind direction of 0°, this ratio was found to be about 2.5 at 120 m.

4. SIMPLE ANALYTIC MODELS

A number of comparatively simple models of building affected dispersion have been proposed in the literature. Perhaps the earliest was that due to Gifford (4), in which an additional area, related to the building frontal area, was added to the plume cross-sectional area, as expressed by $\sigma_y \sigma_z$. This model only predicted ground level, centre-line concentrations, whereas later models modified standard gaussian formulations so as to predict the full concentration field. In these, the plume dimensions, σ_y and σ_z, were individually modified to allow for the presence of the building; e.g. by adding amounts related to the building dimensions (5) or by using virtual origins (6). In the latter, an effective source height was also introduced, though it was the same for all emissons at, or below, roof level. A modified form of the virtual origin model was put forward by Jones (7) in which a near-wake model was added, and its predictions linearly patched to the virtual source model in the region out to 5 building heights. Huber and Snyder (8) also divided the downwind region into a number of zones and used different models in each, terminating in a virtual source model. Their formulation purports to treat any source below, on, or above roof level, whereas the other models are restricted to sources at and below roof level. A brief discussion on the performance of these models is presented below.

5. DISCUSSION

5.1 Intercomparison of Full Scale and Model Results

It is evident from Table I that the full scale results should be comparable with or embraced by the wind tunnel results for 0° and 45°. However, note that in making these comparisons it will be necessary in some cases to multiply the model scale concentrations by 0.72 in order to renormalise them with respect to U_h rather than U_e (see Secton 3.1). Upon making this correction it can be seen that the plume sections at full and model scales (Figs. 6 and 12) show very good qualitative and quantitative agreement. As already stated this case represents one of the better examples; the agreement obtained overall has not yet been properly assessed but it appears to be generally 'good to satisfactory'. Where differences occur they may be the result of natural variability, i.e. each full scale result represents just one sample from the ensemble averaged behaviour which is derived at model scale. Of course, differences may also be introduced by experimental problems as discussed previously.

Some indication of the variability which is obtained at full scale is given by Fig. 13 where the ground level and elevated peak concentrations linked by a vertical line are plotted against range for all cases in which lidar measurements were available (N.B. the elevated peak is given by the upper point). The source position 7 result is entered twice since it behaved in some respects like a point source emission; the two ranges used therefore refer to the building centre and the emission point as origins. Also shown are the ground level centre line concentrations measured at model scale for release positions 2 and 3, and for wind directions 0° and 45°. The comparison shows that the three results for position 3, which were consecutive releases on the same day, give good agreement whilst the seven results for position 2, which were obtained on four different days, show more scatter and a tendency towards lower concentrations than observed at model scale. The latter might be the result of steadier wind conditions being available at model scale.

For comparison with analytic model predictions the modified virtual source model in (7) may be used, this being typical of its kind. This recommends the following virtual source characteristics for the gaussian dispersion region beyond about 5 building heights: $\sigma_z(o)$ = 1/3 building height, $\sigma_y(o)$ = maximum ($\sigma_z(o)$, 1/3 building width) and source height = $\sigma_z(o)$. The model also recommends the use of the Pasquill-Gifford values for σ_y, σ_z but with σ_y ammended to include an additonal term, linearly dependent on travel time, to allow for the effects of large eddies. The ground level centre line concentration versus range behaviour predicted by this model is shown as a pecked line in Fig. 13. The prediction is seen to show good general agreement with both full and model scale results although, once again, the variability trend in the full scale results is toward lower than predicted values.

An indication of plume growth with travel is provided by Fig. 14 in which the half heights and half widths of the integrated vertical and integrated cross wind profiles from the full scale data are plotted, these being the respective dimensions at which the plume concentration has fallen to one half of its respective peak value. Also shown are the corresponding results predicted by the gaussian component of the above virtual source model. Significant disagreement is evident, the half width being overpredicted and the half height underpredicted. Integrated profiles are not yet available from the wind tunnel data. However, corresponding dimensions derived from the centre line vertical profile and the ground level cross wind profile are included in the figure. The variation with source position and wind direction is quite noticeable, this being due to the rapid changes in profile shape with range which could occur. Nevertheless, in spite of this variability it is evident that there is a general agreement between both full and model scales. The apparent failure of the analytical model to predict plume dimensions is therefore substantiated and possibly reflects the inability of simple analytical models to predict both centre line concentrations and plume dimensions.

5.2 Performance of Analytic Models

For a more critical assessment of the performance of analytic models the results of wind tunnel measurements should be used. These do not suffer from the inherent variability of measurements made in the atmosphere and, additionally, the models can be applied in the knowledge that the basic spreading laws for the undisturbed flow are precisely known. Figure 15 shows the model predictions as applied to Oldbury Power Station for a wind direction of 0°, the shape being idealised to a block 54 m high, 150 m wide and 50 m deep. Where uncertainty existed, it was assumed that the wind speed appearing in model formulations was U_{10}, and the most conservative parameter values were adopted. Most of the predictions agree quite well with one another, excepting the Huber-Snyder model which seems only to give adequate predictions if a small effective source height is used for all emissions, rather than the actual source height. For comparison purposes, the modified virtual source model in (7) was also applied to a wind direction of 90°, though it must be said that the models are not formulated to take proper account of variations of dispersion behaviour with wind directon. In comparison with Figs. 9 and 10, it is clear that the models underpredict concentration levels at short range and become progressively more realistic as the downwind distance increases. The worst discrepancies at short range occur for ground level emissions at or near the wake edge, since these tended to produce plumes that were not at all thoroughly mixed by the wake flow and, hence, retained some of the structure of emissions advected directly from source to receptor. Beyond about 10 h downwind, any of the models could be used to give a fair estimate of concentraton levels. These conclusions concerning the adequacy of analytic models should be read in conjunction with the recent review of these matters by Fackrell (9).

6. CONCLUSIONS AND RECOMMENDATONS

This project has enabled the development and application of an experimental technique for three dimensional mapping of building influenced plumes at full scale. Measurements made at an operational nuclear power station over a limited range of wind directions have demonstrated agreement with the predictions of a complementary wind tunnel study. Comparisons with the predictions of a standard analytic model based on virtual source assumptions have shown agreement with the ground level centre line concentraton behaviour, but disagreement over the lateral and vertical dimensions of the plume. It appears that the requirement to predict reasonable centre line ground level concentrations cannot necessarily be matched to the prediction of plume dimensions.

The wind tunnel study permitted an extensive range of release positions and wind directions to be investigated and enabled a more critical assessment of simple analytic models to be made. Close to the building plume behaviour was complex and produced large (typically ×4 at 150 m range) changes in the ground level centre line concentration with varying source position and wind direction. The analytic models do not predict this but only attempt to represent the bulk behaviour. However, beyond about 10 building heights range the plume behaviour simplifies sufficiently for acceptably accurate predictions to be achieved.

The dispersion of a passive emission from a building in neutrally stratified conditions as considered here, probably represents one of the simplest engineering tasks placed before the wind tunnel modeller. Hence, the observed agreement between full and model scales, although rewarding, is not altogether surprising. It is suggested that full scale validation studies should now be extended to consider more difficult modelling problems such as those including thermal or density difference effects and the effects of atmospheric stability.

Concerning the development of analytic models, an immediate task is to improve the prediction of plume dimensions, perhaps by reconsidering the specification of the virtual source parameters.

7. ACKNOWLEDGEMENTS

The authors gratefully acknowledge Dr C.D. Barker's contributions to project administration and for organising the SF_6 measurements aspects of the field programme.
This work is published by permission of the Central Electricity Generating Board.

8. REFERENCES

1. THUILLIER, R.H. and MANCUSO, R.L. (1980). Building effects on effluent dispersion from roof vents at nuclear power plants. EPRI Report NP-1380
2. ROBINS, A.G. (1979). The development and structure of simulated neutrally stable atmospheric boundary layers. J. Industrial Aero., 4, 71-100
3. ROBINS, A.G. (1975). Experimental model techniques for the investigation of the dispersion of chimney plumes. Proc. Inst. Mech. Engrs., 189, 4/75, 44-54
4. GIFFORD, F.A. (1960). Atmospheric dispersion calculations using the generalised Gaussian plume model. Nucl. Saf., 2, 56-59
5. FERRARA, V. and CAGNETTI, P. (1980). A simple model for estimating airborne concentrations downwind of buildings for discharges near ground level. Proc. CEC Seminar 'Radioactive releases and their dispersion in the atmosphere following a hypothetical reactor accident', RISO, April 1980, Denmark
6. BARKER, C.D. (1982). A virtual source model for building wake dispersion in nuclear safety calculations. CEGB Rept. TPRD/B/0072/N82
7. JONES, J.A. (Chairman) (1983). Models to allow for the effects of coastal sites, plume rise and buildings on dispersion of radionuclides and guidance on the value of deposition velocity and washout coefficients. NRPB Rept. R157
8. HUBER, A.H. and SNYDER, W.H. (1982). Wind tunnel investigation of the effects of a rectangular shaped building on dispersion of effluent from short adjacent stacks. Atmos. Environ., 16, 2837-2848
9. FACKRELL, J.E. (1984). An examination of simple models for building influenced dispersion. Atmos. Environ., 18, 89-98

Figure 1   Oldbury Power Station (looking East)

Figure 2   Site plan showing source positions and defining wind direction

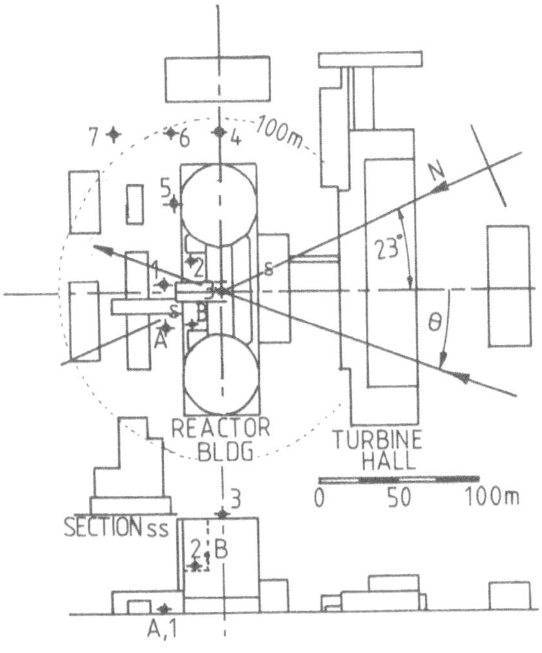

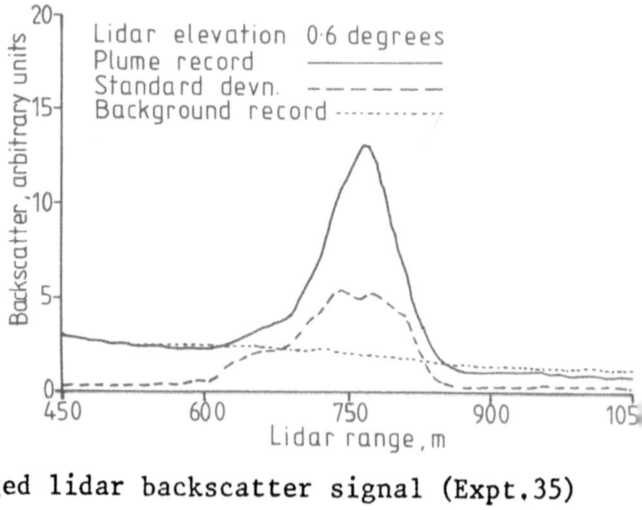

Figure 3   Averaged lidar backscatter signal (Expt.35)

Figure 4 Site plan showing bag sampler positions, plume direction and lidar sight line (Expt.35)

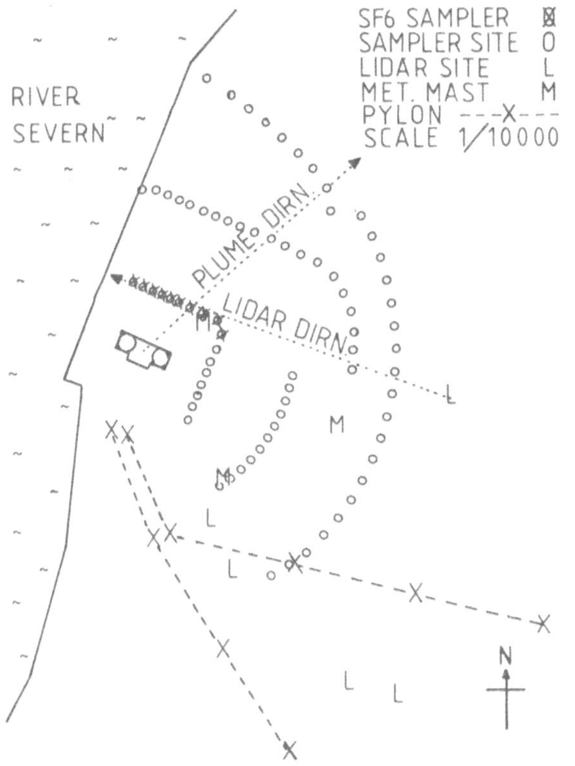

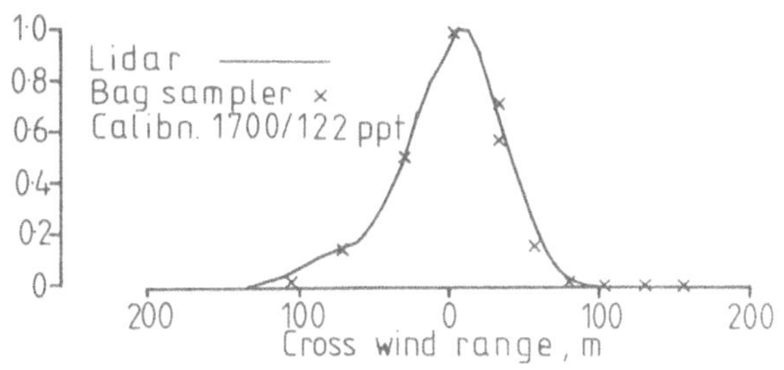

Figure 5 Comparison of lidar and bag sampler plume profile data at 150m downwind range (Expt.35)

Figure 6 Calibrated lidar plume section and intermittencies (i.e. % of shots seeing plume) at 150m downwind range (Expt.35)

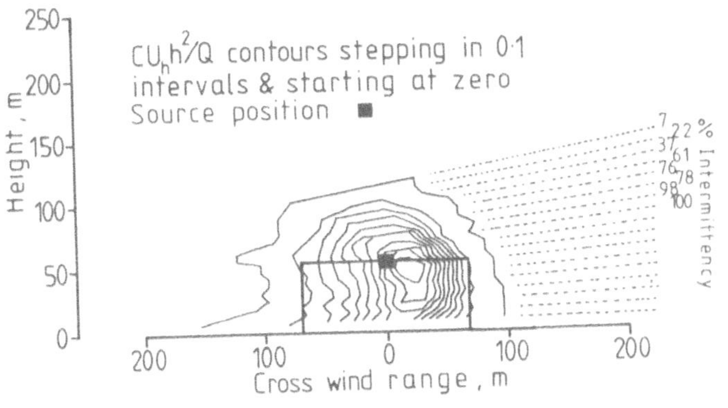

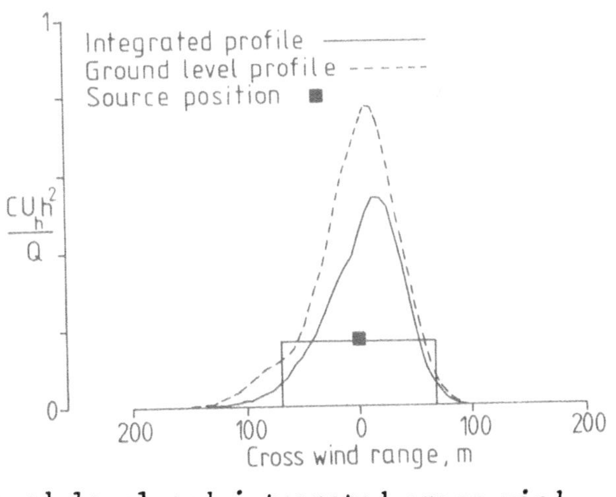

Figure 7 Comparison of lidar ground level and integrated cross-wind plume profiles at 150m downwind range (Expt.35)

Figure 8 Integrated vertical plume profile and fractional intermittency (i.e. fractional plume occurence) versus height at 150m downwind range

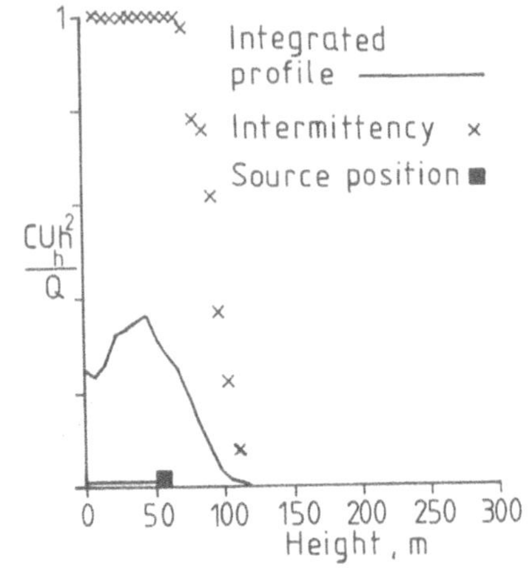

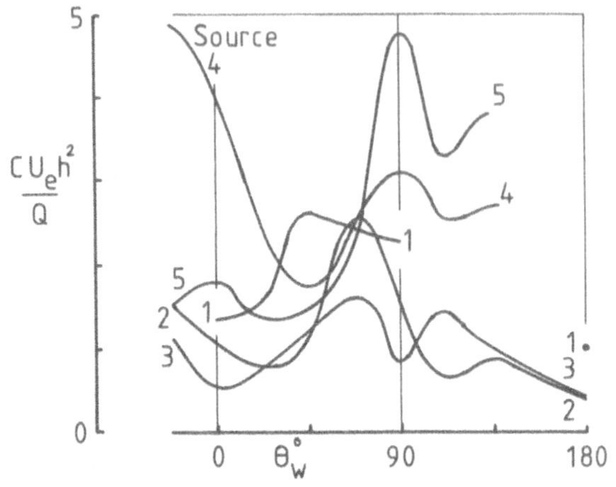

Figure 9 Ground level concentration versus wind direction at 150m downwind range for sources 1 to 5

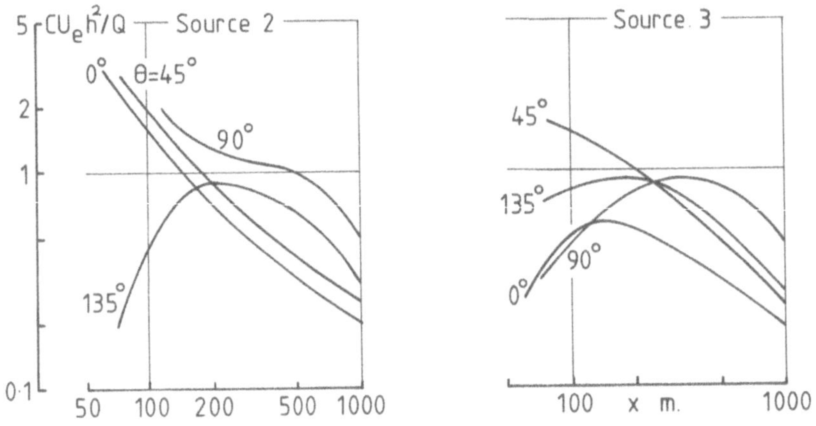

Figure 10 Ground level concentration versus range for sources 2 and 3 at different wind directions

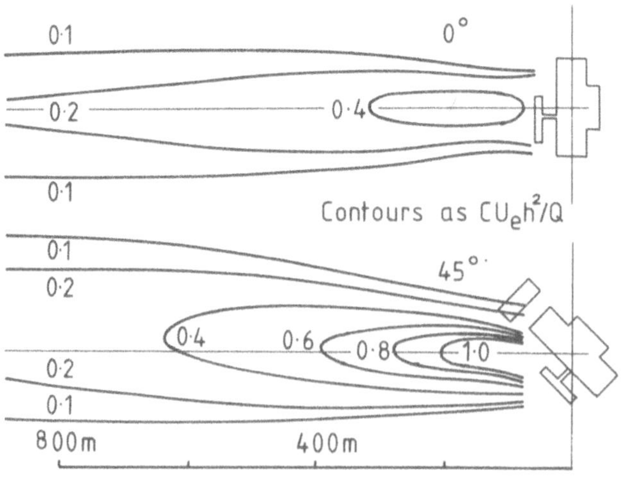

Figure 11 Ground level concentration contours for source 3 at 0 and 45 wind directions

Figure 12 Model scale plume section for source 3 at 150m and 45 wind directio

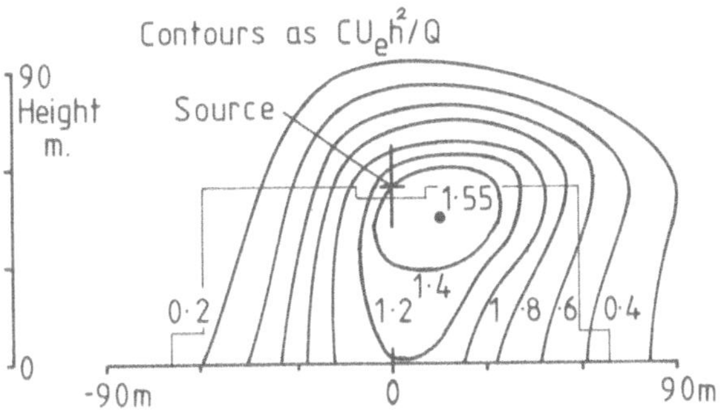

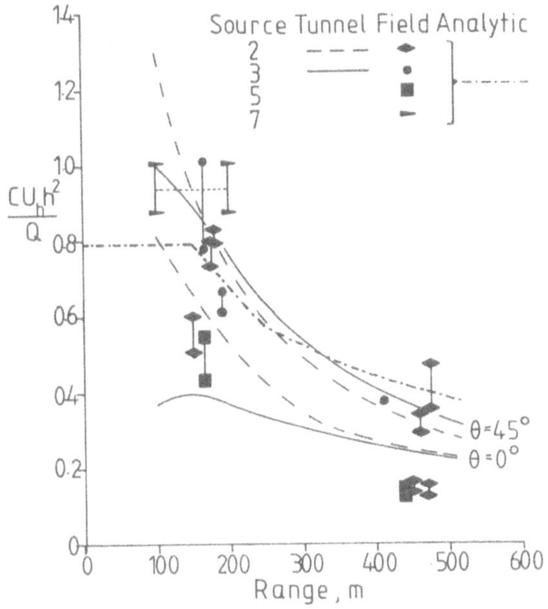

Figure 13 Comparison of plume peak concentrations versus range for full scale, model scale and analytic results

Figure 14 Comparison of plume halfwidth and halfheight versus range for full scale, model scale and analytic results

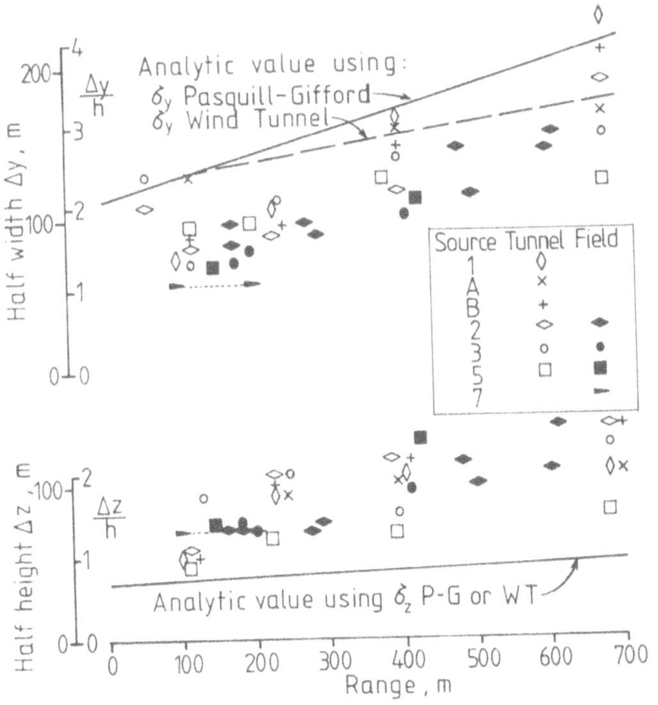

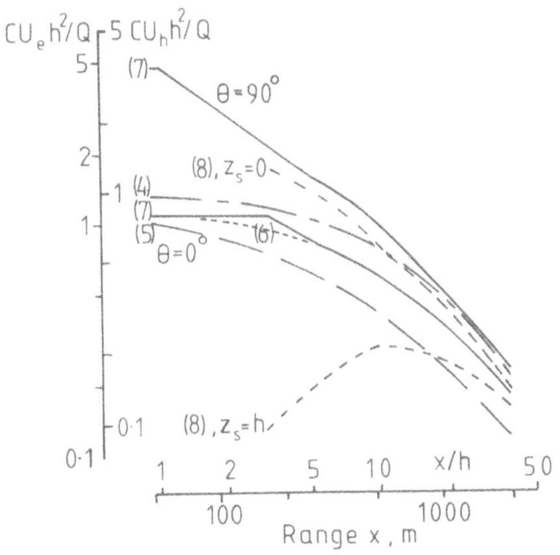

Fgure 15 Some analytic model predictions of centre-line ground level concentration versus range (see References for model identification)

Discussion

Question: B. Tolley - CEC/BXL
What are the continuation activities of CEGB following on from contract 2C?

Answer: P.M. Foster - CEGB/CERL

A few more experiments may be made in order to extend the current data set for neutrally stratified conditions. The full scale plume mapping technique will also be used for validation studies in support of the stratified flow modelling facilities currently being built by the C.E.G.B.

Question: J.G. Kretzschmar, Janssen Pharmaceutica

In your dual tracer release, did smoke and SF_6 disperse in the same way, especially in the vertical direction close to the source?

Answer: Foster - CEGB/CERL

Provided care is taken to avoid effects at source due to the large relative density of the SF_6 gas, and noting the small size of the smoke particulate (i.e. 1µm-2µm), one would expect the two tracers to disperse similarly. Agreement between the cross-wind profiles measured at ground level supports this assumption. However, since SF_6 measurements were confined to ground level only, the tracer profiles could not be compared in the vertical direction. Nevertheless, the assumption that the smoke disperses like a gas is indirectly confirmed by the agreement obtained between full and model scale measurements of the vertical plume section.

Parallel Sessions Research Area C

The release and distribution of radioactive fission products in the atmosphere following a reactor accident

Session C2

Chairman: J.P. Schon (EC.-CENTR/Lyon)

Session secretary: B. Tolley (CEC/BXL)

PREDICTING LIFT-OFF OF MAJOR SELF-HEATING RELEASES UNDER THE INFLUENCE OF A BUILDING

R.W. BENODEKAR, A.J.H. GODDARD and A.D. GOSMAN
Department of Mechanical Engineering,
Imperial College of Science and Technology,
London, SW7 2BX.

Summary

Dispersion in the vicinity of a building is governed by the complex, 3-D recirculating turbulent flow field around it. The present investigation, carried out under the CEC programme, has aimed at developing a general computer method to predict the flow and dispersion around buildings with due consideration to the building shape, type of release and atmospheric stratification. This paper describes the methodology of numerically predicting 3-D turbulent flow and dispersion around buildings. Predictions are obtained for the cases of rectangular prismatic buildings in normally incident thick boundary layers. Results of computations for flow and dispersion for passive and buoyant releases are compared with the measurements. The new method represents a marked advancement in 3-D computing of bluff body flows, particularly in respect of representation of a buoyant plume. Predictions are qualitatively in good agreement with the very complex features revealed by experiment.

Nomenclature

B_D	width of solution domain
c	concentration
$c_1, c_2, c_\mu, \sigma_k, \sigma_\epsilon, \varkappa$	turbulence constants 1.44, 1.92, 0.09, $\varkappa^2/[(c - c_1)c^2]$, 0.4187
g	acceleration due to gravity
G	mechanical generation of turbulent kinetic energy
G_B	generation of k due to buoyancy
hxbxl	height x width x length of the obstacle
H_D	domain height
i_∞	free stream turbulence intensity
k	turbulent kinetic energy
L_u, L_d	lengths of upstream and downstream computational boundaries
M	molecular weight
p	pressure
Q_{in}	rate of emission of effluent
Re	Reynolds number
R_f	flux Richardson number
T	temperature
u,v,w	mean velocity in x,y and z directions
u_i	mean velocity in direction x_i
w_*	friction velocity
x_i,x,y,z	Cartesian coordinates
x,y,z	lateral, vertical and longitudinal directions
δ	boundary layer thickness
ϵ	dissipation of turbulent kinetic energy
μ	molecular viscosity
μ_t	turbulent viscosity
μ_{eff}	effective viscosity
ρ	density
σ	turbulent Prandtl number

1. INTRODUCTON

This paper describes research aimed at developing general computer method for predicting the near-field disposal of releases from a reactor building given the local atmospheric conditions, shape of the building and the nature of the release. The complex flow field around a building, influenced by the various possible atmospheric conditions, interacts with the process of plume lift-off from the ground. While estimated risk is only weakly dependent on details of plume rise modelling [1], due to eventual depletion by rain, whether or not the plume leaves the vicinity of the ground is clearly of significance [2]. Even in the case of open, level terrain the criteria for lift-off is uncertain and models do not exist [2, 3] to treat the lift-off problem when the release may be mixed in a turbulent building wake. In a building wake additional uncertainties arise due to interaction of the plume with the recirculation zone, with building vortices and with the enhanced turbulence. Does lift-off become an ever less well-defined process and will there always be a residual part left near ground level?

The present project was begun through an awareness that simple models such as Gaussian which have been developed for lift-off and plume rise over smooth terrain cannot automatically be extended to releases in the vicinity of a building. Wind tunnel measurements have an important role but may only cover a certain range of conditions. The present investigation set out to develop a detailed numerical procedure, embodying a turbulence model to obtain solutions for the detailed velocity, turbulence, surface pressure and concentration fields and to compare these with available data. When fully developed it is anticipated that it will be used to extrapolate to cases of interest which may be outside the conditions amenable to experiment, as well as leading to a deeper physical insight into the process involved.

2. METHODOLOGY OF 3-D PREDICTIONS

2.1 Introduction

Turbulent flow and dispersion around a bluff body poses many problems with respect to both numerical and turbulence modelling. Many predictions of separated flows have been attempted in the past using the finite volume method of solution. The conventional approach of using hybrid differencing [4] and k-ϵ turbulence model [5] has been found to underpredict the gross features of the recirculation zone behind 2-D obstacles [6,7] due to the numerical diffusion associated with the upwinding of convective terms (see e.g. Raithby [8], Castro [9]) and to the inadequacy of the turbulence model to account for local anisotropy and important strain effects such as those resulting from streamline curvature and buoyancy [10].

In the present study we have reduced numerical diffusion through the use of Quadratic Upwind Differencing (QUD) originally applied by Leonard [11] to 1D and 2D problems. A flux blending technique developed by Gosman and Peric [12] has been used to suppress the nonphysical wiggles which may be generated by the QUD scheme. The resulting scheme is referred to as BQUD. The QUD scheme has been used here for 3D calculations with, if found necessary, recourse being made to BQUD.

The standard k-ϵ model of turbulence has been used with the modifications to represent the effects of buoyancy. Due account is taken of the effect of turbulence anisotropy and the stratification caused by buoyancy, on the diffusion coefficients for pollutant concentration.

Efficiency of the solution procedure has been improved by implementation of the PISO algorithm [13] along with the highly implicit Stone's Solver [14] replacing the convenctional SIMPLE algorithm and tridiagonal matrix line solver, thereby reducing the required c.p.u. time by a factor of three to four.

2.2 Governing Equations

The governing equations for steady state 3D turbulent flow and effluent dispersal can be written in Cartesian tensor notation as follows:

Continuity equation:

$$\frac{\partial \rho u_i}{\partial x_i} = 0 \qquad (1)$$

Momentum equation:

$$\frac{\partial \rho u_j u_i}{\partial x_j} = -\frac{\partial p}{\partial x_i} + \frac{\partial}{\partial x_i}\left[\mu_{eff}\left(\frac{\partial u_i}{\partial x_j} + \frac{\partial u_i}{\partial x_i}\right)\right] - g_i(\rho - \rho_a) \qquad (2)$$

where $\mu_{eff} = \mu + \mu_t$ (3)

Concentration equation:

$$\frac{\partial \rho u_j c}{\partial x_j} = \frac{\partial}{\partial x_j}\left[f_j \frac{\mu_{eff}}{\sigma_c} \frac{\partial c}{\partial x_j}\right] \qquad (4)$$

Equation of state:

$$p = \frac{1}{M} \rho R T \qquad (5)$$

Turbulence model

$$\mu_t = \rho c_\mu k^2/\epsilon \qquad (6)$$

Equations for turbulence kinetic energy, k and its dissipation rate, ϵ (Launder and Spalding [5]):

$$\frac{\partial \rho u_j k}{\partial x_j} = \frac{\partial}{\partial x_j}\left[\frac{\mu_{eff}}{\sigma_k} \frac{\partial k}{\partial x_j}\right] + (G + G_B) - \rho\epsilon \qquad (7)$$

$$\frac{\partial \rho u_j \epsilon}{\partial x_j} = \frac{\partial}{\partial x_j}\left[\frac{\mu_{eff}}{\sigma_\epsilon} \frac{\partial \epsilon}{\partial x_j}\right] + c_1(1 + c_3 R_f)(G+G_B) - c_2\frac{\epsilon^2}{k} \qquad (8)$$

where $G = \mu_t \frac{\partial u_i}{\partial x_j}\left(\frac{\partial u_i}{\partial x_j} + \frac{\partial u_j}{\partial x_i}\right)$

$G_B = g_i \, \mu_t/\sigma_c \, 1/\rho \, \partial\rho/\partial x_i$

According to the practices of Rodi [15], the following values are employed for coefficient c_3:

$c_3 = 0$ for $\partial\rho/\partial x_i > 0$ (unstable stratification)

$= 1$ for $\partial\rho/\partial x_i \leq 0$ (stable stratification)

The values of other coefficients appearing in the turbulence equations are given in the nomenclature. The flux Richardson number is given by

$$R_f = \frac{-G_B}{(G + G_B)} \qquad (9)$$

The non-isotropic diffusivity formulation as in [16] is used here as follows. We write the vertical and longitudinal diffusivities as in the standard turbulence model (by taking $f_y = f_z = 1$ in Eq.(4)). For passive releases, the lateral diffusivity is taken twice that of the vertical (i.e. $f_x = 2$) which is approximately the ratio of lateral to vertical turbulent normal stresses near ground level (see e.g. measurements of Castro and Robins [17] for thick boundary layer). For buoyant releases which are carried to greater heights a more realistic variation of the above ratio with height is specified. We also specify the variation of σ_c as a function of R_f as suggested by Turner [18]:

$$\sigma_c = 0.7\,(1 - R_f)^2 / (1 - R_f/0.15) \tag{10}$$

Upper and lower bounds are specified on σ_c, viz. 2.0 for the most stably stratified and 0.5 for the most unstably stratified regions within the plume.

2.3 Boundary Conditions

The form of calculation domain is illustrated in Fig.1. Shown are three views of the 3-D cuboid obstacle of dimensions hxbxl, in turbulent boundary layer, with its face normally incident to the approach flow. Because of the flow symmetry, only one half of the symmetry section is considered for the calculation. Flow boundary conditions described in Ref. [19] for 2-D predictions are applicable to 3-D calculations as well. Ground roughness is simulated by the use of log-law of velocity for rough wall, following Hinze [20].

The concentration source is specified over the areas where the effluent mass is injected as a distributed release. Special treatment is necssary to reduce the initial dilution if effluent release is from a point source, as in the case of a stack release. The treatment is based on calculating the concentration analytically, using a Gaussian plume model, until the plume is spread over a few computational cells from where it is picked up as an area source for the numerical calculations. Diffusivity and longitudinal velocity at the source location, obtained from the flow calculations, are used in Gaussian model.

2.4 Numerical Solution Procedure

In order to derive the finite volume analogue of the partial differential equations described in Section 2.2, we consider a rectangular 3-D grid in Cartesian coordinates over the computational domain shown in Fig.1. Eqs.(2),(4),(7) and (8) are integrated over the control volumes around the grid nodes by introducing centred approximations for the diffusion terms. The QUD scheme is used to approximate the convective fluxes by quadratic upstream interpolation. Discretised equations are obtained to relate the value of a generalised variable at a grid node to its neighbouring values. The resulting set of algebraic equations are solved by iterative solution procedure.

The QUD scheme contains negative coefficients which sometimes introduce non-physical wiggles in the solution. The problem is overcome by making recourse to BQUD which uses a blend of higher order QUD and lower order upwind differencing schemes in such a way that the resulting solution satisfies the local bounds.

A "deferred corrector" approach is used which reduces the 13 point module discretised equation due to QUD to a 7 point one as far as the solver is concerned. A highly implicit Stone's solver [14] which is less sensitive to the departures from diagonal dominance of the coefficients (the latter is the sufficient condition for convergence of an iterative procedure) is used in the present solution procedure.

The overall solution procedure uses a staggered grid system as in the TEACH code [21] in which the velocity components are stored midway between the pressure and scalar storage locations. The present method of solution uses the PISO algorithm (Pressure Implicit Split Operator) algorithm which is a two stage predictor corrector procedure to deal with the coupling between velocity and pressure variables. Its application to 2D calculations has been described in Ref.[19]. Further modifications have been incorporated in the PISO procedure to include the effects of density coupling on Eq.(2) and (4).

2.5 Differences from Original 2-D Methodology

Predictions of flow around 2-D obstacles were obtained with the aim of improving the accuracy of the conventional numerical solution procedure as in the TEACH code [21] and have been reported in Ref.[19]. The Bounded Skew Hybrid Differencing (BSHD) was used in 2-D to reduce numerical diffusion. However, when the BSHD scheme was derived in 3-D, it involved a 27 point module whereas the standard HD scheme required only a 7 point module. Hence the BSHD scheme proved to be unwieldy in 3-D and was therefore abandoned in favour of QUD, requiring only a 13-point module. The accuracy of 2D predictions had been further improved by employing modifications to the k-ϵ model based on the streamline curvature correction and preferential dissipation treatment suggested by Leschziner and Rodi [10]. These corrections have not been attempted in 3-D calculations because of their complexities in 3-D.

3. RESULTS OF COMPUTATIONS

3.1 Introduction

The predictions of flow over 2D surface mounted ribs were obtained as the first phase of the present investigation and have been reported separately [19]. It was found that the combination of the BSHD scheme with its improved accuracy and the efficient PISO algorithm resulted in a predictive method which is accurate, econominca1 and robust. 3-D predictions were carried out based on the experience gained from the 2D predictions.

Predictions in 3-D were obtained for steady state negligible momentum releases in neutral flow for two configurations of a rectangular building normally incident to the approach flow. One corresponded to detailed wind tunnel measurement of flow and of dispersion around a cube in a thick boundary layer. In the second case wind tunnel concentration measurements were simulated, for buoyant emissions from a rectangular model building in a simulated boundary layer. The various geometrical, flow and dispersion parameters for the two cases are listed in Table 1.

3.2 Predicted Flow Around a Cube

Here the flow predictions correspond to the wind tunnel measurements of Castro and Robins [17]. A non-uniform grid was arranged in decreasing density away from the solid boundaries. It was ensured that first grid line near the obstacle was within 5 percent of the obstacle height and the expansion factor was less than 1.3 in all three directions. A grid of 20x22x34 (14,960 points) was used in the predictions; further grid refinement was not carried out because of computer time limitations.

Fig.2 shows predicted longitudinal velocity profiles in the centre plane at three longitudinal locations, along with the data. Predictions agree very closely with measurements except in the recirculation zone where the maximum negative velocity is overpredicted by about 30 percent. The maximum predicted length of the recirculation zone downstream of the cube as inferred from the numerical results is 1.5 times the cube height, which agrees with the measured value.

Surface pressure coefficients on the exposed surfaces of the cube are shown in Fig.3. Although the qualitative trend has been correctly predicted, there are significant quantitative differences, especially on the front of the cube, where the maximum pressure is overpredicted by about 30 per cent and the minimum is underpredicted by about 20 percent. Also, the predicted location of the maximum lies more towards the top edge of the cube. On the top face, the predicted pressure rises continuously from the front to the rear edge, unlike the data, which has a constant pressure region for about 25 percent of the top surface, where a recirculation zone was detected. The predictions show no separation. The pressure does not recover as much as the measurements at the rear and side faces of the cube.

Predicted velocity vector plots in the longitudinal plane near the plane of symmetry and the cross sectional plane nearly two-thirds way down the front face of the cube are given in Fig.4(a) and (b). Fig.4(a) clearly shows the recirculation zones in the front and rear of the cube. On the front face the stagnation point is indicated at $y/h = 0.85$ which divides the impinging flow into upward and downward directing flow regions. The separation is not resolved at the top leading edge as in the experiment. The streamlines undergo a deflection up to a height of about 2.5 h. Fig.4(b) shows the presence of the horse shoe vortex formed by vorticity stretching of the upstream flow [22]. An intriguing flow structure is observed on the lower half of the side, which shows an upward flow meeting the downflow of the horse shoe vortex and the presence of a small eddy in the separation zone on the side face, rotating in the opposite sense to the main horse shoe vortex. The flow near the ground level is pointing inwards, which may be due to lower vector velocities and high suction in the cavity zone immediately downstream of the cube.

3.3 Dispersion of Passive Point Source Release

Robins and Castro [23] made additional measurements of the concentration of passive material dispersed in the near wake of the cube. Here the predictions are obtained for a modelled stack of height of 1.5 times the cube height and located at the roof centre. The plume was traced initially by the analytical solution up to $z/h = 1.3$. Fig.5(a) and (b) show comparison with the data of non-dimensional concentration contours in the longitudinal plane near the plane of symmetry and horizontal plane near ground level. The figure shows excellent agreement in the wake region. The dispersion in the recirculation region is overpredicted which may be partly attributed to the coarse grid used at the source height which cannot resolve the peak in the profile. The plume is also seen to descend quicker to the ground overpredicting the maximum ground level concentration by about 30 percent.

3.4 Dispersion of a Buoyant Plume in a Neutral Boundary Layer

Predictions were obtained for flow and dispersion around a rectangular prism as investigated by Hall et al [24] and Hall* [25]. A grid of 20x22x30 (= 13,200 prints similar to that used for the previous case) was used in predictions. Only concentration measurements were available for comparison with prediction.

Predictions are reported here for four cases of releases; two release rates (hence different buoyancy parameters) and two positions of release (i.e. top and rear surface) in each case, although Hall has made a more extensive range of measurements. In case A the release rate of helium is very small, hence the buoyancy of the release can be neglected and considered to be a passive release. The buoyant release case G as in [24] corresponds to an equivalent heat release of 80 MW at a windspeed of 3 m/s.

Predictions for all four cases are presented in the form of centre line ground level concentrations in Fig.6. Predicted profiles for non-buoyant and buoyant cases are in reasonably good agreement with the data

*The present authors wish to express their appreciation for being provided with access to the as yet unpublished data.

for the release from the rear face. When the source is on the top face, the non-buoyant case results show an underprediction of about 30 percent. For the buoyant release, a different form of the downstream variation is predicted. Instead of monotomic decrease with distance, as in the measurements, the predictions show an initial increase and a subsequent decrease. however, the average level of ground concentration matches the measurements satisfactorily.

Concentration contours in the cross-sectional planes at two locations downstream of the building, are presented in Figs.7 and 8 respectively for non-buoyant and buoyant releases from the rear surface. The contours are qualitatively in good agreement with the data. Fig.7 shows that·the horizontal and vertical spread as well as the maximum concentration agree fairly well. Predictions for the buoyant plume in Fig.8 show that the bifurcated plume structure is also in reasonable agreement although discrepancies in vertical spread exist.

The predicted flow field for case G with source at the rear of the building is presented in Fig.9(a) and (b). A completely different picture of the wake flow is observed in the vertical slice of the flow (Fig.9(a)). the buoyant gas mixed with air in the cavity region tends to raise the flow indicating no recirculation region at all near the plane of symmetry. Cross sectional vector plot shown in Fig.9(b) clearly demonstrates the buoyancy induced vortices which are responsible for bifurcating the plume.

3.5 Comments on Error Sources

Velocity predictions for case (1) are in good agreement with the measurements except in some isolated regions, such as flow separation near the top leading edge of the cube, which is not resolved by the QUD scheme using a grid of 20x22x34. Discrepancies in the pressure distribution are partly attributed to the absence of fine details in the predicted velocity field. Inadequacies in the turbulence model and the three-dimensionality of the flow which may introduce numerical diffusion in using the QUD scheme along the grid directions, will also be responsible for errors in the calculations.

A simplified empirical approach has been used in treating nonisotropic diffusion coefficients as described in Section 2. It would seem appropriate to use, instead, an algebraic-stress treatment (perhaps similar to Gibson and Launder [26] and include all the various secondary production terms associated with the three-dimensional flow field. The quantitative discrepancies for buoyant releases could be due to modelling errors in applying k and ϵ model to buoyant flows and the empirical relation for σ_c given by Eq.(12) which need further research and testing. more detailed turbulence and flow measurements are required to accompany concentration measurements in testing the improved models.

3.6 Comments on Cost Effectiveness

The computations were carried out on IBM3081 computer. Calculations for the highly buoyant case were performed using a grid of 20x22x30 (= 13,200 points). The converged solution required 60 iterations, 25 mins of c.p.u. time and a core of 400 k words. The new methodology therefore represents a substantial advancement in 3D computing of bluff body flows especially with a buoyant plume over the previous full scale predictions such as [27].

4. CONCLUSIONS

The predictions for turbulent flow and dispersion of passive and buoyant releases were compared with the available measurements and were found to be in satisfactory agreement. The prediction procedure was tested successfully for very complex flow fields; highly buoyant releases from a modelled building. Some discrepancies were observed in the predictions; further improvement is likely to require the improvement of the turbulence model. Further development of the model will permit the treatment of non-orthogonal approach flow, atmospheric stratification, non-rectilinear building shapes, source momentum, self-heating and transient release effects. In order to complement these theoretical developments, there is a clear need for improved measurements in which both flow and concentration fields are adequately characterised. A properly validated methodology may be applied to problems of practical significance in risk assessment, for example in developing criteria for lift-off in the presence of buildings.

ACKNOWLEDGEMENTS

The authors are grateful to D.J. Hall of the Warren Spring Laboratory for access to as yet unpublished data. Computing resources for this project have been made available by the U.K. Science and Engineering Research Council under grants GR/B/62761 and GR/C/36698.

REFERENCES

1. NEA Report (1983). International Comparison Study on Reactor Accident Consequence Modelling, Sept.1983.
2. UNDERWOOD, B.Y et al. (1983). Review of Specific Effects of Atmospheric Dispersion Calculations, Report to CEC XII/917/83 EN.
3. JONES, J.A. (1983). The Fifth Report of a Working Group on Atmospheric Dispersion, NRPB-R 157.
4. SPALDING, D.B. (1972). A novel difference formula for differential expressions involving both first and second derivatives, Int.J.Num. Meth. in Eng., 4, 551-559.
5. LAUNDER, B.E. and SPALDING, D.B. (1974). The numerical computation of turbulent flows. Comp.Meth. in Appl.Mech. and Eng., 3, 269-289.
6. VASILIC-MELLING, D. (1976). Ph.D.Thesis, Imperial College, London.
7. McGUIRK, J.J., TAYLOR, A.M.K.P. and WHITELAW, J.H. (1981). The assessment of numerical diffusion in upwind-differencing calculations of turbulent recirculating flows. Paper at 3rd Symp. on Turbulent Shear Flows, Univ.Calif., Davis, U.S.A.
8. RAITHBY, G.D. (1976). A critical evaluation of upstream differencing applied to problems involving fluid flow. Comp.Meth. in Appl. Mech. and Eng., 9, 75-103.
9. CASTRO, I.P., CLIFFE, K.A. and NORGETT, M.J. (1981). Numerical predictions of the laminar flow over a normal flat plate. A.E.R.E. Harwell, Rep.HL.81/666.
10. LESCHZINER, M.A. and RODI, W. (1981). Calculation of annular and twin parallel jets using various discretisation schemes and turbulence model variations. Trans.ASME, 103, 352-360.
11. LEONARD, B.P. (1979). A stable and accurate convective modelling procedure based on quadratic upstream interpolation. Comp.Meth. in Appl.Mech. and Eng., 19, 59-98.
12. GOSMAN, A.D. and PERIC, M. (1983). Private communication.
13. ISSA, R.I. (1982). Int.Rep., Dept.Min.Resources Eng., Imperial College, London.
14. STONE, H.L. (1968). Iterative solution of implicit approximation of multidimensional partial differential equations. SIAM J.Numer.Anal., 5, (3), 530-559.

15. RODI, W. (1979). Influence of buoyancy and rotation on equations for the turbulent length scale. Paper at 2nd Symp. on Turbulent Shear Flows, London.
16. BERGELES, G. (1976). Ph.D.Thesis, Imperial College, London.
17. CASTRO, I.P. and ROBINS, A.G. (1977). The flow around a surface-mounted cube in uniform and turbulent streams. J.Fluid Mech., 79, (3), 307-335.
18. TURNER, J.S. (1973). Buoyancy effects in fluids. Cambridge Univ. Press.
19. BENODEKAR, R.W., GODDARD, A.J.H., GOSMAN, A.D. and ISSA, R.I. (1983). Numerical prediction of turbulent flow over surface mounted ribs. ASME Paper, No.83-FE-13, presented at Fluids Eng.Conf., Houston, Texas, June.
20. HINZE, S.D. (1959). Turbulence. McGraw-Hill.
21. GOSMAN, A.D. and IDERIAH, F.J.K. (1976). Int.Rep., Dept.Mech.Eng., Imperial College, London.
22. HUNT, J.C.R. (1971). The effects of single buildings and structures. Phil.Trans.Roy.Soc., London, 268, 457-467.
23. ROBINS, A.G. and CASTRO, I.P. (1977). A wind tunnel investigation of plume dispersion in the vicinity of a surface mounted cube - II. The concentration field. Atmos.Env., 11, 299-311.
24. HALL, D.J., BARRETT, C.F. and SIMONDS, A.C. (1980). Wind tunnel model experiments on a buoyant emission from a building. Rep.No. LR 355 (AP). Warren Spring Laboratory, U.K.
25. HALL, D.J. (1983). Private communication, September
26. GIBSON, M.M. and LAUNDER, B.E. (1976). On the calculation of horizontal turbulent free shear flows under graviational influence. ASEM, J.Heat Transfer, 98(C), p.81.
27. HIRT, C.W. and COOK, J.L. (1972). Calculating three-dimensional flows around structures and over rough terrain. J.Comp.Phys., 10, 324-340.

TABLE I. GEOMETRICAL, FLOW AND DISPERSION PARAMETERS

Parameter, Dimension	Case (1) Castro and Robins [17] Robins and Castro [23]	Case (2) Hall et al [24] Hall [25]	
		Case A	Case G
h(m) x l(m) x b(m)	0.2 x 0.2 x 0.2	0.167 x 0.167 x 0.333	
δ,m	2	?	
Model scale	1/300	1/300	
Reference velocity w_r m/s	$w_\infty = 7.5$	$w_{y = 0.1\ m}$ 1.17	0.35
Re	10^5	1.3×10^4	3.4×10^3
i_∞	0.0005	0.0073	0.0053
w*	0.4875	0.1	0.03
y_o	4.4	1.5	1.5
Type of source	point (stack release)	area	area
ρ_g kg/m³	1.205	0.166	0.166
location	top centre ($y_s = 0.3$ m)	top or rear	top or rear
Q_{in} l/min	19	1.0	31

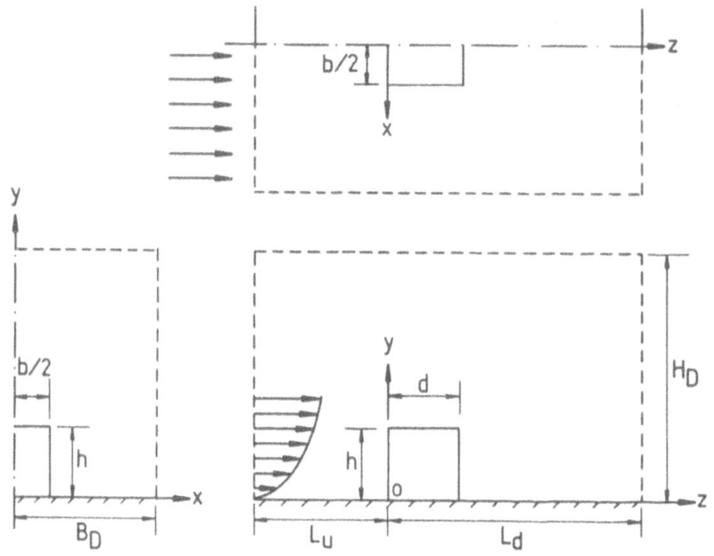

Fig 1. 3-D Rectangular Building and Domain of Calculation

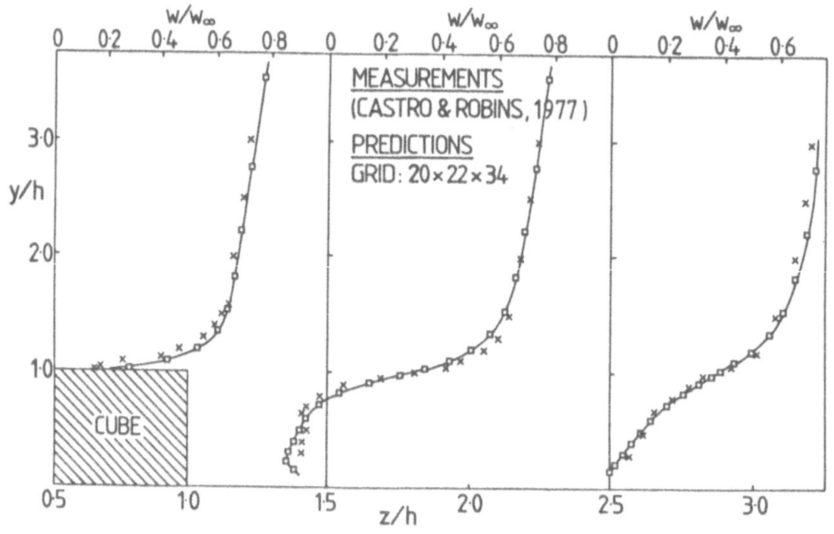

Fig 2. Mean Longitudinal Velocity Variations in the Wake (x/h = 0.025) of a Cube

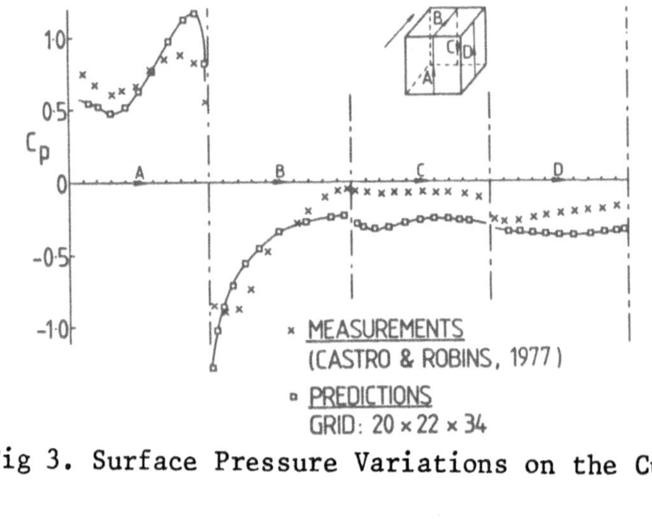

Fig 3. Surface Pressure Variations on the Cube

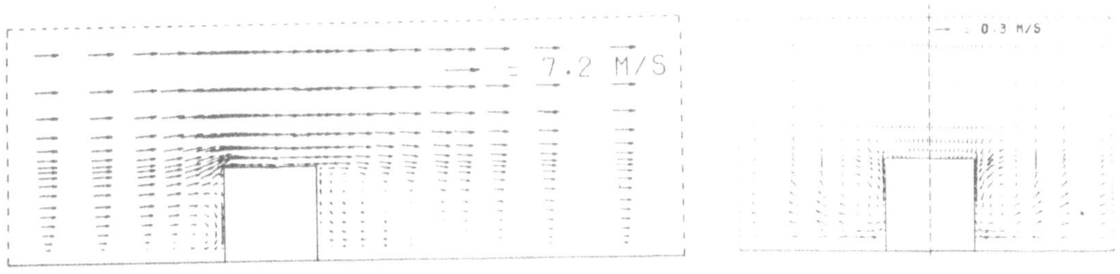

(a) (b)

Fig 4. Predicted Flow Field around the Cube

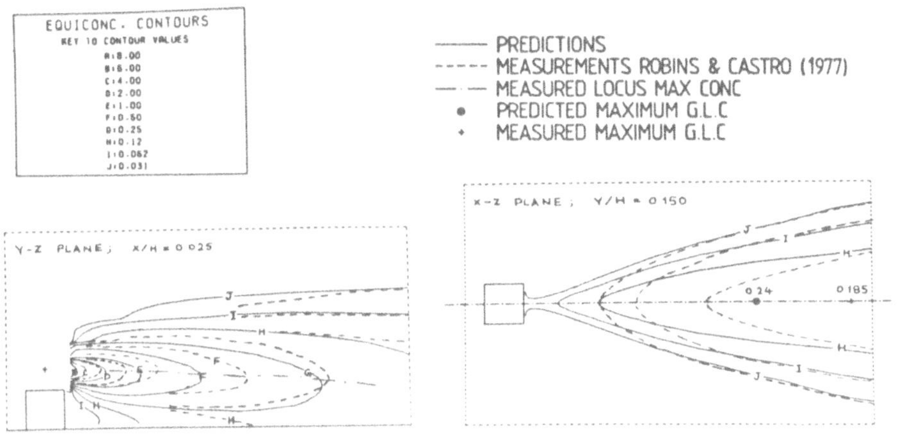

(a) (b)

Fig 5. Concentration Contours $(cw_{\infty}h^2/Q_{in})$: $y_S/h = 1.50$

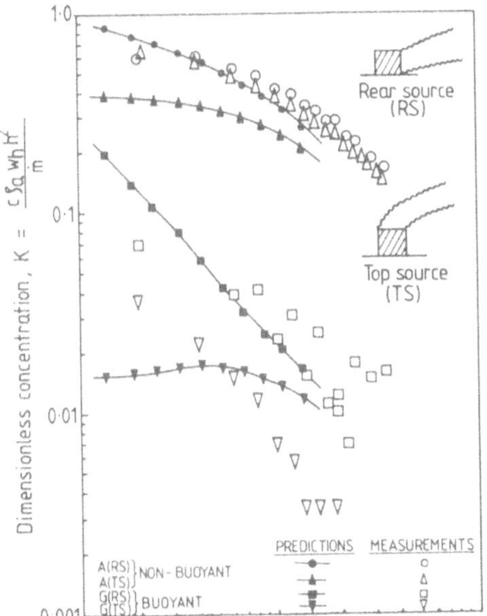

Fig 6. Comparison of Predicted Centre-line Ground Level Concentration with the Measurements of Hall [25]

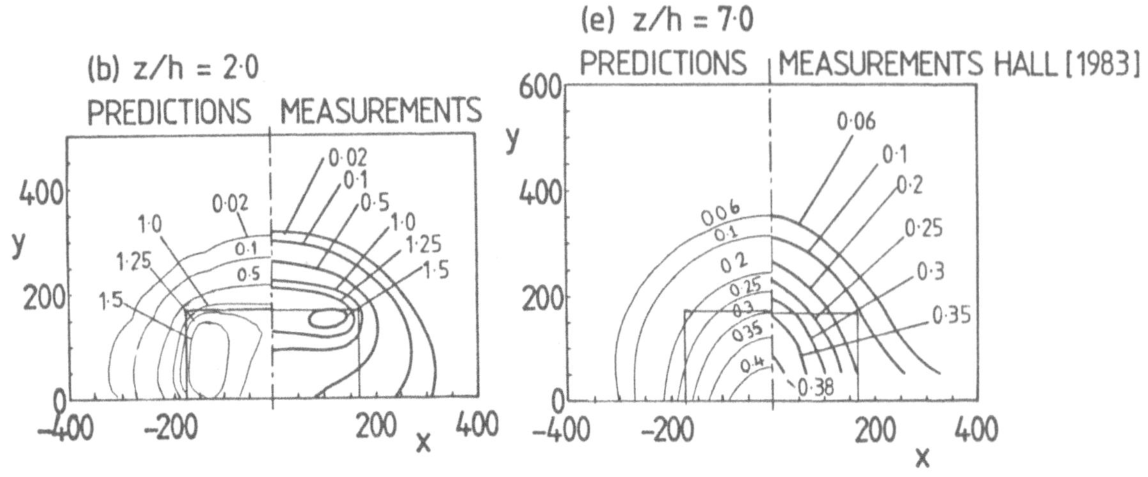

Fig 7. Comparison of Concentration Contours (c $\rho_a w_h h^2/\dot{m}$)

Case A (Rear Source)

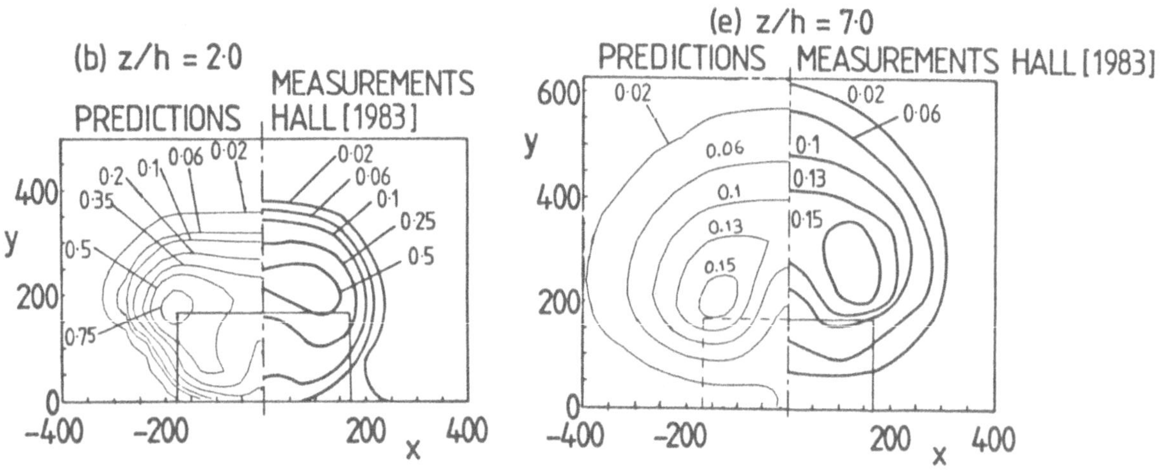

Fig 8. Comparison of Concentration Contours (c $\rho_a w_h h^2/\dot{m}$)

Case G (Rear Source)

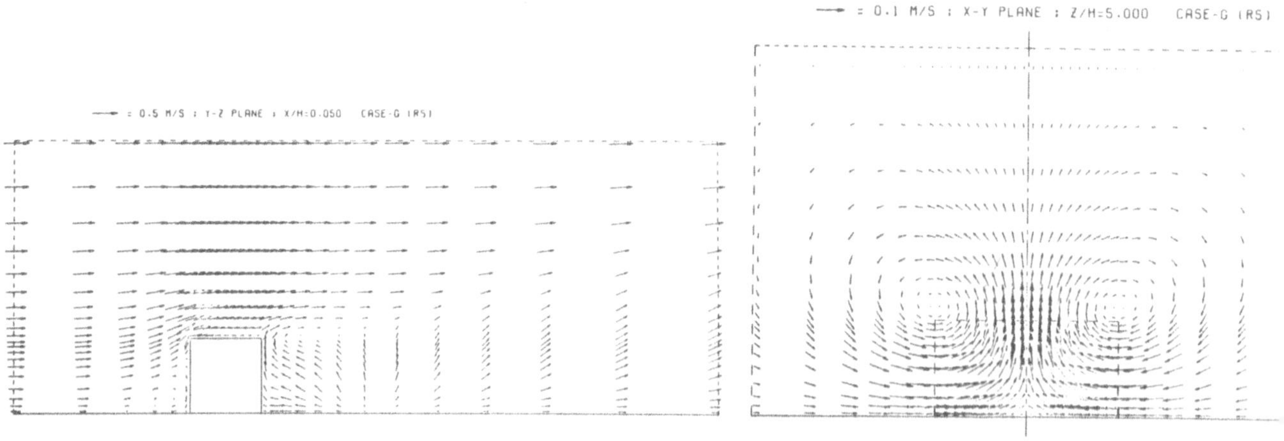

(a) (b)

Fig 9. Predicted Flow Field around Rectangular Building with Buoyant Release (Case G (Rear Source))

Discussion

Question: A. Hauguel - EDF/Chatou

What is the CPU time for a standard computation?
Could you give more details on the formulation you use for production terms in the k equation?

Answer: Goddard - Imp. Col./London

A typical CPU time was 25 minutes on an IBM 3081.
Production terms in the k equation consist of production due to mechanical turbulence and due to buoyancy. The former is the production term in the standard k-e model. The latter is related to density (or temperature) fluctuations and is modelled by the gradient hypothesis to be proportional to the vertical density gradient. The details are given in the paper.

Question: Schon, Ec. Centr./Lyon

Do you intend to introduce multiple scales in the turbulence modelling?

Answer: Benodekar/Goddard - Imp. Col./London

We have not considered the need for this, but such a development may well be required for applications to the real atmosphere.

DEVELOPMENT OF AN INTEGRAL MODEL OF A RADIOACTIVE JET RELEASED ACCIDENTALLY FROM A NUCLEAR REACTOR

D. GRAND and A. BADRE
Commissariat à l'Energie Atomique
Institut de Recherche technologique et de Développement Industriel
Centre d'Etudes Nucléaires de Grenoble

Summary

This paper concerns the theoretical study of a hot jet with internal heat generation issued vertically into a horizontal cross-flow. The study focuses on the initial region of the jet, where the initial momentum and buoyancy of the jet are still influent. The main objective of this study is to predict the height reached by the jet in order to give initial conditions to large scale polluant transport models. The model MICAR developed by CEA is based upon an integral formulation : The conservation equations are integrated in a cross-section through the jet. It requires an empirical model for the mass flow of the entrained fluid.
Parametric studies are done with the model. Attention is focused on the trajectoreis and on the dilution of a polluant contained in the emission. External parameters which have been covered are :
. Internal heat generation,
. Ratio of wind velocity to jet velocity,
. Ambient stratification.

1. INTRODUCTION

Most of the existing models for the evaluation of the dispersion of airborne radioactivity in nuclear power plant accident situations use a Gaussian model. Briefly, one can say that the Gaussian model depends on a particular solution of the diffusion equation. In this solution for a point-source continually emitting at some height, some exponential terms appear in which the diffusivities in the lateral and vertical directions are taken into account. This atmospheric dispersion theory depends on the fact that the atmospheric release has lost its individuality and is passively transported by the ambient flow. This assumption does not apply in the initial region of the release, also called the near field. This is why in the atmospheric gaussian dispersion model one has to define a fictious source located upstream of the real source, at some height above it. The purpose of this study is to provide a model for the release in the near field in the case of a heated release with internal heat sources can be used as initial conditions for atmospheric gaussian dispersion models.

2. DESCRIPTION OF THE MODEL

This work is concerned with the behaviour of gaseous jets discharged at arbitrary angles to an infinite ambient through a single circular orifice. The flow of the ambient atmosphere is horizontal. For the trajectory to be two-dimensional, the initial velocity of the jet $\vec{U}_0$ and the velocity $\vec{U}_\infty$ of the cross-flow are assumed to belong to the same vertical plane |see fig. 1|. While it grows, the jet is subjected to different regimes, and sometimes, two regions are distinguished : the near-field and the far-field.

The near-field is defined as being the zone of the jet nearest its exit. In this region the dynamics of the jet plays an important role and the jet induced turbulence is the primary cause of the dispersion. In this region, two zones are distinguished :

A first initial zone just after the orifice. In this part of the jet a shear layer develops at its boundaries where there is mixing with the ambient fluid while at the core the jet remains undisturbed by the external influences. The length of this zone is small (a few initial diameters). This region is called "the zone of flow establishment". The jet bends over and its inclination tends towards the horizontal.

Then the jet enters the far-field where the atmospheric turbulence prevails entirely being solely responsible for the dispersion of the plume.

This work is restricted to the zone of established flow, in the near-field. The following simplifying assumptions are made :

- The flow is steady in the mean.
- The fluid is Newtonian.
- The flow is fully turbulent : molecular fluxes are neglected with respect to turbulent ones.
- The Boussinesq approximation applies to the fluid : all physical properties other than density ρ are assumed to be constants and ρ is assumed to be constant except in the buoyancy term.
- The curvature of the jet and its consequences are neglected.
- The pressure distribution is hydrostatic.
- The flow is axisymmetric : the mean flow variables are symmetrically distributed around the jet axis.
- The boundary layer approximation can be made : the mean transverse velocity in the jet is much smaller than the axial component of the velocity and transverse gradients are much greater than longitudinal ones.

These assumptions made, the method followed here is the integral approach. It was initially introduced by Morton et al (1956). It is based on the temporal and spatial integration of the fundamental balance equations of Fluid Mechanics (fig. 2).

The transport equations included in the model are :

- continuity
- momentum (horizontal and vertical projections)
- energy
- concentration of a radioactive contaminant
- moisture (concentration of vapor)

The complementary assumptions needed for the closure of the mathematical formulation concern mainly :

- the distributions of the physical variables (velocity, temperature,...) inside a cross-section : one assumes gaussian distributions,
- the mass flow entrained by the jet or entrainment function.

The entrainment function is a concept well fitted to the integral method. It is supposed to represent in a reliable way the lateral fluid flux into the jet from the ambient. The entrainment depends on factors such as the local conditions due to buoyancy and momentum, the initial conditions, the ambient values etc,... A review of existing entrainment functions has been made from which it was concluded that Hirst (1971) and

Wu and Koh (1977) proposals were the best fitted to the present model.

3. PREDICTIONS OF INTERNALLY HEATED BUOYANT JETS ISSUED IN A CROSS FLOW

The purpose of this study is to show the influence of the most prominent parameters such as the initial angle of exit conditions, the ambient conditions (wind and temperature profiles), humidity (atmospheric one as well as that of the plume), and of course, the heat generated by the fission products. The jets considered will all be taken in unsaturated conditions. Also, the radioactive jet is supposed to be emitted at a height of 25 m, and the surrounding is supposed to be flat. The ambient velocity is either uniform or following the logarithmic law.

Radioactive power efflux from the source is varied from 0 to 16 250 W/s.

Different computed trajectories are shown in the following figures. The influence of initial conditions is illustrated in fig. 3 where the angle of the initial velocity is varied.

In fig. 4, only the stability of the atmosphere is varied (2 stable conditions (curves (1) and (2), 1 neutral atmosphere (3), 1 instable atmosphere (4)).

Figs 5 and 6 show the influence of the internal heat sources upon the trajectory in two cases of atmospheric stability (neutral ans stable).

REFERENCES

1. This study has now been reported in European Applied Research Reports. (1983). Volume 5, number 1, pp 1-120 (EUR 8662).
2. HIRST, E.A. (1971). Analysis of Round, Turbulent, Buoyant Jets Discharged to Flowing Stratified Ambients, Oak Ridge National Laboratory, Report n° ORNL-4685, UC-48.
3. MORTON, B.R., TAYLON, A.G. and TURNER, J.S. (1956). Turbulent Gravitational Convection from Maintained and Instantaneous Sources". Proc. Roy. Soc. London, A234, 1-23.
4. WU, F.H.Y. and KOH, R.C.Y. (1977). Mathematical Model for Multiple Cooling-Tower Plumes. W.M. Keck Laboratory of Hydraulics and Water Resources. California Institute of Technology, Pasadena, California, Rep. N° KH-R-37.

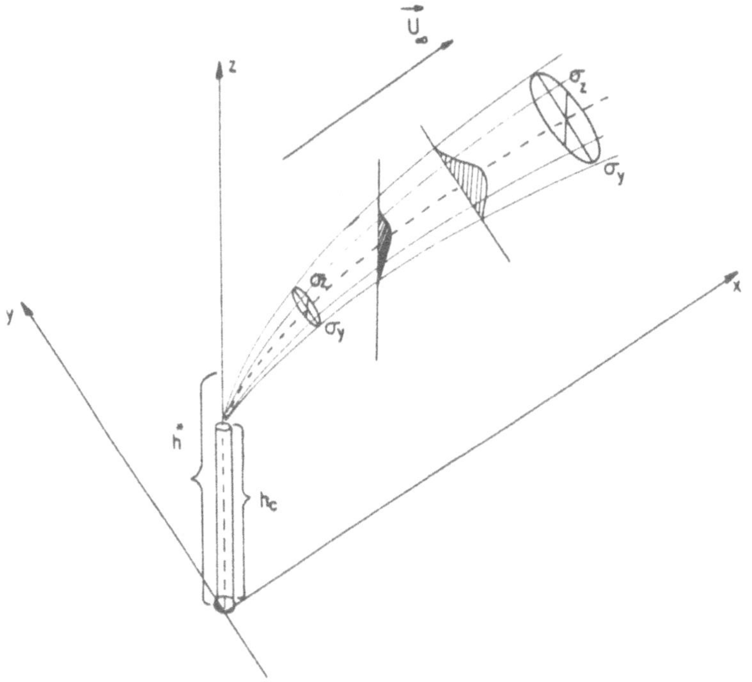

Fig. 1 Description of the release

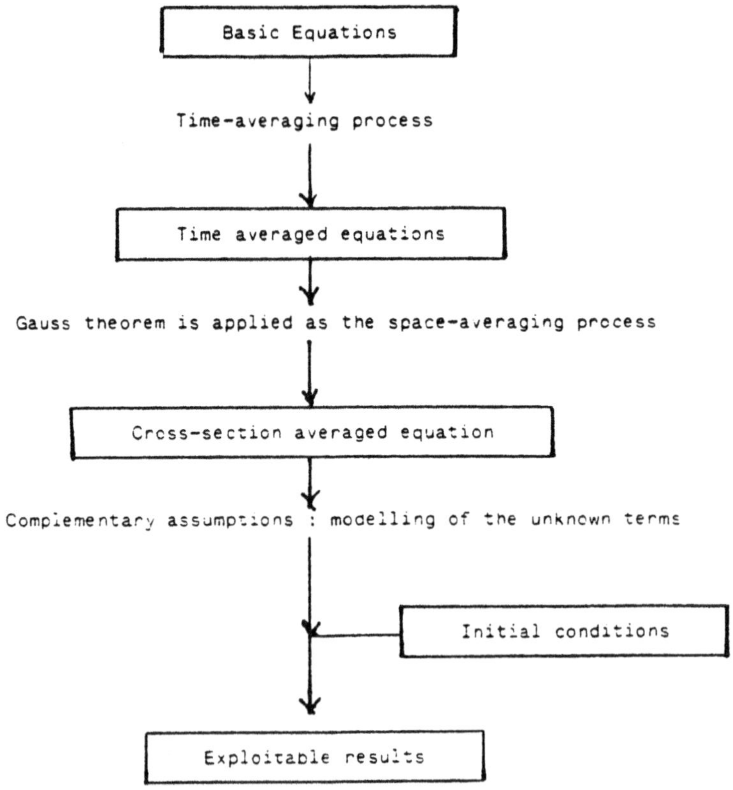

Fig. 2 Derivation of the integral model

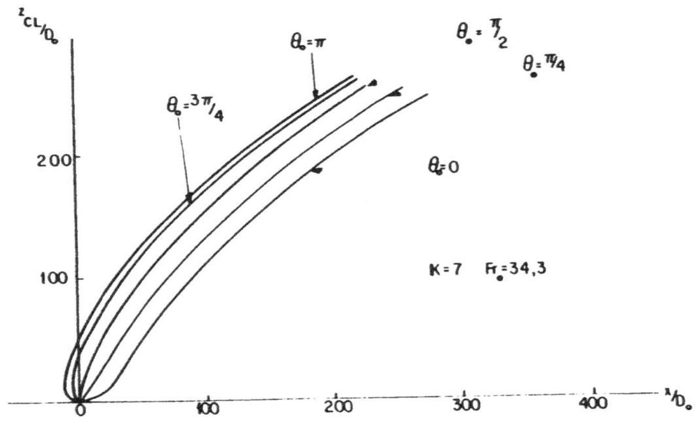

Fig. 3 Influence of initial conditions

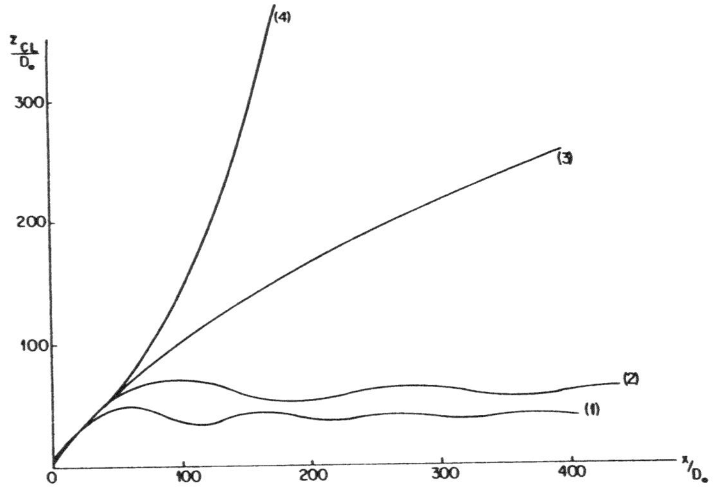

Fig. 4 Influence of the atmospheric stability

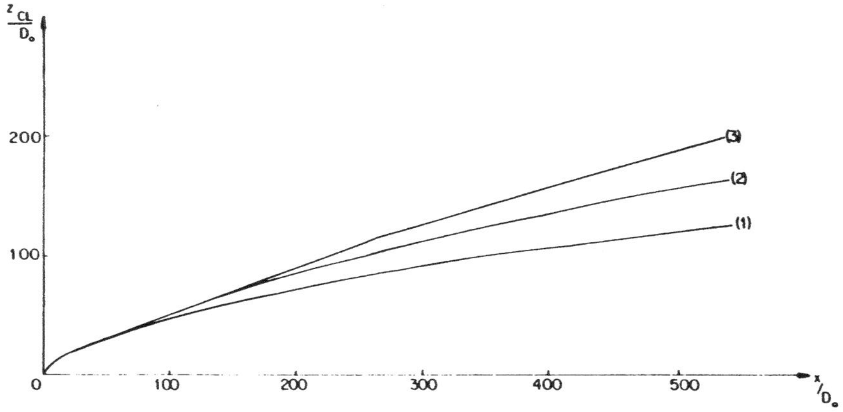

Fig. 5 Influence of internal heat generation
(neutral atmosphere

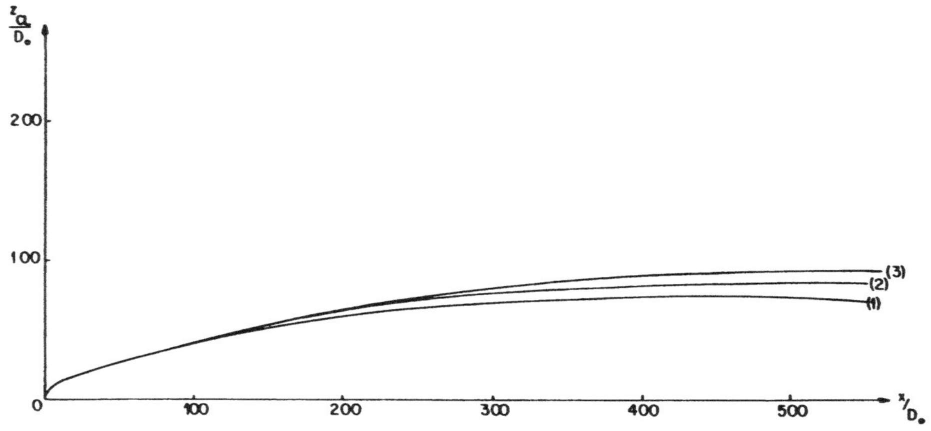

Fig. 6 Influence of internal heat generation
(stable atmosphere)

Discussion

Question: P.M. Foster - CEGB/CERL

Briggs and other authors have included a virtual mass term into the vertical momentum equation for their plume rise formulation. Was this done in the present formulation?

Answer: D. Grand - CEA/CEN-G

This term is an empirical attempt introduced by Briggs to take into account the drag force exerted by the wind upon the jet. We did not include it in our study. It does not seem to produce much difference in the cases for which we compared our predictions to Briggs' ones.

Question: J.P. Schon - Ec. Centr./Lyon

To what are due the irregularities of the stable case?

Answer: D. Grand - CEA/CEN-G

The jet rises first over its equilibrium position and oscillates around it. This is simular to lee waves. However, one cannot ensure that the integral model gives a quantitative description of these waves.

Question: J. Paesler - KFK

Did you compare your results with the results of Mr. Briggs? And do you have a comparison here on a sheet?

Answer: D. Grand - CEA/CEN-G

In this thesis, A. Badre compared the MICAR model to experimental results on other integral models: Briggs (1969), Hirst (1971), Schatzhan (1979) for example.

DEVELOPMENT OF A PROTOYPE MESOSCALE COMPUTER MODEL INCORPORATING TREATMENT OF TOPOGRAPHY

H. M. APSIMON, A. J. H. GODDARD, K. KITSON and M. FAWCETT.
Imperial College, London

Summary

More sophisticated models are now available to simulate dispersal of accidental radioactive releases to the atmosphere; these use mass-consistent windfields and attempt allowance for site-specific topographical features. Our aim has been to examine these techniques critically, develop where possible, and assess limitations and accuracy. The resulting windfield model WAFT uses efficient numerical techniques with improved orographic resolution and treatment of meteorological conditions. Time integrated air concentrations, dry and wet deposition are derived from TOMCATS, which applies Monte-Carlo techniques to an assembly of pseudo-particles representing the release, with specific attention to the role of large eddies and evolving inhomogeneous rainfields. These models have been assessed by application to hypothetical releases in complex terrain using data which would have been available in the event of an accident, and undertaking sensitivity studies. It is concluded that there is considerable uncertainty in results produced by such models; although they may be useful in post-facto analysis, such limitations cast doubt on their advantages relative to simpler techniques, with more modest requirements, during an actual emergency.

1. INTRODUCTION

In recent years more sophisticated models, such as ARAC (1) or ARIES (2), have been developed to predict dispersion and deposition of radioactivity released to the atmosphere in the event of an accident. These attempt to allow for particular topographical characteristics. Their implementation requires substantial investment in computer facilities and supporting meteorological data networks surrounding each nuclear installation. However the atmospheric dispersion models included have limited representation of the physics involved. The aim of the present work has been to examine these models critically, improve where possible, and assess their limitations and accuracy.

The predictions required are spatial distributions of time-integrated atmospheric concentrations (T.I.C.) and activity deposited on the ground by dry and wet processes out to distances 100 km from the source. The problem may conveniently be divided into two parts. First, there is prediction of 3-dimensional windfields advecting the release over the region, making optimum use of the limited wind data likely to be available in the area at the time of the accident. For this purpose the windfield model WAFT (Windfields Adjusted For Topography) was developed - see Section 2. Secondly there is transport, dilution and depletion of radioactivity in this windfield and deposition on the ground. Monte-Carlo methods have been devised to simulate such dispersion, embodied in the TOMCATS model (Topography Oriented Monte-Carlo Atmospheric Transport Simulation) described in Section 3.

The methods developed have been widely tested including assessment by application to hypothetical releases over real terrain using meteorological data which would have been available if an accident had occurred at the time. These case studies are discussed in Section 4 and illustrate the conclusions of Section 5 on the suitability of such models for accident assessment. A full description of the work is given in (3).

2. Derivation of Windfields - WAFT

The WAFT model yields 3-dimensional windfields in the lowest 1 or 2 kilometres of the atmosphere (the boundary layer in which pollutants released at low level are transported) out to 100 km or so from the source. For such an area there might be say 20 surface winds reported, and possibly limited information on winds aloft (eg radiosonde). Geostrophic or synoptic winds can also be useful to indicate general airflow over the region. The aim is to make optimum use of such data to define the wind components (u_{ijk}, v_{ijk}, w_{ijk}) at a 3-dimensional rectangular grid of points spanning the region, allowing for the fact that air will have to flow over or round hills and other orographic features. WAFT is similar to MATHEW(4) incorporated in ARAC (1), but has we believe several substantial advantages.

WAFT is composed of 2 distinct stages. First an initial guess at the windfield (u_{ijk}^{c} , v_{ijk}°, w_{ijk}°) is obtained by using the wind data to interpolate at each gridpoint above ground. Imagining the grid as an assembly of cells, it is then possible to estimate fluxes across each cell face, and hence fluxes in and out of each cell. In general these fluxes will not balance violating mass conservation. Moreover it will often be implied that winds flow through sloping ground. The second stage therefore makes small adjustments ($\Delta\hat{u}_{ijk}$, $\Delta\hat{v}_{ijk}$, $\Delta\hat{w}_{ijk}$) to the initial flow field such that mass conservation is achieved, subject to the boundary condition that flow is tangential at the ground. More specifically this revised windfield is chosen to minimise the weighted sum of adjustments squared

$$S = \sum_{ijk} \alpha x_{ijk}\, \Delta\hat{u}_{ijk}^2 + \sum_{ijk} \alpha y_{ijk}\, \Delta\hat{v}_{ijk}^2 + \sum_{ijk} \beta_{ijk}\, \Delta\hat{w}_{ijk}^2 \qquad [1]$$

subject to a linear equation of constraint for each unburied grid cell representing mass conservation.

2.1 The calculational mesh and specification of orography in WAFT:-

Since the effects of a pollutant plume encountering a hill can depend quite critically on how closely the streamline flow takes the plume to the surface, it is important to represent ground contours accurately. In MATHEW (4) the ground is crudely treated like an assembly of bricks, each grid cell lying entirely above or below ground. In WAFT exact ground heights are used at the corner edges of each vertical column of cells; this introduced problems to solve due to partially submerged cell walls. However the added resulting complexity is well justified by better flow fields - see figure 1 comparing the maximum error in the derived windfield beyond various radii, for simulated flow over a hemisphere calculated with WAFT and MATHEW. WAFT windfields are an order of magnitude more accurate close to the surface, and a factor 3 more accurate at larger distances.

2.2 Method of solution:-

The efficiency and reliability of numerical procedures used was recognised as being fundamentally important. The minimisation of S in equation [1] may be achieved using the method of Lagrange multipliers (5) resulting in a fairly standard problem; namely that of solving a large number of simultaneous equations typically 20,000, one for each grid cell. However each equation involves few quantities since each grid cell has at most 6 faces through which flow can take place. Written in matrix form

$$\underline{A} \cdot \underline{x} = \underline{B} \qquad [2]$$

where $\underline{A}$ is a sparse matrix with a banded structure. The use of Incomplete Cholesky Conjugate Gradient (ICCG) iterative methods (6) introduced a great improvement both in efficiency and reliability, as illustrated in figure 2.

2.3 Representing physical conditions in the atmosphere

By varying the weighting factors αx_{ijk}, αy_{ijk}, β_{ijk} the mass adjusted windfield may be controlled. Thus by using high values of β_{ijk} in selected regions of the grid a mass-consistent windfield is obtained with relatively small alteration to vertical wind components in those regions. Hence by suitable specification of weighting factors the windfield may be forced to reflect physical atmospheric conditions such as stability and temperature inversions. The treatment of such phenomena in MATHEW (4) has been severely criticized (7). The greater flexibility of spatial adjustment of weighting factors in WAFT is an advantage here.

Thus flow over and round a hill in WAFT is based on "Froude number", which relates the kinetic energy of air upwind to the potential energy required to rise displacing air at the top of the hill. In stably stratified conditions with denser air near the ground the airstream near the top of the hill still manages to climb over, but at lower levels the air is forced to flow round rather than over the hill (see figure 3). This behaviour has been reproduced in WAFT by suitably defining the weighting factors but requires atmospheric stability data. Also a complex region has to be rather arbitrarily divided into separate hills (although objective methods of doing this have been devised).

Elevated inversions can also have an important effect on windfields forcing flows round rather than over hills introducing large windshears. In WAFT they are treated as flat impenetrable ceilings to airflow, their deformation over mountainous terrain being very difficult to predict. The behaviour of WAFT, and control using the weighting factors, has been extensively tested using theoretical and laboratory flows.

3. Dispersal in the windfield and deposition - TOMCATS.

The TOMCATS model represents a radioactive release as an assembly of particles using Monte-Carlo techniques to simulate turbulent displacements. Associated with each particle is a "mass" parameter representing the probability that the particle is still airborne and has not been lost by decay or deposition. Each particle is followed in a series of time steps with 3 superimposed displacements in each time step. First there is a translation along a streamline due to advection; secondly a semi-random displacement with horizontal and vertical components depending on the turbulence field, and retaining correlation with previous time steps according to the appropriate Lagrangian time scales for persistence; thirdly there may be additional horizontal displacement due to larger horizontal eddies comparable in scale perhaps to hills and gaps between them and introducing fluctuations into the advecting windfield.

Monte-Carlo methods were preferred to the "particle-in-cell" model ADPIC (8) in ARAC (1) since ADPIC suffers from the same problems as K diffusion models. A single diffusivity function can not represent the whole spectrum particularly of horizontal turbulence, and estimating concentration gradients is difficult when material becomes dilute. TOMCATS thus has several advantages. It avoids numerical diffusion also, but with inhomogenious turbulence artificial accumulation of particles can occur in regions of low turbulence unless suitable compensation is made. In TOMCATS this trend was overcome by introducing an additional "drift velocity" to counteract the clustering tendency.

$$v_{drift} = \Delta t.\ 1/2.\underline{\nabla}\ \sigma^2 \qquad [3]$$

where Δt is the length of the time steps for particle tracking and σ defines the turbulent intensity. (Note that equation [3] differs from formulae previously quoted (9) in the factor 1/2, but has been carefully checked and justified in practice.) Turbulence profiles and Lagrangian time scales have been chosen following Hannah (10). Good agreement between results from TOMCATS and from traditional Gaussian plume treatments has been established over flat terrain where the latter are applicable.

To study the effect of large horizontal eddies in hilly terrain the "impinging eddy" model has been developed. This model assumes a random assembly of such large eddies frozen into the flow field and advecting with it. Pollutant particle trajectories are perturbed as they move in and out of the larger eddies, and initial correlation between particle motion depends how closely the particles are released in space and time. Thus the model may be used to investigate sensitivity to large eddies for different release durations and sampling times. In the particular case studies considered however extra lateral spreading due to enhanced windshear in complex terrain tended to dominate large eddy effects.

Dry and wet deposition are also simulated in TOMCATS. Dry deposition is calculated statistically by estimating a probability of deposition proportional to the dry deposition velocity each time a particle impinges on the ground as a result of turbulent motion. Great care was taken in satisfying streamline prediction particularly in tangential flow close to the ground. Accurate streamline prediction is in any case important for advection terms in particle displacements. Wet deposition may either be estimated in the usual way assuming a uniform removal rate by washout with instantaneous deposition on the ground, or by specifying spatially and temporarily varying rainfall resulting in inhomogeneous wet deposition and possible "hot spots". Suitable data on precipitation fields is now available from modern radar networks.

4. Applying WAFT/TOMCATS to hypothetical releases in real terrain

To assess the applicability and accuracy of WAFT/TOMCATS over complex terrain several case studies were conducted over the hilly Bristol Channel in the UK and the mountainous Vosges in France. Dispersion of hypothetical releases from nuclear installations was simulated in a variety of atmospheric conditions. Meteorological data from the area which would have been available in the event of an accident was used, not a special dense experimental network as is sometimes used in model validation studies. Particular emphasis was placed on streamlines within the flow since these are so important for pollutant dispersal. Sensitivity tests were conducted on such factors as siting of data points, and the choice of interpolation scheme used both horizontally in the surface and field and for vertical wind profiles. These case studies are fully described in (3).

Figure 4 illustrates typical results, showing streamlines originating at different heights (m) from Hinckley Point and Oldbury power stations. The dotted line indicates the upper level trajectory when a different interpolation scheme (cubic spline instead of inverse square) was used in the initial windfield specification. It is very disturbing that such large differences can occur just due to the choice of mathematical technique. There is much more wind shear introduced by the Welsh mountains for the more northerly trajectories from Oldbury. This leads to a much wider plume as is shown in figures 5 and 6 giving isopleths of TIC for a 1 Ci release of I^{131} over 1/2 an hour from Oldbury and Hinckley Point respectivley. Using a Gaussian plume treatment both would have the same lateral spread corresponding to Pasquill category D conditions. Effects of wind shear can also lead to rather different areas of contamination by dry and wet deposition as illustrated in figures 7 and 8. Dry deposition patterns depend on concentrations in air near the surface, whereas wet deposition may occur with material borne aloft over areas with negligible contamination of surface air.

5. Summary and conclusions - applicability of complex models in emergencies

Current emergency procedures vary between countries in assessing accident situations. Some emphasize measurements rather than numerical models, but the latter have a potential advantage of forecasting ahead if reliable forecast winds were available. A great difficulty in real accident situations is that the source terms are likely to be very uncertain anyway. At present there are two very different approaches to numerical modelling for emergencies. The first involves simple Gaussian plume of puff models based on meteorological conditions at the source, requiring minimal computing and operator skills. The uncertainties introduced by changes of wind direction etc. are obvious. By contrast the more complex approach adopted in the current work attempts to allow for topography. However it implies substantial investment with extensive requirements for data, computing and trained staff. In developing WAFT/TOMCATS we have, we believe introduced several substantial improvements compared with previous models (1, 2). It is encouraging that an independent evaluation (7) has recognised several of the same deficiencies that we have attempted to overcome. But do the more complex methods produce significantly better predictions than simple Gaussians?

Unfortunately we have had to conclude that this may not be so. A very severe criticism is that the operational methods such as the choice of mathematical interpolation scheme can introduce differences in the windfield as large as the adjustments made for mass consistency. Even disregarding this we have found many other drawbacks. For example data from badly sited stations can spoil windfield predictions. Spurious vertical motion may be generated to achieve mass consistency; this may arise due to limits of accuracy on wind measurements. Better resolution of critical orographic areas may be required. Vertical wind profiles are very important and will vary considerably in space and time; adequate data is unlikely to be available. Shallow winds (eg katabatic winds or sea breezes) evolve in time; they can make surface data misleading with regard to winds aloft and complicate vertical profiles. Internal boundary layers due to changes in surface are not resolved. Inversions are very important but their deformation over complex terrain is not possible to predict at present, and must affect windfields. Stable conditions also have a marked effect, and in light winds several sequential windfields are necessary. Meteorological data is then likely to be very uncertain and material may meander about giving extended exposure periods.

We thus conclude that at present simple procedures in emergencies based on Gaussian puffs or plumes seem the best option, but with an appreciation that a plume may be bent or spread far wider laterally; also that wet deposition may occur even where ground level concentrations are not detected. (The most useful addition to basic data on low level wind at the site is probably wind measurements at higher levels to pick out marked variations in the vertical wind profile). On the other hand the more complex WAFT/TOMCATS type of approach could be useful in post factum analysis when there is leisure to assess the uncertainties, or in training on rehearsal scenarios since they will produce typical unanticipated departures from the simple Gaussian treatment.

REFERENCES

1. DICKERSEN, M. et al (1974). A Concept for Atmospheric Release Advisory Capability. Lawrence Livermore Laboratory unnumbered report.
2. CNEN-DISP. (1981). ARIES - First Progress Report. RT1-AMB.
3. APSIMON, H. M., KITSON, K., FAWCETT, M., GODDARD, A. J. H. (1984). Development of a Prototype Mesoscale Computer Model Incorporating Treatment of Topography. Final Report CEC Contract SR 014 UK.
4. SHERMAN, C. (1978). A mass-consistent model for windfields over complex terrain. Journal Applied Meteorology <u>17</u>. P 312-319.
5. COURANT and HILBERT. (1953). Methods of Mathematical Physics Vol 1. Published by J. Wiley & Sons.
6. MEIJERINK, J. A. and VAN DER WORST. (1977). An iterative solution method for linear systems of which the coefficient matrix is a Symmetric M-Matrix. Mathematics of Computation, <u>31</u>. No137. P 148-162.
7. LLEWELLEN et al. (1982). The evaluation of MATHEW/ADPIC as a real time dispersion model. NUREG ICR-2199. ARAP report no 442.
8. LANGE, R. (1973). ADPIC - A 3-dimensional transport-diffusion model for the dispersal of atmospheric pollutants and its validation against regional tracer studies. Journal Applied Meteorology. <u>17</u>. P320-329.
9. JANICKE, L. (1981). Particle Simulation of Inhomogenious Turbulent Diffusion. Proc. 12th. NATO/CCMS. Meeting Air Pollution Modelling and its Application.
10. HANNA, S. (1981). Diurnal variation of horizontal wind direction fluctuations in complex terrain at Geysers. Cal. Boundary Layer Meteorology <u>21</u>. P207-213.

ACKNOWLEDGEMENT

We should like to express our thanks to Dr Van der Worst for his help in suggesting and implementing the ICCG techniques.

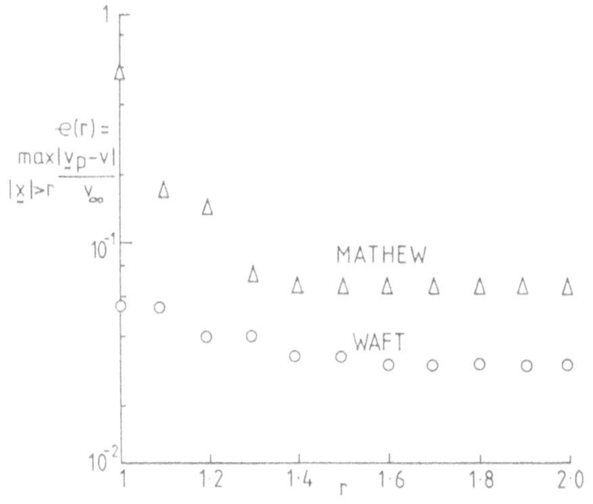

Figure 1. Comparison of MATHEW and WAFT for flow over a hemisphere. Maximum error beyond given distance r radius.

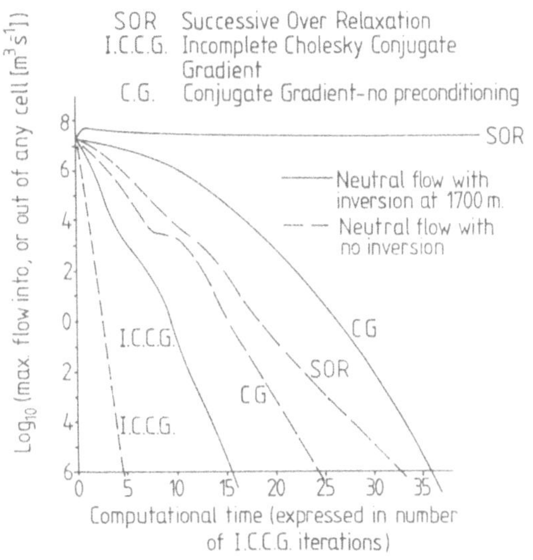

Figure 2. Comparison of different methods of solution in WAFT.

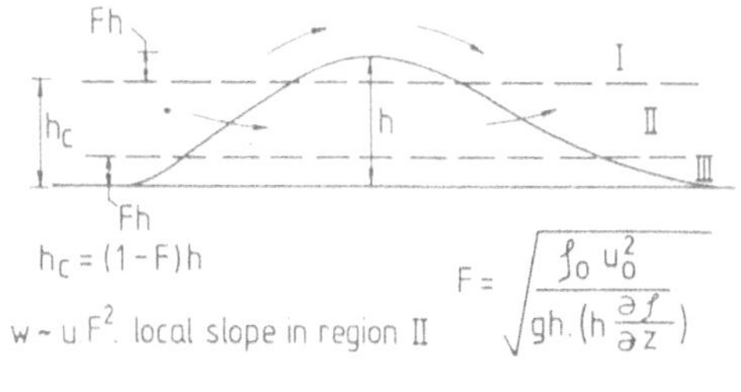

Figure 3. Regimes in flow over a hill in stable conditions

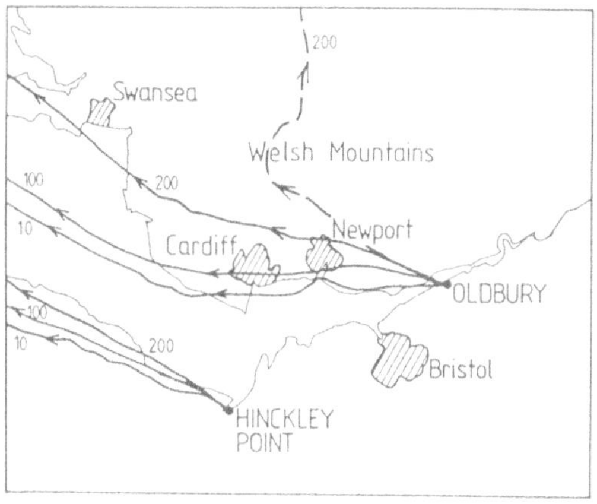

Figure 4. Streamlines from Hinckley Point and Oldbury. 6th February 1976 9.00 hours originating at 10, 100 and 200 m.
(Arrows plotted every ½ hour. Dotted line indicates effect of using cubic spline instead of inverse square interpolation.

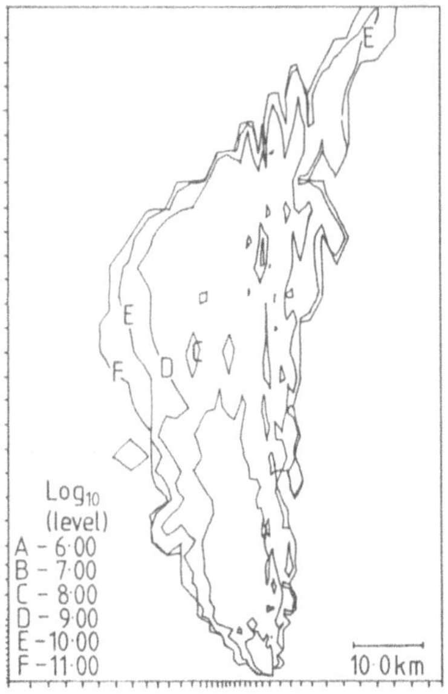

Fig 5. Isopleths of TIC for hypothetical 1 Ci release of I^{131} over ½ hr from Oldbury. ($Ci\ sm^{-3}$) Feb 6th 1976.

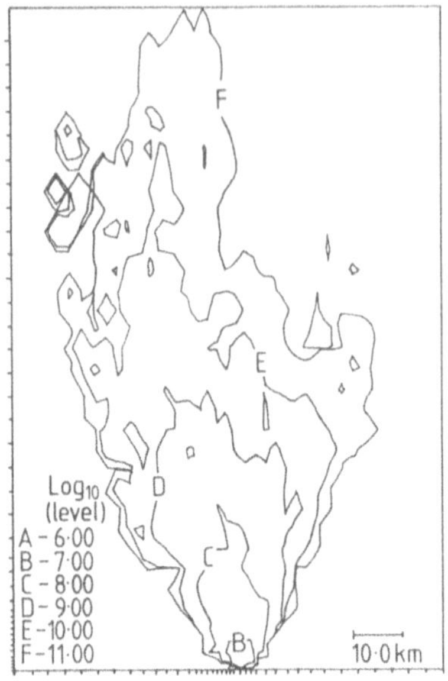

Figure 6. Isopleths of TIC for hypothetical 1 Ci release of I^{131} over ½ hr from Hinckley Point. ($Ci\ sm^{-3}$) Feb 6th 1976.

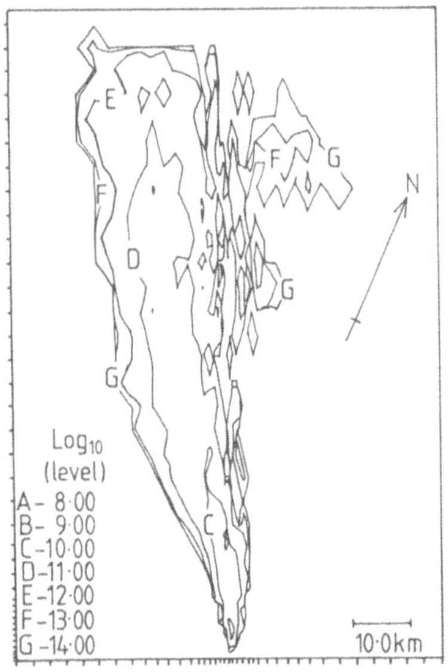

Figure 7. Pattern of dry deposition. Hypothetical release of Cs^{137} from Hinckley Point. Jan 29th 1976.

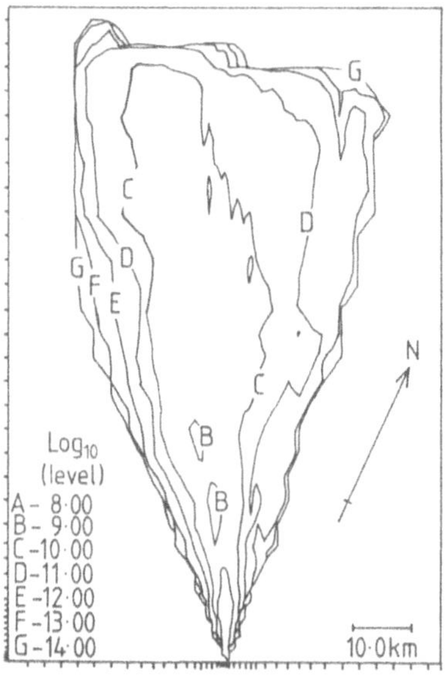

Figure 8. Pattern of wet deposition. Hypothetical release Cs^{13} from Hinckley Point. Jan 29th 1976.

Discussion

Question: F. Desiato - ENEA/Roma

1) Could you say some more words about the treatment of topography in the model?

2) In which way is the wind field interpolated in the vertical?

3) Which values of the Lagrangian time scale T_L are used by the diffusion code?

Answer: H. ApSimon

1) The height of terrain is specified on each vertical grid line. Within a vertical column the heights at the 4 corner edges are used to define the surface as 4 triangles meeting at the central point at the average height. This complicates the flux calculations in mass consistency since all walls may be just partly blocked and submerged. However, the greater complexity is justified by the better surface wind fields.

2) In the case studies radiosonde data was used to specify the vertical wind profile but also other wind profiles were tried to test sensitivity. These were based as theoretical formulations e.g. log law or power law. What worried us was how sensitive predictions would be to vertical wind profile specification, about which little data is likely to be available in an emergency accident situation.

3) Our specification of turbulence profiles and Lagrangian time scales was based on Hanna's work.

Question: Van Ulden - KNMI

The British meteorological office is developing a mesoscale forecast model. Do you think such a model is better able to provide 3-dimensional windfields than Waft or Mathew?

Answer: H. ApSimon - Imp. Col./London

The non-hydrostatic models have a better representation of the atmospheric physics involved. Nevertheless there are still limitations on windfield prediction in complex terrain due to grid resolution of orographic features.

Question: C.J. Wheatley - UKAEA/SRD

1) What computing resources are required for your calculations - both time and size of memory?

2) Does the drift velocity correction you make to compensate for regions of low intensity turbulence alter the physics or merely bias the sampling and hence particle trajectories to provide quicker convergence?

Answer: H. ApSimon - Imp. Col./London

1) The windfield prediction is much more economical than the Monte-Carlo simulation. It takes less than 5 min CPU on a CDC 6600. (Depending on such factors as stability). The dispersal of the release in the windfield depends on how many particles are used to simulate the release. Taking typically 10000 particles, computing time for high to moderate winds would be about 30 mins.

2) The aim of the drift velocity correction is to compensate for artificial accumulation of particles in regions of low intensity turbulence.

Question: J.M. Quinault - CEA/CEN Cad

Have you any problems to correlate scales of space and time: for adjustment of wind in a large area are you obliged to use mean values of wind, the time of averaging depending on space scale?

Answer: H. ApSimon - Imp. Col./London

The data available for the wind model will typically be for example hourly averaged surface wind observations from synoptic stations, etc.
Windfields are updated hourly in simulating dispersal of a release.

Question: G. Fraser, CEC/Lux

What thought has been given as to how environmental monitoring data (airborne concentrations etc.) might be used to modify the model predictions and improve the "fit" should one wish to use it in real time in an accident situation?

Answer: H. ApSimon - Imp. Col./London

This is a general question for emergency procedures. How should one combine measurements and modelling-whatever modelling techniques are used. There are so many uncertainties and parameters involved in the complex models that I find it difficult to know quite where one would start in model optimisation to fit concentration patterns.

FINITE ELEMENT THREE-DIMENSIONAL FORMULATION OF A VARIATIONAL PROCEDURE OF WIND FIELD ADJUSTMENT OVER A DOMAIN OF COMPLEX TOPOGRAPHY.

J.Y. CANEILL and J.P. GRANIER
Electricité de France
R. ROSSET and P. RACHER
LAMP, Université de Clermont II

Abstract.

The aim of the paper is to describe a finite element method providing optimal adjustment of a given interpolated wind-field, in order to get a non divergent flow over topography. We describe the mathematical method, and we include a discussion concerning empirical parameters taken into account in the model formulation. An exemple of application is presented, using data from a mesoscale experiment in the mid-Rhine Valley.

1. INTRODUCTION

Rapid progress has recently been made in analysing and forecasting meteorological phenomena at local and regional scales. This progress is liable to at least three main causes (2. BROWNING 1980) : first avaibility of weather observations of high resolution from radars and satellites ; secondly improvement and scope of mesoscale numerical models and thirdly new and more efficient procedures of data processing and analysis.
In the following, our study concerns the third point quoted namely development and testing of a finite element variational procedure of objective analysis applied to observed wind field over a domain of complex topography.
In other terms, our objective is to define a method of physical interpolation of the wind field measured at a set of irregularly spaced stations in a hilly or mountainous region. The incompressible continuity equation is selected among the meteorological equations as a strong physical constraint to be included in a variational procedure (4. SASAKI 1970), allowing for the estimation of the three wind components u,v,w at any point in the domain, whatever its topography. For that purpose, a finite element scheme, a relatively new technique in formulating geophysical fluid dynamics problems, has been retained, in conjunction with a variational procedure in order to insure an optimal description of orography.

After presentation of the variational procedure and its finite element formulation, we display some examples of application of that procedure, after having discussed the importance of the specification of some empirical parameters we have to take into account in the mathematical formulation of that problem.

1. The variational procedure

Under the constraint of mass conservation, the general formulation of variational objective analysis previously reported by (5. SHERMAN (1978)) has been adopted.
Let us consider the domain Ω of frontier Γ in $\mathbb{R}^3$ where Γ_1 is the no flow - through portion of the boundary and Γ_2 its flow through part. - $\vec{V}$ (u,v,v) is the analyzed wind field constrained to verify :

$$\begin{cases} \nabla \cdot \vec{V} = 0 & \text{in } \Omega \quad (1) \\ n \cdot \vec{V} = 0 & \text{on } \Gamma_1 \quad (2) \end{cases}$$

V_0 being some interpolation of the observed wind field (in some stations), we define the functional J as :

$$J(\vec{V}) = \frac{1}{2} \int_{\Omega} (\vec{V} - \vec{V}_0)^t [W] (\vec{V} - \vec{V}_0) \, d\Omega + \frac{\beta}{2} \int_{\Gamma_2} n (\vec{V} - \vec{V}_0)^2 d\Gamma_2$$

where (W) is a weighting matrix expressed as :

$$[W] = \begin{bmatrix} W_H & 0 & 0 \\ 0 & W_H & 0 \\ 0 & 0 & W_V \end{bmatrix}$$

W_H, (resp W_V) being characteristic horizontal (resp.vertical) wind velocities and β a weighting parameter for the contribution of the integral bearing upon Γ_2 .
Thus posed, the variational objective analysis problem reduces to an optimization problem under constraint in the following form :

Determine the wind field $\vec{V}_1$ such as :

$$J(\vec{V}_1) = \operatorname*{Inf}_{\vec{V} \,:\, \nabla \cdot \vec{V} = 0 ,\; \vec{n} \cdot \vec{V} = 0 \text{ on } \Gamma_1} J(\vec{V})$$

In order words, we want to determine the wind field obeying relations (1) and (2) and fitting at best the initial wind field $\vec{V}_0$, in the generalized least - square sense (3)
This is a classical optimization problem, which is solved by introducing a Lagrangian. Stationnarity of the latter enables us to write the following Euler equations :

$$
\begin{cases}
\vec{\nabla}\cdot\left[[W]^{-1}\vec{\nabla}\lambda\right] = -\vec{n}\cdot\vec{V}_0 & \text{in } \Omega \quad (4)\\
\vec{n}\left[[W]^{-1}\vec{\nabla}\lambda\right] = -\vec{n}\vec{V}_0 & \text{on } \Gamma_1 \quad (5)\\
\beta\vec{n}\cdot\left[[W]^{-1}\vec{\nabla}\lambda\right] + \lambda = 0 & \text{on } \Gamma_2 \quad (6)\\
[W]\left(\vec{V}-\vec{V}_0\right) = \vec{\nabla}\lambda & \text{in } \Omega \quad (7).
\end{cases}
$$

where λ is the Lagrange multiplier. Solving (4) with boundary conditions (5) and (6) allows us to get the final wind field with (7).

2. Finite element formulation.

The objective analysis procedure above realizes three dimensional adjustment of the mesoscale wind field incorporating topographic forcing ; this adjustment obeys equations (4) to (7) which have to be solved after selection of a discretization basis. A finite element partition of the domain Ω has to be retained. These three-dimensional elements have been defined with a progressively decreasing resolution in the vertical (higher resolution near the ground) together with a surface and horizontal resolution proportional to topographic gradients or increased in regions of particular interest (secondary circulations).
Discretization of variables is then formulated in terms of piecewise interpolation functions within each element, together with nodal values. Thus, on the basis of the N nodes ensuring gridding of the domain Ω , the initial wind field $\vec{V_0}$ and the unknowns (wind field V and the Lagrange multiplier λ) is written at any point P as :

$$
\begin{cases}
\vec{V}_0(P) = \sum_{i=1}^{N} \phi_i(P)\,\vec{V}_{0i}\\
\vec{V}(P) = \sum_{i=1}^{N} \phi_i(P)\,\vec{V}_i \qquad (8)\\
\lambda(P) = \sum_{i=1}^{N} \psi_i(P)\,\lambda_i
\end{cases}
$$

with $\phi_i(P)$ and $\psi_i(P)$ formally designing values of the global basis functions for $\vec{V_0}$, $\vec{V}$ and λ at node i - V_{0i}, V_i and λ i are respectively the values of these variables at this same node i.
According to a weighting - residual procedure of the Galerkin type, one gets a weak integral formulation of the problem, equivalent to the system of equations (4) to (7).
That leads to a system of algebraic equations, solved Ly an interative method. (see 3. CANEILL and al. 1984).

For specification of the elements we have choosen a compromise between a detailed description of the wind field and reduced computational costs in using a 8 node isoparametric hexaedron element with tri linear basis functions). (see 3. CANEILL and al. 1984).

3. Sensitivity of the procedure to the specification of empirical coefficients - An estimation of K.

Sensitivity of the results of the objective analysis procedure has been tested against the values of the empirical parameters and $K = \frac{W_H}{W_V}$. We will discuss here, only the latter point. The ratio

$K = \frac{W_H}{W_V}$ controls relative contributions of flow around and over mountains. As a result of the sensitivity analysis, it has been found that decreasing K favours flow around the relief, while increasing it produces predominant overcrossing effects. (cf : 3. CANEILL and al. 1984).

Due to its effect on the airflow over mountains, one basic physical ingredient in K is the atmospheric static stability. In order to calibrate in some sort the value of K relevant in a particular situation, one has made use of the vertical influence height H_0 defined by 1. BUZZI et al (1977).

This height is a function too of static stability as well as of characteristic topographic horizontal scales. It is given by :

$$H_o = L_T \frac{f}{N} (1 - R_o^2)^{\frac{1}{2}}$$

where L_T refers to the above characteristic orographic scale, f being the Coriolis parameter, $R_0 = \frac{\overline{V}}{f L_T}$ is the flow Rossby number in which $\overline{V}$ is a typical horizontal velocity and N being the Brunt-Vaissala frequency :

$$N^2 = \frac{g}{\theta} \frac{\partial \theta}{\partial z}$$

This height of influence can be calculated from basic meteorological parameters relevant of the situation to be studied. Given, now, the topography of the site, it is possible to derive using the model, a calibration curve (H versus K) where H is the mean height where $W(H) = e^{-1} W(z_s)$, specific of the site. Thus it is possible to determine an average value for K, using H_0 and this curve, for a general meteorological situation. By this way it is possible to take into account the physical as well as the numerical nature of the parameter K.

4. Examples of application : The Rhine Valley.

This area is of great interest for mesoscale models validations for mainly two reasons :

1) the topography of the domain exhibits two characteristic features (Vosges mountains and Black forest) (See Fig. 1).
2) a field experiment was organized by E.D.F. and the French Met'Office in October 1976.

From this field experiment (extra radiosoundings, plane cross-sections), the air flow in the lower layer of the atmospheric has been characterized according to the synoptic fluxes, leading to four principal regimes for the period of the experiment :

- West/South-West synoptic flow (moderate or strong),
- South synoptic flow,
- North-East synoptic flow.

We can summarize some particular physical features :

- Channeling effect is observed within the valley for South-West and North-East regimes (moderate wind),
- Important horizontal gradients are noted for South-West flow,
- A secondary circulation zone is observed between Belfort and Bâle for weak South regime.

We apply here our variational objective analysis scheme, to one of the situation encountered during the field experiment : 6th October 1976.

We use surface measurements (20 stations) as well as three radiosoundings (Nancy-Payerne-Stuttgart) at 12 TU, to initialize the vertical structure of the flow and surface winds (see the full network on figure 2).

The main characteristics of the simulation are :

- Mesh sizes $\Delta x = \Delta y = 10$ km
- Variable vertical meshing

The mean static stability of the atmosphere is provided by the Nancy radiosounding. Choozing L_T : 200 km for a characteristic length scale of the topography, and U = 10 m/s, leads to H_0 = 1290 m so that we retain for the factor K : K = $0,85.10^{-3}$ using the methodology described in the previous section.

The results are quite satisfactory. As it can be seen on figure 3, the channeling effect of the valley is quite well reproduced in the lower part of the atmosphere.

The vertical velocity field shown on figure 4 ,exhibits a natural behaviour and has been compared with results of a more complex model (primitive equation model) with a rather good agreement.

At last one can observe that the west part of the domain is not strongly perturbed. This fact is confirmed by comparison of the initial radiosoundings in Nancy, and its adjusted value on figure 5.

5. Conclusion.

A finite element model has been presented, to calculate a non divergent flow over complex topography from measurements, using a variational procedure. It is found that care must be taken in specifying some empirical parameters inherent to the formulation. The results obtained with the model on a real case are satisfactory, allowing to depict typical circulations in the Rhine Valley. This methodology seems promising for analysing the structure of the flow over complex thermal power plant sites.

REFERENCES.

1. BUZZI, A. and S. TIBALDI 1977 : Initial and frictional effects on rotating stratified flow, over topography. Q.J.R.M.S, Vol 103, pp 135 - 150.
2. BROWNING, K.A. 1980 : Local weather forecasting. Proc. Roy. Soc London, A 371, 179-211.
3. CANEILL, J.Y. and al 1984 : CCE report - Contract n° 015 SRF.
4. SASAKI, 1970 : Some basic formalisms in numerical variational analysis. M.W.R., 98, pp. 875-883.
5. SHERMAN, C. 1978 : A mass consistent model for wind fields over complex terrain. J. Appl. Meteorology Vol 17, pp 312-319.

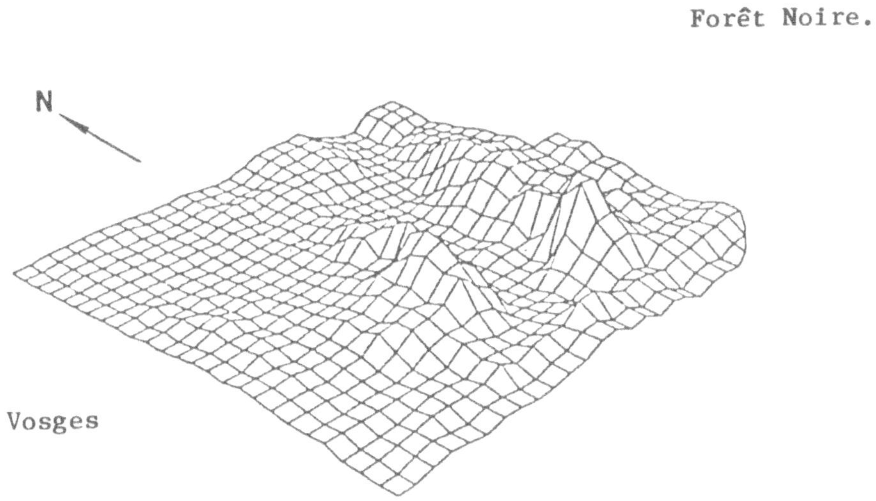

Fig. 1. **Topography of the Mid-Rhine Valley area.**

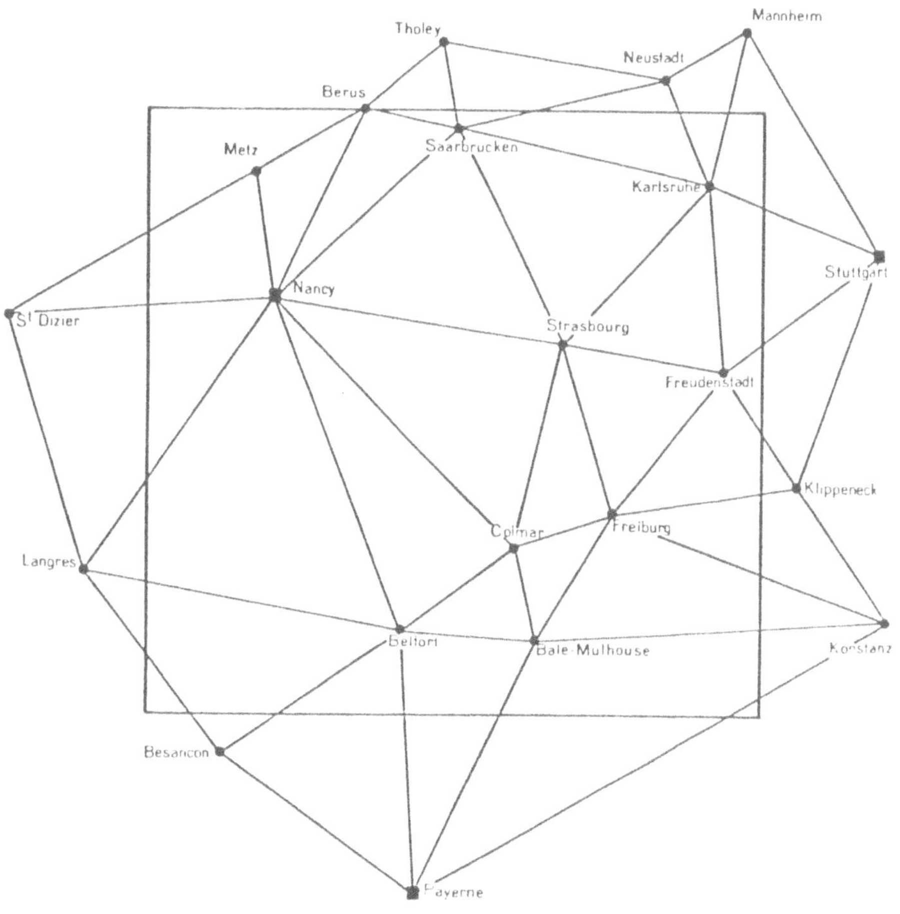

Fig. 2.a. **Network of wind-surface stations used in the calculation.**

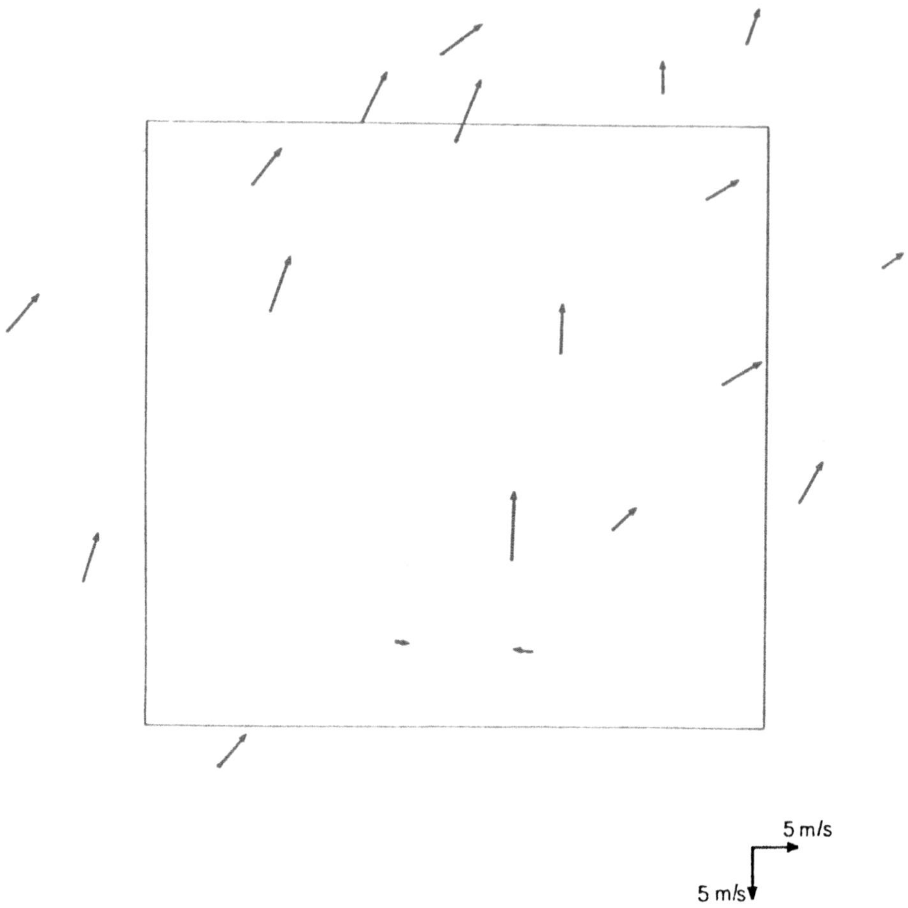

Fig. 2.b. Measured wind at surface stations (6/10/76 - 12h TU).

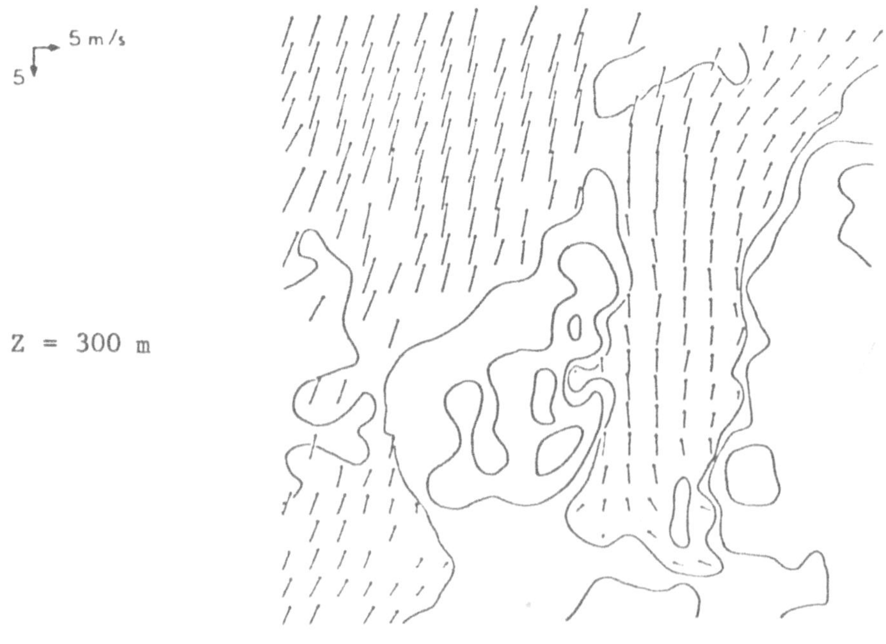

(a) Z = 300 m

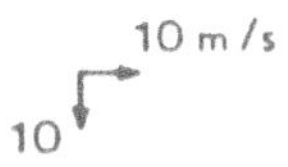

(b) Z = 700 m

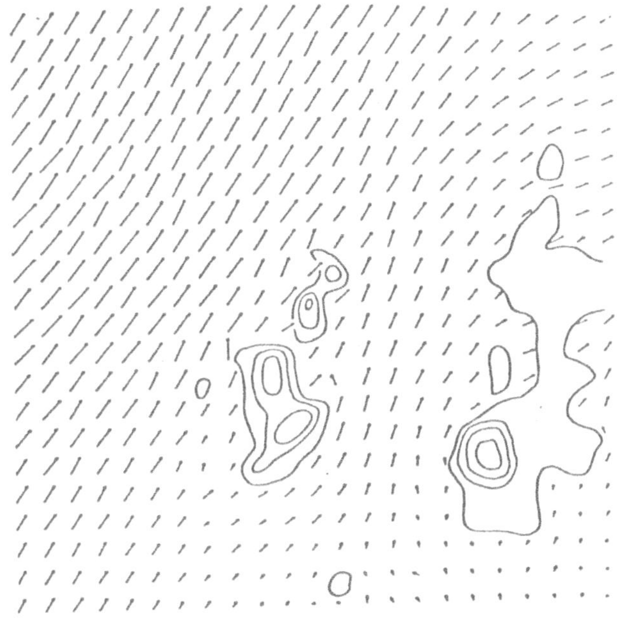

__Fig. 3.__ Cross sections of the horizontal wind-field at different altitudes.

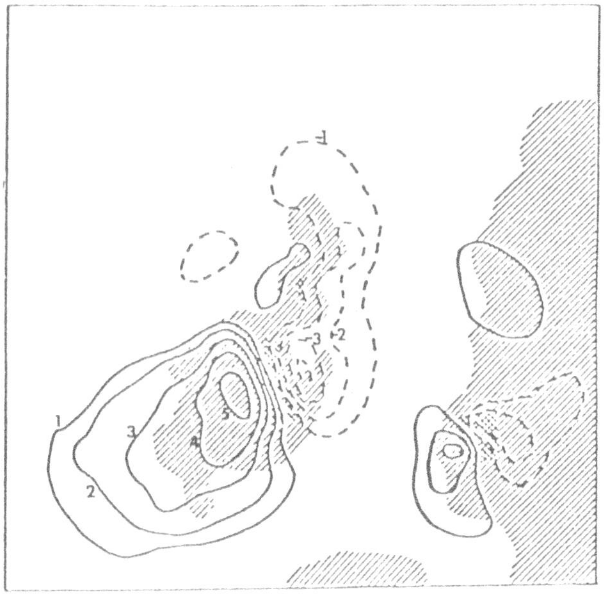

__Fig. 4.__ Vertical velocity field at Z = 1500 m.
(isovalues : $\pm$ 1, $\pm$ 2, $\pm$ 3, $\pm$ 4, $\pm$ 5 cm/s).

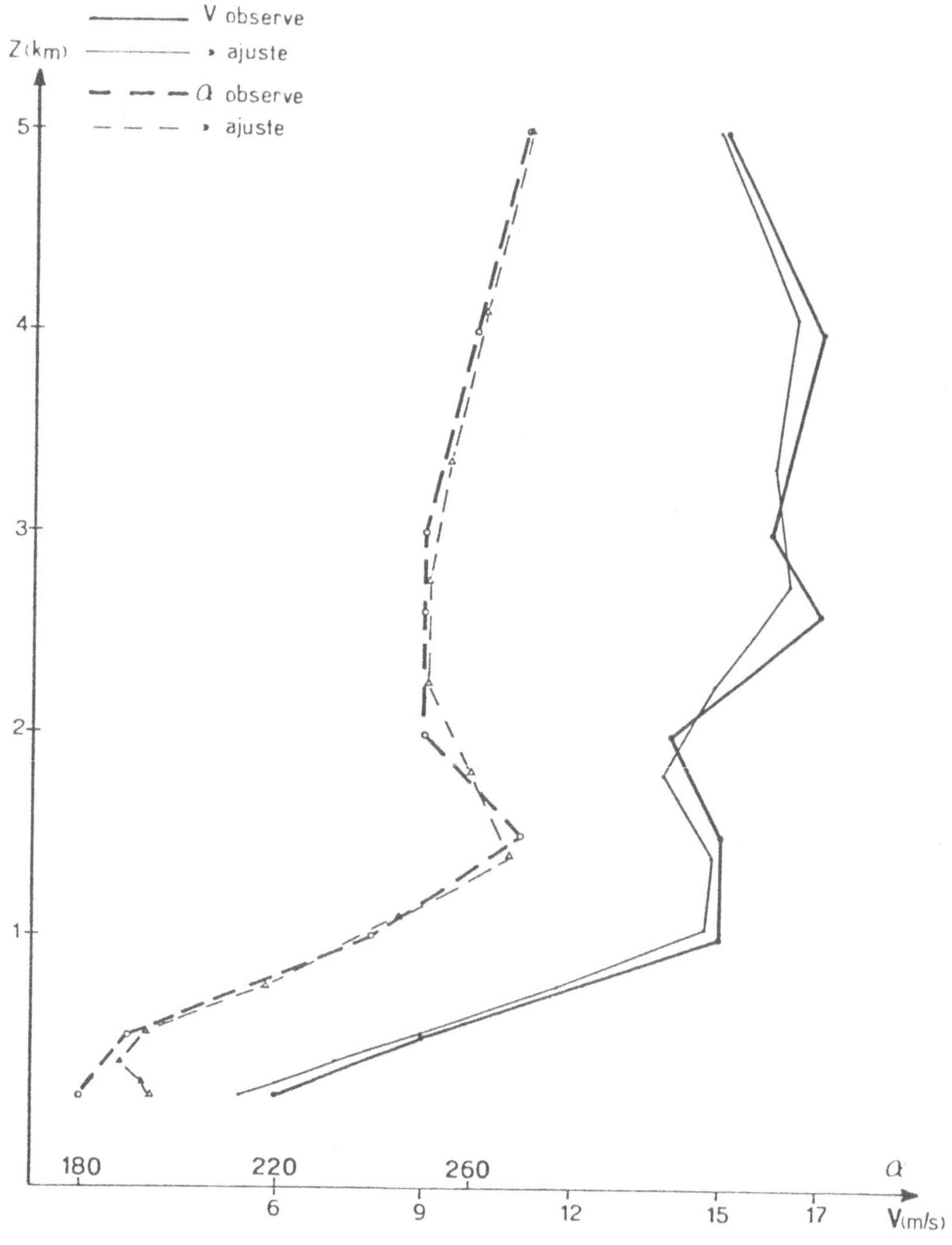

Fig. 5. Vertical structure of the horizontal wind-field at Nancy Station.

Discussion

Question: B. Crabol, CEA/CEN-G

1) How was determined the curve relating the value of K to the height of the topography?

2) Do you think that pronostic models could help the understanding of the influence of the stability on the wind field in the case of complex terrain?

Answer: J. Caneill - EDF/Chatou

1) Using the model with different values of K= PH/PW, we look in each case at the height where the vertical velocity is decreasing at some fixed rate. So H is plotted versus K or log K.

2) I think that more complex modelling techniques (non-hydrostatic models with thermodynamics) could help in the specification of the empirical parameter K, because they contain, a priori, the necessary physics to understand phenomena of interest here.

Question: Van Ulden - KNMI

1) Have you considered transforming surface winds to higher level winds by using boundary layer scaling laws?

2) Do you think it is useful to do such a thing in the context of wind analysis.

Answer: J. Caneill - EDF/Chatou

1) Not in the framework of this study. This is in fact a possibility, which is right to mention, and that can be used, in absence of other information. However, and this concerns the second question too, in the case of complex terrains it may be doubtful to use such scaling laws that are generally established in regular condition.

Question: C.J. Wheatley - SRD/UKAEA

1) Should the time averages of the measurements of the velocity field be related to the time of travel of the plume?

2) Do you agree with the conclusions made at the end of the previous paper with regard to the future development of this type of mesoscale modelling?

Answer: J. Caneill - EDF/Chatou

1) As the final computed wind-field (mass -adjusted) represents an average of the wind on typical elements used in the numerical procedure, one has to perform a time-average of the measurements. This time average may depend on the meteorological situation (fluctuations low winds, strong winds), or the plume travel time.

2) I agree in part with the conclusions of the previous talk in the sense, that in complex terrain sites, there are important effects that one cannot take into account in that model: (drainage winds, radiation cooling, sea breeze effects) and which may be important in emergency procedures. However, for analysis of experimental data this kind of model is very useful to get information on the behaviour of the flow in space.

Question: G. Fraser - CEC/Lux

It has been pointed out that the mesh size can be optimized in accordance with the complexity of the topography. The number of measuring points for horizontal and vertical input data necessary to obtain "good" results will presumably increase with the terrain complexity. It would appear, therefore, that the optimized mesh size could also serve as a measure of the number of measuring points necessary. Could some idea be given of the number of necessary points per unit mesh area which would result from this relationship?

Answer: J. Caneill - EDF/Chatou

I see in fact two points. The possibility with the finite element technique to refine the mesh is viewed in terms of a adjustment of the measured data to the topography. But it is true, that if you increase the resolution on some particular features of the domain (for instance narrow valleys), it is necessary to include pertinent measurements related to the phenomena you want to take into account (valley breeze for instance). That point has to be investigated, I think, using well-documented experiments on site, with the aid of more complex models.
Recently, we found, by the application of such a model to the Fos.Sur.Mer experiment 1983, that adjustment did not bring much more information than the measured field (37 stations + 2 sodars). But further work is needed to give a definite evaluation.

DESCRIPTION OF A RECENT NORDIC MESOSCALE DISPERSION EXPERIMENT OVER A LAND-WATER-LAND AREA (THE ØRESUND EXPERIMENT)

S.-E. GRYNING
Meteorology Section, Risø National Laboratory
DK-4000 Roskilde, Denmark

Summary

The experimental set-up is described together with some results from a project that was carried out to investigate the nature of the turbulence and atmospheric dispersion over a mesoscale region with warm land and cold water surfaces. The measuring campaign was carried out as a joint effort of 15 institutions located in the Nordic countries (Denmark, Finland, Norway and Sweden), in Belgium, Germany and in the Netherlands. The experiment took place in the Øresund region from May 15 to June 15, 1984. Use was made of a large proportion of the available meso- and micro-meteorological instrumentation in the Nordic countries including six three-component doppler sodars. The atmospheric dispersion process in the area was investigated by carrying out tracer experiments as well as by tetroon-flights. The tracer was released from about 100 meters height. The release took place over a period of several hours for each experiment. When released the tracer was advected over the Øresund water surface for about 20 km. When the tracer plume reached the opposite coast it was sampled at ground level by sampling units positioned in an arc close to the water front as well as inland. Measurements of instantaneous tracer concentrations were carried out from a van, from a boat, and from an airplane. The project performs a thorough mapping in time and space of the meteorological parameters that control the atmospheric dispersion process. These measurements, in addition to the results from the tracer experiments, will be available for general scientific use.

1. INTRODUCTION

Increasing concern over air pollution on the mesoscale (5-200 km) has created a growing need for reliable model calculations of the atmospheric dispersion on this scale. Because mesoscale meteorology is controlled by those parameters that have often been neglected on a smaller scale, the extrapolation of the validity range to include the mesoscale of models that have been developed and evaluated from a micro-meteorological scale basis cannot readily be justified except under exceptional conditions. Therefore, experimental verification of the calculations of the atmospheric dispersion on the mesoscale is essential in order to establish the credibility of the models and the environmental assessment based on model simulations. Contrary to the micro-meteorological scale, where numerous well-qualified experiments exist for evaluating of dispersion models, only few investigations of the dispersion on the mesoscale have been made.

It is characteristic for the mesoscale meteorology that conditions rarely are stationary and homogeneous as often assumed on the microscale. This implies that a thorough mapping in time and space of the meteorology will be a required input for mesoscale models. Consequently, experiments will be very instrument demanding, costly and cumbersome to be carried out.

The coastal zone is one of the mesoscale sceneries that for a long time has attracted attention. The Øresund-region represents such a scenery. It is of specific interest partly because this region has a very high population density and partly because a number of power plants are situated along the coast. This configuration gave the motive for an investigation of the local meteorology and dispersion processes in the area. Also, the Øresund-region is well suited for such an investigation from a scientific point of view because the coast on both sides of the sound is rather straight and parallel, and the land area is nearly flat.

2. FRAMEWORK OF THE ØRESUND-PROJECT

The project is named after the sound between Denmark and Sweden. The purpose of the experiment was to study the atmospheric flow, turbulence and dispersion over the region under conditions of warm land and a cold water surface. The temperature difference between the two affects the flow and disperson through several mechanisms two of which will be mentioned :1) it tends to produce different atmospheric stability in the air over the water and the land with an interface between them, and 2) it tends to cause a difference in the air pressure above the land and water surfaces which will influence the flow field.

The planning of the project was carried out by institutions from the Nordice countries with NORDFORSK (Nordic co-operative organization for applied research) acting as project co-ordinator. The following institutions from the Nordic countries participated in the measuring campaign; from Denmark: The Air Pollution Laboratory of the National Agency of Environmental Protection, the Danish Meteorological Institute, and the Risø National Laboratory; from Finland: the Finnish Meteorological Institute and the Institute of Marine Research; from Norway: the Norwegian Institute for Air Research; from Sweden: Chalmers University of Technology, the Department of Meteorology at Uppsala University, Studsvik Energiteknik AB, the Swedish Meteorological and Hydrological Institute, the Swedish Military Weather Service and the Swedish National Defence Research Institute. The following institutions from non-Nordic countries participated in the measuring campaign; from Belgium: S.C.K./C.E.N. Nuclear Energy Research Center; from Germany: Kernforschungszentrum Karlsruhe; and from the Netherlands: the Free University of Amsterdam.

The measuring campaign took place in the period May 15 to June 15, 1984. In the period June 4-10 a special intensive measuring programme was carried out.

3. SITE DESCRIPTION

A map of the experimental site is shown in fig. 1. The Øresund-region is well suited for investigations of boundary layer dynamics and dispersion processes. The land east of Øresund is rather flat, hilly farmland, reaching a height of 100 m 5-10 km inland. On the western side of Øresund lies Copenhagen and its suburbs; this area is very flat with a high surface roughness due to the urban area. During the experimental period, the water surface temperature of Øresund was a few ^{o}C less than the land temperature during daytime.

4. METEOROLOGICAL MEASUREMENTS

The geographic locations where the meteorological measurements were carried out, shown in fig. 1, represent the meteorological conditions 1) over land, upwind of the coast line, 2) at the coast line, 3) in Øresund, that is over the water surface, 4) at the downwind coast line and finally, 5) downwind of Øresund, inland. The instrumentation and measurements included:

1) Three-component Doppler sodars. This instrument measures the wind speed and direction in the lowest several hundred meters in the atmosphere. Six of these instruments were available, the locations 4, 7, 8, 10, 11 and 13 in fig. 1 were equipped with a sodar. The sampling time was 10 min at Kastrup, Furulund, Borlunda and 20 min at Gladsaxe, Middelgrunden and Barsebäck. This time was chosen in dependence of the noise level of the environment and the auxiliary equipment of the sodars. The sodars were of the make "Sensitron". Prior to the experimental campaign, all the sodars were put into operation near the 120 m meteorology mast at Maglarp, and measurements of wind speed and direction from the mast and the individual sodars were compared. The comparison was satisfactory for all of the sodars.

2) Radiosondes. They were launched from Jægersborg (situated close to Charlottenlund) and Borlunda every 6 hours during the entire period, and every 3 hours during the intensive period and tracer experiments. During tracer experiments, mini-sondes (pressure, temperature and wet-bulb temperature) and sometimes the so-called "Sprengersondes" (temperature only) were launched from a fishing boat every 1 1/2 hours from various locations in Øresund.

3) Microwave radiometer for 52.8 - 58.8 GHz. This instrument measures temperature profiles every 20 minutes in the absence of clouds. Profiles are derived with support of a priori data from previous radiosoundings. The microwave radiometer was installed at Borlunda.

4) Turbulence measurements at three levels in the 100 m mast at Barsebäck (hot wire, temperature and humidity) during a great deal of the measuring period, and at 115 m height at Gladsaxe (cup, vane, vertical propeller; three component wind velocity fluctuations) during tracer releases.

5) Profiles of wind, temperature and in some cases humidity and pressure (one level only). These measurements were made at the tall masts (~ 100 m) at Risø, Barsebäck and Maglarp, and at small masts (~ 15 m) at Sjælsmark, Margretheholm, Avedøre, Charlottenlund, Saltholm (two masts, one at the east and one at the west coast, both very close to the water), Barsebäck and Furulund. The sampling time was 10 min. From the profile data the turbulent fluxes of sensible heat and momentum near the ground can be estimated by applying current flux-profile relationships.

On a less regular basis pilot balloon measurements were performed at Barsebäck and measurements with a tethered balloon up to 300 m at Barsebäck and Charlottenlund.

During the intensive period, June 4-10, the research vessel "Aranda" carried out measurements of water and air temperature, infrared reference measurements and wind direction, speed and temperature at several heights.

"Aranda" circulated between five positions in Øresund, performing measurements over a period of one hour at each position.

A research airplane participated during some of the tracer releases. The airplane programme included east-west traverses at varying heights while making measurements of dissipation and temperature. The purpose of the flights was to determine the height of the internal boundary layers arising from the water and land surfaces. The aircraft programme also included a south-north traverse over Øresund at a constant pressure surface (~ 900 mb) in order to determine the baroclinity over the area.

Satellite images of the area were received every day and stored in digital form when appropriate. In combination with the infrared reference measurements from "Aranda" these images give information of the skin surface temperature in Øresund.

A study of the transformation of the air as it moves from sea to land took place at Barsebäck. Two masts with profile instrumentation were used, one near the coast and the other about 200 m inland. During selected periods flux measurements were carried out at these masts by use of sonic-anemometers.

5. DISPERSION AND ADVECTION MEASUREMENTS

The atmospheric dispersion process in the areas was investigated by carrying out tracer experiments. Sulphurhexafluoride was used as tracer. The tracer was released from about 100 meters height. As release point at the Swedish coast the meteorological mast at Barsebäck was used. This mast is situated about 1/2 km from the coastline such that the tracer at easterly wind was advected over the water surface for about 20 km. When the tracer plume reached the Danish coast it was sampled at ground level by sampling units positioned in an arc close to the waterfront, i.e. before the tracer actually had been influenced by the land surface and further sampled in an arc about 2 km inland such that the dispersion of the tracer has been influenced by the land surface. At westerly winds, the mast at Gladsaxe was used as tracer release point. The mast is situated about 6 km from the coastline. The tracer was sampled in Sweden in an arc along the coast and one about 2-4 km inland. A number of ~ 25 tracer sampling units was placed in each arc at positions separated by about 1.5° as seen from the release point. In some of the experiments tracer sampling was also carried out in more than two arcs. The consecutive tracer samples were taken in a way such that the total averaging time was one hour. For a typical tracer experiment tracer release started at 9:00, tracer sampling started at 12:00 and stopped at 13:00; however, deviations from this timetable took place. All indications of time throughout the paper are given in local summer time, 12:00 GMT = 13:00 MET (mean European time) = 14:00 local summer time.

Also measurements of instantaneous tracer concentrations were carried out. A van equipped with a gas chromatograph drived in the period between start of tracer release and sampling around in the area searching for the tracer plume. Direct information about the plume location was very important, because the wind field in the area usually was not straight. At the start of tracer sampling, the van was parked in the plume and measured instantaneous tracer concentrations once per minute. Typically, the plume drifted away from the position of the van after roughly 1/2 hour.

A research airplane participated in some of the experiments. The airplane performed traverses in the middle of the Øresund perpendicular to the plume, measuring consecutive 1/2-s averaged tracer concentrations. The traverses were made at typical heights of 50, 100 and 150 m. In one of the experiments, tracer concentrations were measured from a fishing boat while making traverses in Øresund. The institutions concerned with tracer measurements intercompared SF_6-standards during the campaign.

The advection in the area was investigated during the intensive week with tetroon-flights. A tetroon is a balloon designed to fly on an isopycnic surface after release. The tetroons were tracked by radar. They were launched near Barsebäck and were flown at a height of about 650 m. The tetroons that were released during the tracer experiments were equipped with transponders and were tracked for more than 60 km. A corner reflector was used for tetroon flights performed beyond the periods of tracer releases, which reduced the tracking distance to about 50 km.

6. THE EXPERIMENTS

A total of 9 tracer esperiments were carried out during the experimental campaign. One of the experiments failed because the tracer plume did not pass over the set-up of tracer sampling units. Of the remaining eight experiments, two were carried out during the so-called intensive phase (June 4-10). Below is given a short description of characteristic features of the experiments. The description of the meteorology is based only on 30-min routine observations (METAR) from Kastrup Airport, because the analysis of the meteorological measurements has not yet been concluded. Characteristic times for the tracer experiments are given in Table 1. The performance of the instruments that were in operation during the experimental campaign is illustrated in Table 2.

Experiment May 16. The tracer release was carried out from Barsebäck at south easterly wind. Due to a shift in wind direction the plume passed almost parallel to the coast and was detected in Helsingør, 40 km from the point of release. Only about 2/3 of the time-averaged tracer concentration distribution was measured because a part of the plume passed over the water between Helsingør and Helsingborg. The cloud cover was 2/8 stratus with the base at 300 m in combination with an overlying 5/8 cirrus. Wind speed was around 5 m/s.

Experiment May 18. Tracer release from Gladsaxe at a wind from the southwest with a wind speed of about 4 m/s. The plume was measured between Landskrona and Helsingborg. The sky was nearly cloudless, the observations at Kastrup gave 1/8 cumulus with the cloud base at 800 m.

Experiment May 22. Tracer release from Barsebäck at an easterly wind, such that tracer sampling took place in the northern part of Copenhagen. The wind speed was fairly high, that is about 10 m/s. Cloud cover was reported as 1/8 cumulus with the base at 1100 m.

Experiment May 29. The wind was from the east; therefore the tracer was released from Barsebäck. The wind speed at Kastrup was reported to be 8 m/s. In the morning 1/8 stratocumulus was reported at 9:00, but the clouds disappeared and the sky was cloudless during the whole tracer measuring period.

Experiment May 30. This was the only tracer experiment that failed. The wind direction was 90° at 9:00 when the tracer release started, but changed gradually throughout the period of tracer release to 160° at 13:00, which brought the plume outside the set-up of the tracer sampling units.

Experiment June 4. This was the first tracer experiment during the intensive phase of the campaign. The wind was steady east such that the tracer release was carried out from Barsebäck with sampling in the northern part of Copenhagen. The wind velocity was 7-8 m/s, and the sky was nearly cloudless.

Experiment June 5. This was the last of the two experiments during the intensive phase. This also was the experiment with the most intensive meteorology measurement programme. The tracer release was from Barsebäck with tracer sampling in the northern part of Copenhagen. The wind speed was about 7 m/s. The cloud cover at Kastrup Airport was reported to be 1/8 cumulus with a 1 km base in combination with 4/8 altocumulus with base at about 3.5 km. The temperature was 22°C. Figure 2 shows the dissipation (measure of turbulence) over the area. Examples of the one-hour-averaged tracer concentration profiles are shown in fig. 3. Figure 4 shows examples of instantaneous tracer measurements carried out by an airplane 11 km from the point of tracer release. Figure 5 shows instantaneous tracer measurements as a function of time 25 km from the point of tracer release and figure 6 shows the trajectories estimated from tetroon flights.

Experiment June 12. Experiment at westerly wind with tracer release from Gladsaxe. The wind velocity was as low as 4 m/s and consequently the wind direction fluctuated considerably. Cloud cover was reported to 3/8 cumulus with base at 1 km. The tracer plume was detected by the van at about 11:30. The van was able to follow the plume until 13:15 when the plume disappeared. However, the tracer plume was again detected by the van after 14:00. Time-averaged tracer sampling was carried out between 13:00 and 14:00. The analysis of the 3×20-min averaged tracer concentrations showed a well-established plume between 13:00 and 13:20, but then the plume completely disappeared in the 13:20 - 13:40 and 13:40 - 14:00 runs. A huge cumulus cloud was observed over the measuring area during the tracer sampling. It is possible that this cloud split the plume.

Experiment June 14. During the morning a cold front passed over the area. Immediately behind the front the wind direction was 320°, but the direction changed gradually during the next hours to 280-290°. When we decided on the set-up of the tracer sampling units, this change in wind direction was taken into account, and the experiment was delayed until the plume was covered by the tracer-sampling units. The wind direction was very fluctuating during the shifting period. The wind speed was about 5-7 m/s. During the tracer sampling the cloud cover at Kastrup Airport was reported to 3/8 cumulus with a 800 m base, in connection with 5/8 stratocumulus at 1600 m.

ACKNOWLEDGMENTS

It is a pleasure to acknowledge the assistance the project received from the Danish Post- and Telecommunication, The Directorate of Civil Aviation in Denmark and Sweden, the Danish Air Force, Gentofte Kommune and the Barsebäck Nuclear Power Plant. The never failing interest of Mari-Mai Lagus, NORDFORSK, is fully appreciated. The experiment was made possible by grants or support from: in Belgium: S.C.K./C.E.N. Nuclear Energy Research Center; in Denmark: the Danish Technical Research Council, the Ministry of Energy, and the National Agency of Environmental Protection; in Finland: the Finnish Meteorological Institute and the Institute of

Marine Research; in the Netherlands: the Free University of Amsterdam; in Norway: Norwegian Institute for Air Research; and in Sweden: National Institute of Radiation Protection, Ringhals Nuclear Power Plant, Studsvik Energiteknik AB, Swedish Environmental Protection Board, and the Swedish Military Weather Service. Further support to the project was received from NORDFORSK and the Commission of the European Communities.

TABLE I

Characteristic times for the tracer experiments

Experiment	Place of tracer release	Starting time tracer release	Starting time tracer sampling	Stopping time tracer release	Stopping time tracer sampling
16 May	Bb	10:30	14:30	15:30	15:30
18 May	Gl	9:30	13:20	14:10	14:20
22 May	Bb	10:00	12:00	13:00	13:00
29 May	Bb	9:00	12:00	13:00	13:00
30 May*	Bb	9:00	12:00	13:00	13:00
4 June	Bb	9:30	12:00	13:00	13:00
5 June	Bb	9:00	12:00	13:00	13:00
12 June	Gl	9:30	13:00	13:45	14:00
14 June	Gl	11:15	14:00	14:55	15:00

* Tracer outside sampling unit set-up.
Bb: Barsebäck; Gl: Gladsaxe. Time is local summer time.

TABLE II

Measuring periods.

Date May/June	15	16	17	18	19	20	21	22	23	24	25	26	27	28	29	30	31	01	02	03	04	05	06	07	08	09	10	11	12	13	14	15
METEOROLOGY																																
Doppler Sodars:																																
Gladsaxe	*	*	*	*	*	*	*	*	*	*	*	*	*	*	*	*	*	*	*	*	*	*	*	*	*	*	*	*	*	*	*	*
Kastrup	*	*	*	*	*	*	*	*	*	*	*	*	*	*	*	*	*	*	*	*	*	*	*	*	*	*	*	*	*	*	*	*
Middelgrunden	*	*		*	*	*	*	*	*	*	*	*			*	*	*	*	*	*	*	*	*	*	*	*	*	*	*	*	*	*
Barsebäck	*	*	*	*	*	*	*	*	*	*	*	*	*	*	*	*	*	*	*	*	*	*	*	*	*	*	*	*	*	*	*	*
Furulund	*	*	*	*	*	*	*	*	*	*	*	*	*	*	*	*	*	*	*	*	*	*	*	*	*	*	*	*	*	*	*	*
Borlunda	*	*	*	*	*	*	*	*	*	*	*	*	*	*	*	*	*	*	*	*	*	*	*	*	*		*	*	*	*	*	*
Small masts:																																
Sjælsmark	*	*	*	*	*	*	*	*	*	*	*	*	*	*	*	*	*	*	*	*	*	*	*	*	*	*	*	*	*	*	*	*
Margretheholm	*	*	*	*	*	*	*	*	*	*	*	*	*	*	*	*	*	*	*	*	*	*	*	*	*	*	*	*	*	*	*	*
Avedøre	*	*	*	*	*	*	*	*	*	*	*	*	*	*	*	*	*	*	*	*	*	*	*	*	*	*	*					
Charlottenlund	*	*	*	*				*	*	*	*	*	*	*	*	*	*	*	*	*	*	*	*	*	*	*	*	*	*	*	*	*
Saltholm west	*	*	*	*	*	*	*	*	*	*	*	*	*	*	*	*	*	*	*	*	*	*	*	*	*	*	*	*	*	*	*	*
Saltholm east	*	*	*	*	*	*	*	*	*	*	*	*	*	*	*	*	*	*	*	*	*	*	*	*	*	*	*	*	*	*	*	*
Barsebäck	*	*	*	*	*	*					*	*	*	*	*	*	*	*	*	*	*	*	*	*	*	*	*	*	*	*	*	*
Barsebäck, NL	*	*	*	*	*	*	*	*	*	*	*	*	*	*	*	*	*	*	*	*	*	*	*	*	*	*	*	*	*	*	*	*
Furulund	*	*	*	*	*	*	*	*		*	*	*	*	*	*	*	*	*	*	*	*	*	*	*	*	*	*			*	*	*
Tall masts:																																
Risø	*	*	*	*	*	*	*	*	*	*	*	*	*	*	*	*	*	*	*	*	*	*	*	*	*	*	*	*	*	*	*	*
Barsebäck		*		*		*	*	*	*	*	*	*	*	*	*	*	*	*	*	*	*	*	*	*	*	*	*	*	*	*	*	*
Maglarp	*	*	*	*	*	*	*	*	*	*	*	*	*	*	*	*	*	*	*	*	*	*	*	*	*	*	*	*	*	*	*	*
Radiosonde:																																
Jægersborg	*	*	*	*	*	*	*	*	*	*	*	*	*	*	*	*	*	*	*	*	*	*	*	*	*	*	*	*	*	*	*	*
Borlunda	*	*	*	*	*	*	*	*	*	*	*	*	*	*	*	*	*	*	*	*	*	*	*	*	*	*	*	*	*	*	*	*
Øresund		*		*											*	*					*	*			*				*		*	
Microwave-radiom.:																																
Borlunda													*	*	*	*	*	*	*	*	*	*			*							
Turbulence:																																
Gladsaxe		*		*				*							*	*					*	*							*		*	
Barsebäck		*		*		*	*	*	*	*	*	*	*	*	*	*	*	*	*	*	*	*	*	*	*	*	*	*	*	*	*	*
Tethered balloon:																																
Charlottenlund																					*		**	**								
Barsebäck															*	*	*				*	*							*		*	
Pilot balloons:																																
Barsebäck																					*	*							*		*	
"Aranda" (ship):																				*	*	*	*	*	*	*	*					
Airplane:															*						*	*							*			
TRACER SAMPLING																																
Time-averaged:																																
coast arc		*		*				*							*						*	*							*		*	
inland arc		*		*				*							*						*	*							*		*	
other arcs								*													*	*									*	
Instantaneous:																																
van		*		*				*							*	*					*	*							*		*	
airplane																					*	*							*			
fishing-boat																						*										
Tetroon-flights:																					*	*		*	*							

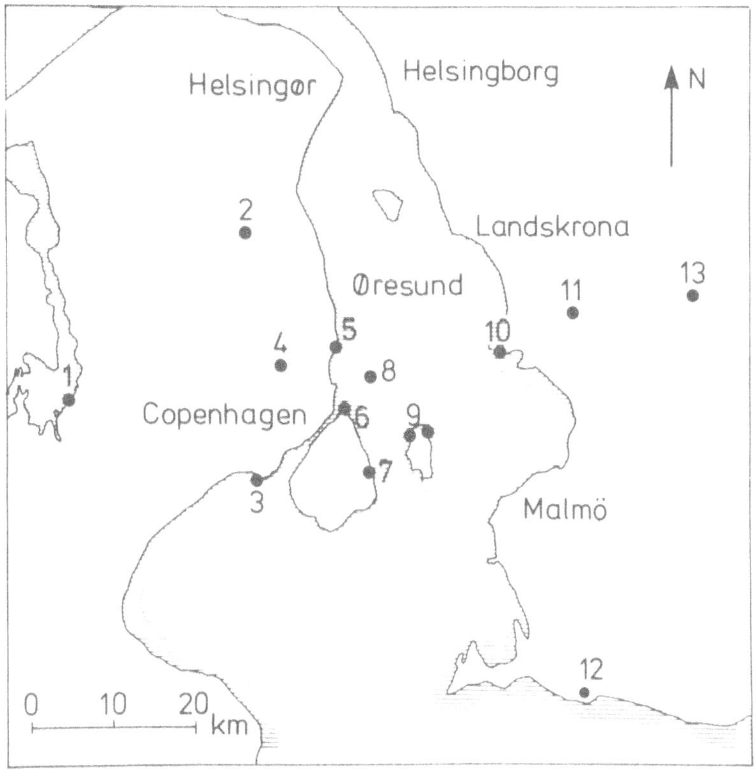

Fig. 1. The geographical locations where the meteorological measurements were carried out. The positions refer to 1) Risø, 2) Sjælsmark, 3) Avedøre, 4) Gladsaxe, 5) Charlottenlund, 6) Margretheholm, 7) Kastrup Airport, 8) Middelgrunden, 9) Saltholm, 10) Barsebäck, 11) Furulund, 12) Maglarp and 13) Borlunda. Some of the cities in the area are shown.

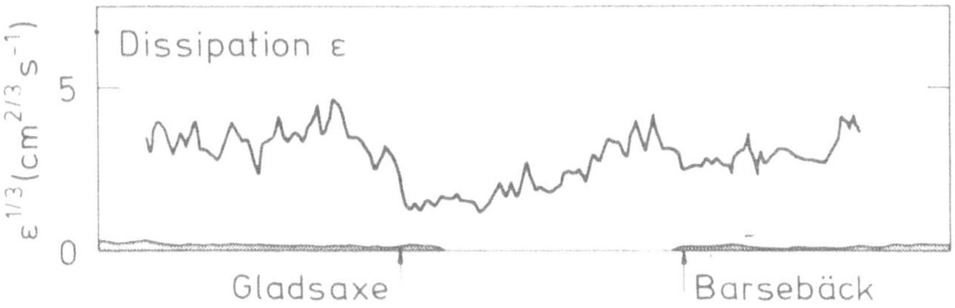

Fig. 2. Turbulence measured by an airplane in a traverse at 270 m height over the Øresund-region on June 5 1984. Turbulence was measured by a dissipation rate sensor. The wind was easterly and the gradually lowering of the turbulence level as the air passes over the Øresund water surface can be observed as well as the abrupt increase in turbulence when the airplane is in the internal boundary layer of Copenhagen.

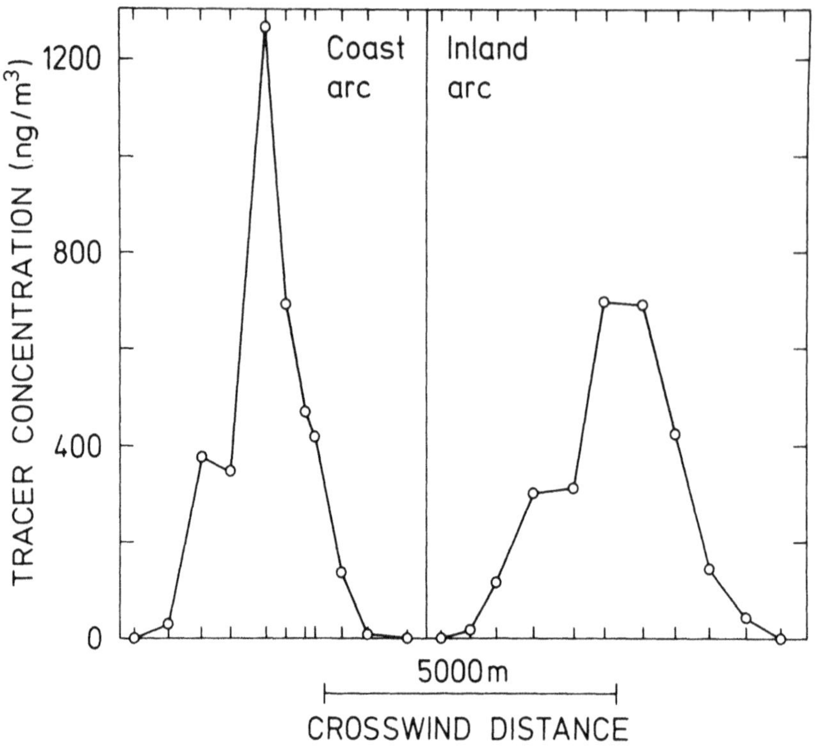

Fig. 3. Two (out of four) of the measured tracer concentration profiles, averaged over 60 min for the experiment of 5 June 1984; the coast arc is 22 km and the inland arc is 25 km from Barsebäck.

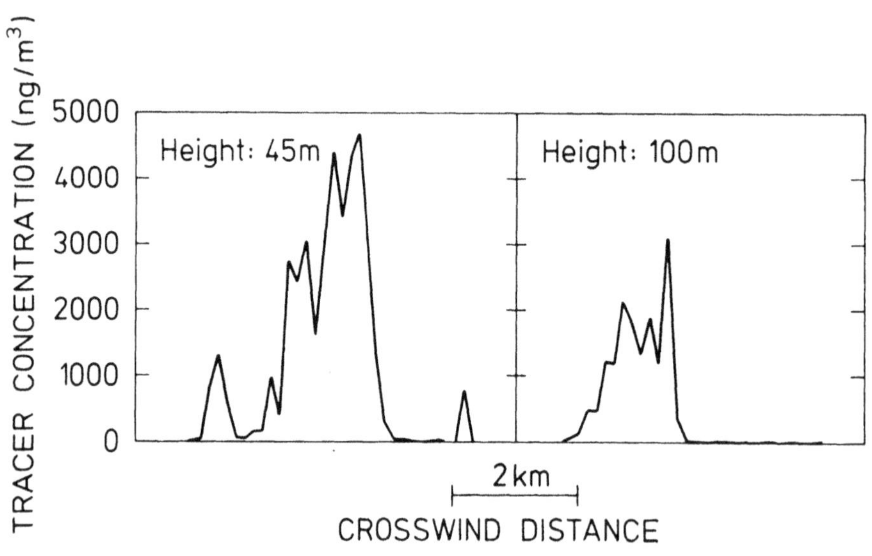

Fig. 4. Tracer concentration profiles measured by an airplane over Øresund 11 km from Barsebäck; from the experiment of 5 June 1984.

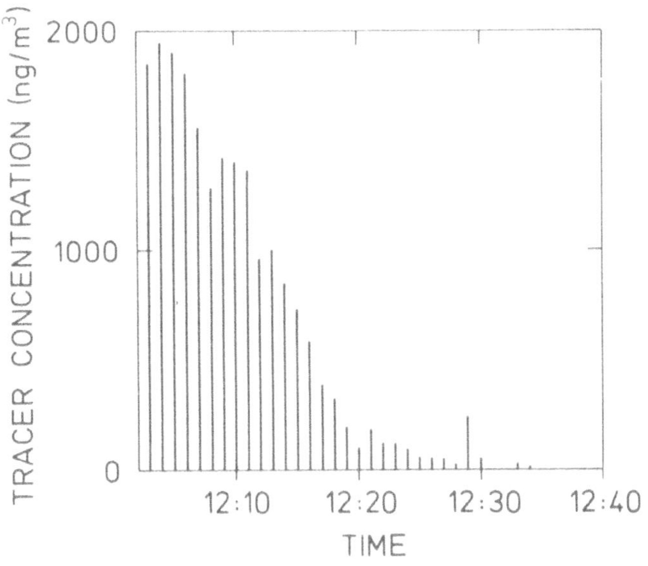

Fig. 5. Instantaneous tracer concentrations measured once per minute at a location 25 km from Barsebäck, from the experiment of 5 June 1984.

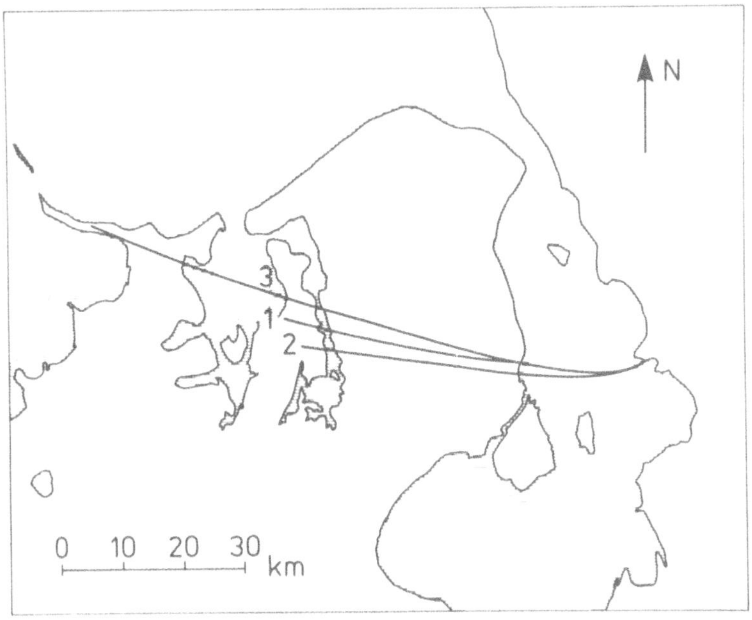

Fig. 6. Trajectories of the tetroon flights on 5 June 1984. Flight No. 1 started 8:38, mean flight level was 500 m; flight No. 2 started 10:05, mean flight level 700 m; and flight No. 3 started 11:35, mean flight level 750 m.

Discussion

Question: B. Tolley - CEC/BXL

Was it possible to carry out more than one experiment with variation of only one variable, for ex. wind speed?

Answer: S.E. Gryning - Risø N.L.

Some of the experiments were made under "rather similar" conditions. As an example it can be mentioned that the sky was near cloudless and the wind speed approximately 8 m/s during the experiments that were carried out on 29 May, 4 June and 5 June.

Question: J.A. Hertig - EPF/Lausanne

Did the group make some humidity profile measurement near the sea surface in order to control the heat and buoyant fluxes?

Answer: S.E. Gryning - Risø N.L.

Humidity profiles on the sea were obtained from mini-sondes that were launched from a fishing boat.

Question: J.P. Schon - Ec. Centr./Lyon

Are there numerical models planned by your staff?

Answer: S.E. Gryning - Risø N.L.

Yes, simulations of the experiments are planned by groups in Denmark at the Risø National Laboratory , and the Air Pollution Laboratory of the Environmental Protection Agency in Sweden by Uppsala University and the Swedish Meteorological and Hydrological Institute; and in Norway by the Norwegian Institute for Air Research.

Parallel Sessions Research Area C

The release and distribution of radioactive fission products in the atmosphere following a reactor accident

Session C3

Chairman: H.M. ApSimon (IMP.COL/London)

Session secretary: B. Tolley (CEC/BXL)

WIND TUNNEL MODELING OF TURBULENT DIFFUSION OF POLLUTANT PUFFS

J.P. SCHON**, O. DEBBARH*,
P. MEJEAN*, R. MOREL*, J.P. GRANIER***

*Ecole Centrale de Lyon, **Université Saint Etienne, *** E.D.F. Chatou

Summary

Statistical characteristics on turbulent concentration fields are given for different limiting conditions in a wind tunnel simulation of atmospheric flows. The data is obtained by a laser visualization associated to digital image processing.
The cases studied concern plumes and puffs emitted from a source at a given height in the boundary layer.
Different thermal stratifications have been used, and the effect of a clift on the concentrations has been investigated.
The instantaneous concentration fields due to puffs have been determined at two given instants after release. The evolution of some geometrical characteristics of these puffs could so be computed. All the measurements showed clearly that the method is of interest for the study of unsteady releases in the simulation of accidental situations.

1. INTRODUCTION

The present paper corresponds to the main results obtained during a C.E.E. study. A detailed description of the methods used and of all the data which have been derived can be found in SCHON et Al. (1984).

The goal of the study was the measurement of instantaneous concentration fields in a wind tunnel simulation of plumes and puffs. One of the limits of the use of wind tunnels to simulate dispersion in the atmosphere is the difficulty to determine the concentration fluctuations. It is necessary to get the concentration fluctuations because of the time scale ratio between the wind tunnel and the atmosphere is the same than the length scale ratio. Several methods have been developed to get information on the concentration fluctuations (FACKRELL-1980, LA RUE-1977). The visualization technique which was used here has the advantage to give the concentration field at a given instant in a plane. That allows the computation of spatial characteristics of the concentration field.

We will first describe the experimental set up and then give them some characteristic results of the work.

2. EXPERIMENTAL SET-UP

2.1. Wind tunnel

The experiments have been made in the E.D.F. - E.C.L. atmospheric wind tunnel built at the Ecole Centrale de Lyon with technical and financial support of E.D.F. - Division M.A.P.A. A detailed description of the characteristics of the wind tunnel can be found in FAVERGER (1977) and SERRES (1978). The test section of the wind tunnel is 15 m long, 3.7 m wide and the height can be varied from 2 to 2.5 m to adjust the longitudinal pressure gradient.

The reference of the longitudinal coordinate is taken at the end of the boundary layer generation plate (7 m). The external velocity used in this study is 2 m/s.

The wind tunnel is equiped with classical hot wire systems and a sophisticated traversing mechanism. All the data acquisition system is run by a PDP 11/34 mini-computer.

2.2. Plume and puff generation system

The quantitative visualization technique is based on particle illumination so the generation of particles has to be well defined. The particles are obtained by burning incence. Their mean size is .8 μm. An electronic system allows the controlled action of valves to generate puffs and the synchronization of the cameras.

2.3. Flow visualization

The particles which are present in the flow are illuminated by a sweeping laser beam. The sweeping is obtained by a set of rotating mirrors (figure 1).

The light diffused by the particles is a function of the size and the concentration of the particles and is recorded on a 400 ASA film processed at 1600 ASA. The process is not simple and the relationship between the particle concentration and the response of the film has been carefully analysed, (BALINT-1982).

2.4. Image digitizing

The photographic films are digitized by a vidicon camera associated to a SPATIAL DATA EYECOM II system. Each image is defined by 512 x 640 Pixels with 256 gray levels for each pixel.

2.5. Calibration procedure

The images are digitized by sets of 40 to 250 images. For each set, uniform images are recorded and digitized and are used to compute correction images which are used to get images without optical distorsions.

The relationship between gray level and concentration is obtained by computing a "mean image" and by comparison of the gray level on this image with the mean concentration measured by a ROYCO 225 particle counter (figure 3). The calibration curve allows to compute "instantaneous" images with a linear relationship between concentration and gray level.

The procedure has to be respected because of the strong non linearity of the photographic process. On figure 2, the mean image of a jet is given, the figure 3 shows the image obtained after the whole calibration and correction procedure.

3. EXPERIMENTAL RESULTS

3.1. Experimental conditions

All the results were obtained with an outside velocity of 2 m/s. The boundary layer was about 30 cm thick at the reference point. The relative positions for the different emission are given on figure 4. As for the puffs no mean concentration can be measured to calibrate the system, each puff experiment has to be associated to a plume experiment. Photographs are taken at two given instants after puff release (figure 4).

3.2. Examples of results

The figure 5 shows the instantaneous concentration field of a plume taken with the second camera in the case of an unstable flow. One notices that small regions show quite high concentrations.

The mean concentrations show much lower values as shown for example in the inversion case for the two positions of the puff (figures 6 and 7). The great variability of the concentrations is confirmed by the RMS values of the concentration fluctuations as well for the plumes (figure 8) than for the puffs in the stable case (figure 9). For the step case the results show clearly that the dispersion is enhanced by the presence of the obstacle. The probability of occurence of a concentration level higher than a given one can be computed as shown by figure 10.

4. CONCLUSION

Because of the limited space available only few results could be presented here. However, it is clear that the method is of great interest to determine the statistical characteristics of the dispersion of plumes and puffs.

REFERENCES

1. BALINT J.L., (1982), Application d'une méthode de visualisation laser et de traitement d'images à l'étude de la dispersion dans les écoulements turbulents. Thèse, Lyon.
2. FACKRELL J.E., (1980), J. Phys. E. Scient. Inst. 13.
3. FAVERGER J.P., (1977), Etude et réalisation d'une soufflerie de simulation atmosphérique, Mémoire C.N.A.M., Lyon.
4. LA RUE J.C., LIBBY P., (1977), Measurements in the turbulent boundary layer with slot injection of helium. Phys. Fluids, Vol. 20, n° 2.
5. SCHON J.P. et al. (1984), Atmospheric diffusion of puffs simulation in a wind tunnel. Final report C.E.E.-E.D.F.-E.C.L., Project 9C/Contract SR 017 F (G), February 1984.
6. SERRES E. (1978), Etude de la simulation en soufflerie des basses couches de l'atmosphère. Thèse, Lyon.

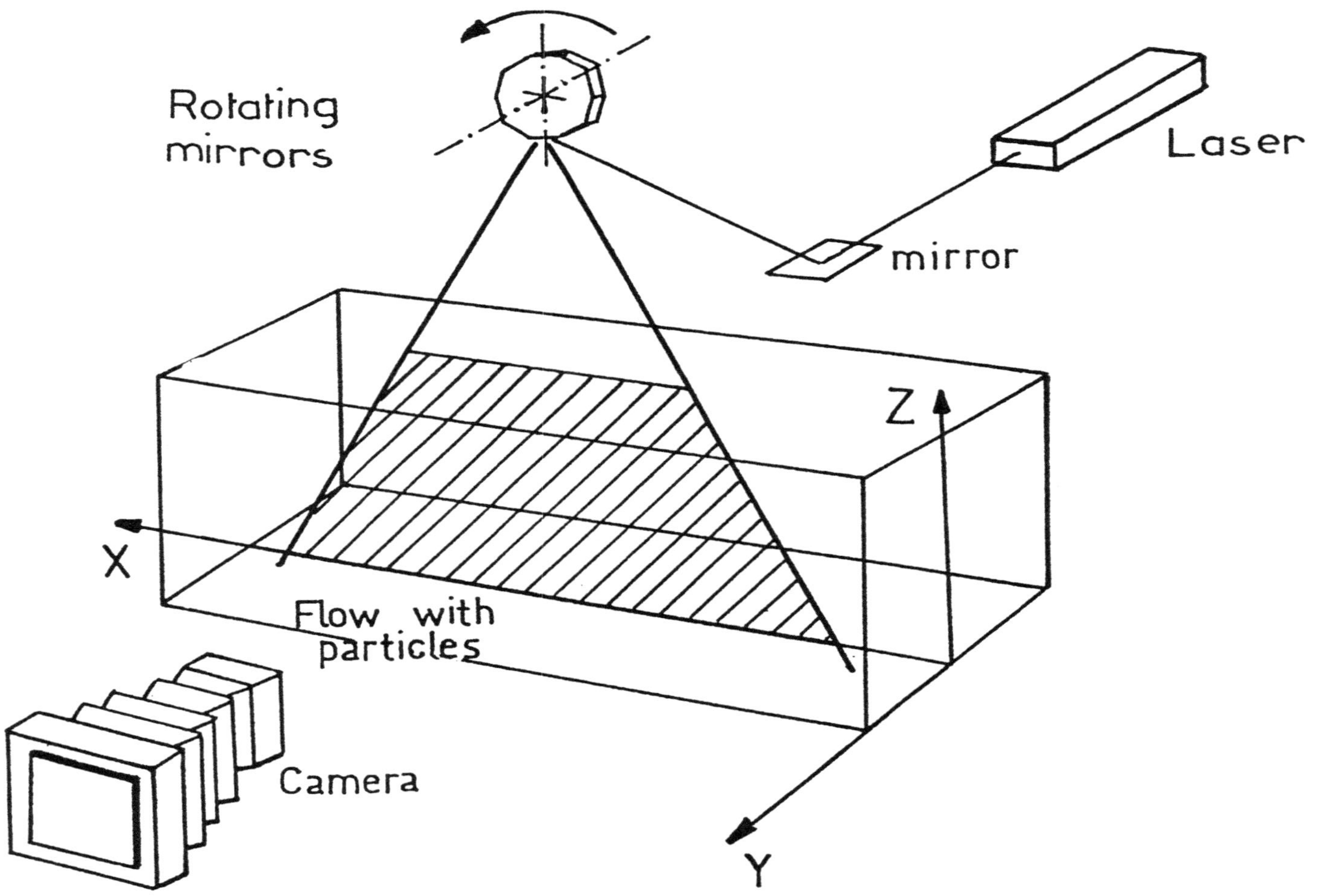
Rotating
mirrors
Laser
mirror
Z
X
Flow with
particles
Camera
Y

Figure 1.

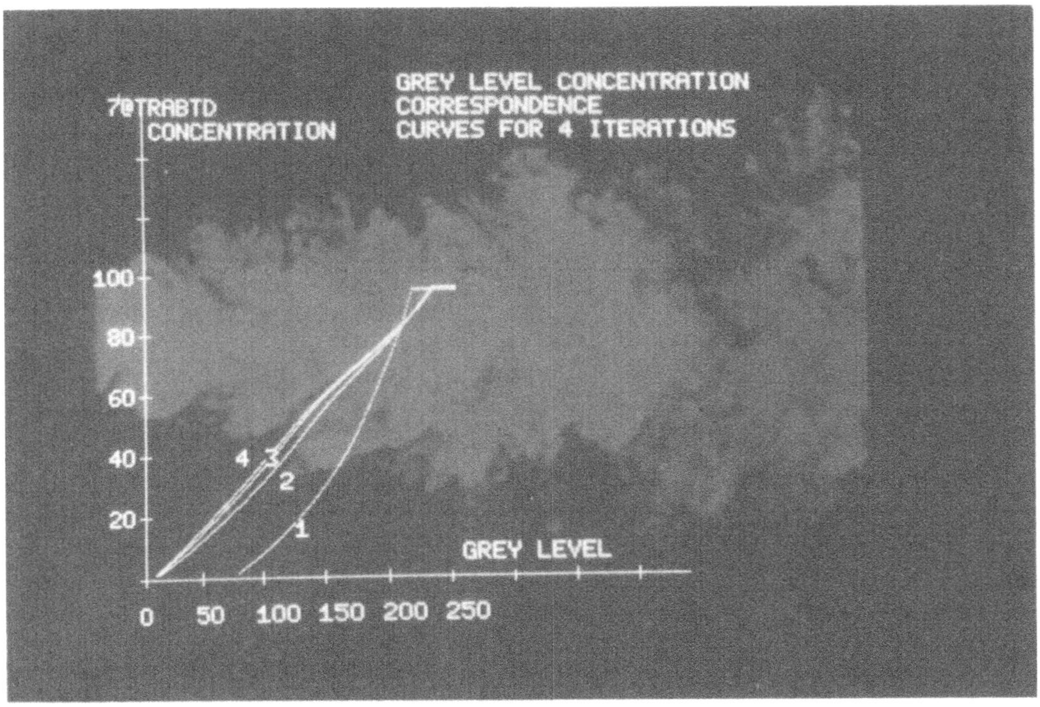

Figure 2.

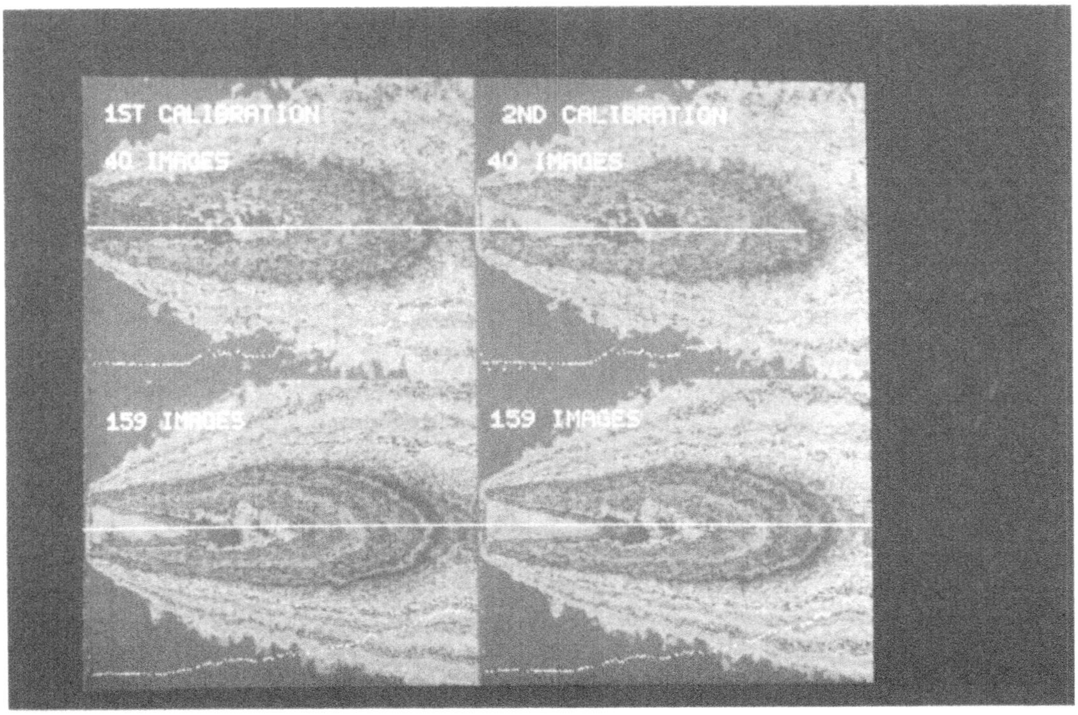

Figure 3.

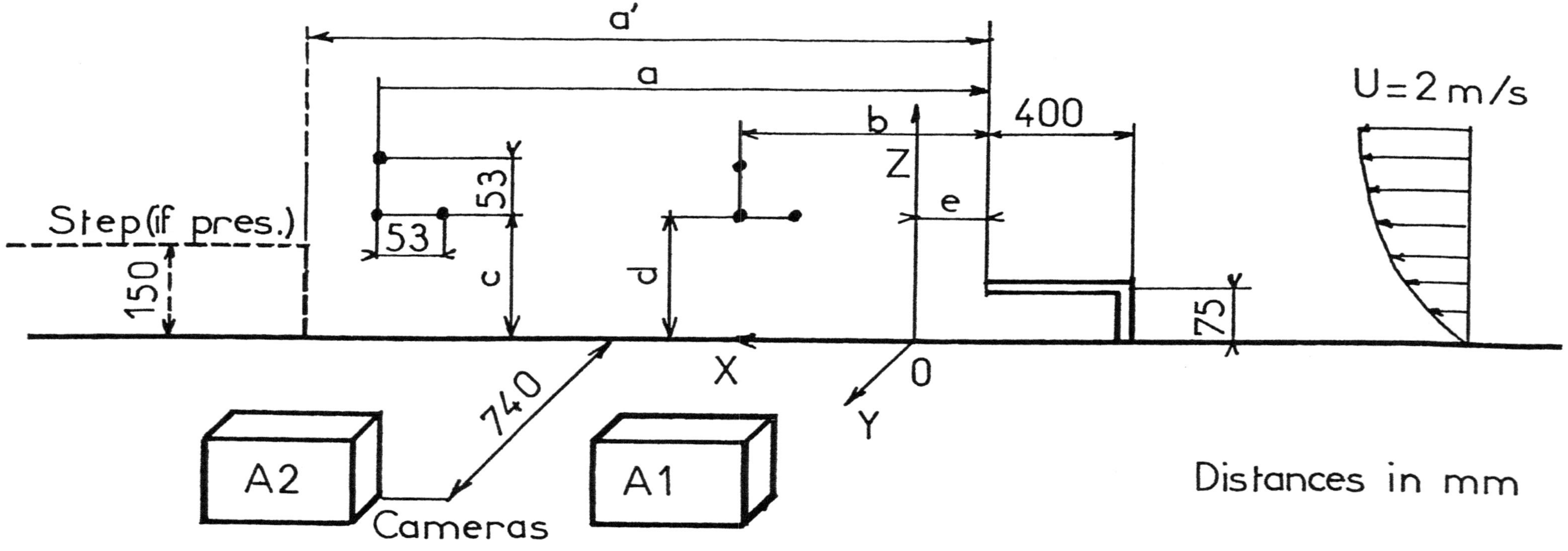

EXPERIMENT	a, a'	b	c	d	e	Delay 1(s)	Delay 2 (s)	Release (s)
Neutral	520	370	108	158	400	0.1	0.22	0.03
Unstable	520	370	108	158	400	0.1	0.22	0.03
Stable	565	395	158	150	-1260	0.1	0.29	0.03
Inversion	565	400	158	155	40	0.1	0.27	0.03
Step	790	420	—	180	40	0.16	0.40	0.03

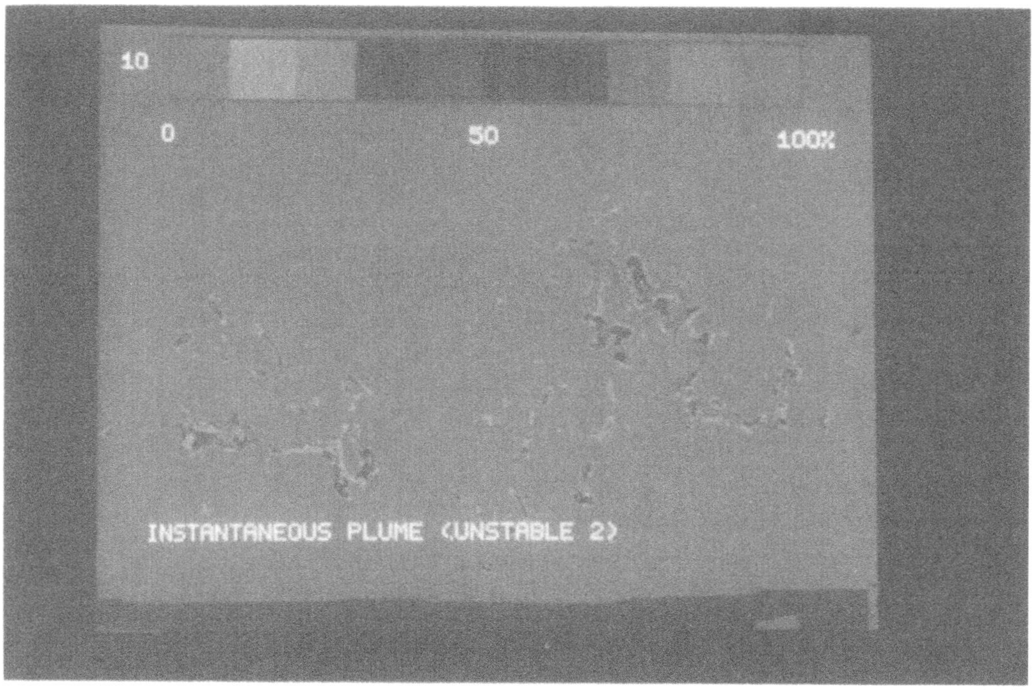

Figure 5.

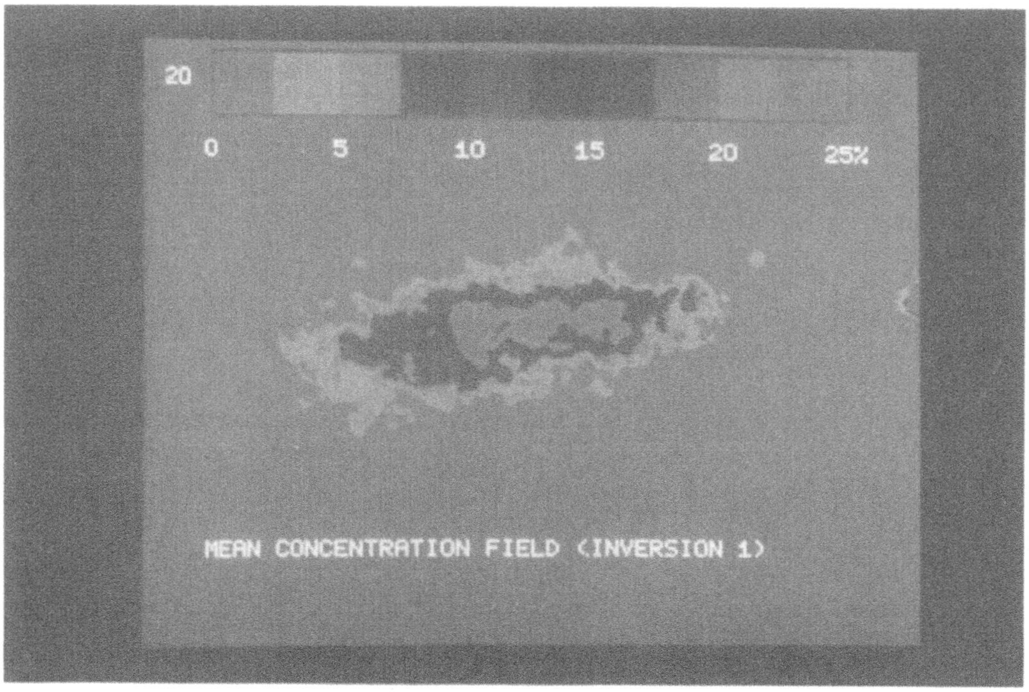

Figure 6.

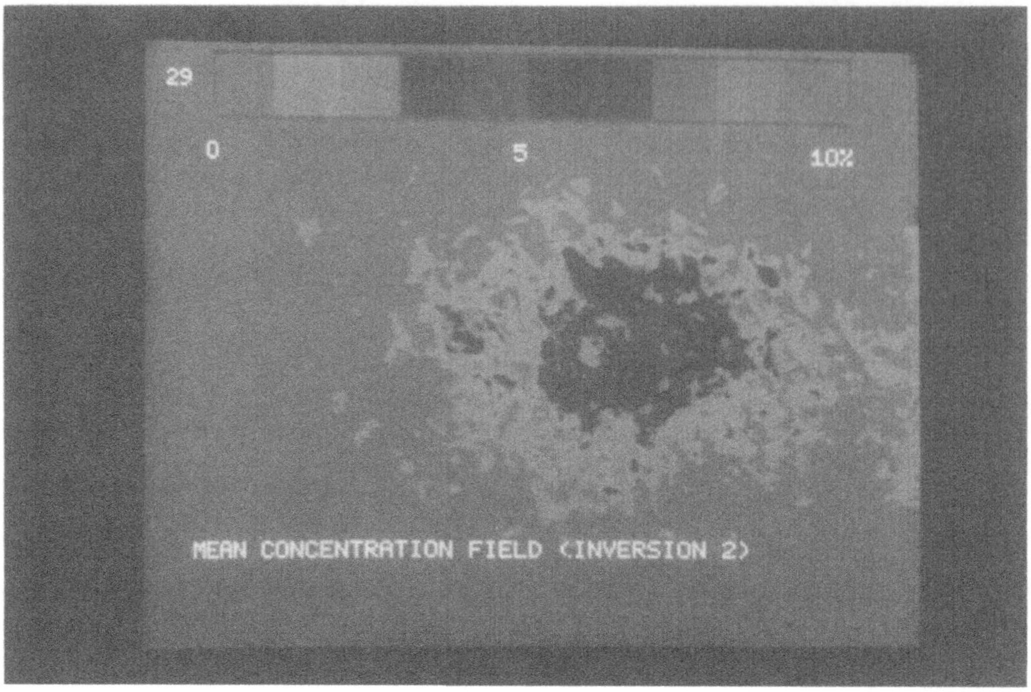

Figure 7.

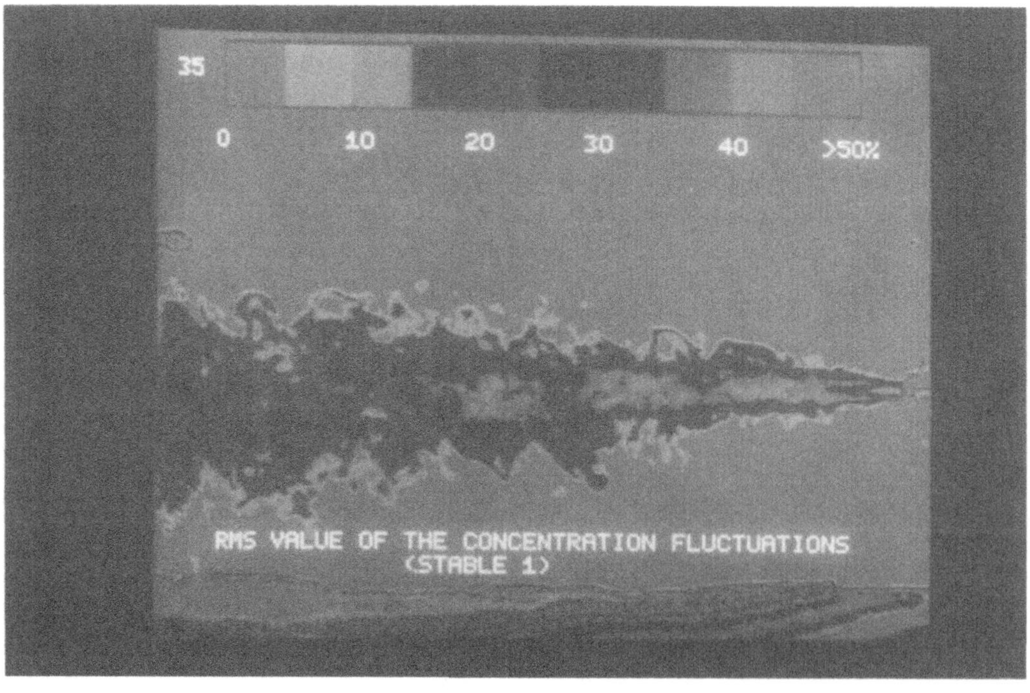

Figure 8.

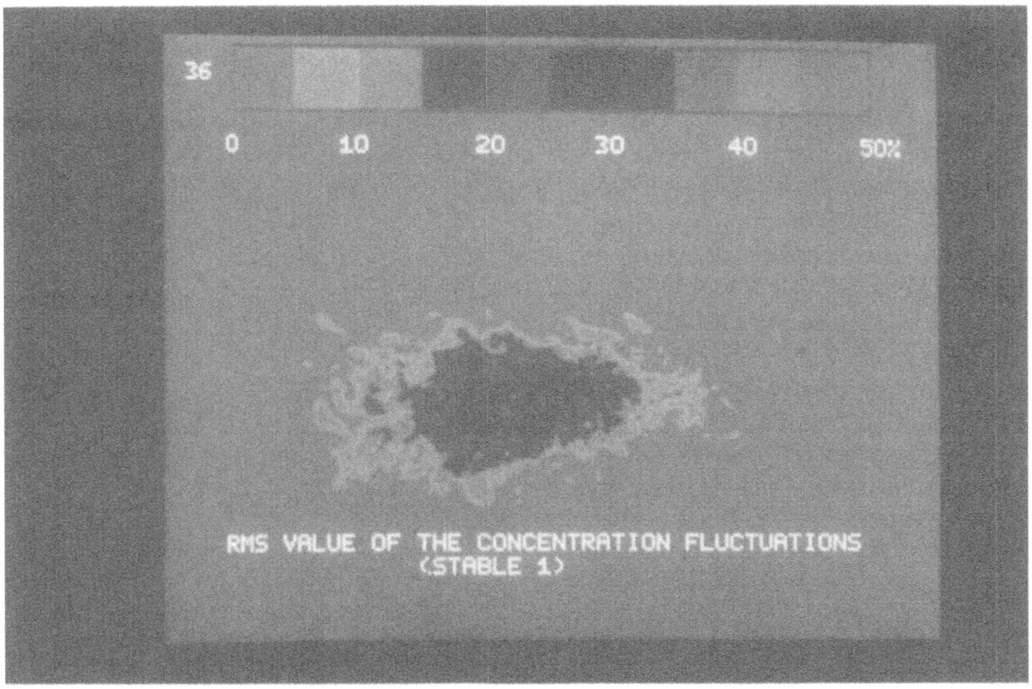

Figure 9.

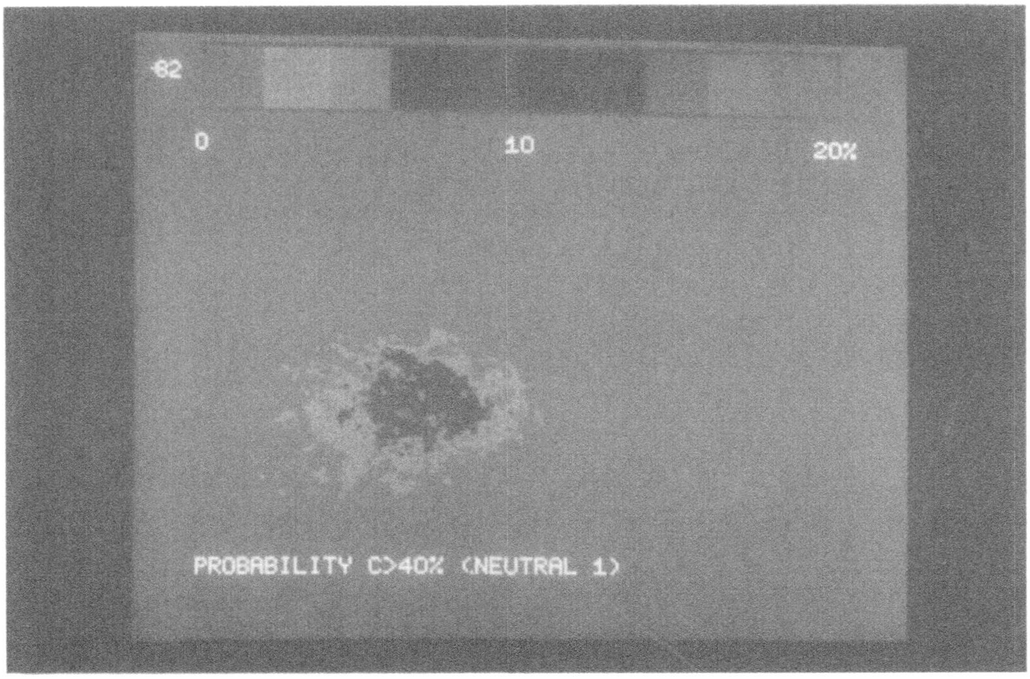

Figure 10.

Discussion

Question: J.M. Quinault - CEA/CEN-Cad

In your pictures, the turbulence looks very homogeneous, do you think that the wind tunnel represents all the variability of real turbulence in the atmosphere?

Answer: J.P. Schon

In our experiments we simulate only small scale turbulence so the effect of large scale perturbations are not taken into account (Crabol has studied this problem). The results are pessimistic compared to the atmosphere.

Question: H. ApSimon - Imp. Col./London

Your laser techniques are most impressive for flow visualisation. The ability to analyse instantaneous concentrations suggests application to concentrations relative to explosive or flammable units. Are there any difficulties in simulating for example dense gas dispersal, and retaining a representative concentration pattern from the particle distribution?

Answer: J.P. Schon - Ec. Centr./Lyon

There is only a technical problem for seeding the heavy gas. We are doing some trials to perform such experiments. The main shortcoming of the system is the parasitic reflexion near the walls which prevents making concentration measurements near the walls.

PHYSICAL MODEL OF THE DISPERSION OF A RADIOACTIVE CONTAMINANT IN THE ATMOSPHERE ABOVE A HEAT ISLAND

J.A. TOLY
Commissariat à l'Energie Atomique
Institut de Recherche technologique et de Développement Industriel
Centre d'Etudes Nucléaires de Grenoble

Summary

This project deals with the impact of surface heating in urban areas on the dispersion of contaminants in the atmosphere. The atmospheric boundary layer is simulated in a water flume. Ground heating is applied locally reproducing the heat flux of an urban region. Fission products for which internal heat source is neglected are simulated by horizontal plumes of sodium hydroxyde. The main results are the characterization of an internal boundary layer downstream of the leading edge of the heated ground, comparisons of the concentration distributions of pollutants with and without surface heating. A transposition of the results, expressed in terms of global parameters, enables information on the heat island effect due to urban regions on the dispersion of contaminants in the atmosphere to be obtained. The conclusion is that, though surface heating has a smaller impact on dispersion than roughness changes, it cannot be locally neglected.

1. INTRODUCTION

Surface heating results in generating buoyancy forces above an urban complex. These are expected to change the dispersion process and modify the concentration distribution of pollutants down to ground level. The main aim of this study is to obtain information on this "heat island" effect through physical modelling in a water flume. For this purpose, the typical size of the region to be modelled - at least 10 km - has to be one order larger than those commonly simulated in physical models describing the immediate vicinity of polluted sources - up to 1 or 2 km.

The atmospheric flow is modelled by water circulation in a rectangular flume. So, simulation of neutral atmospheric conditions only requires an isothermal flow. The geometry of the flume implies a 2 D approach. Furthermore the simulation does not account for Coriolis effects. This limits the extension of the simulated region to less than 10 km, which is about the size of a city and its immediate surroundings. The history of the pollutant before it reaches the city is out of the scope of the study. Terrain is considered as flat in topography and is simulated by a flat horizontal steel plate. However gravel is placed on the floor so as to obtain an aerodynamically rough flow. As a consequence, the surface roughness and type of ground of the region upstream from the city are imposed.

The "heat island" itself is obtained by heating the plate locally. The flux is determined to represent the transition between land and an urban region. Practically, the heating effect is combined with a mechanical effect due to the sudden change in surface roughness when the flow passes over a city. This latter effect is in fact stronger the smoother the surrounding land is (water or grass). On the contrary, when the surface roughnesses of the rural area and the city can be compared (forested region) thermal heating becomes a major factor.

More especially, as poor experimental and field information exists on the subject, it seemed to us of interest to look first at these effects separately. This consideration leads us to simulate the heated region by keeping a constant surface roughness consistant with both land and urban regions.

Concerning the pollutant itself we admit that the fission products, emitted tens of kilometers upstream, have lost their own momentum and are convected by the atmospheric boundary layer flow. Their dispersion depends essentially on the characteristics of the surrounding atmosphere. So the fission products are looked on as passive, and simulated by injection of an acid or a basic solution. The evolution of the concentration with distance is then related to pH measurements.

2. EXPERIMENTAL EQUIPMENT

The experimental work was carried out in the water flume of the Service des Transferts Thermiques du C.E.A. in Grenoble. The closed-circuit flume has a 0.5 x 1 x 12 m test section. After an initial steel section of 9 m, the floor consists of a 1 m long copper plate which can be heated up to 20° above the ambient water temperature. 2 m are left to prevent constriction effect at the end of the flume.

The copper plate is heated by a hot secondary water circuit beneath the plate. The temperature distribution all along the plate can be maintained approximately constant (by means of six meanders) (variations below 1°C).

Two lateral vertical glass panels enable flow to be visualized and laser anemometry measurements to be carried out in the last 3 meters.

The flow rate ranges between 50 and 250 m^3/s.

The side-wall boundary layers are each about 20 cm thick, thus leaving a working width of 60 cm modelling the approaching atmospheric flow, which is a turbulent two-dimensional shear flow.

To ensure aerodynamically rough flow, calibrated gravel is dispersed all along the floor (between 8 and 15 mm).

The transposition factors are the following :
length scale = 1/5000
temperature scale = 5
velocity scale = 1/100

3. EXPERIMENTAL RESULTS AND DISCUSSION

The flow approaching the city is characterized mainly by velocity and turbulence distributions in the vertical direction.

Close to the surface (but well above the viscous layer) the profiles approach the expected logarithmic form.

$$\frac{\bar{U}}{u_*} = \frac{1}{k} \ln \frac{z - d}{z_o} + B$$

The distance d which accounts for the displacement from zero level due to the model roughness is taken as 0.9 cm.

The values of u_* and z_0 were experimentally determined and tested against the critical Reynolds number criterion.

3.1 Flow over the heat island

One might expect heating of the surface to affect the velocity field with increasing instability. ARYA's observations did not indicate this trend to be very significant. Ours confirm these results, all the more so as the simulation of an urban "heat island" only needs a slight heating of the plate (a typical estimation of Ri_δ made with a 1°C temperature difference across the model boundary layer is - 0.07 which classifies the experimental runs in stability group V). In contrast to the mean velocity, fluctuation intensities increase all along the heated plate.

3.2 Temperature profiles

We can define an internal boundary layer I.B.L. where fluctuations over 0.2°C in r.m.s. values are still existing. The nature of these fluctuations is characterized by temporary hot bursts rising from ground surface.

3.3 Comparison between the hot and the neutral cases

"Heat island effect".

At ground level the two situations are alternatively dominant. Along the first 150 mm the heat island participates to an increase of the pollution level. Afterwards, the neutral case is more unfavourable as far as x : 250 mm, then a new hot dominated situation reoccurs.

At z/δ = 0.18, the situation is reversed. Those general factors are even more accentuated when we consider the r.m.s. fluctuations.

These results illustrate the complexity of the evolution of the temperature field in the presence of the "heat" effect.

They prove the difficulty encountered to obtain good quantitative pH measurements and though we could not obtain the accuracy we hoped all these results enlighten in a relatively good way the mechanism involved when the destabilizing effect due to the heat momentum remains relatively slight, as in the case of a classical "heat island".

The general conclusion would be that if the "heat island" effect is not normally very important, it cannot be neglected.

4. CONCLUSION

This experimental campaign has confirmed the ability to simulate the "heat island" effect due to urban excess heating on the dilution of a passive contaminant in a water flume.

Two important limitations, principally due to the reduction scale of 1 : 5000 must however be noted :

- The difficulty of taking roughness changes properly into account as the necessity of obtaining an aerodynamically rough flow was incompatible with a detailed representation of the real surface (canopy layer...).
- The rather limited range of model velocities satisfying both the Reynolds and Froude criteria.

The main results concern :

a - The characterization of an internal boundary layer.

b - The influence of heating on the dispersion of a single, continous point source of pollutant.

REFERENCES

1. PASQUILL, F. (1974). Atmospheric diffusion. Ellis Horwood, 2d edition.
2. SNYDER, W.H. (1972). Similarity criteria for the application of fluid models to the study for air pollution meteorology. Boundary layer Meteorology 3, 115-134.
3. SCHLICHTING, H. (1968). Boundary layer theory. 6th edition, MC Graw Hill.
4. CHAUDHRY, F.H. and CERMAK, J.E. (1971). Wind tunnel modelling of flow and diffusion over an urban complex. Contract N° DAAB07-68-C-0423, Colorado State University. Fort Collins, Colorado 80 521.
5. CRABOL (1978). Contribution à l'étude de la simulation en laboratoire de transferts de masse en atmosphere neutre. Thèse Paris VI.
6. ARYA, S.P.S. (1975). Buoyancy effects in a horizontal flat-plate boundary layer. J. Fluid Mech. 68, 321-343.
7. SCHÖN, DEBBARH, MEJEAN, PERROUS and GRANIER. Atmospheric diffusion of puffs. Simulation in a wind tunnel. Project 9C.

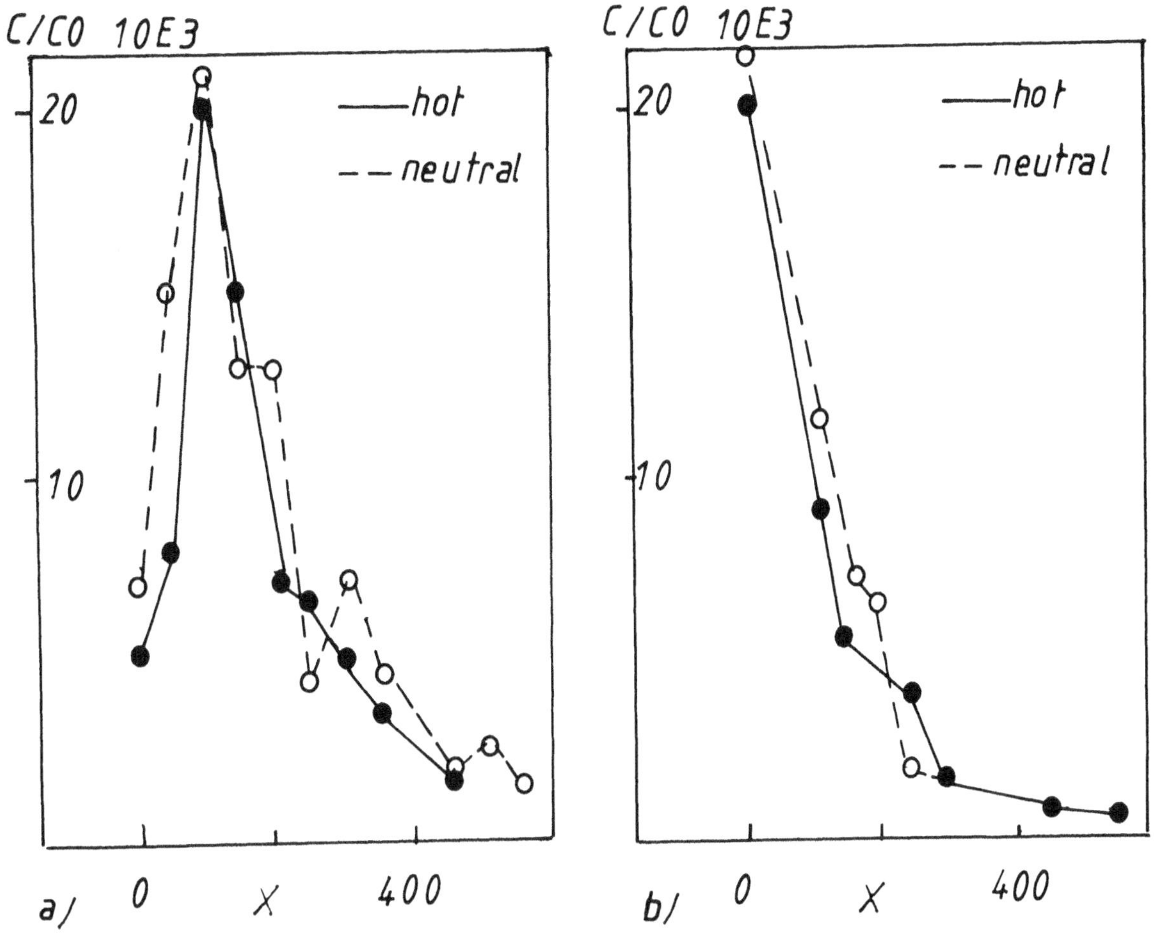

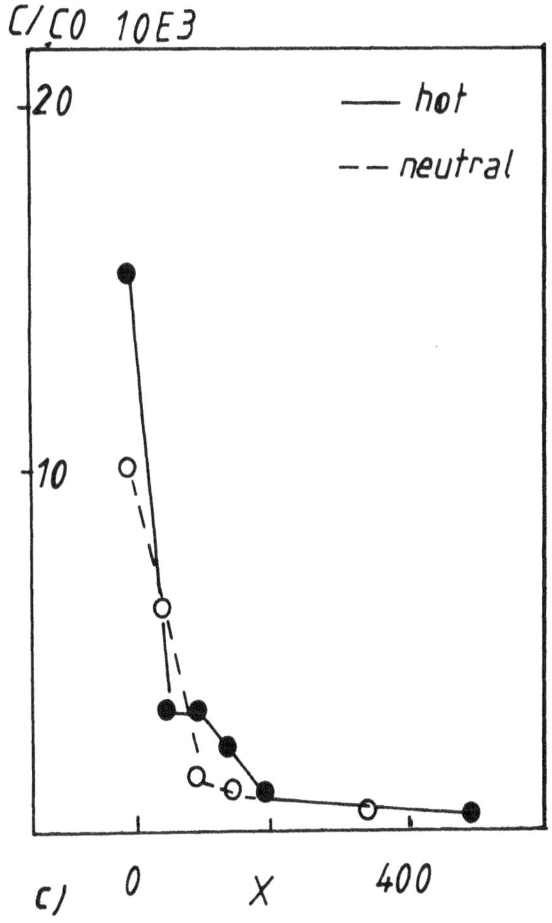

DILUTION ALONG X AXIS

a - z = 15 mm

b z = 25 mm

c z = 45 mm

injection

z_i = 45 mm

x_i = - 150 mm

Discussion

Question: J.A. Hertig - EPF/Lausanne

What is represented by the first transparency sheet that you have presented?

As you have two lengths as axis, what is the meaning of the blue lines?

Answer: J.A. Toly, CEA/CEN

The blue lines simulate an artificial boundary of the polluted cloud. The chosen criterion is a partition of the space in regions where the concentration levels are above or below 10 ‰ of the source level.

Question: H. ApSimon - Imp. Col./London

Just a comment. Your conclusions on the relative importance of a heat island and changes in surface roughness apply to 2-dimensional flow, and would be different in a fully 3-dimensional situation.

Answer: J.A. Toley, CEA/CEN-G

In the case which has been simulated, Coriolis effect was not considered. Curvations in the transverse direction were neglected, which limits the length to less than 10 km in the horizontal.
A simulation of a 3-dimensional situation would certainly give different results. But it would be of interest to simulate the Coriolis effect too, to give a real improvement in this field.

WIND TUNNEL SIMULATION OF ATMOSPHERIC DISPERSION IN STABLE CONDITIONS AT A REAL SITE

J.-A. Hertig, P. Liska
Ecole Polytechnique Fédérale Lausanne

F. Ladhari
Ecole Centrale de Lyon

M. Ayrault, J.P. Schon
Université St-Etienne

SUMMARY

The flow visualisation technique developed at Ecole Centrale de Lyon (contract no. 017-SRF) has been used for the study of a plume dispersion on a site simulating the experiment field work by CRABOL (contract no. 010-SRF).
Very stable situations could be reached by cooling the surface of the model with liquid nitrogen. Quasi-laminar flows were obtained with these conditions, while the neutral and slightly stable cases performed fully turbulent flows. Results on the velocity and concentration fields are given and compared with the field data.

INTRODUCTION

In the present study the physical modelling is used in order to simulate transport and dispersion of a plume and to compare the plume behaviour in neutral and very stable condition on a real site.

The technique of flow simulation developed by the Institute of Energy Economics and Power Engineering (IENER), (Hertig and Liska, 1983) is able to reproduce the three dimensional flow pattern in stable condition on a topographic model. These flows are gravity driven through temperature changes on the regional topography. For the present study this technique is associated with a turbulent flow created in a classical boundary layer wind tunnel.

The measurement of the concentrations which is another difficulty, is done by the technique developed at "Ecole Centrale de Lyon" by Schon et al 1984. This technique allows to measure the mean, rms, and some high order moments of the concentration in a section of the flow by means of plan visualisation associated to digital image processing.

The combination of these two techniques is very promising; it will offer the possibility of simulating and measuring dispersion and transport of pollutant in complex terrain with very complicated flows. Some aspects of flow simulation and plume behaviour are presented in the present paper. The image processing results will be described in a future communication.

SIMILARITY

As exact similarity conditions based on dimensional analysis are not applicable, we are obliged to use a approximate similarity. It has been shown by several authors, Cermak (1976), Meroney (1976), Snyder (1981), Schon (1974), Mery (1969), Rey (1981) that acceptable similarity is possible for the atmospheric boundary layer phenomena if the Reynolds number is high enough and if the stratification is not too strong. Moreover, the effect of Coriolis acceleration can be relaxed if the distance is less than 5 to 10 km. In this case the similarity is related to a turbulent Reynolds number instead of the Reynolds number; some criteria such as Rossby, Froude, Euler, Eckert numbers are neglected and some others such as Prandtl, Schmidt numbers are respected for their turbulent signification.

This type of similarity is a slight extension of the turbulent boundary layer concept of the atmosphere.
More recently Hertig (1983) has treated the problem of geometrical and temperature scale distorsion in order to authorize the reduction of horizontal scale in such a way that the edges of the basin can be included in the model.

It is obvious that this sort of simulation (turbulent to laminar) is not correct for the plume dispersion simulation at short distance, but it may be a good approximation of the simulation of transport of pollutant.

In order to check this affirmation and the limits of the simulation at low speed, three different similarities will be used in this experiment:

1. Turbulent boundary layer similarity of the atmospheric boundary layer in neutral atmosphere.
2. Extreme similarity case of drainage flow simulation with strong stratification and laminar flow ($Re_{t/R} = Re/_M$).
3. Combination of the turbulent boundary layer and strong stratification, laminar and turbulent flow in the vicinity of the surface.

DESCRIPTION OF THE EXPERIMENT

The experiment is based on a 1/1000 scale model of a site near Cadarache (France) in which some field trial was done (Crabol 1983). The model is made of a thin aluminium sheet (0,1 mm thick) formed on a existing topographical relief. The aluminium sheet is fixed on frames made of four (15 mm thick) wood plates defining topographical sections. Each frame represents a model surface of 1.25 x 0.60 m^2. Full model dimensions are 5.00 x 4,00 m^2.

The model is disposed on 10 insulating boxes containing an aluminium basin and three fans with vertical axis. During the experiment the basins have been filled with liquid nitrogen (liquid air) which is then evaporated and the very cold nitrogen gas is blown by the fan against the lower face of the model aluminium shee. A strong cooling of the model surface results. The intensity of the cooling rate is regulated by means of rotational fan speed variation.

The model and the cooling system are installed in the diffusion test section of the IENER boundary layer wind tunnel, as shown in fig. 1.
The velocity measurement in neutral flow is done by means of a conventional hot film anemometer.
In stratified flow an original low speed anemometer developed at IENER has been used. It is based on a two thermistor probe.

For the concentration measurement an aspiration probe of TSI model 1440-6 has been used.

The measurement of concentration is also made by the visualisation technique developed at "Ecole Centrale de Lyon" and presented by Schon in an other paper.
Generation of particles consists of a incense generator, an accumulator and distributing pipes (Balint (1982)).

The particles are illuminated by a laser beam, the light diffused by the particles is a function of the size of the particles which permits the concentration of these particles to be measured.

The light diffused by the particles is recorded on a Kodak TRI X film processed at 1600 ASA.

After correction of image distorsion, calibration of the digitized images are added for mean and higher order moments of concentration field.

EXPERIMENTAL RESULTS

Flow description

Figure 3 shows the mean velocity profile with an exponent of the power law of about 0.17 and a thickness of about 240 mm for the lower layer.

The profile of turbulence intensity (figure 3) shows a value of 20% extrapolated at ground level, and a value of less than 2% in altitude. This behaviour agrees well with what is expected for a topographical region similar to the Cadarache site, with neutral flow.

The spectrum of turbulence which presents a macro scale of turbulence of 100 mm. This corresponds to 100 m for the most energetics vortexes, and is well in agreement with several field experiments.

To study the behaviour of the flow on the topography, the cooling of the model surface corresponds to three conditions of the wind tunnel velocity flow. These are:

- high speed in altitude (1.2 m/s)
- medium speed in altitude (0.70 m/s)
- pure drainage flow without wind in altitude.

On figure 4 we have the velocity, temperature and direction profiles taken at the point of release. We observe a maximum velocity in the drainage flow of 0.30 m/s which corresponds to about 1 m/s (velocity scale $1/\sqrt{10}$).
The speed of the wind in altitude is nearly 0. On a second profile, taken 0.60 m downstream, we observe a speed increase of about 10% (0.33m/s) and a slight diminution of the stratification.

For evaluating the wind influence in altitude on the drainage we have represented on figure 5a the variation of the thickness of the thermal layer δ_T as a function of the distance along the model. δ_T is defined as the height above ground at which the temperature is 95% of the temperature difference between surface and flow in altitude. Figure 5a shows the case of drainage flow in which the layer is quasi constant. Figure 5b presents the case of a 0.10 to 0.15 m/s wind at altitude. The thickness is also nearly constant.
The intermediate case with a speed of about 0.70 m/s (figure 5c) shows a variation between 50 mm at 1 m distance which is the thickness of highest speed, and 98 mm at 5 m distance which corresponds to the drainage flow. The case of highest speed of 1.20 m/s is shown in figure 5d. In these conditions the layer thickness varies considerably along the plate.

The shape of the longitudinal profile of the thickness plotted in non dimensional form on figure 6 reveals that the influence of the topography increases the thickness in the upper part of the valley for the drainage flow in comparison with the theoretical law in $x^{1/4}$ for the drainage flow. This is due to the lateral flow of the valley. In the case of flow in altitude of 1.20 m/s, the shape differs in the opposite direction in comparison with the drainage flow. The theoretical shape has to be in $x^{1/2}$ and we can see that the evolution gives higher thickness in the low valley than at the beginning.

Results of the concentration measurement with the aspirating probe

The location of the four profiles corresponds to those taken in the field at distance of about 400, 1000, 2000, 3000 mm from the source. The profiles are also located along the line which is illuminated by the laser beam. So these measurements can be compared with the field measurement as well as with the digital image processing results.

Figure 7 shows the resulting concentration profiles in the neutral case (shear flow). The profiles seem agreeing well with what is expected for the dispersion in a turbulent boundary layer.
On the enlargement of the profile at 3000 mm we can see that the concentration presents a slight increase in the highest profile part at about 200 - 400 mm. This may be due to the change of turbulence structure on the internal boundary layer upper limit.

On figure 8 the case of stratified flows is represented at scale 1/2 of these shown on figure 7. Thus a big increase appears in concentration field. It also shows that the thickness of the highest concentration layer is smaller than in the case of neutral flow.

A comparison between these two cases is presented on figure 9. The concentration plotted as a function of the distance along the plane is also compared with the fixed experiment. We have taken the "Pissenlit 11" field trial reported by Crabol (1983).

For the feasability test of the image measurement we have retained two extreme cases which are turbulent neutral flow and quasi laminar stratified flow (fig. 2).

CONCLUSIONS

The part of the study presented here shows that the technique of a strong cooling associated with a turbulent approach flow is well applicable to the simulation of plume transport and dispersion.
A relatively flat terrain such as the small Cadarache valley has an important influence on the flow pattern. This topographical influence is not restricted to the creation of a generalised slope wind but associated with the creation of a valley wind caused by the convergence of the flows. This influence is noticeable and remains under stable conditions even with a strong wind in altitude. This valley wind has also an remarkable influence on the concentration. An increase of about 100 % of the concentration in neutral case was observed for stratified flow.
This increase is similar to this observed for the field trial.
Some preliminary results show that the measurement of concentration by image processing is applicable to the case of stratified flow and that the presence of water dropplets in the flow do not affect too much the results.

REFERENCES

Balint J.L. 1982: Application d'une méthode de visualisation laser et de traitement d'images à l'étude de la dispersion dans les écoulements turbulents. Thèse Lyon.

Cermak J.E. 1975: Application of fluid mechanics to wind engineering. A Freeman Scholar Lecture. Journal of Fluid Engineering, vol. 97.

Crabol B. 1983: Assessment of the dispersion of fission products in the atmosphere following a reactor accident, under meteorological conditions of low wind speeds with or without high temporal and spatial variability in wind speed direction. Progress report no. 4-5, Control CEEA SRF/010/F, January-July.

Hertig J.-A., P. Liska 1983: Simulations of regional atmospheric flows on small scale topographical models. Journal of Wind Engineering and Industrial Aerodynamics 15 (1983) 77 - 89.

Meroney R.N., J.E. Cermak 1967: Wind modelling of flows and diffusion over San Nicolas Island, California. Technical note CSU CER-66-67 RNM-JEC-44.

Mery P. 1969: Reproduction en similitude de la couche limite atmospherique. La Houille Blanche no. 4.

Rey C. 1981: Critère de simulations des échelles de la turbulence atmospheric en soufflerie, colloque "Construire avec le vent", Nantes vol. 2, article VII-3.

Schon J.P. 1977: Contribution à l'étude des couches limites stratifiées en température. Thèse Lyon.

Schon J.P., and Al 1984: Atmospheric diffusion of puffs simulation in a wind tunnel. Technical report, project 9C/contract SR 017 F(G), CEE-EDE-ECL.

Snyder W.H. 1981: Guideline for fluid modeling of atmospheric diffusion. EPA-600/8-81-009.

Fig. 1: 1/1000 scale model in the IENER wind tunnel

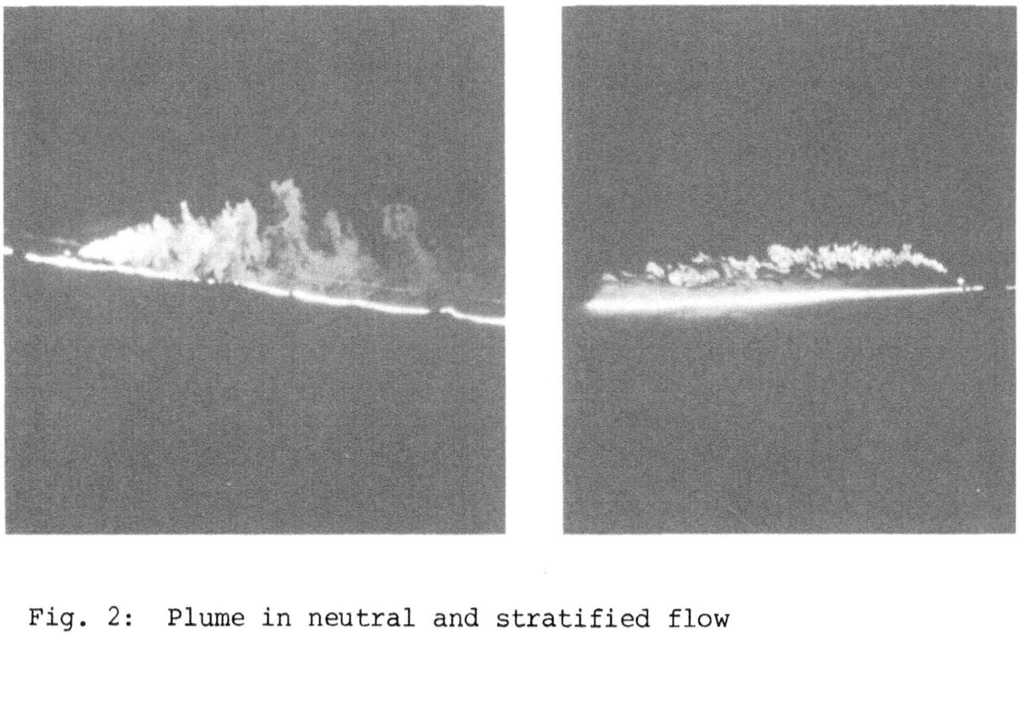

Fig. 2: Plume in neutral and stratified flow

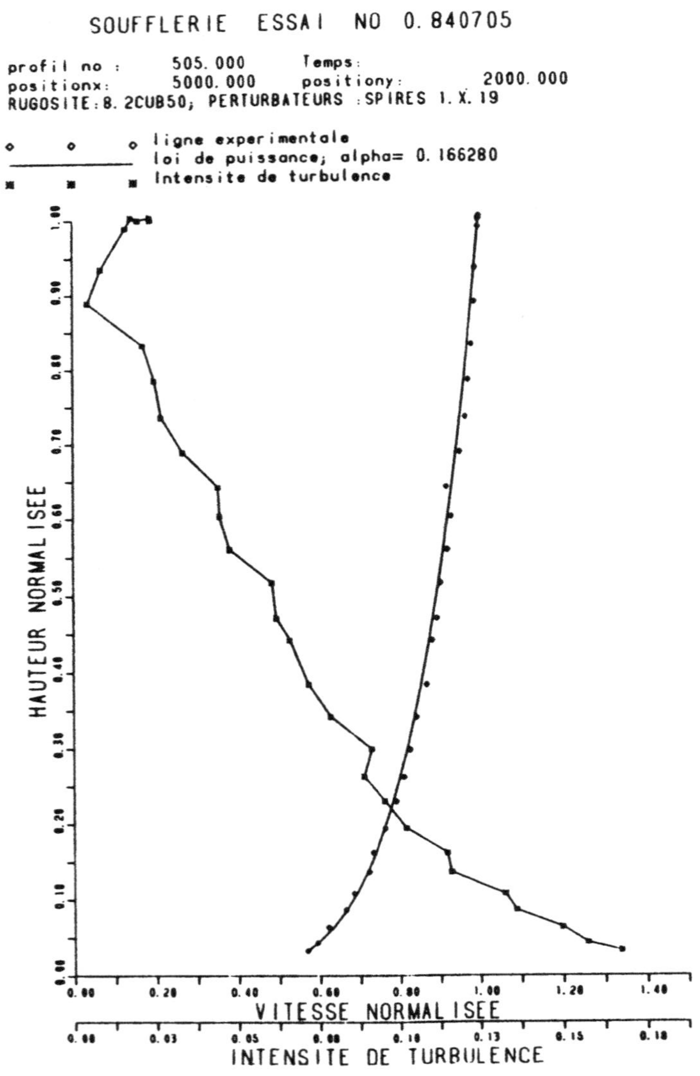

Fig. 3: Turbulent boundary layer flow in neutral condition

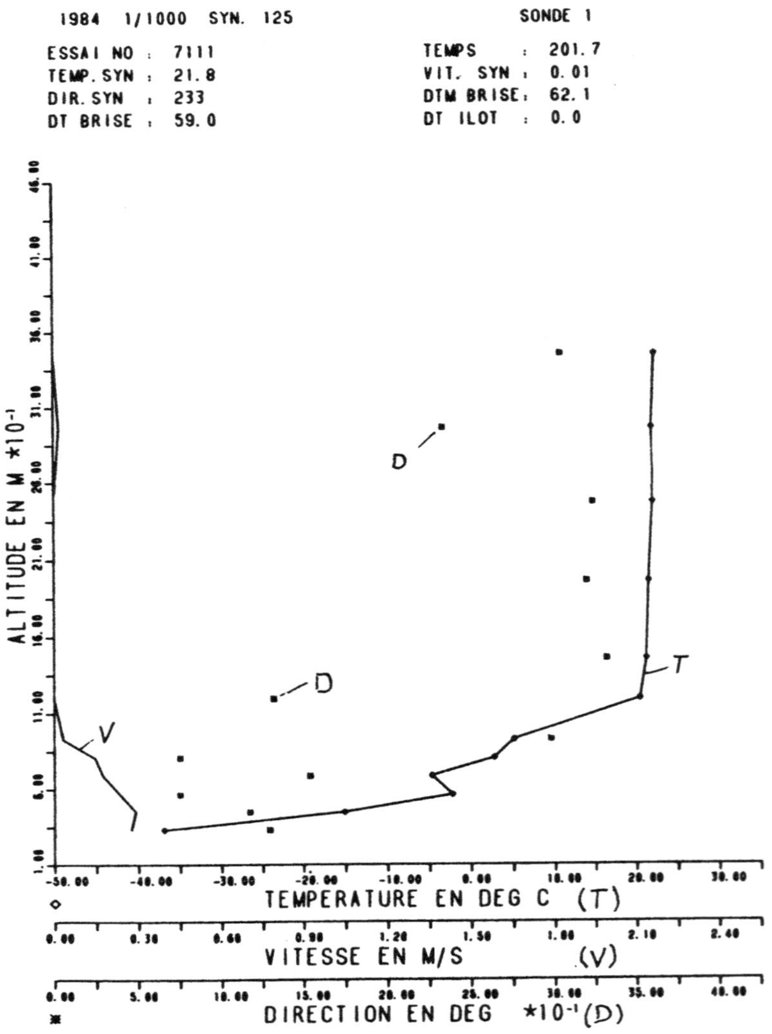

Fig. 4: Drainage flow profile

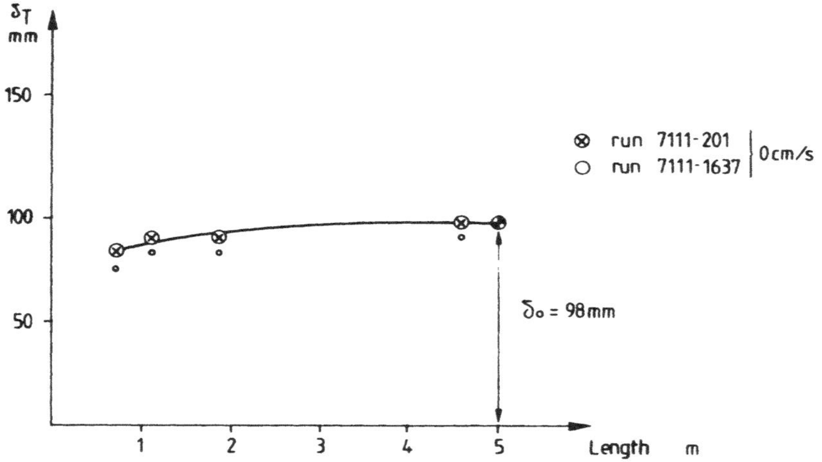

5a: Pure drainage

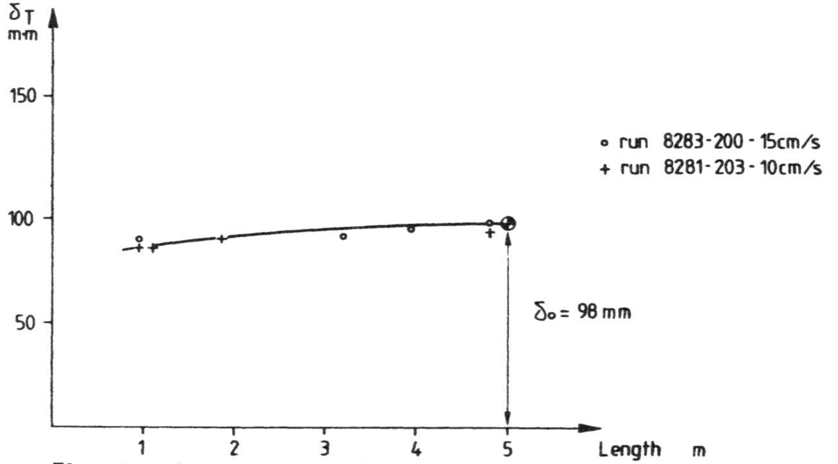

5b: Drainage associated with wind in altitude of 10 - 15 cm/s

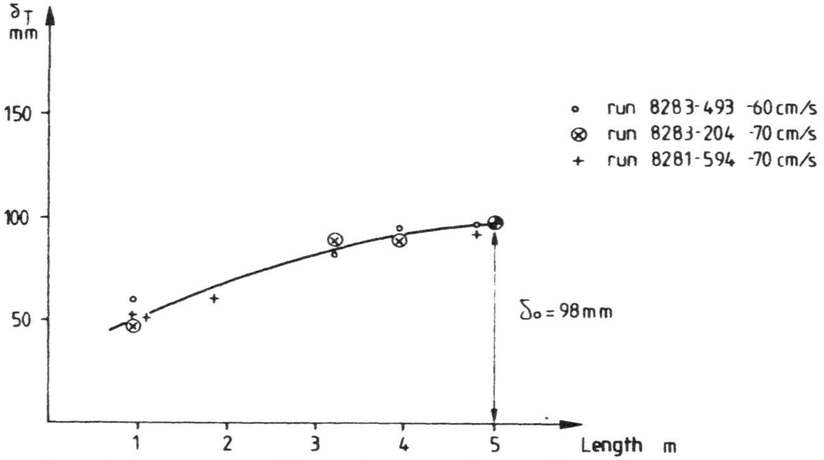

5c: Wind in altitude of 60 - 70 cm/s

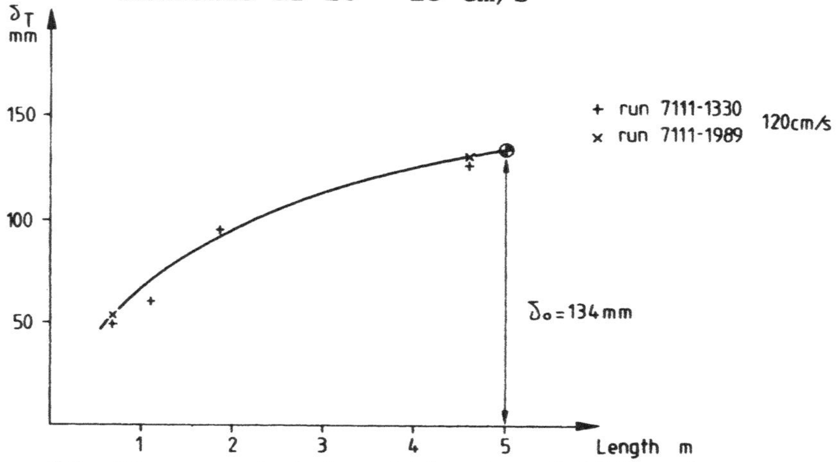

5d: Strong wind in altitude of 120 cm/s

Fig. 5 Thickness of the thermal layer δ_T as a function of the distance along the model L. and of wind speed at altitude.

22.10.84 Hq/Wlo

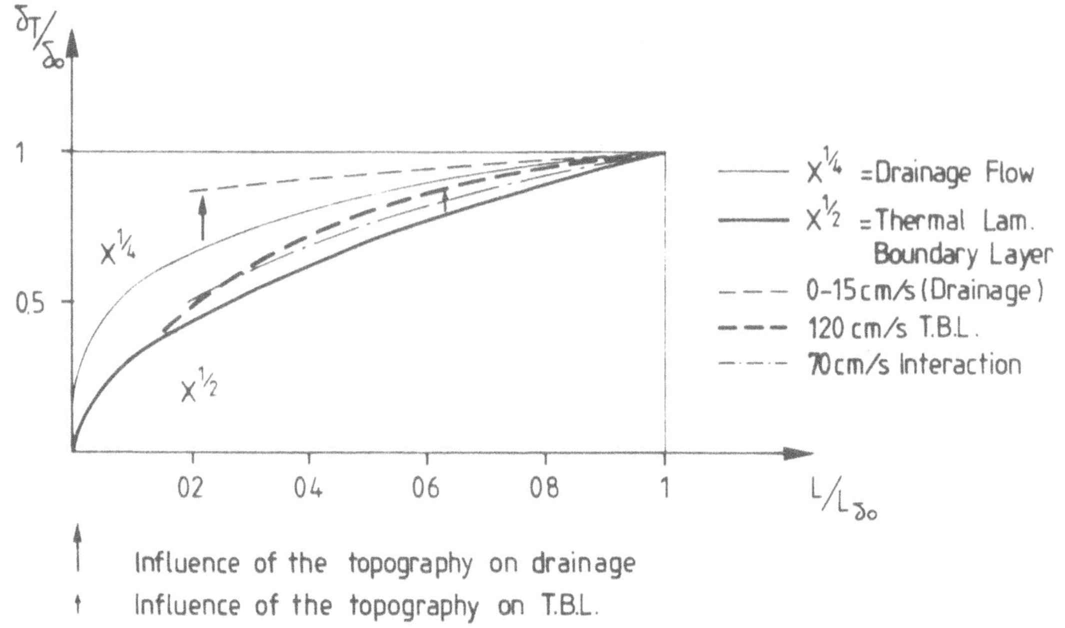

Fig. 6: Evolution of the non dimensional thickness of thermal layer

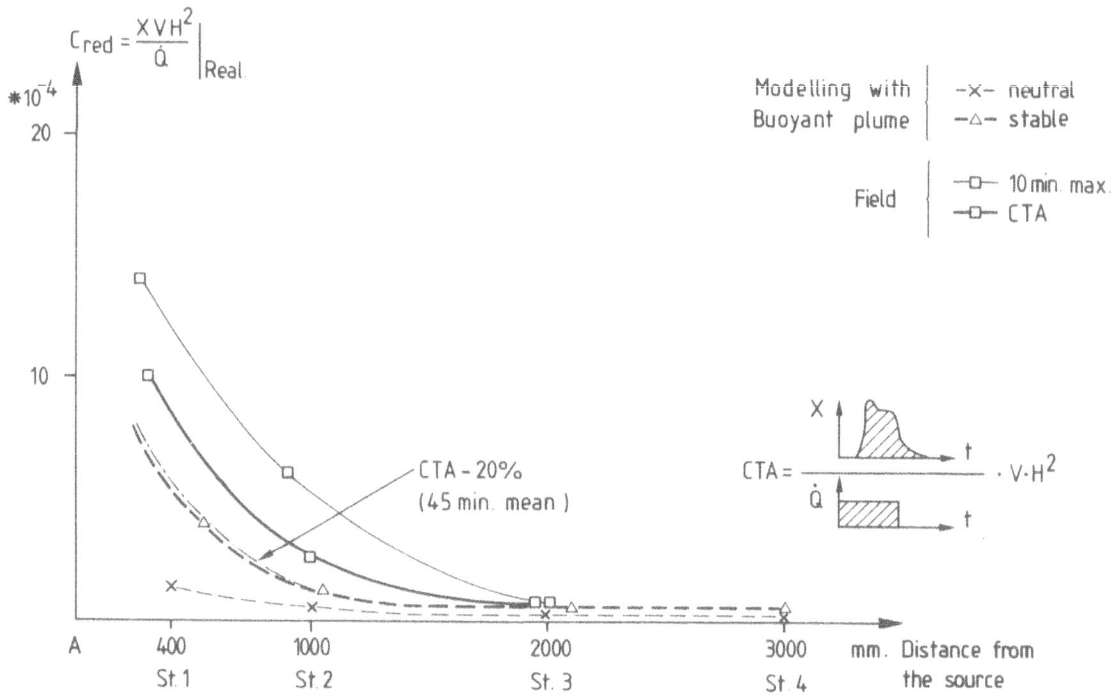

Fig. 9: Comparison with field measurement

IENER N 522101

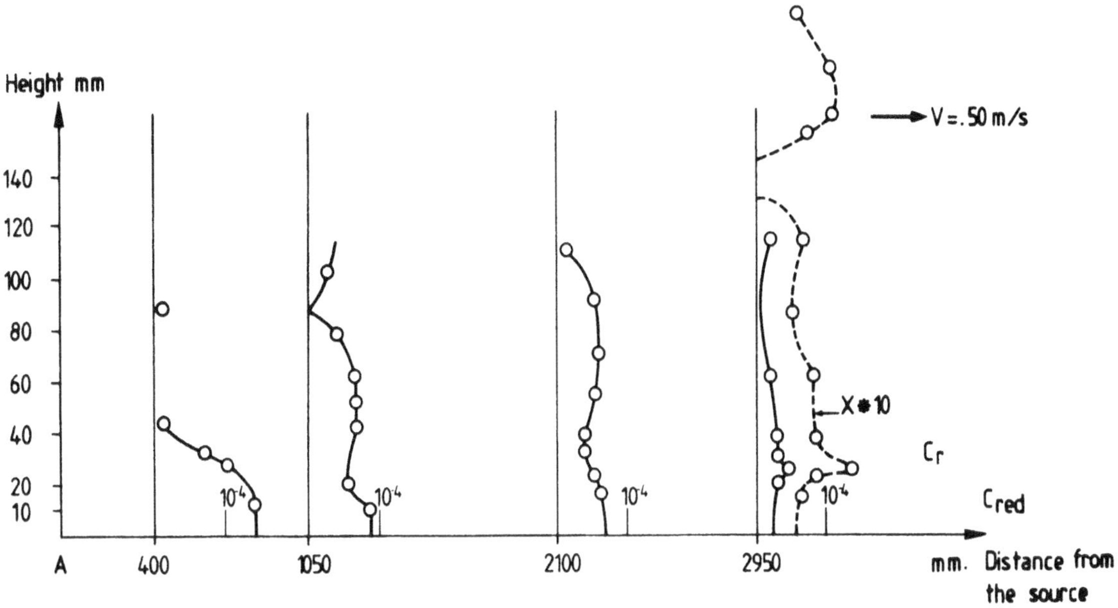

Fig. 7 Concentration profile in the shear flow in neutral atmosphere $\frac{\bar{X}}{Q}$

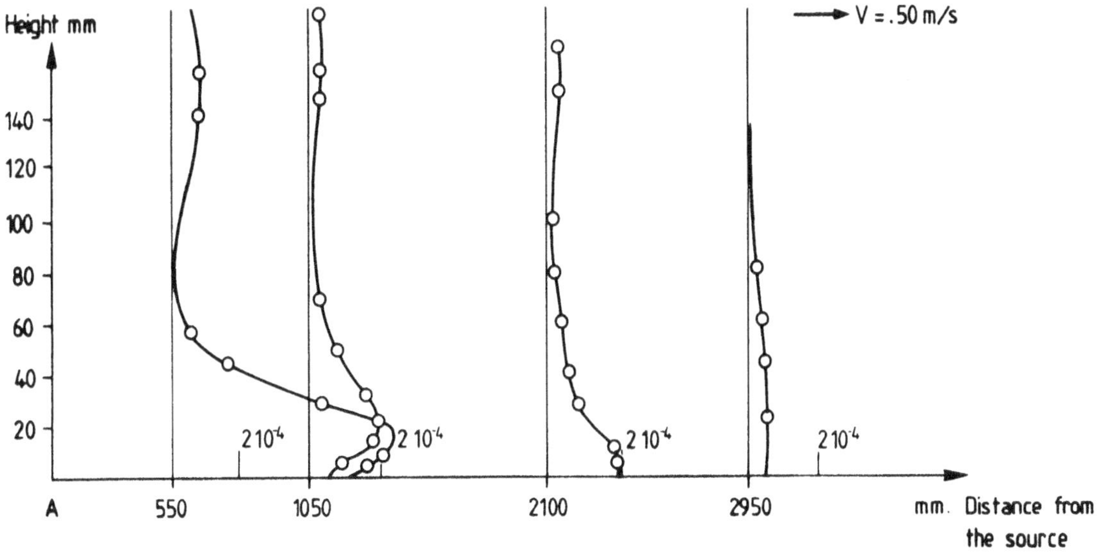

Fig. 8 Concentration profile in the stratified flow $\frac{\bar{X}}{Q}$

IENER N 522102

Discussion

Question: H. ApSimon - Imp.Col./London

Would you like to comment on the importance of humidity for flow characteristics and dispersion in low wind conditions?

Answer: J.A. Hertig, E.P.F./Laussane

In fact, you make mention of another project in which we are involved with the study and simulation of air quality in a town situated on the site of a Swiss lake (Biel). In this study we have observed a drastic influence of humidity on thermal structure and heat radiation. We have found that the presence of moisture creates the possibility of a super adiabatic layer of about 7°C just above the lake because of heat in air by evaporation and by radiation of the moist layer in direction of the sky during night.

Question: B. Crabol, CEA/CEN-Far

You mentioned that, in stratified conditions, you observe a meandering of the plume on the small-scale model while it is not observed in the field experiments. How do you explain it?

Answer: J.A. Hertig - E.P.F./Lausanne

It is a question of time scale. In the field it is true that you observe in neutral flow a time variation of the direction of the velocity due to the high level of turbulence. In stratified flow these variations decrease but the trajectories still varies in the valleys due to the well known variation in feeding of lateral drainage with a period of about one hour.

Comment: J.M. Quinault, CEA/CEN

The experiments of VOGT at Julich confirm your observations concerning very small variation of the plume direction in case of stable weather.

ASSESSMENT OF THE DISPERSION OF FISSION PRODUCTS IN THE ATMOSPHERE FOLLOWING A REACTOR ACCIDENT UNDER METEOROLOGICAL CONDITIONS OF LOW WIND SPEED

B. CRABOL
CEA/IPSN/DAS
FONTENAY-AUX-ROSES/FRANCE

Summary

The aim of the study is the assessment of the dispersion in a low wind speed situation and the validation of the computer code ICAIR3 by means of SF6 tracing experiments carried out on the CADARACHE site under different stability conditions. The results show clearly some characteristic features of the dispersion. In particular, high concentrations are found in the experimental field several hours after the end of the release. Large differences of the plume width are observed depending on the atmospheric stability. The flow seems well organized under stable conditions, probably in relation with a topographic effect (CADARACHE is situated in a valley), while there is a much larger spread out of the plume in neutral or unstable conditions. A reasonable agreement with the values predicted by the calculation code is found for the maximum concentration.

1. INTRODUCTION

The purpose of the contract is the study of atmospheric dispersion in a low wind speed situation and the validation of the ICAIR3 computer code by means of the "PISSENLIT" experiments carried out on the CADARACHE nuclear research center under different stability conditions.

In different aspects, low wind speed situations need further studies in order to assess in a realistic way the atmospheric dispersion. Only few experiments have been carried out in this type of situation and there is a need in this domain. The numerical modelling of this situation is difficult notably because of the heterogeneity and non-stationarity of the wind field as often observed.

The contract, including both theoretical and experimental work, has been divided into three phases :

- acquisition of real meteorological sequences with the aid of 4 stations located on the CADARACHE site.
- Sensitivity study of the ICAIR 3 code to the acquisition parameters (sampling time of the wind speed and direction, number of stations taken into account in the calculation).
- Experimental validation of the code . This phase consists in the realization of 20 tracer experiments on the CADARACHE site under different stability conditions and the interpretation of the results with the computer code ICAIR 3.

We present hereunder the calculation hypothesis of the code and the main results obtained in the different phases.

2. PRESENTATION OF THE MODEL ICAIR 3

The method of calculation has been the subject of previous publications (1) . It notably includes a Gaussian solution to the diffusion equation . The basic principle is to schematize a real release by a sequence of instantaneous release of separate puffs. The assessment of consequences is made by adding, at the observation point, the contributions of all the puffs.

The trajectory of each puff is calculated by using the observed variations of the wind field in space and time, supposing that the puff center follows the wind speed and direction measured, at the instant of the calculation, on the nearest mast. The standard deviations of the diffusion of each puff are function of the travel time and are presented on figure 1.

3. RESULTS OF THE SENSITIVITY STUDY

The sensitivity study of the model has been carried out by means of real meteorological sequences collected on the CADARACHE site during a 6 month-period. Each sequence includes data of wind velocity and direction simultaneously obtained at the 4 measuring stations settled on the site (figure 2) . The minimum sampling time of the data was 2 min. The height of the measuring point was 10 m or 30 m above the ground-level. The atmospheric stability is determined by the value of the vertical temperature gradient measured on the mast n°3.

The sensitivity study has been carried out by varying:

1) the number of measuring stations that are taken into account (from 1 to 4),
2) the sampling time of the meteorological data (from 2mn to 1h).

3.1 Sensitivity to the number of measuring stations

The full results are presented in (3).

The following conclusions may be drawn : it is noticed important differences in the results according to whether the station n°3 (cf. figure 2) intervenes or not in the calculations. The reason seems to be a clear decrease of the wind velocity observed at this station compared to the first 2 stations. It entails a significant slowing-down of the puffs, and an increase of the concentrations.

Under the conditions of calculation considered here, the 4 measuring stations are not of the same importance for the assessment of transfers. It is noted that there is no difference between the results obtained with the data from the station n°1 or from the 2 first stations. On the other hand, the 4st station does not bring generally much change. It follows that there are mainly two wind regimes on the site characterized by the measurements at stations n°2 and 3 for example.

3.2 Sensitivity to the sampling time of the meteorological data

The calculations have been carried out by using the meteorological data sampled over a time varying between 2 min and 60 min. This study brings out a very clear difference in the turbulent structure of the atmosphere under stable and neutral conditions with low winds on the CADARACHE site.

Under stable conditions, little influence of the value of the sampling time is observed .The flow is channelled by the valley and the variations of the meteorological data are small and take place at rather low frequencies. It is showed that under these conditions a short sampling time is unnecessary as it provides little significant information for a better assessment of the dispersion.

Under neutral conditions, it is observed on the contrary an important sensitivity of the results to the value of the sampling time. The horizontal flow seems much less organized than in the previous case and contains higher frequencies of significant energy. A short sampling time is necessary to take into account in a determinist way in the calculation these variations.

4. EXPERIMENTAL VALIDATION

Up to now, 13 tracer experiments have been carried out (7 in stable conditions, 6 in neutral conditions). The experimental conditions are reported table I. We present hereunder the experimental procedure, the main results and their interpretation illustrated on two selected experiments.

4.1 Experimental procedure

4.1.1 Gas release, samples and analysis

About 40 kg of sulphur hexafluoride (SF_6) are released punctually into the atmosphere at a constant rate during one hour. Gaseous samples are taken by pumping during a determined period of time (typically 10 or 20 min) by means of an autonomous battery-powered apparatus, which is capable of procuring automatically up to 10 consecutive samples. The analysis of the samples is made by chromatography in the gaseous phase. The threshold of measurable concentration is 10^{-7} g/m^3.

4.1.2 Meteorological instrumentation

The wind speed and direction are measured simultaneously on the masts n° 1,2,3 (figure 2) and averaged every two minutes. The measurements are carried out at height of 10 m or 30 m according to the release height : 2 m or 45 m. The atmospheric stability is determined by measuring the vertical temperature gradient between 2 and 100 m on the mast n° 3.

4.1.3 Observation points

About 32 observation points are laid out along three cross-sections A,B,C approximately located at 500, 1000, 2000 m from the release point. The sampling occurs during a maximum of six hours, the duration of each sample being 10 min during the two or three first hours and 20 min after.

4.2 Results and interpretation

4.2.1 Stable atmosphere

The results of the experiment PISSENLIT 11 (cf. table I) carried out at night under stable conditions and the interpretation with the ICAIR 3 computer code are presented here.

4.2.1.1 Presentation of the results

An example of result is presented figure 3. It shows the evolution of the concentration versus time at a point of the first cross-section.
A concentration peak is observed shortly after the beginning of the release, then a slow decrease of the concentration. Concentrations inferior to 10^{-7} g/m^3 are attained 2 hours after the end of the release. Then a second concentration peak is noticed 2 hours later. It is generally observed on all the sampling points (4).

The low decrease of the concentration is coherent with what can be expected in a low wind speed situation. The longitudinal diffusion of each puff occurs at a speed ($d\sigma_x$ /dt) which is of the same order as its travel speed (mean wind speed). Thus the puff continues to contribute at a given point a long time after its mass center passes this point.

It is obvious in this case that the modelling of the transfer with a plume model (in which the longitudinal diffusion is neglected) is not realistic (3).

4.2.1.2. Comparison with the ICAIR 3 code

Figure 4 presents an example of the comparison of the theoretical and experimental values of concentration at a given instant after the beginning of the release on the first cross-section.

Good agreement can be observed at short transfert time after the beginning of the release. In particular a very good prediction of the highest concentration in the center of the plume is to be noted.

Beyond that, it seems that the dispersion observed in the field is weaker than expected. The experimental plume is much narrower than the theoretical one ; the value of the concentration in the center of the plume is under-estimated by the model. These observations are true for the three cross-wind sections.

This difference can be explained by a reduction of the horizontal turbulence, at least in a certain range of the atmospheric spectrum, linked with a combined effect of the topography of the site and of the high stability conditions. This result confirms the conclusions of the sensitivity study (cf §. 2).

Some tests have been performed to fit the experimental and theoretical data. They essentially show that the original transverse standard deviation had to be reduced by a factor 4,5, at long transfer times (cf. figure 1).

For the higher observation times (> 10800s), the second concentration peak (cf. figure 3) is difficult to obtain by the model even with a modification of the set of standard-deviations. It seems that it could only be explained by a change in the advection term (wind speed) in the calculation. Transport velocity had to be supposed to have been lower than the one measured in the experiment. No obvious explanation can be given for this feature.

An interesting point is the comparison of the experimental and calculated atmospheric transfer coefficient (A.T.C.) - in $s.m^{-3}$ - defined as the integrated concentration over the time divided by the released mass, at the different observation points. The results are presented table II. There is a rather good agreement concerning the maximum values on the axis of the plume. However, on the extremities of each cross-section, the experimental values are higher than the calculated ones. The first ones are mainly due to the high values of concentrations at long time intervals after the end of the release while the calculation does not reflect this behaviour (see hereabove). This result shows that these high values of concentrations are of relative weak importance compared with the maximum values found on the axis of the plume.

4.2.2 Neutral or unstable conditions

The results of the experiment PISSENLIT 13 (cf. Table 1) are presented here. The mean wind speed and direction during this experiment are similar to those observed during the previous experiment carried out in stable conditions. This similarity permits a direct intercomparison between the results.

4.2.2.1 Presentation of the results and comparison with those of the previous experiment

An example of result is presented figure 5 . One notices that the maximum concentration is observed between 20 and 30 min after the beginning of the release. The decrease of the concentration is then much more rapid than the one noticed in stable conditions (the background is reached approximately one hour after the end of the release). The widths of the plume in the two experiments have been compared . The plume in neutral conditions is much wider than in stable conditions . This is in agreement with the conclusions of the sensitivity study (cf. 3.2) where it was pointed out that the flow in stable conditions is much well organized than in neutral or unstable conditions. As previously noted, this behaviour is probably linked with the topography of the site. This effect is much more important than it is generally expected by authors, (5) for example, in the case of a flat terrain.

4.2.2.2 Comparison with the code ICAIR 3

The comparison of the experimental results with those of the code ICAIR 3 is presented figure 6 . One notices that the maximum concentration on each cross-section (except perhaps on the last one) is overestimated by the calculation. Similarly, the width of the plume is underestimated. The theoretical lateral spread of the plume, i.e. the value of σ_y , is small compared with the observed one. This result, opposite to the one obtained in stable conditions (cf. 4.2.1.2) enhance the considerations mentioned hereabove (cf. 4.2.2.1). The values of the experimental and theoretical A.T.C. are reported table II. The comparison leads to similar results, i.e. the model over estimates the values of the maximum A.T.C. on each cross-section and underestimates the other values.

5. COMMENTS

The clearest point which can be pointed out from the results hereabove seems to be the strong effect of the atmospheric stability on the characteristic of the flow. Both the sensitivity study (cf § 3) and the results of the tracing experiments (cf § 4) show that, in neutral or unstable conditions, there are large variations of the wind direction and a great lateral spread-out of the pollutant on the site. In stable conditions, the flow seems on the contrary well organized and there is a much weaker transverse dispersion . The difference between the two cases is largely more important than expected by different authors , in the case of a flat terrain. The reason for this seems to be the topography of the site. The "meandering" of the plume, as often observed in the case of a flat terrain, and which contributes to make the lateral dispersion not, or weakly, dependant on the stability conditions, seems to be strongly lessened by the topography in the case of a stable atmosphere. It follows in this case an unexpected organized flow with few energy at a low frequencies and a weak dispersion at long transfer times.

An other noticeable point is the low decrease of the concentration after the end of the release. Especially in stable conditions, significant concentrations are found several hours after the release of gas has ended. This effect may partly be explained by the fact that the rate of growth of a pollutant cloud in the longitudinal direction is of the same order as the velocity of the mass center of the cloud. This effect is numerically simulated with a puff model. The simulations we have performed with the code ICAIR 3, while they reflect this tendancy, show a much weaker effect compared with the observed one. No clear reason to this can be given. A possible explanation could be the recirculation of the pollutant on the experimental site, due to a flow upon the drainage flow and of opposite direction to it.

The comparison of the maximum concentration at ground between experiment and model in stable conditions shows a rather good agreement up to relatively long time intervals before the effect mentioned hereabove is observed. In neutral conditions, the model slightly overestimates these concentrations. The effect at long time intervals, while not as pronounced as in stable conditions, is also observed here.

The comparison of the atmospheric transfert coefficients on the axis of the plume shows a good agreement in stable conditions and a slight overestimate by the model in neutral or unstable conditions.

6. CONCLUSION

This study permits to point out some characteristic features of the atmospheric dispersion in a low wind speed situation. However, some of these features are probably linked with the topographic particularities of the experimental site. An obvious combined effect of the topography and of the stability conditions on the atmospheric dispersion is put forwards.

This study shows the ability of a gaussian puff model to assess the consequences of a pollutant release in the case of low wind speed. The comparison of the results of the tracing experiments and of the code ICAIR 3 shows generally a reasonable agreement with regard to the values of concentration and time-integrated concentration on the axis of the plume in stable conditions and a slight over-estimate by the code in neutral or unstable conditions.

REFERENCES

1. A. DOURY (1976). Une méthode de calcul pratique et générale pour la prévision numérique des pollutions véhiculées par l'atmosphère. Rapport CEA-R-4280 (rév. 1).

2. A. DOURY (1980). Les limites d'utilisation des modèles panaches aux courtes distances et par vents faibles. Séminaire CEE sur les rejets radioactifs et leur dispersion dans l'atmosphère à la suite d'un accident hypothétique de réacteur, RISØ (Danemark).

3. B. CRABOL (1983). Contrat CEEA n°SRF/010/F. 4ème rapport d'avancement.

4. B. CRABOL (1983). Contrat CEEA n°SRF/010/F. 5ème rapport d'avancement.

5. F. PASQUILL (1974). Atmospheric Diffusion (second edition). Ellis Horwood.

I am grateful to IPSN/DERS/SERE, particularly to M. G. DEVILLE-CAVELIN, who carried out the experimental part of this contract.

EXPERIENCE N°	RELEASE HEIGHT (M)	MEAN WIND SPEED (M/S)	MEAN WIND DIRECTION	STABILITY CONDITION
8	45	1.5	S.E.	S *
9	45	2.0	S.E.	S
10	45	2.1	S.E.	S
11	2	1.5	S.E.	S
12	45	1.9	S.E.	S
13	2	2.1	S.E.	N/U **
14	2	2.3	S.E.	N/U
15	2	3.6	N.W.	N/U
16	2	2.9	N.W.	N/U
17	2	2.4	N.W.	N/U
18	2	2.4	N.W.	N/U
19	2	1.1	S.E.	S
20	2	0.2	S.E.	S

* stable

** neutral or unstable

CONDITIONS OF THE TRACING EXPERIMENTS

TABLE I

TABLE II

A Cross section	measured	$7.1.10^{-7}$	$1.1.10^{-6}$	$8.2.10^{-7}$	$8.8.10^{-5}$	$4.7.10^{-4}$	$1.4.10^{-4}$	$8.8.10^{-7}$	$8.7.10^{-7}$	8.10^{-7}
	calculated	7.10^{-12}	$1.3.10^{-11}$	$1.9.10^{-11}$	$3.1.10^{-8}$	2.10^{-5}	$1.7.10^{-4}$	$1.8.10^{-5}$	$2.5.10^{-9}$	$1.4.10^{-11}$
B Cross section	measured	$2.5.10^{-7}$	$7.1.10^{-7}$	$7.3.10^{-7}$	$8.3.10^{-7}$	$1.2.10^{-4}$	$1.3.10^{-4}$	$4.5.10^{-7}$	$9.4.10^{-7}$	$9.6.10^{-7}$
	calculated	$9.7.10^{-12}$	$5.5.10^{-11}$	$3.9.10^{-10}$	$7.8.10^{-8}$	$1.7.10^{-5}$	6.10^{-5}	$1.4.10^{-5}$	$3.6.10^{-7}$	$2.4.10^{-11}$
C Cross section	measured	1.10^{-6}	$8.3.10^{-7}$	$7.1.10^{-7}$	$6.2.10^{-6}$	$3.6.10^{-5}$	$9.8.10^{-5}$	$1.1.10^{-4}$	4.10^{-5}	$4.4.10^{-6}$
	calculated	$1.3.10^{-7}$	$8.4.10^{-7}$	$2.8.10^{-6}$	$8.3.10^{-6}$	$1.4.10^{-5}$	$1.7.10^{-5}$	2.10^{-5}	$2.4.10^{-5}$	$1.7.10^{-5}$

A Cross section	measured	-	$3.7.10^{-7}$	$7.7.10^{-8}$	$9.5.10^{-7}$	$1.5.10^{-6}$	$3.9.10^{-6}$	$4.7.10^{-6}$	$6.6.10^{-6}$	
	calculated	$2.8.10^{-10}$	$2.9.10^{-10}$	$1.6.10^{-7}$	$6.5.10^{-6}$	$9.8.10^{-6}$	$3.9.10^{-5}$	$2.1.10^{-5}$	$2\ .10^{-6}$	
B Cross section	measured	$2.4.10^{-8}$	$1.9.10^{-8}$	$5.1.10^{-8}$	$1.7.10^{-7}$	$1.6.10^{-6}$	$1.1.10^{-6}$		$1.1.10^{-6}$	$1.4.10^{-6}$
	calculated	$1.5.10^{-9}$	$2.8.10^{-9}$	$3.6.10^{-9}$	$2.8.10^{-8}$	$7.5.10^{-7}$	$5.5.10^{-6}$	$1.2.10^{-5}$	$7.8.10^{-6}$	$6.8.10^{-7}$
C Cross section	measured	$7.4.10^{-8}$	$7.6.10^{-8}$	$9.1.10^{-8}$	$1.1.10^{-7}$	$2.1.10^{-7}$	$2.5.10^{-7}$	$1.2.10^{-7}$	2.10^{-7}	6.10^{-7}
	calculated	$3.6.10^{-8}$	$4.6.10^{-8}$	$6.1.10^{-8}$	$7.5.10^{-8}$	$9.3.10^{-8}$	$1,2.10^{-7}$	$1.5.10^{-7}$	$1.5.10^{-7}$	$3.9.10^{-8}$

TABLE III

Comparison of experimental and theoretical values of Atmospheric Transfer Coefficients at the different points of each cross-section

TABLE II : experiment PISSENLIT 11 ; TABLE III : PISSENLIT 13

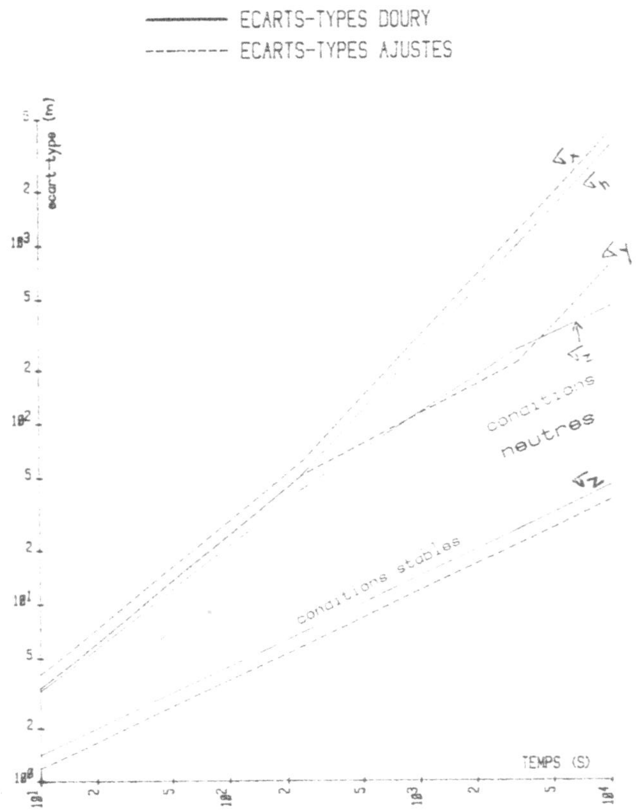

FIGURE 1

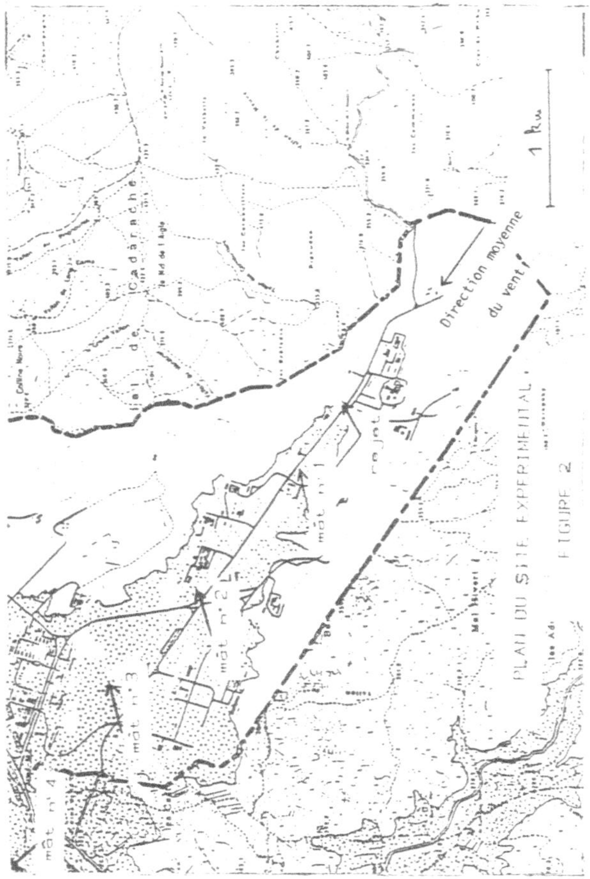

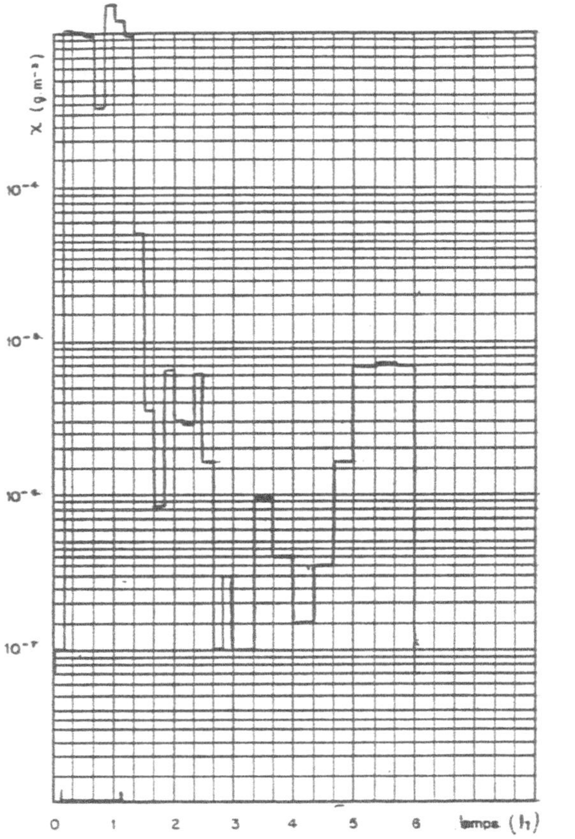

Concentration en fonction du temps au point A6
Experience PISSENLIT 11

FIGURE 3

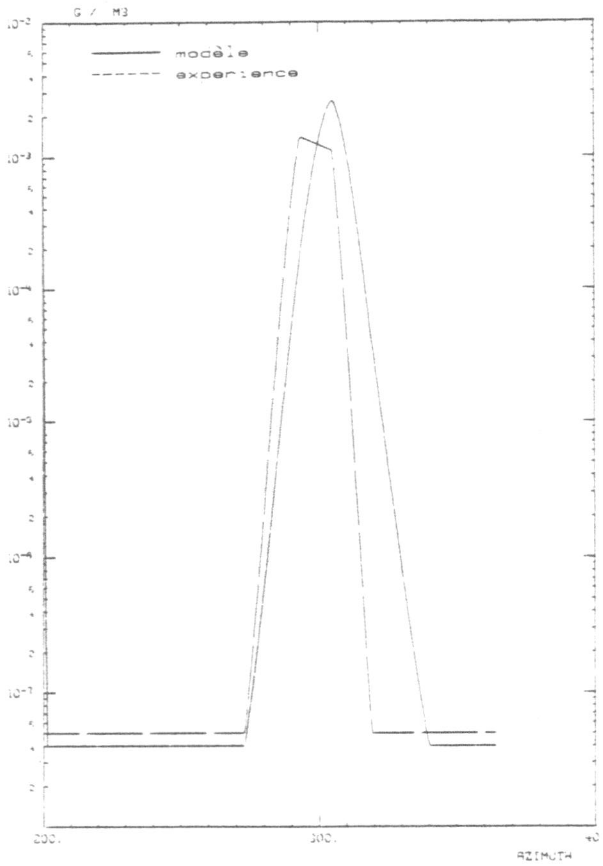

Comparaison des concentrations theoriques et experimentales entre les instants 500-1200 s.
Coupe A, Pissenlit 11

FIGURE 4

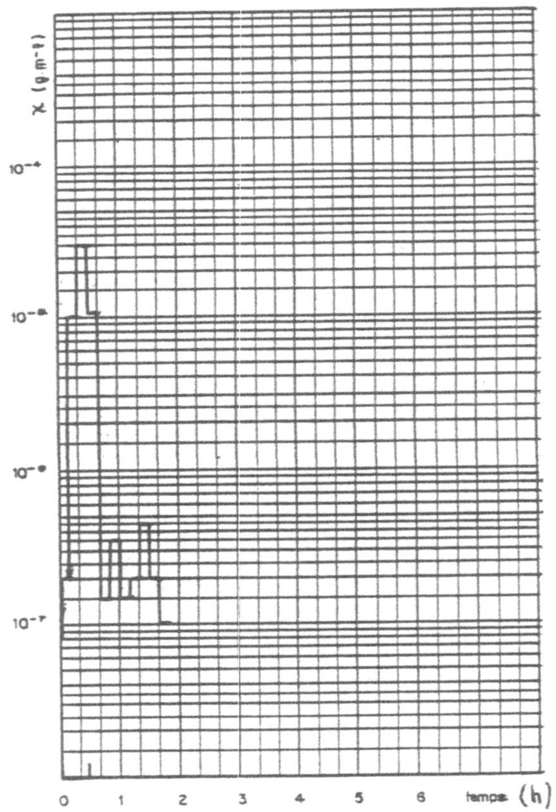

Concentration en fonction du temps au point B8
Experience Pissenlit 13

FIGURE 5

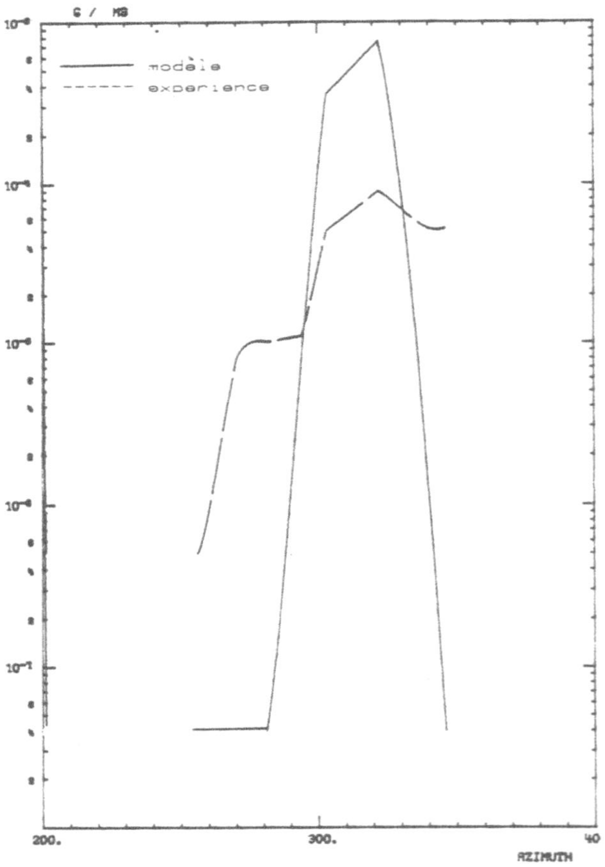

Comparaison des concentrations theoriques et experimentales entre les instants 600-1200 s.
Coupe A. Pissenlit 13

FIGURE 6

Discussion

Question: B. Tolley - CEC/BXL

When will the further results be available?

Answer: B. Crabol - CEA/CEN/Far

End of October.

Question: H. ApSimon - Imp. Col./London

The site is well wooded. Is there any possibility with the shallow nocturnal flows that material penetrates into the forest canopy and subsequently seeps out slowly contributing to the tendency of material to linger in the vicinity.

Answer: J.A. Hertig - E.P.F./Lausanne

From physical modelling we have found that there is another possible explanation of this secondary augmentation of concentration. We have observed a small local depression in the highest part of Cadarache Valley, just above the release point. In this depression of terrain there is the creation of a cold air lake in which part of the release must enter in accumulation.
So, a part of the SF_6 emitted is entrained directly by drainage flow and is observed first and after that the second part recycled from the cold air lake basin may be observed later because of the longer trajectory and of the retention time.

Comment: B. Crabol - CEA/CEN-Far

Yes, it is one of the possible explanations to the low decrease of concentration observed in the experimental field.
Question: J. Paesler - KFK

In your Puff model you use σ_y-Parameters being independent of stability for the single puffs.
Further your model gets the time varying wind direction as an input. So why can't your model simulate the higher angular spread in instable situations. Is it because you do not resolve the spatial wind direction fluctuations?

Answer: B. Crabol - CEA/CEN-Far

It can be explained partly because the model does not resolve completely the spatial wind direction fluctuations, partly because large horizontal fluctuations of periods less than 2 min (the sampling time of meteorological data in the experiments) are not taken into account in the calculation.

Parallel Sessions Research Area C

The release and distribution of radioactive fission products in the atmosphere following a reactor accident

Session C4

Chairman: J.G. Kretzschmar (Janssen-Pharm.-Beerse)

Session secretary: B. Tolley (CEC/BXL)

LOCAL SCALE ATMOSPHERIC DIFFUSION AT A COASTAL SITE IN THE PRESENCE OF BREEZE EFFECT

P. CAGNETTI, V. FERRARA, A. PELLEGRINI
ENEA - PAS Department - Environmental Sciences Division - Rome

Summary

The aim of this work is the characterization, from the thermal and anemological point of view, of the lower layers of the atmosphere at a coastal site, affected by breeze circulation, in order to set up possible diffusion models for accidental releases of airborne materials, both of short and prolonged duration. Five inland meteorological campaigns, starting from April '82 (Apr., Jul., Oct. 1982, Jan. 1983), have been carried out. An appropriate extension of the research allowed the execution of two more campaigns in the open sea (Apr., Jul. 1983), utilizing the oceanographic ship "Bannock" kindly supplied by CNR. The analysis of the data showed the development of a well defined IBL during on-shore flow only in Spring and Summer, while an inversion layer was detectable aloft independently of the season (provided that an anticyclonic situation was present). According to those relevant features a simple diffusion model has been developed for short duration releases at local scale. Finally, the analysis and elaboration of the data collected on site by a meteorological automatic station, allowed the extension of the model to prolonged releases.

1. INTRODUCTION

The aim of this work is the physical characterization of the lower layers of the atmosphere at a coastal site, during breeze circulation, in order to set up possible diffusion models for accidental releases of airborne materials, both of short and prolonged duration.

The work has been carried out in two phases. The first phase, aiming at choosing the site, organizing meteorological campaigns, collecting and pre-elaborating data, consisted of 4 seasonal campaigns, starting from April 82 (Apr., Jul., Oct. 1982, Jan. 1983). From a preliminary analysis of collected data, we recognized the opportunity to carry on further off-shore campaigns. An appropriate extension of the research program allowed us to execute two more campaigns in the open sea (April, July 1983) utilizing the Oceanographic Ship "Bannock" kindly supplied by C.N.R.. The second phase provided the elaboration of diffusion models, based on experimental data elaboration, in order to evaluate the local scale impact on environment and man of accidental releases from LWR Reactors.

1.1 Synoptic meteorological conditions during the campaigns

The Summer and Winter campaigns as well as the two above-sea campaigns, were in practice dominated by high-pressure fields at synoptic scale and by associated weak air circulation at ground level. This latter fact allowed the increase of local scale phenomena, the most typical of which is the breeze circulation. Moreover, those local phenomena had often a prevailing character with regard to synoptic scale motions and, hence, the probability of significant observations was elevated.

1.2 Methodologies and instrumentation

The map (fig. 1) shows two of the main measurement stations, located at approximately 1 and 6 km from the coastline, and three of the mobile stations which have been occasionally used. In September 1981 a complete and automatic meteorological station was installed on the site (loc. "Centrale"). Every 10-minutes interval it records, on magnetic cassettes, the values of wind at 10 m height, of temperature and relative humidity, rainfall rates, solar radiation, atmospheric pressure at ground level, according to the WMO standards. As far as the equipment for thermic and anemological data assessment, at upper levels (up to 1,200 m approximately), is concerned, a Tethersonde system was used. It consists of a tethered ballon radiosonde apparatus, whose ground station was connected on line to an HP-85 system (mini-computer +plotter+printer+mass-storage). The measures obtained by the radiosoundings are: wet and dry bulb temperature, atmospheric pressure, wind velocity and direction. The data collected were automatically recalled from the mass storage and elaborated, in order to obtain graphs of wet and dry bulb temperature, wind velocity and direction, as a function of height. Wind velocity and direction measurements at upper levels were as well obtained by means of pilot balloons; consequently, zenith and azimuth, recorded at specific time intervals, were elaborated by means of appropriate software. The purpose was to have graphs of wind velocity and direction as a function of height, and horizontal projection of the pilot path.

2. SHORT DURATION RELEASE

In order to set up local scale diffusion models, in presence of breeze, for short duration releases, it is essential to verify experimentally the space-time trend of some basic parameters. It is well known that the essential conditions for the setting up of breeze regime are: synoptic situation characterized by high pressure and low wind speed, at ground level; at local scale the presence of an adequate difference in temperature between over land surface air and sea surface water (generally about 2-3 oC). Parameters to be considered are: the anemological- thermic structure at lower layers, its relationship with the distance from the coastline and the intensity of incoming solar radiation. The final elaboration of the data collected by vertical soundings, over the various campaigns, is taken into consideration in the following sections, a simple local scale diffusion model being proposed.

2.1 Elaboration of data collected during inland and above sea campaigns

Taking as a basis the individual soundings, some elaborations have been carried out in order to describe synthetically the space-time trend of isothermal lines on a period of time during which it has been possible to have continuous soundings of the atmosphere. Since our purpose tends to obtain, from that elaboration, data relative to the diffusion properties of atmosphere, the layers of more intense stability have been underlined (thermal inversion layers); they are given by the dashed areas shown in fig. 2, which is given as an example. The isothermals of the Summer and Winter campaigns present some peculiar characteristics: in addition to the usual development of nocturnal inversion at ground level (related to the absence of clouds) they show an inversion layer aloft which is always detectable and which is , unaffected by diurnal cycles. The base of this stable layer oscillates from 400 to 700 m, showing no significant differences between the two main measurement stations (A and B in fig. 1). A relationship between the presence of a synoptic anticyclonic situation and the presence of the stable layer aloft, seems to exist: therefore this layer is detectable in typical situations which are favourable to breeze formation. Yet, is has to be clarified whether the presence of such inversion is related to the onset of breeze, or both phenomena are consequences of the larger scale situation. The following section will deal with part of this problem, considering as well the soundings carried out offshore in April and July 1983 (the latter shown in fig. 3). Note that in high pressure conditions, an inversion layer is as well detectable offshore; its base oscillates between sea surface and a few hundreds meters. The graph at the bottom of fig. 3 shows the temperature difference between water and air close to the sea surface, as a function of time.

2.2 Analysis of data

In analyzing the data of the vertical thermal structure, during the hours of maximum on-shore flow intensity (from 1 p.m. to 5 p.m.), there emerged some difficulties in the determination of IBL height, even if various methodologies were used, whilst the presence, almost constant, either by day and by night, of a stable layer aloft is thought to be determinant for the formulation of a model.

Two basic methods in order to assess the IBL height have been considered: firstly, we tried to identify a ground-based layer characterized by a superadiabatic lapse rate; secondly we used a method based on the observation of the vertical trend of relative humidity, either as a check of the results of the first method, or as an alternative to those particular cases in which the first method was not suitable for a certain identification of the IBL height. In details, the criterium used in the second method stems from the hypothesis that on-shore flow is originally formed by saturated or near-to-saturation air: hence, as the inland fetch increases, relative humidity diminishes, an such decrease reaches higher and higher layers of atmosphere, as a consequence of heat turbolent flow upward, peculiar to the IBL growth. During the Summer campaign these methods allowed us to determine IBL height in the majority of the soundings made during the first hours of the afternoon (2-5 p.m.), when sea breeze was well established. Fig. 4 supplies mean values and standard deviations at 1 and 6 km from the coastline. In Autumn, the IBL mean value diminishes of a factor of 1.5-2, due to the difference in the incoming solar radiation. In the Winter campaign, although breeze is present in several cases, we could not detect IBL presence from sounding data. Similar results were obtained during the Spring campaign; however, in this case, it should be remembered that synoptic conditions were not favourable for breeze.

2.3 Proposal of a simple model for short duration releases in presence of breeze circulation

In order to explain the factors which bring about this almost constant inversion aloft, the soundings made off-shore in anticyclonic conditions have been studied. In open sea as well we detected the constant presence of a stable layer whose base fluctuates between sea surface and a few hundred meters of height. It appears logical to link the presence of this inversion layers with the anticyclonic situation, i.e. with the subsidence phenomenon which, in open sea, seems to affect the air layers close to the surface as well. It has been found that the inversion base tends to be higher over the land than over the sea and, again, rises as the inland distance from the coastline increases. Furthermore, in the majority of the observations, this stable layer lies within the on-shore flow, during both day and night (breeze return flow in the latter case); while, in the case that this condition is not verified, a strong wind shear occurs both in direction and intensity.

In order to set up a diffusion model for short releases in the presence of breeze, we have to evaluate both the horizontal and vertical diffusion parameters. As far as horizontal dispersion is concerned, we recognized the necessity to carry on further experiments utilizing airborne tracers, because the breeze could be regarded as a steady state wind, with limited lateral fluctuations about the mean direction, and thus the horizontal dispersion parameter may result lower than that estimated in the same insolation and wind speed conditions, at least when the breeze flow is well developed. Until different estimates will be set up in the future following tracer experiments, we shall assume, for our purposes, sigma-y (horizontal dispersion parameter) values corresponding to F and B category during off-shore and on-shore flow respectively.

No problem seems to arise, for the vertical dispersion parameters (sigma-z), when considering the land breeze by night, as the atmosphere is stable from ground level upward, and generally wind speed is quite low: hence, dispersion parameters proper to the F category can be undoubtedly used. On the contrary, during the day, vertical dispersion is favoured by the growth of IBL, while it is certainly limited by the inversion aloft. Fig. 5 shows that IBL height (180 m in this case) can be lower than that of the inversion base: however, as concerning diffusion, it is more important to know the latter one, because, as it is shown in the same fig., the layer between the top of IBL and the inversion base is quite neutral, and therefore it reduces but not inhibits upward diffusion. Hence, if we know the probability Pi that the inversion base height is lower than a particular value, we can infer a series of ground level concentration values Ci, each one related to the value of probability Pi: so that Pi represents the probability of occurrence of a concentration greater than Ci.

Let H_{IBL} be the height of IBL, and σ_z^D the value of sigma-z proper to category D: the following expression is assumed to describe the height of the mixed layer H_{MIX} to be utilized in the model:

$$H_{MIX} = H_{IBL} + \sigma_z^D \cdot \sqrt{\pi/2} \qquad (1$$

where both H_{IBL} and σ_z^D are functions of downwind distance.

Let H_{INV} be the height of the inversion base: it should always result H_{MIX} lower than H_{INV}. Hence, the usual fumigation formula can be used:

$$\chi = [Q \cdot \exp(-y^2/2\sigma_y^2)] / (H_{MIX} \cdot \sigma_y \cdot U \cdot \sqrt{2\pi}) \qquad (2$$

where - χ is the mean ground level concentration; - Q is the tracer release rate; - U is the mean surface wind speed. The fig. 6 shows the results obtained by means of equation (2, using the value of sigma-y in cat. B, the mean value of IBL height in Summer, and the mean value of H_{INV} obtained by averaging between 1 and 6 km values: fig. 6 shows therefore the values of concentration whose probability of occurrence, in sea-breeze conditions, is greater than, respectively, 10 %, 20 %, 30 % and 50 %. The same calculation was executed for Autumn, obtaining similar results. Absolute values of concentration are higher in Winter, due to the reduced thickness of IBL, and to the height of the inversion base, which is lower. It should also be remarked (see fig. 6) that the higher the inversion base is, the greater the distance is at which the inversion itself affects ground level concentration values. It could finally be concluded that the values of concentration nearby the source are mainly affected by the IBL thickness, while the inversion aloft begins to affect the concentration values at greater distances directly depending on the height of the inversion base.

3. PROLONGED RELEASE

Once defined diffusion conditions, and, hence, the model to be utilized for calculating ground level air concentrations in the presence of breeze effects for short releases, the extension of such a model to prolonged releases requires a study about wind direction persistence and variability up to 24 h.

Therefore we utilized the data collected by the automatic meteorological station on-site, elaborated in a specific way, as shown in the next section.

3.1 Elaboration and interpretation of data collected on-site

The data collected automatically on magnetic tape every 5 minutes were elaborated to obtain hourly mean values as well as minimum and maximum recorded values, over a 24 hours cycle. In particular, in order to obtain the 24 h hodograph in the presence of breeze flows starting from hourly mean values, we have analysed the values of wind direction referred to the same hour of the day all over a period of 5-10 days and then calculated the corresponding mean values of such hourly values. In this way we have a representation wind vector for each hour in well developed conditions of breeze flows, smoothed by stochastic fluctuations, and consequently a complete hodographic diurnal cycle, representative of the whole period considered.

During July we have analysed 13 consecutive days (12-24 July 1982), whose results are shown in fig. 7, where the corresponding hodograph is plotted. In the same way we have also considered the following periods: 15-22 May 82; 13-16 March 82; 7-14 October 81. Analysing the hodographs of the different seasons it can be concluded that in Summer the on-shore and off-shore flows are anyway persistent during day and night respectively, while a fast wind direction rotation is present at intermediate hours (during sunrise and sunset). In this case the wind speed firstly decreases and then increases again in the opposite direction. During Winter, on the other hand, the breeze wind persistence is consistent only for off-shore flows, while the wind direction rotation appears to be slow during intermediate hours and nearly constant during daytime. Therefore, if we consider the on-shore flow, we can state that for prolonged releases (up to several hours) the horizontal dispersion owing to the wind rotation is quite high during Winter, and quite low during Summer, at least in central daytime hours.

3.2 A simple model for evaluating ground level concentrations following a prolonged release in the presence of breeze

We have seen in sec. 2.3 a methodology for evaluating air concentrations χ (1 h) following a short release ($\leq$ 1 hour) in the presence of breeze. If the release occurs during daytime, in the presence of on-shore flow, the time averaged air concentration (χ(T) over averaging time T) decreases as a function of averaging (or sampling) time T considered, according to a function R(T) given by:

$$R(T) = \chi(T)/\chi(1) \qquad (3$$

which depends on two factors: - a) the horizontal dispersion parameter sigma-y referred to a short release (it is assumed that Pasquill cat. B is constantly present during on-shore flow, because as the incoming solar radiation increases the wind speed increases too); - b) the wind rotation and wind speed variability hour by hour.

In order to calculate the R(T) function it has been chosen a reference angle of 20 degrees about the prevailing direction of on-shore winds and estimated for each direction θ (considered degree by degree) and for each sampling time T (starting with 1 h and up to 13 h) the mean and the maximum value of R(T). Those values have been obtained among all the values calculated by considering periods of T consecutive hours starting from every hour after the breeze onset (usually at 8-9 o' clock) and up to T hours before the breeze setting-down. For every direction θ and for a fixed value of T, the expression for R(T) is:

$$R_\theta(T) = \chi(T)/\chi(1) = (U/n)\sum_{i=1}^{n} \exp\left[-\tfrac{1}{2}(Y_i/\sigma_y)^2\right]\cdot U_i^{-1} \qquad (4$$

where - $Y_i = x\cdot\Delta\theta_i$, x being the downwind distance; - $\Delta\theta_i$, the angle between the considered direction and the direction corresponding to the occurrence of release during the i-th hour. It can be noted that R(T), calculated as the average value over the $R_\theta(T)$ values, depends on sigma-y value referred to a short release, and on wind speed and direction (U_i, θ_i), if it is assumed that the mixing height remains nearly constant during the on-shore flow. The mean value of $R_\theta(T)$ and its dispersion over the 20 θ-values have been evaluated for the different periods of year described in the previous section. The same calculation has been carried out for the off-shore flow, considering also the sigma-y value referred to a Pasquill cat. F.

The following observations could be made:

1) On-shore flow: - a) During the warmer periods of the year (May-July) it is reasonable to assume a persistence of the on-shore flow of about 6-7 hours, at least during the warmer hourrs. In this time interval R(T) can be expressed as follows:

$$\overline{R}(T) \propto T^{-1/3} \quad ; \quad R_{max}(T) \propto T^{-1/5} \quad , \; T \leq 7 \text{ hrs} \qquad (5$$

For T > 6-7 h, R(T) is proportional to T^{-1} , i.e. no overlapping of different releases exists anymore. In other words the wind rotation cancels the effect of wind persistence. - b) During the periods of the year corresponding to intermediate seasons (April, October) the mean on-shore flow persistence is limited to 2-4 hours, during which is again valid an expression for R(T) similar to the previous one. - c) During Winter periods the wind direction regularly rotates and, hence, there is no overlapping for the whole day. In this case it is reasonable to propose an expression for R(T) proportional to T^{-1}. Finally, once $\chi(T)$ over T hours is calculated by means of $\chi(1)$ and R(T), it is sufficient, when the release duration is prolonged for many days, to multiply the $\chi(T)$ values obtained over one day by the number of the days considered for the release duration. But this calculation is valid only assuming that in the following days the same conditions can occur for the on-shore flow as in the first day. Analogously the off-shore flow can be treated when considered separately.

2) Off-shore flow: as expected, being sigma-y for F cat. quite low, a small rotation of the wind direction is enough for eliminating whatever overlapping. Thus, in the majority of the situations, one can assume $R(T) \propto T^{-1}$. It is worth observing that, however, there are situations of well defined off-shore flow, during which the persistence period ranges between 7 and 10 hours, independently of any season. As representative values for the proposed model it can be assumed, in the case of off-shore flow, the following equations:

$$\overline{R}(T) \propto T^{-2/5} \quad ; \quad R_{max}(T) \propto T^{-1/3} \qquad (6$$

4. CONCLUSION

Data collected during the 4 experimental campaigns at Montalto site and above-sea campaigns on Bannock Oceanographic Ship (vertical soundings) have been analyzed as well as data collected by the automatic meteorological station set-up on the site itself. The presence of IBL was identified better during intermediate seasons and during Summer, whereas during Winter such an identification seems quite impossible. However the relevant results obtained through the vertical soundings carried out both inland and off-shore during meteorological situations favourable to the breeze development (such as anticyclonic conditions associated with light ground level winds), have shown the nearly constant presence of an inversion layer in the lower atmosphere. The base of this inversion layer was located sometimes even at sea surface (when negative values of the difference between sea water and air temperature were established) and anyway at heights above ground or sea level of about 400-500 meters, This occurred in practice every time of year independently of the seasonal cycle. Taking account of these observations a reliable model at local scale has been developed, in order to evaluate air concentration at ground level following a short release of airborne pollutant to the atmosphere in the presence of on-shore breeze winds. Moreover, utilizing the data collected by the automatic meteorological station, it has been possible to estimate the mean value of the air concentration for prolonged releases lasting T hours both in the presence of on-shore and off-shore flows. Consequently also the alpha-values (assuming $\chi(T)/\chi(1) = T^{-\alpha}$) have been respectively deduced for the two latter cases. It is to be pointed out that the difference referred to different seasons is only related to the more or less prolonged persistence of the on-shore breeze wind direction. In particular, for on-shore flows, the wind persistence is longer (about 8-11 hours) during Summer, while it decreases down to 3-4 hours during intermediate seasons and, finally, approaches to 0 during Winter. On the other hand the off-shore breeze wind persistence remains approximately constant about the mean value of 10 hours and this occurs in a way uncorrelated to the seasonal cycle.

REFERENCES

1. CAGNETTI, P., FERRARA, V. (1980). Le Transport et la Dispersion de Rejets Radioactifs dans l'Atmosphere a Moyennes et Grandes Distances: Evaluation des Concentrations dans l'Air et des Depots au Sol. Vol. II, Contrat d'etude 1124-5-78 L/V, DG V, CCE
2. GRYNING, S.E., LICK E. (1980). Medium-Range Dispersion Experiments Downwind from a Shoreline in Near Neutral Conditions. Atm. Env., Vol. 14, pp.923-931
3. KERMAN, B.R., MICKLE, R.E., PORTELLI, R.V., TRIVETT, N.B., MISRA, P.K. (1982). The Nanticoke Shoreline Diffusion Experiment, June 1978-II. Internal Boundary Layer Structure. Atm. Env., Vol. 16, pp. 423-437
4. LYONS, W.A., COLE, H.S. (1973). Fumigation and Plume Trapping on the Shores of Lake Michigan during Stable Onshore Flow. J. Appl. Meteor., Vol. 12, pp. 494-foll.
5. NIXON, W., GHOBADIAN, A. (1983). A Simulation Study of Atmospheric Turbulence over the Southern Coast of Long Island during On-Shore Flow. UKAEA Report SRD R268, Saf. and Rel. Dir.
6. PASQUILL, F., SMITH, F.B. (1983). Atmospheric Diffusion. Ellis Horwood Limited, Chichester
7. RAYNOR, G.S., SETHURAMAN, S., BROWN, R.M. (1979). Formation and Characteristics of Coastal Internal Boundary Layer during Onshore Flows. Bound. Layer Meteor., Vol. 16, pp. 487-514
8. VAN DER HOVEN, I. (1967). Consequences of Activity Release: Atmospheric Transport and Diffusion at Coastal Sites. Nucl. Saf., Vol. 8, pp. 490-499

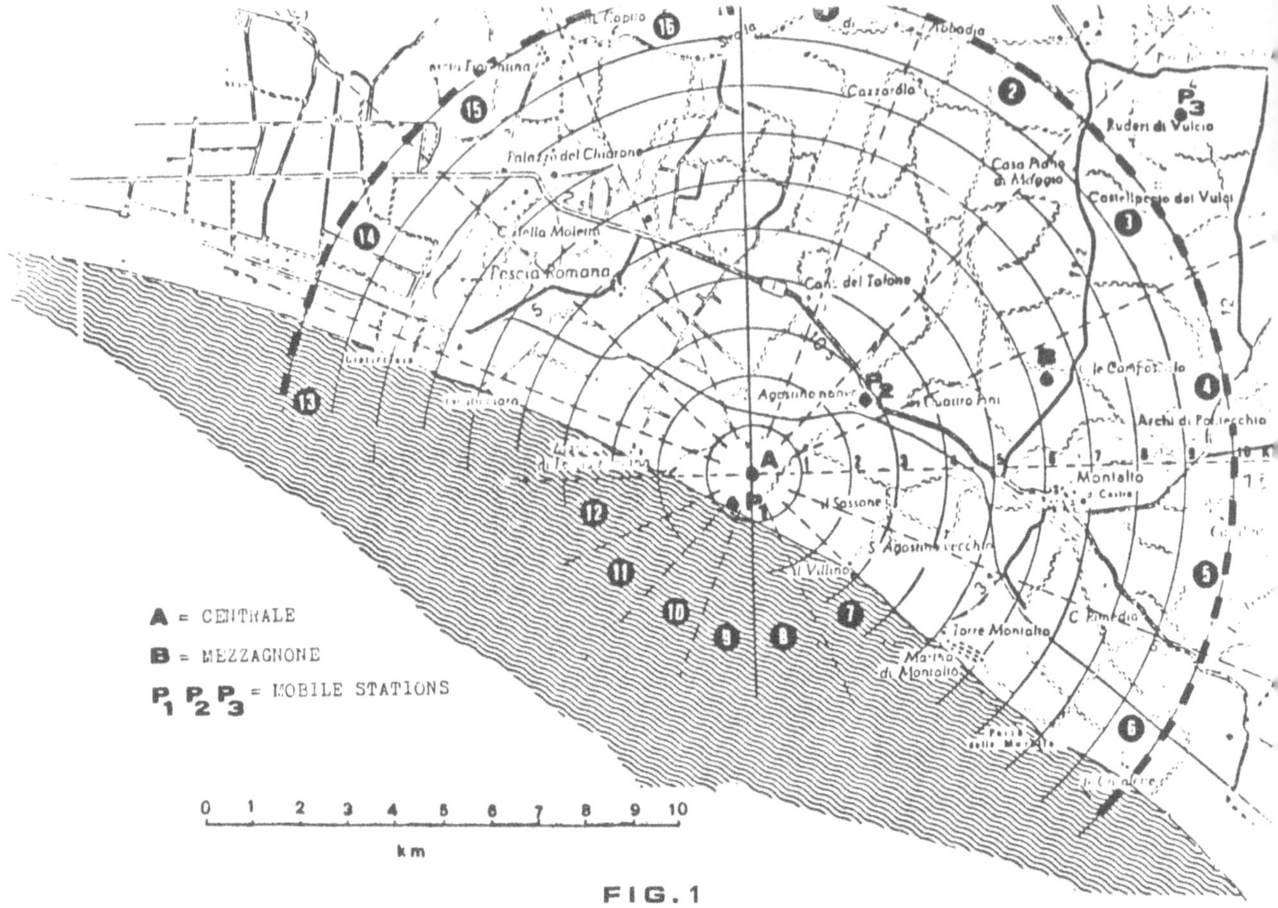
A = CENTRALE
B = MEZZAGNONE
P1 P2 P3 = MOBILE STATIONS
0 1 2 3 4 5 6 7 8 9 10
km
Palazzo del Chiarone
Pescia Romana
Cazzarola
Casa Piana di Mefaggio
Ruderi di Vulcia
Castellaccio del Vulci
Cant. del Tafone
Agostino nuovo
Quattro Pini
Archi di Pontecchio
Montalto di Castro
Il Sassone
S. Agostino vecchio
Il Villino
Torre Montalto
Marina di Montalto

FIG. 1

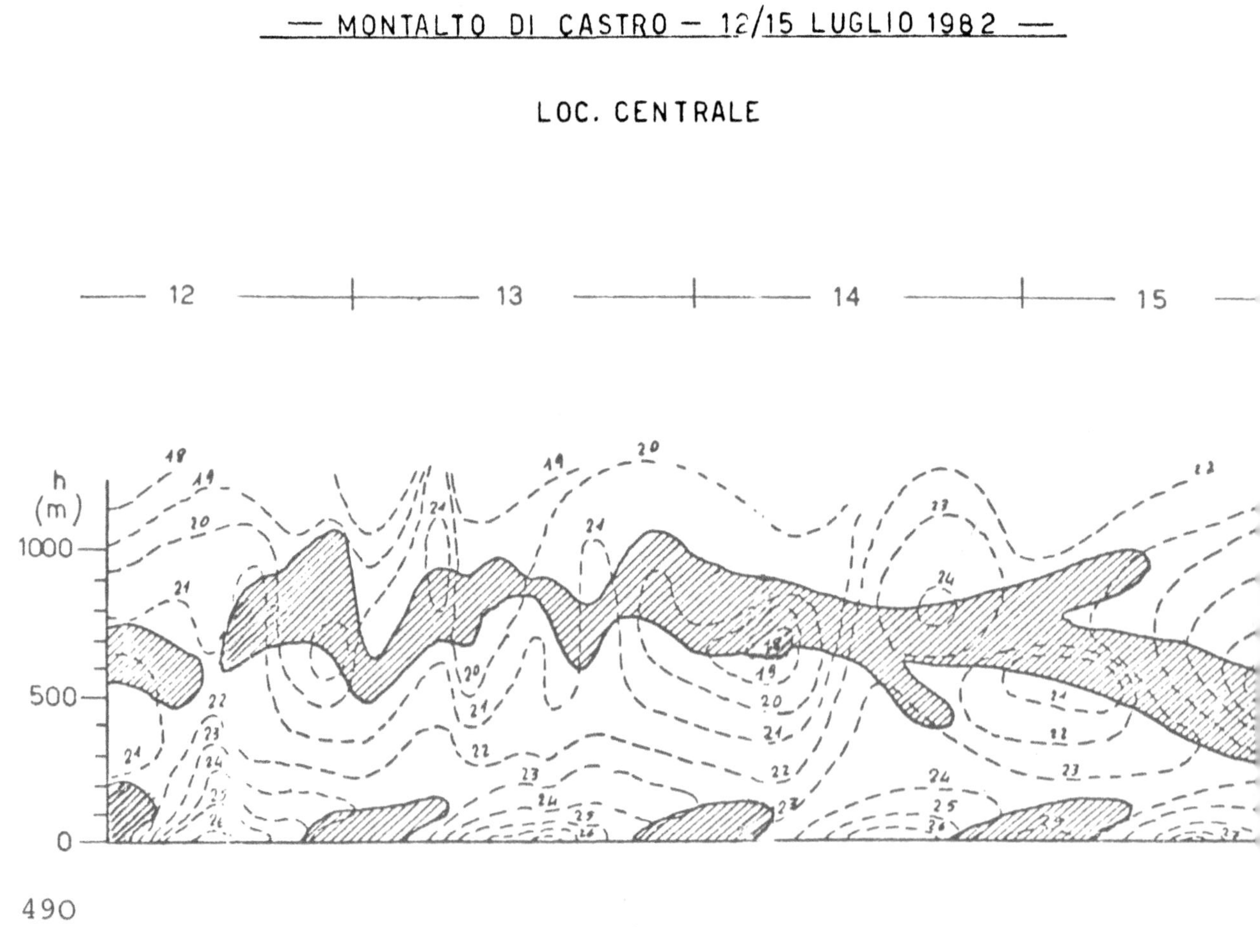
— MONTALTO DI CASTRO — 12/15 LUGLIO 1982 —
LOC. CENTRALE
12
13
14
15
h
(m)
1000
500
0

FIG. 2

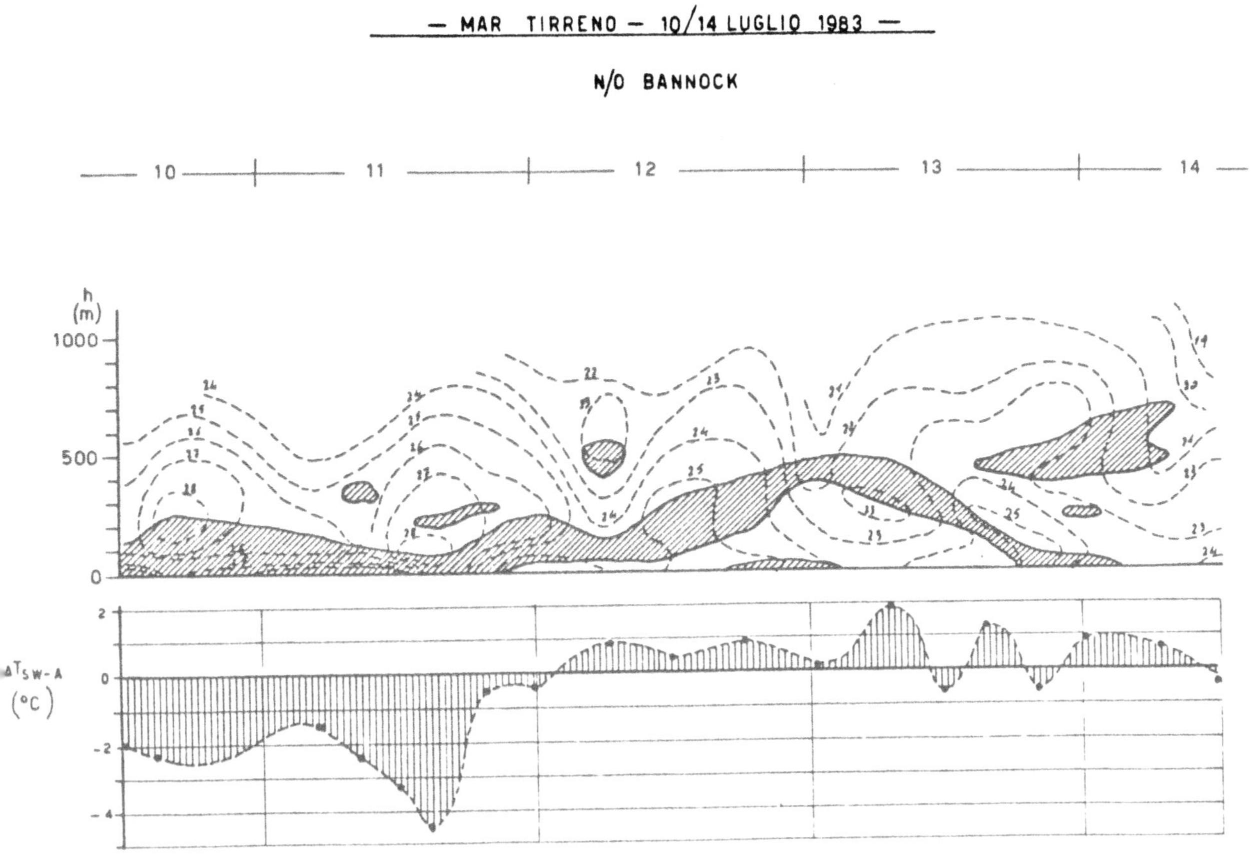
— MAR TIRRENO — 10/14 LUGLIO 1983 —
N/O BANNOCK
10
11
12
13
14
h (m)
1000
500
0
ΔT SW-A (°C)
2
0
-2
-4

FIG. 3

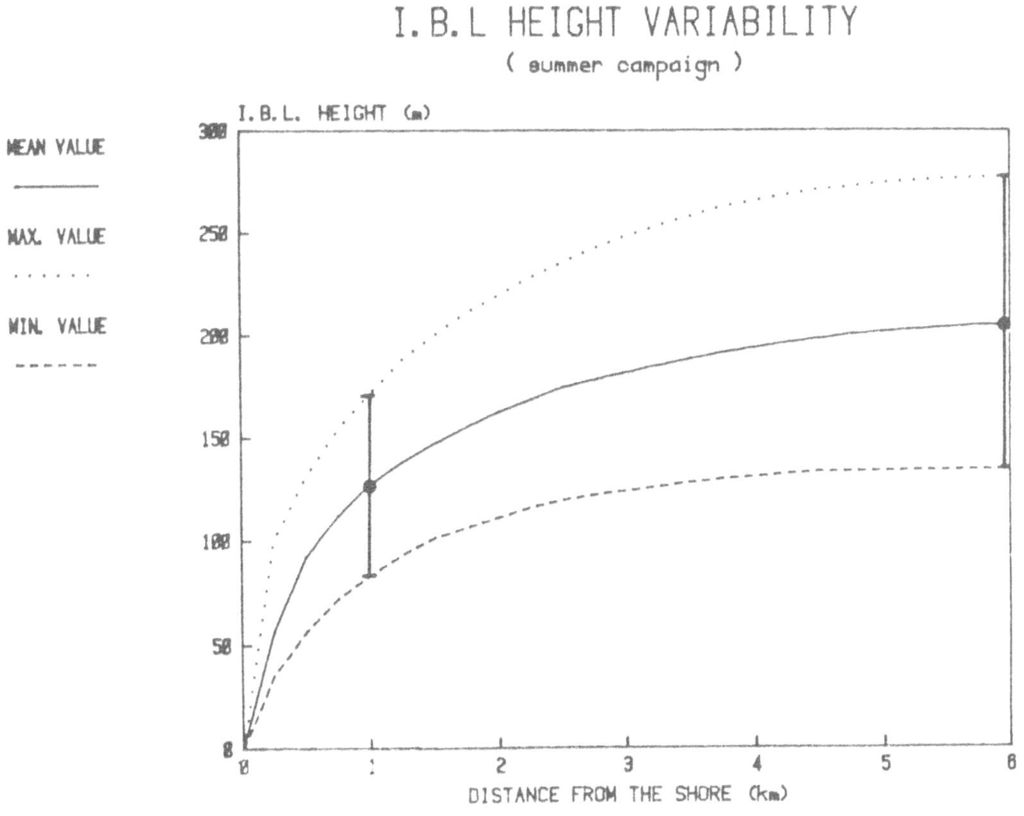
I.B.L HEIGHT VARIABILITY
(summer campaign)
I.B.L. HEIGHT (m)
MEAN VALUE
MAX. VALUE
MIN. VALUE
300
250
200
150
100
50
0
0
1
2
3
4
5
6
DISTANCE FROM THE SHORE (km)

FIG. 4

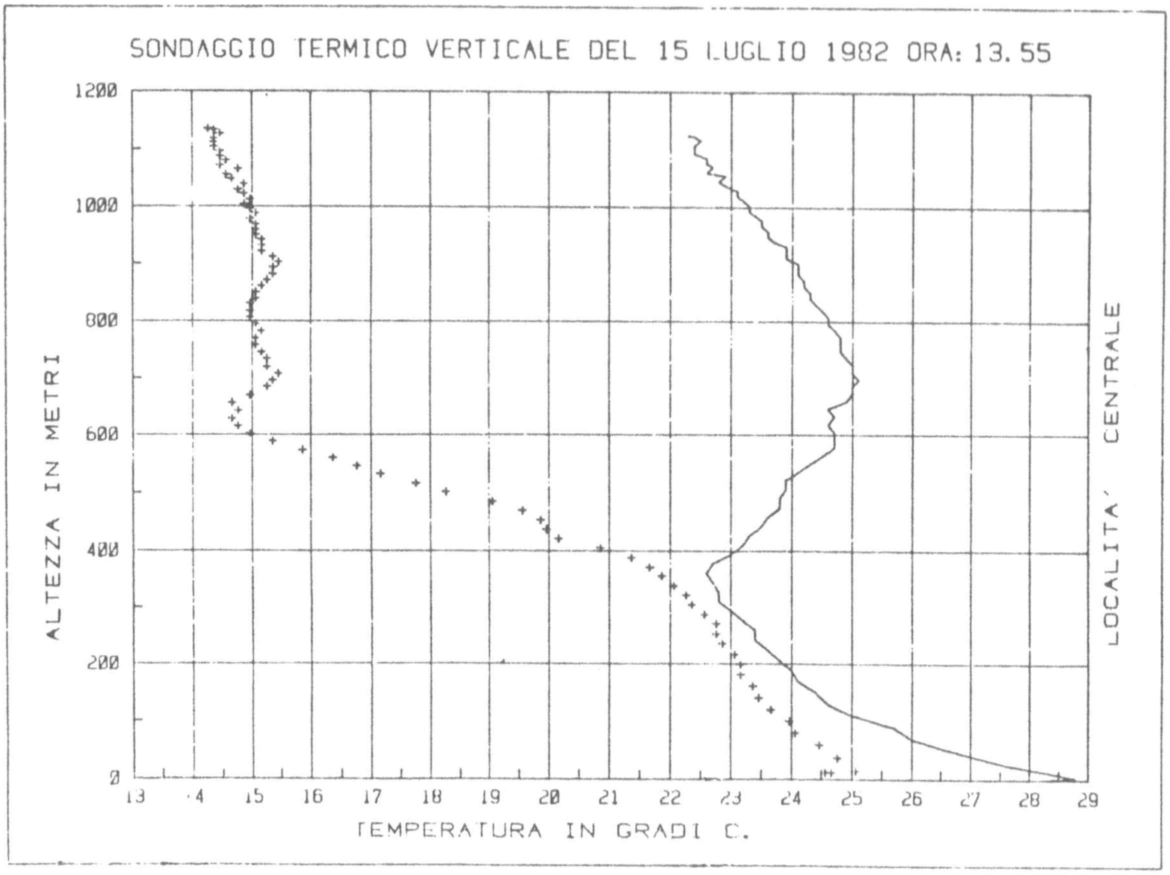
SONDAGGIO TERMICO VERTICALE DEL 15 LUGLIO 1982 ORA: 13.55
ALTEZZA IN METRI
TEMPERATURA IN GRADI C.
LOCALITA' CENTRALE
1200
1000
800
600
400
200
0
13
14
15
16
17
18
19
20
21
22
23
24
25
26
27
28
29

FIG. 5

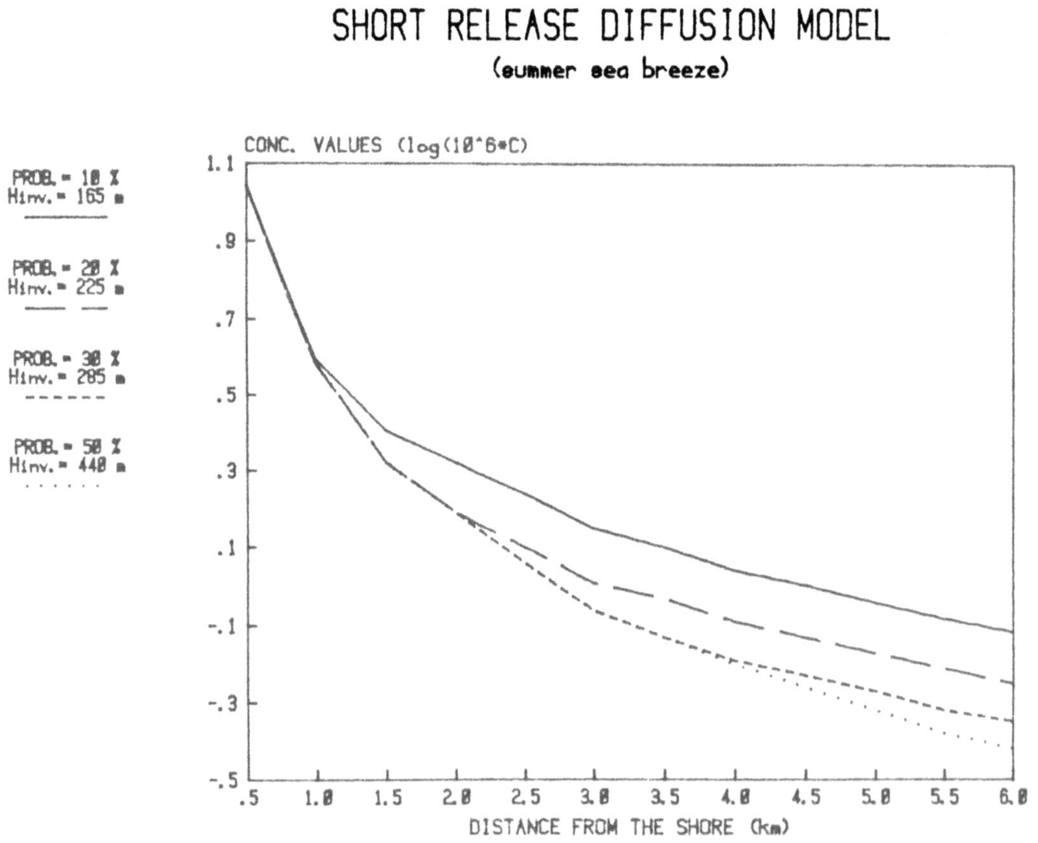
SHORT RELEASE DIFFUSION MODEL
(summer sea breeze)
CONC. VALUES (log(10^6*C)
PROB. = 10 %
Hinv. = 165 m
PROB. = 20 %
Hinv. = 225 m
PROB. = 30 %
Hinv. = 285 m
PROB. = 50 %
Hinv. = 440 m
1.1
.9
.7
.5
.3
.1
-.1
-.3
-.5
.5
1.0
1.5
2.0
2.5
3.0
3.5
4.0
4.5
5.0
5.5
6.0
DISTANCE FROM THE SHORE (km)

FIG. 6

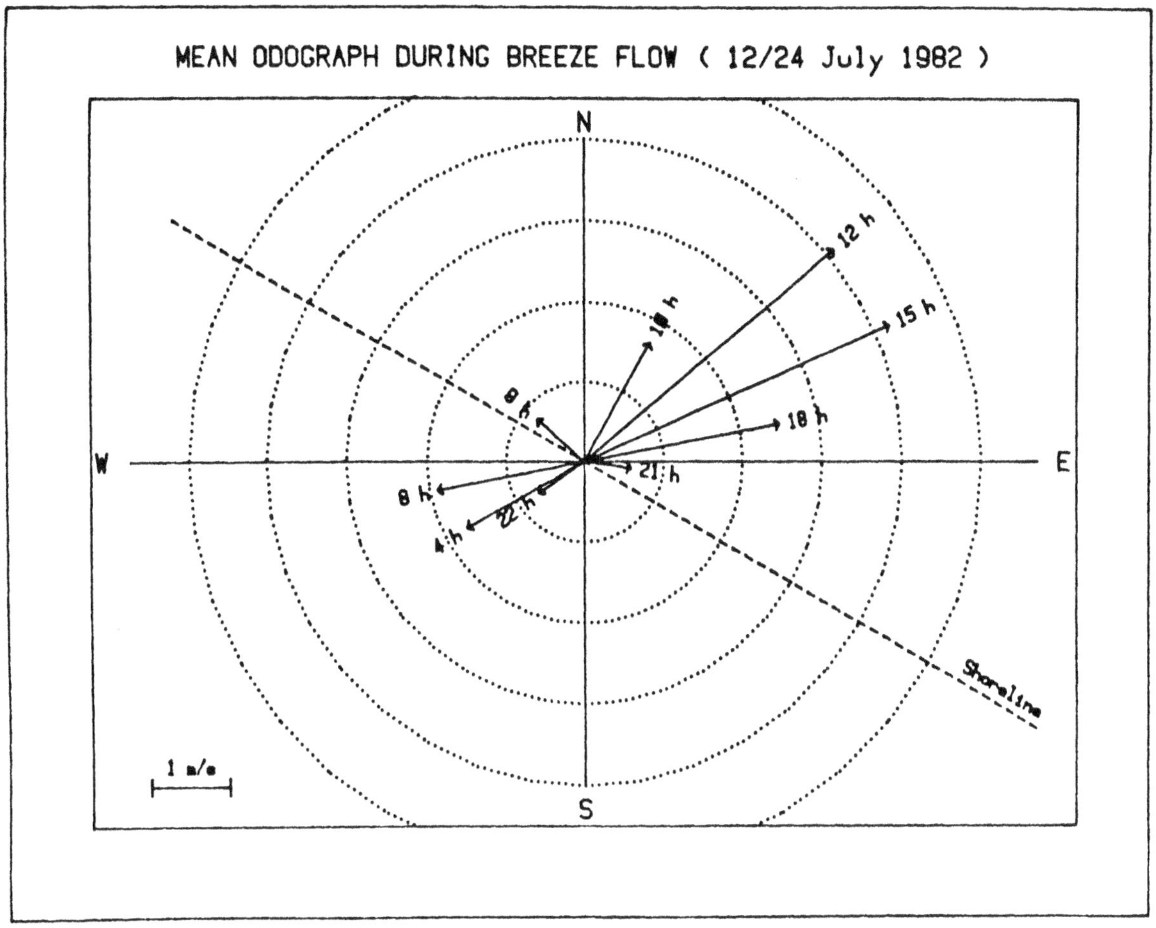
MEAN ODOGRAPH DURING BREEZE FLOW (12/24 July 1982)
N
S
W
E
12 h
15 h
18 h
21 h
22 h
4 h
8 h
Shoreline
1 m/s

FIG. 7

Discussion

Question: F.B. Smith - Brit. Met. Office

My question is in 3 parts. First, do your data on the height of the internal boundary layer agree with the predictions of the various available models?
Secondly, can you see a response in the height at your fixed monitoring site as the fetch increases with the turning of the sea-breeze-wind direction?
Thirdly, do you understand why this wind-turning occurs? Classical theories link it to a Coriolis effect but the upwind fetch over the sea within the sea-breeze circulation is only some 10-20 km. Scarcely long enough for Coriolis forces to operate.

Answer: P. Cagnetti/A. Pellegrini - ENEA/Casaccia

1) They agree quite well with other experimental data available in the literature, but we have not compared our data with the results of any particular model, because we have not found a clear correlation between IBL height and other parameters, such as solar radiation intensity, surface wind speed, etc.

2) As Mr. Pellegrini has said, the turning of the wind in the intermediate hours (dawn and sunset) is very fast during summer, so that in this case the fetch for a fixed point inland during on-shore flow does not vary significantly. In winter the rotation is very slow, but in this case we have found that IBL does not develop.
3) We have accepted, for our purposes, the explanation of clockwise rotation in terms of Coriolis effect, as stated by classical theories; on the other hand, it was beyond the aim of our research to investigate the physical mechanisms that give rise to the breeze circulation and cause its peculiar characteristics. In this sense, I would like to recall that our purpose in this work was to introduce a deeper experimental knowledge into a simple diffusion model; we thought that this approach was the only realistic one in order to provide a calculation algorithm "easy to handle".

Question: J.A. Jones - Nat. Rad. Prot. Board

You have indicated in your final report that σ_y in the sea breeze may be nearer to that for category E than for category B. How does this affect the value of the ratio R.

Answer: P. Cagnetti - ENEA/Casaccia

The value of R(t) depends evidently on σ_y evaluations, so that it will be important in the future, perhaps in the programme of the indirect actions, to schedule some diffusion experiments with tracers on the coast to obtain realistic values of σ_y during on-shore-flow. Actually, if σ_y proper to cat. E is taken (as it has been found in some preliminary tracer experiments carried out at Montalto site in collaboration with the C.C.R. Ispra), the level of concentration for short releases (t=1h) is increased but the R(T) function will approach T^{-1} i.e. this means that the lateral dispersion for a short release will be in this case so small that there will be little or no overlapping at all of tails, from successive hourly releases (prolonged release, lasting T hours); when taking into account the hodograph.

Question: J.M. Quinault - CEA/CEN-G

I do not understand the significance of R(T). Can you explain it again to me? Is it a measured or calculated parameter?

Answer: A. Pellegrini/P. Cagnetti - ENEA/Casaccia

R(T) represents the relationship between the mean concentration value of a sampling time T and for a time of 1 hour. It has been calculated from a Gaussian model for a short duration release (t = 1 h) and uses the results obtained on the site with persistant see breeze effect, see fig. 7 of the paper.

SENSITIVITY, APPLICABILITY AND VALIDATION OF BI-GAUSSIAN OFF- AND ON-LINE MODELS FOR THE EVALUATION OF THE CONSEQUENCES OF ACCIDENTAL RELEASES IN NUCLEAR FACILITIES

B. VANDERBORGHT, J. KRETZSCHMAR and I. MERTENS
S.C.K./C.E.N. Studiecentrum voor Kernenergie
Boeretang 200, B-2400 Mol, Belgium

Summary

The use of the IFDM and CAERS models of the S.C.K. for the simulation of the atmospheric dispersion of routine respectively incidental releases has been validated in two phases. In task 1 the sensitivity of the results of the bi-Gaussian model to different input parametrinations has been discussed. In task 2 the model results are experimentally validated by comparing the calculations with measurement results of 11 tracer dispersion experiments.

1. INTRODUCTION

Since about two decades, several mathematical models for the simulation of the dispersion of pollutants in the atmosphere under various less or more complicated conditions, have been developed. Some of them have been experimentally validated (1,2) and some are used for environmental impact assessment studies. It can be stated that, for source-receptor distances up to about 20 km, the majority of the applications of atmospheric dispersion models is based on the bi-Gaussian dispersion formula. The disadvantage of the theoretical limitations of the theorem is counter-balanced by the fact that only for the bi-Gaussian model the necessary input parameters are usually available.

The availability of the necessary model input parameters is a major consideration in the decision to use the bi-Gaussian model for the simulation of atmospheric dispersion in emergency situations in case of an accident or incident at a nuclear facility. The objective of the research contract between the CEC and the S.C.K./C.E.N., was to investigate the applicability of this approach.

The work was subdivided into two main tasks :

Task 1 : to investigate the sensitivity of the model results to the use of different turbulence typing schemes and dispersion parameter sets.

Task 2 : to validate the model results by comparing the calculated immission with the measurements of 10 tracer dispersion experiments.

The granted length of this paper does not allow for an illustration of the results and conclusions of the project. It will therefore be focused on the major conclusions. More information is published in references (3-6).

2. INFLUENCE OF TURBULENCE TYPING SCHEME AND DISPERSION PARAMETER SET

2.1. Experimental

With the meteorological measurements on the 120 m meteo tower at the S.C.K./C.E.N., Mol and the climatological observations at the airport of Kleine Brogel, 20 km E of Mol, a data base for three consecutive years (1977 to 1979) was assembled. It contained all meteo parameters necessary to determine on an hourly basis the atmospheric dispersion conditions using the procedures of several commonly used turbulence typing schemes.

Using this meteo data base, the atmospheric stability class was determined on an hourly basis using the following 10 schemes that have been published in the literature :

- the S.C.K./C.E.N. system based on the vertical temperature gradient and the horizontal wind-speed
- six systems based on the work of Pasquill : the original Pasquill approach based on synoptical observations (SYN) ; the procedure of the Nuclear Regulatory Commission (NRC) using the vertical temperature gradient ; the system proposed in the Probabilistic Risk Assessment Procedures Guide (PRA) based on the vertical temperature gradient and the horizontal wind-speed ; the adaption of Turner using the cloud cover and height and the solar angle ; the NRPB system developed by Smith and Hosker and finally the evaluation of the Prairie Grass experiments by Klug.
- two classification schemes developed by Vogt in Jülich during many tracer dispersion experiments. One turbulence classification system is based on synoptical observations, the other on the vertical temperature gradient and the wind-speed.
- in contrast to the previous systems, containing 6 or 7 classes, the French Doury system discerns only poor and normal dispersion conditions.

Following the determination of the atmospheric stability class the horizontal and vertical dispersion parameters (σ_y and σ_z) were determined on an hourly basis using the dispersion parameter sets corresponding to the proper turbulence typing scheme. The results of the Pasquill Synoptical classification was furthermore combined with the dispersion parameter sets of Pasquill-Gifford, Briggs open country and Briggs urban area.

In order to compare the results of the different alternative combinations of turbulence typing scheme-dispersion parameters, systematic calculations for different averaging times (8h, 16h, 3d and 26d), receptor orientations and source-receptor distances and for a ground level release and a release at an effective height of 100 m were done by means of the bi-Gaussian Immission Frequency Distribution Model (IFDM) of the S.C.K./C.E.N.

2.2. Results and Discussion

The first step in the intercomparison of the different systems, is the deduction of the atmospheric stability class starting from the same meteorological data. The results of the 10 different turbulence typing schemes were bilaterally compared by making co-frequency matrices where the frequency of simultaneous occurrence of one stability class in a given scheme with the occurrence of a class in another scheme is indicated. The obtained co-frequency matrices were surely not of the diagonal type and the scatter within each matrix was large and scheme dependent.

This means that a given meteo situation can be classified in a certain stability class by one scheme but into another class by another scheme. The differences can be systematic or stochastic. The majority of the bilateral combinations resulted in less than 50 % agreement on the stability class. Combinations that resulted in more than 50 % agreement were on one side : S.C.K./C.E.N., Pasquill NRC and Pasquill PRA, and on the other side : Pasquill SYN, Turner, Vogt SYN and NRPB. The three systems in the first group use the vertical temperature gradient as a main input parameter. The four systems in the second group are based on synoptic observations.
Even if the stability classification is as rough as simply : stable, neutral or unstable, the different systems obtain different frequencies of occurrence as is illustrated in table 1.

Table 1 - Frequency of occurrence of stable, neutral and unstable situations using different classification procedures on the same meteo data base

<table>
<tr><th>System</th><th>% stable</th><th>% neutral</th><th>% unstable</th></tr>
<tr><td>S.C.K./C.E.N.</td><td>60</td><td>23</td><td>17</td></tr>
<tr><td>Klug</td><td>30</td><td>30</td><td>40</td></tr>
<tr><td>Pasquill SYN</td><td>27</td><td>47</td><td>26</td></tr>
<tr><td>Pasquill NRC</td><td>47</td><td>42</td><td>11</td></tr>
<tr><td>Pasquill PRA</td><td>44</td><td>29</td><td>27</td></tr>
<tr><td>Turner</td><td>28</td><td>59</td><td>13</td></tr>
<tr><td>NRPB</td><td>22</td><td>72</td><td>6</td></tr>
<tr><td>Vogt SYN</td><td>26</td><td>55</td><td>19</td></tr>
<tr><td>Vogt GRAD</td><td>26</td><td>41</td><td>33</td></tr>
<tr><td>Doury</td><td>47</td><td colspan="2">53</td></tr>
</table>

The questions is whether these important differences in stability classification have a large influence on the immission calculations.

Therefore the cumulative frequency distribution of the concentrations were calculated for receptors at 0.5 km, 1 km, 2 km, 4 km and 8 km in 8 directions around the source, on an hourly basis.

The scatter of the results of the different sets is high, as well for the lower as for the higher percentiles and for the maxima.
Although the relative difference between the schemes varies with the specific combination receptor - statistical parameter - averaging time, some general conclusions remain valid.

- Due to the narrow plumes in Klug's system, the higher percentiles yield the highest concentrations of all sets and the cumulative frequency distributions drops steeply to the lower percentiles.
- The three Pasquill-turbulence typing schemes (SYN, NRC and PRA) and the Turner scheme, combined with the Pasquill-Gifford dispersion parameter set yield fairly similar cumulative frequency distributions, despite the fact that the schemes showed a marked difference in stability classification.
- Also Vogt SYN and Vogt GRAD yield very similar cumulative frequency distributions. This means that the choice of the dispersion parameter

set, and the mutual differences between these sets, influence the yearly statistics much more than the differences between the stability classification schemes.

- Klug's scheme almost systematically gives the highest values, followed by Turner and the different Pasquill systems in combination with the Pasquill-Gifford dispersion parameters. Also Pasquill SYN in combination with Briggs open country belong to this group. Smith-Hosker's scheme generally gives lower results and is quite often at a somewhat intermediate position in the ensemble of twelve combinations. Further down are the results of S.C.K./C.E.N., followed by both Vogt systems, not drastically different from each other in their end results, and finally Pasquill SYN in combination with Briggs urban area. For the extreme levels, especially for the shortest averaging times, Vogt, Briggs urban area and Smith-Hosker sometimes change positions as a function of the specific combination.
- Doury's results are specific in the sense that they are highest, or belong to the group of highest results, for the yearly average levels and move gradually to the group of lower or lowest values with decreasing averaging time and/or increasing percentile level to give the smallest short-term maxima in certain receptors.
- Depending upon the receptor, differences in calculated yearly averages can be as much as a factor 8 while one to two orders of magnitude were noted for individual extremes.
- For the 100 m-release comparable differences below a factor 8 for the yearly average levels were noted, although the differences in the maximum short-term levels were much smaller than for the ground-level releases, at least when not taking into account the position where the maxima in the different combinations occurred.
- Concentration-distance curves are indeed quite different from one system to another and wind-direction as well as averaging time dependent. The grouping of the end results noted already for the ground-level releases persisted but in a reverse way. When comparing the maxima for a given statistical parameter, occurring somewhere between 300 m and 8 km from the source, the following ranking from smallest to highest can be made in a first approximation : Doury, Klug, Turner and the Pasquills with as well the PG-curves as Briggs' open country dispersion parameters, Smith-Hosker, S.C.K./C.E.N., Vogt SYN and Vogt GRAD, and Pasquill SYN with Briggs urban area. As pointed out before this ranking is not absolute and many less or more pronounced variations do actually occur.

This study clearly illustrated that the appropriate choice of a specific turbulence typing scheme, and the corresponding set of dispersion parameters in the bi-Gaussian dispersion formula, is critical when impact assessment studies are undertaken for a specific installation at a given site.

The outcome here has already been given by B. Turner in his 1979 critical review of atmospheric dispersion modelling where he noted "There is a significant need to plan and execute additional field experiments to improve our ability to estimate dispersion. In order to do this a significant multi-disciplinary field effort is required".

3. MODEL VALIDATION

3.1. CAERS

The sensitivity of the bi-Gaussian model to its input parameters has been investigated using the statistics of the calculations over one-year periods in different receptors. The calculations have been done with the Immission Frequency Distribution Model : IFDM. The second task of the project was to validate the bi-Gaussian theorem for the simulation of individual periods using tracer dispersion experiments. The comparison of calculated with measured immission must not be done on a statistical but on an individual basis, and thus the calculations should not be done with the IFDM code. Another computer code was therefore elaborated at the S.C.K./C.E.N. with the goal to simulate individual periods with the bi-Gaussian theorem or with a bi-dimensional puff trajectory model with tri-Gaussian diffusion of the puffs for the case that the initial assumption of wind-direction persistency should not be fulfilled. The computer code, incorporating both models and the 12 different turbulence typing schemes and dispersion parameter sets (described in § 2), developed at the S.C.K./C.E.N. is called CAERS : Computer Aided Emergency Response System.

3.2. Experimental

The experimental data, necessary to validate the performances of a mathematical model, are : the emission conditions, the meteorological parameters during the experiment and the resulting concentration field. In order to obtain this information, eleven tracer dispersion experiments were carried out at the S.C.K./C.E.N. Sulfurhexafluoride (SF_6) tracer was released at 2 m height at a rate of about 3 g/s during 2 to 3 hours.

Air was sampled at 20 to 35 locations up to 7km downwind the source during 2 or 3 consecutive periods of 30 minutes. The 30 minutes average SF_6 concentrations were analysed afterwards. During the experiment instantaneous SF_6 concentrations were determined using a SF_6 analyser in a measuring van. Cross-wind profiles of the instantaneous SF_6 concentration and variation of the SF_6 concentration as a function of time were obtained.

In order to validate the bi-Gaussian model, whenever possible, a Gaussian curve was fitted through the SF_6 ground-level concentrations, measured on an arc more or less perpendicular on the average plume axis.The parameters of the Gaussian profiles are : the horizontal and vertical dispersion (σ_y and σ_z) and the direction of the plume axis. The wind-speed also influences the height of the Gaussian fit ; it was obtained from meteorological measurements in the area. However the accuracy of the estimation of the vertical dispersion is hampered by the accuracy of the knowledge of the wind-speed, influencing the actual SF_6 plume. Figures 1 to 3 give for a typical experiment (not the best nor the worst) the schematic representation of the emission and sampling network, with the measured SF_6 concentration as a bar with the length proportional to the measured concentration pointing from the sampling place to the emission site ; the Gaussian curve fitted through the experimental points, with the parameters of the fit in the figure caption ; and the scatter diagram between the measured concentrations and the concentrations calculated with CAERS and the S.C.K./C.E.N. turbulence typing scheme.

Most of the measured cross-wind ground-level concentration profiles could be well fitted with a Gaussian curve. 41 profiles from 11 experiments were used for the model validation. The maximum concentration of a profile was deduced from the maximum of the fit rather than the maximum measured concentration, because the real maximum can fall between two sampling points.

When measurements are done during two or three consecutive half-hour sampling periods, under apparently equal atmospheric conditions, the correspondingly measured dispersion parameters can be almost equal but can also differ up to 50 % one from the other. The same is true for the maximum concentration at a given distance. The variability of the measured dispersion under similar atmospheric conditions must be held in mind, when the model values are compared with the measured dispersion parameters.

3.3. Model validation

The measured horizontal and vertical dispersion parameters of the 41 cross-wind profiles were compared with the values obtained with the S.C.K./C.E.N. turbulence typing scheme dispersion parameter set. The average ratio of measured over model dispersion was 0.82 $\pm$ 0.22 for the horizontal and 1.25 $\pm$ 0.65 for the vertical. The precision of the experimental estimation of σ_z is worse because the vertical concentration profile could not be measured directly. The difference between model and measured dispersion is statistically not significant.

The, from a health safety point of view, important simulation of the maximum concentration has been successful with the S.C.K./C.E.N. system : the average ratio of calculated over measured maximum concentration is 1.02 $\pm$ 0.65 for the 41 observations.

When the measured concentrations of a single sampling period are compared with the corresponding concentrations calculated with the S.C.K./C.E.N. system by means of a scatter diagram and a least square linear fit, a very good correlation is usually obtained : the average correlation coefficient is 0.92 and the average slope is 1.03. An important aberration of the slope from unity usually corresponds with a large difference between σ_z used in the model and σ_z derived from the measurements.
When the data from all experiments are collected in one scatter diagram, situations of different days, that are obviously not physically correlated, are mixed in one scatter diagram resulting in a decrease of the correlation coefficient (r = 0.80) and slope (a = 0.55) of the linear fit.

The measured dispersion parameters have been compared with the values obtained with the 12 turbulence typing scheme-dispersion parameter sets, by means of the fractional error.
For the horizontal dispersion parameter the average fractional error (i.e. the accuracy) is the smallest for the Pasquill based system (-13 % to +15 %), it is about 20 % for the S.C.K./C.E.N. system while the Vogt and Doury systems overpredict the measured σ_y with 40 to 60 %. The scatter of the results (i.e. the precision) is best for the S.C.K./C.E.N. system (20 %), and ranges from 30 to 50 % for all the others.

The accuracy and precision of the vertical dispersion parameters are the best for the S.C.K./C.E.N. system. The Pasquill based systems underestimate the vertical dispersion with 30 to 80 % while the scatter of the results is high too (50 to 80 %). Vogt overestimates σ_z with 50-60 % and Doury underestimates σ_z with 80 %.

The variations of the plume under non steady emission and meteo conditions are well simulated by the puff trajectory model.

4. CONCLUSION

The sensitivity analysis of the bi-Gaussian model in the IFDM code to the origin of the input parameters has been demonstrated to be very

important. The tracer experiments have shown that the bi-Gaussian model in the CAERS code, using the S.C.K./C.E.N. turbulence typing scheme, can simulate short-term concentration levels very well.
It is clear that for a reliable use of mathematical models, the dispersion parameters must be optimized for typical topographical textures and that they cannot be simply used for other areas.

The fiability of CAERS and IFDM allows to conclude that mathematical models for the simulation of atmospheric dispersion can give a significant contribution to the management and the interpretation of air pollution concentration measurements in emergency and routine immission situations.

ACKNOWLEDGMENTS

This work was partly supported by CEC contract 028-SRB in the framework of the Indirect Action Research Programme on the Safety of Thermal Water Reactors.

REFERENCES

1. VANDERBORGHT, B. and KRETZSCHMAR, J. (1984). A literature survey on tracer experiments for atmospheric dispersion modelling studies. Atm. Env., in press.
2. VANDERBORGHT, B., MERTENS, I. and KRETZSCHMAR, J. (1983). Comparing the calculated and measured aerosol concentrations and depositions around a metallurgical plant. Atm. Env. 17, 1687-1701.
3. KRETZSCHMAR, J., MERTENS, I., DE BAERE, G. and VANDERVEE, J. (1983). Influence of the turbulence typing scheme upon the cumulative frequency distributions of the calculated relative concentrations for different averaging times. CEC-EUR 8478 EN (1982)
4. KRETZSCHMAR, J., MERTENS, I. and VANDERBORGHT, B. (1984). Development and evaluation of the possibilities of on-line dispersion models for use in an emergency plan of a nuclear installation. CEC document XII/371/84-EN
5. KRETZSCHMAR, J. and MERTENS, I. (1984). Influence of the turbulence typing scheme upon the cumulative frequency distributions of the calculated relative concentrations for different averaging times. Atm. Env., in press.
6. VANDERBORGHT, B. and KRETZSCHMAR, J. Atmospheric tracer dispersion experiments. 3rd European Symposium : Physico Chemical Behaviour of atmospheric pollutants, Varese (I), 10-12 April 1984.

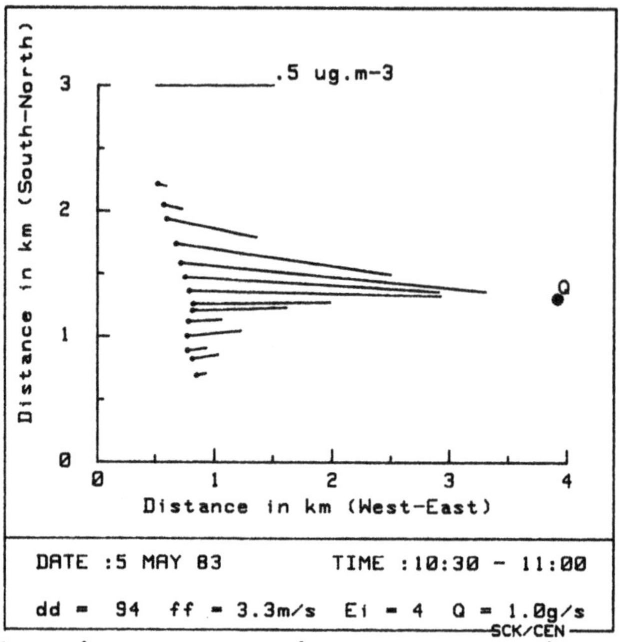

Figure 1 : Schematic representation of a cross-wind SF_6 ground-level concentration profile

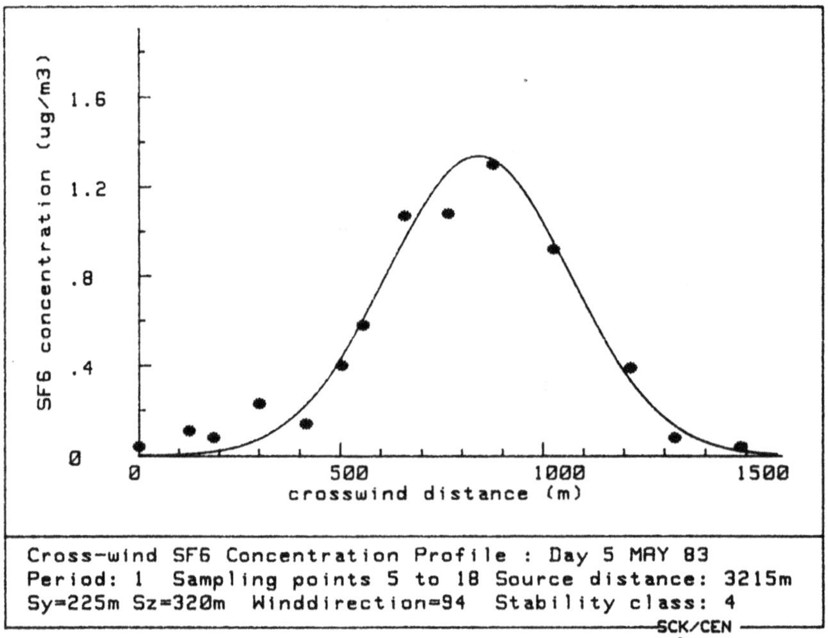

Figure 2 : Gaussian fit through the SF_6 concentrations measured on a line perpendicular to the wind-direction

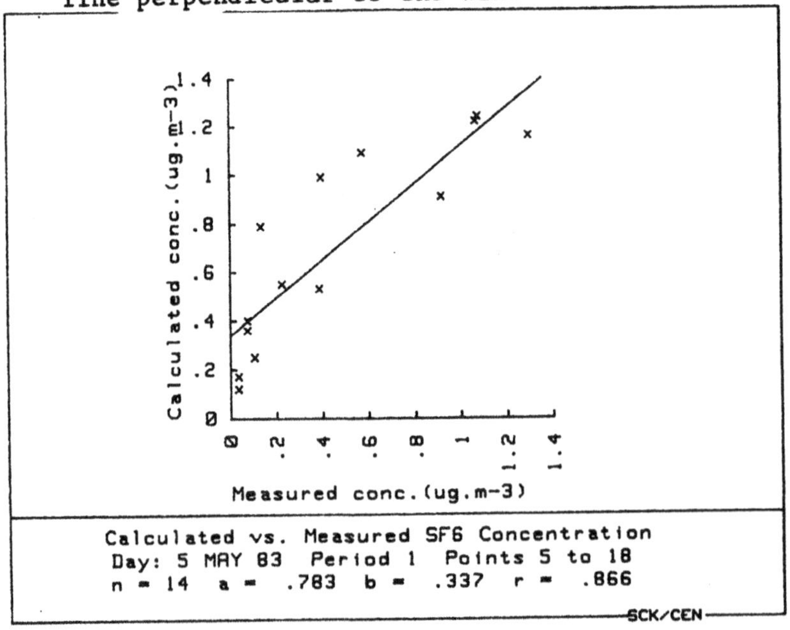

Figure 3 : Scatter diagram of calculated versus measured SF_6 concentration

Discussion

Question: U. Bäverstam - NIRP (SSI)

1. You said the "classification errors" were site specific. Are there better and worse kinds of sites referring to classification?
2. Do you think it worthwhile to perform a similar classification comparison for a coastal site (for emergency planning)?
3. Do you think one should avoid stability classes in emergencies, using for example variation in wind direction only?

Answer: B. Vanderborght - SCK/CEN-Mol

1. Not only the results of the stability classification procedures but also, and especially, the dispersion parameterisation are site specific. There is no one "standard site" nor "standard procedure" that should yield reliable results in all situations.
1 and 2. There is no doubt that for valuable impact assessment studies at least different "typical sites" should be specified, resulting as such in a combination system "site classification - atmospheric stability classification dispersion-parameter set". The idea is gaining approval by people involved with atmospheric dispersion work, but the practical realisation can only be expected from a multinational coordinated research programme including field experiments.
This consideration is valid for impact assessment studies for routine emissions and surely also for emergency situations.
Obviously "coastal site" should be a topic in the "site classification".
3. In emergency situations, the variation of wind direction is a valuable procedure to estimate horizontal dispersion. I don't think it is as valuable for estimating the vertical dispersion, where the vertical temperature gradiant is more useful.

Question: P. Foster, CEGB/CERL

What was the overall accuracy of your tracer experiments, e.g. including both source strength and measurement techniques?

Answer: B. Vanderborght - SCK/CEN-Mol

The emission rate of SF_6 gas can be measured with a rotameter, a gasmeter or by weighing. By proper control of all physical parameters the rate determination by the three methods agreed within 1.5%. The accuracy of the concentration measurements has been the subject of contract 032 SRB and shown to be somewhere between 3 and 10%.
Maybe the determination of the windspeed at plume height is the point with the worst accuracy, and also the estimation of this accuracy is not straightforward.

INTERLABORATORY COMPARISON OF SF_6 DETERMINATION USED WITHIN THE FRAMEWORK OF TRACER RELEASES.

B. VANDERBORGHT
S.C.K./C.E.N.
Boeretang 200, B-2400 Mol, Belgium

Summary

An international round robin exercise was organized to investigate the accuracy of the determination of trace amounts of SF_6 in air samples, with the possibility of a large scale joint European atmospheric tracer dispersion experiment in mind.

Experimental

The S.C.K./C.E.N., Mol Belgium, was the coordinator of the experiment. It made 6 samples of SF_6 diluted in air in the 10 ppt to 10 ppb concentration range. The preparation procedure was based on the dilution of a small amount of pure or diluted SF_6 in a large volume of pure air under atmospheric pressure. The volume of the first amount was accurately determined by weighing the water content of the syringe, used for the injection of SF_6 gas in the larger volume, which was obtained with a "mercury-ring cylinder" with accurately known inner diameter and height. The final samples were stored in PVC plastic bags under atmospheric pressure.

The samples were analysed by the following laboratories : CEN (F), CERL (UK), JRC (I), KEMA (NL), LHW (NL), NILU (N), APL (DK), WDB (FRG) and SCK (B).

All participants used gas chromatography with electron capture detection for the analysis. The following calibration procedures were used :

- 3 step semi-dynamical dilution, atmospheric pressure : SCK
- static dilution, atmospheric pressure : APL, WDB
- exponential dilution, atmospheric pressure : CERL, NILU
- static dilution, increased pressure : JRC, KEMA, LHW
- static dilution with partial pressures : CEA

The diversity of the calibration procedures increases the value of the accuracy test.

Table 1 gives the analysis results of the 6 intercalibration samples, as they were reported by the participants. The second column gives the concentrations as they were calculated from the dilutions made at SCK.

Table 2 gives the ratio of the measured to the calculated concentrations.

Table 1 : Analysisresults of the participants (concentrations in ppt)

sample number	calculated	SCK	CEN	CERL	JRC	KEMA	LHW	NILU	APL	WDB
1	9,84	-	-	16,5	89	11,2	25	5,6	7,4	-
2	39,4	34,8	≃ 21	42,8	470	33,4	39	35	32	-
3	148	135	56	148	180	107	111	140	120	-
4	605	555	412	567	520	405	403	586	460	300 ±30
5	2642	2544	2223	2360	2100	1677	1915	2550	2000	2100 ±200
6	8752	-	8687	7410	8340	-	7507	7250	6700	7800 ±500

Table 2 : Ratio of measured over calculated concentrations

sample	SCK	CEN	CERL	JRC	KEMA	LHW	NILU	APL	WDB
1	-	-	1,68	(9,04)	1,14	(2,54)	0,57	0,75	-
2	0,88	(0,53)	1,09	(11,9)	0,85	0,99	0,89	0,81	-
3	0,91	(0,38)	1,00	1,22	0,72	0,75	0,95	0,81	-
4	0,92	0,68	0,94	0,86	0,67	0,67	0,97	0,76	(0,50)
5	0,96	0,84	0,89	0,79	0,63	0,72	0,97	0,76	(0,79)
6	-	0,99	0,85	0,95	-	0,86	0,83	0,77	(0,89)
average	0,92±0,03	0,84±0,16	1,08±0,31	0,96±0,19	0,80±0,21	0,80±0,13	0,86±0,15	0,78±0,03	

Discussion

- From the analysis results of the SCK after the return of the samples could be concluded that the JRC samples 1 to 3 have been seriously contaminated. It can be questionable whether the other samples 4 to 6 have been damaged too. The JRC analysis results must therefore be interpreted with extreme care.
  Here and there, another sample can have been damaged : LHW 1 and 2, KEMA 1 and NILU 6. It is however uncertain whether the damage has occured before or after the analysis by the lab.
  APL and WDB have analysed and returned their samples very late (≃3 months after synthesis). The bags seem to have lost about 20 % of the initial SF_6 concentration.
  The other samples seem to have withstood the trip without concentration change.

- CEN and WDB normally operate in a higher concentration range.
  The samples 1 to 3 are close to or below their detection limit.
  The accuracy of the analysis is rather poor for the lower concentrations but improves with increasing concentration.

- The results of CERL and NILU agree fairly well with the calculated concentrations.

- The results of SCK, CERL and NILU are however slightly smaller than the calculated concentrations. This could mean that the intercomparison samples show a slight positive error of a few procent.

- The results of APL, KEMA and LHW are systematically and significantly lower than the calculated concentrations.
  From the results after return can be deduced however that the APL samples have lost SF_6 due to aging. The comparison of APL and SCK analyses (table 3.5) is indeed very good.
  This is not the case for KEMA and LHW.

- Considering the limitations from the first remark, the results of JRC agree within some 20 % with the calculated concentrations.

- Summarizing can be stated that :

  a. SCK, CERL, NILU are in very good agreement.
  b. also APL seems to be in agreement with SCK, CERL and NILU if the aging of the samples is considered.
  c. KEMA and LHW underestimate the calculated concentrations by about 20 %.
  d. CEN and WDB are used to work in higher concentration ranges and the results of the low concentration measurements in this intercomparison are less accurate.
  e. At least some samples of JRC have been contaminated before analysis. The accuracy of the JRC analyses cannot be evaluated unambiguously.

- The four lab's : SCK - CERL - NILU and APL use more or less different methods for the synthesis of the calibration samples (table 2.1), but they all make the calibration samples under atmospheric pressure. KEMA and LHW on the other hand, make the calibration samples in steel bottles at a pressure up to 150 bar.
  In order to further investigate the possible remaining differences, further intercomparison work has been done between SCK and KEMA, and between SCK - APL and NILU.

The further comparisons with KEMA did not yet reveal the reason of the differences. Work is still going on.

In May and June '84 SCK has visited RISO - APL National Laboratories with the SCK measuring van. SCK and APL have made 5 calibration samples independently from each other. Both calibration curves agreed within 5 %. This confirms the good agreement between SCK and APL in the intercomparison, when aging of the intercomparison samples between November '83 and January '84 was considered.

As a supplementary test, three unknown samples of the Oresund tracer dispersion experiment have been analysed by SCK, RISO -APL and NILU. They agree within 4 %.

Conclusions

- A very good agreement for the determination of trace amounts of SF_6 in air is obtained between SCK, CERL, APL and NILU.

- LHW and KEMA obtain systematically lower results than the four other laboratories. The reason for the difference has not yet been resolved, but it seems that the problem can be in the manufacturing of the mothersamples under high pressure.

- CEN and WDB are used to work in a higher concentration range. Their analysis of trace amounts is less accurate.

- The JRC samples have been contaminated and the accuracy of the JRC analysis cannot be evaluated unambiguously.

- The SF_6 samples in PVC plastic bags are not stable enough over long periods, especially when the ambient temperature is higher than 30° C. Transport of samples in plastic bags over long distances is not advisable when the transport conditions cannot be well controlled.

Discussion

Note: No discussion has taken place on this paper.

Estimates of Uncertainty in Dispersion Modelling

by F. B. Smith, Boundary Layer Research Branch,
Meteorological Office,
Bracknell, Berkshire, RG12 2SZ, U.K.

SUMMARY

Risk assessments associated with releases into the air of any potentially hazardous material (such as radionuclides, toxic or inflammable gases) must involve not only the expected air concentration, deposition or dosage but also the uncertainty involved in such expectations.

This paper briefly examines the errors that arise from the models and their associated parameters, from the numerical methods employed, from the inexactness or inappropriateness of the basic meteorological (and other) input data, and from the inherent variability of the atmosphere arising from its turbulent character.

It goes on to consider how these errors affect the pattern and magnitude of the downwind groundlevel hazards in subtle but often dramatic ways. It emphasises that on-site measurements of meteorological parameters should be made with proper regard to the scales of the advection-diffusion processes involved, and not over some quite arbitrary and unrelated time-scale as is often the case at present.

1. General aspects of uncertainty

A release of any gaseous or particulate species into the atmosphere is subject to many processes which involve uncertainty. The uncertainty may arise from fundamental ignorance (complete or partial) of the physical and chemical processes involved, from a lack of information about the circumstances under which the release took place, from an inability to extract the most information from the input data that is available (perhaps because computer facilities or required time may be limited or non-existent), and from the inherent variability that is always present in a turbulent medium such as the atmosphere. These areas of uncertainty are summarised in Table 1

Generally the greatest area of uncertainty arises from a lack of information concerning the input data to models. The input data can be divided into data on emission characteristics, data on the state of the atmosphere, and data on the underlying terrain. The first of these includes the source strength (how much material has been released in total and how did the release rate vary in time), the buoyancy associated with the release and the velocity of the release-gases (both of which influence the rise of of the plume), where the release into the atmosphere took place (from a stack, from ventilation vents, from near the ground etc), and the physical and chemical character of the release (the particulate size distribution, the decay or reaction rates, etc). This source of uncertainty may be the greatest for accidental releases of radionuclides at nuclear plants, especially during the first few days following the accident until a detailed survey of what actually happened is available. Uncertainties of an order of magnitude may be expected which will therefore swamp all other uncertainties.

Usually uncertainty on the state of the atmosphere will be the next largest. However, how important this will be in practice will depend very much on what aspect, or consequence, of the release has to be modelled.

For example whether or not a fixed community will be,or has been, affected by a plume during a 24 hour continuous release is decided much more easily and reliably than trying to estimate the concentration at a given fixed place at a specified time. In other words the uncertainty depends on whether we are concerned with air concentrations, total dosages or ground depositions, whether we are concerned with specific point values or ranges of values over a large area, whether at a given time, or at any time during a period, or an integrated time-average, and whether we are dealing with a specific release or a possible release some time in the future.

It is obvious that the more data are available, the more facilities are at hand to extract meaningful information from the data, the less specific one is required to be in terms of time and space, the more accurate one can hope to be in ones estimations.

2. Field and laboratory evidence for uncertainty

(i) Winds and trajectories

Perhaps the most fundamental question of all is "Is this receptor affected by the plume or not?", and the answer, whatever it is, automatically provokes a second question "how certain are you that the predicted path of the plume is realistic?". Clarke et al.(1) has studied 32 tetroon releases which have lasted for more than 250km or 12 hours travel time in various Environmental Protection Agency (EPA) field experiments. They compared the trajectories of these tetroons with trajectories determined by various meteorological trajectory models. The model giving the best overall results was the National Oceanic and Atmospheric Administration's Air Resources Laboratory Atmospheric Transport and Dispersion (ATAD) model. With this model, the tetroon trajectory could be expected to be 8 - 10 % of the trajectory length to the left of the diagnosed position, with a standard error of about 25-30% of the trajectory length. In terms of time of travel T the mean separation e between the tetroon position and the diagnostic position was given approximately by:

$$|e| = 0.8\,T \text{ (kms)} \qquad (3 < T < 21\text{hrs})$$

and the standard deviation $\sigma(|e|)$ just slightly less than this. The errors were thought by Clarke et al. to be probably less in well-defined flow patterns and whenever meteorological data (especially radiosonde data) were available more frequently than every 6 hours. In situations when likely plume trajectories need to be forecast ahead for the following hours (for example as soon as a hazardous emission commences) the extra uncertainty associated with forecasting future meteorological states will increase the probable error. The magnitude of the forecast error depends very much on the sophistication of the model used and, in particular, on the details of the boundary layer representation and its resolution. It also depends quite critically on how the gridvalues (which can often show small-scale roughnesses) are spatially smoothed. No proper analysis has yet been done on these errors, perhaps because of the difficulty in formulating a totally satisfactory comparison.

However as a first stage in determining the quality of forecast trajectories based on the output of the Meteorological Office's so-called fine-mesh operational numerical weather prediction model (15 levels in the vertical,horizontal gridlengths at 50°N of 68km north-south and 83km east-west), Maryon and Riggs have recently compared trajectories starting from the same point at the same real time, one based on an initial actual field but subsequently on the model's forecast fields, and the other entirely on forecast fields based on a run starting 12 hours earlier. The following differences were found:

	median	mean	standard deviation	mean distance
after 24hours	91km	106km	76km	600km
after 36hours	174km	219km	141km	1200km

The expected difference is thus about 1/6th of the distance of travel and is therefore comparable to the error found by Clarke between ATAD trajectories and tetroon trajectories. There is clearly room for a lot more basic research in this area.

(ii) Plume rise

In models that have been tuned to give the correct mean plume rise at downwind distances exceeding 800 metres, the expected difference between predicted and observed rise is about 13% of the total plume height, according to field studies by Moore (2). Leahey and Davies (3) have studied plume rise from a flare made visible through the injection of oil into the flame. The rise h' was generally consistent with the theoretical prediction that $h' \propto x^{2/3}$ until final rise was attained. However the correlation coefficient between 777 observed and theoretical plume rises measured over 2-3 minute periods was only 0.74, i.e. the root mean square expected error is roughly 2/3 the r.m.s variation in plume rise over the 777 observations.

Willis and Deardorff (4) have studied plume rise within a laboratory convectively-mixed layer. They show that active convection has a profound effect on plume rise, indicating the need to modify Briggs' (5) well used formula in these conditions. They also studied the effect of a fluctuating horizontal wind but unfortunately their paper fails to give an indication of the resulting scatter.

(iii) Ground-level concentrations

There are so many variables that govern the concentration of a species emitted from an upwind source that it is difficult to define an absolute measure of uncertainty. In general the uncertainty has to be linked to a specific model which contains only a limited number of parameters. The spread of concentrations measured when these parameters are apparently nearly the same yields the uncertainty for that model for that set of parameter-values. But the problem of doing this is clouded by the sensitivity of the model to possible errors in the inferred parameter-values, and discussion of this point will be left till later.

To give but an example of the uncertainties experienced in practice, Crawford (6) has shown that for a specific hour and at a specified receptor point within 10km of the source when the intervening countryside is flat and the meteorological conditions are steady, the ratio of the predicted concentration to the observed, in which the former is determined using a basic Gaussian plume model, can vary from 0.1 to 10 : a rather large measure of uncertainty!

Vanderborght et al. (7) have made measurements of aerosol concentration and deposition around a metallurgical plant in Belgium over a 14-month period, together with reliable data on emission rates and local meteorological conditions. They claim that a bi-Gaussian model provides predictions that give short-term concentrations which are generally within 40% of the observed, and average depositions within 60%. Comparing this with Crawford's result it is clear that the quality and detail of the input data have a marked effect on the output accuracy.

Harrison and McCartney (8) made ground-level measurements of NO_x concentration 1.2km downwind from five 46m high stacks at a fertilizer factory, and found reasonably good agreement with the predictions of a simple Gaussian plume model using conventional Pasquill-type stabilities and σ_z. For all stabilities the predicted values lay within ± 50% of the observed values, but with a tendency for the former to be higher than the latter, especially in unstable conditions.

On a much longer time-scale, Simpson and Jakeman (9) have shown that annual maximum acid levels at 2 stations vary around 3 to 4 over a ten year period, due almost equally to inter-annual variations in the windfield and to variations in emissions and other meteorological factors.

(iv) Concentrations within plumes

Deardorff and Willis(10) have studied both buoyant and non-buoyant plumes within a laboratory convectively mixed layer. They find that concentrations do vary within plumes and that the cumulative frequency distribution of non-zero sampled concentrations is virtually log-normal. When meandering of the plume, and resulting occasional zero concentrations are included, the mean square concentration fluctuation $\overline{C'^2}$ is found to vary very little across the time-mean plume but decreases significantly with downwind distance, partly because a given receptor is more consistently either totally within the plume or out of it. Fackrell and Robins (11) have made similar experiments in a simulated neutral boundary layer. They find $\overline{C'^2}$ is a maximum at $z/h = 0.75$ for a ground level source (h is the mean-concentration half-height) which presumably has a very small contribution from vertical plume meandering. Elevated sources give profiles of $\overline{C'^2}$ with only a single maximum, the height of which approaches that of the ground-level source as the two mean-concentration profiles approach each other at large downwind distances. Again $\overline{C'^2}$ is shown to decrease substantially with downwind distance. The actual magnitude of $\overline{C'^2}$ depends not only on x,y and z but also on the period of sampling.

(v) Plume widths σ_y and σ_z

The variability of width depends again on the way meteorological conditions are parametrized. On a single occasion when all the mean-flow parameters are by definition fixed, σ_y and σ_z will vary in time due to spatial and temporal variations in turbulent energy on many scales. Smith and Readings (12) have looked at a specific single day in unstable conditions in summer and showed that in the micrometeorological part of the spectrum the energy levels in narrow frequency bands (corresponding to averaging times t_a and sampling times $2t_a$) showed considerable scatter: the root mean square variations of the levels were about 60% of mean levels. This degree of variability would also be reflected in plume widths averaged over $2t_a$, at distances downwind from the source where an equivalent Eulerian averaging of t_a seems appropriate.

Moreover when different occasions with apparently the same mean-field characteristics are grouped together, width-scatter is generated not only from the inherent variability in the turbulence but also from more subtle differences between the occasions.

The National Council on Radiation Protection and Measurement NCRP Report No.76 (13) quotes data on lateral width, σ_y, in which widths measured in field experiments over flat terrain are compared with predictions on the basis of a stability classification expressed in terms of the vertical temperature lapse rate. To define stability in this way is questionable, and this is perhaps one reason why the scatter (90% of the points lie within a factor of 4) is so large. Values of width (measured and predicted) found in the experiments carried out in the more complex terrain at Mt. Iron are also given in the Report. Here classification was made, not in terms of lapse rate, but in terms of wind direction fluctuations. Apart from a few very wild outliers, the great majority of points are now within a factor of 2 in the comparison of measured to predicted, showing that in spite of the rougher terrain the parametrization was more realistic.

The Report also quotes Briggs and McDonald's (14) analysis of σ_z inferred from ground-level measurements made in the Prairie Grass Experiment By proper parametrization, in which σ_z/L is expressed in terms of $u_* x/(UL)$ (where L is the Monin-Obukhov length scale and u_* is the friction velocity), the scatter of σ_z is reduced to within a factor of about 2 in both stable & unstable conditions out to $x < 10\ U|L|/u_*$.

Observations of plumes emanating from sources close to the ground on top of buildings suggest that the character of gusts, associated with incursions of faster moving air from greater heights, is very important to the spread at short range. A sudden gust causes a rapid alongwind convergence and a compensating lateral divergence of the plume. The rate of spread can then be linked to the sharpness of the gust "front", an aspect which is not explicitly considered in normal atmospheric diffusion theory, but which is commonplace in aerodynamic theory.

3. Sources of Uncertainty

Table 1 sets out the major sources of error and uncertainty in dispersion estimation. The first and most obvious source is in the nature of the models themselves. They are only simplified models of the real world They are designed to reflect our limited scientific understanding of what is actually taking place, to meet certain specifiable objectives, which may differ from model to model, to be expressible in terms of a limited input-data supply, and to be capable of yielding results with available resources (manpower, time and computer facilities). An individual model may work well within a certain range of conditions for which it was designed but relatively badly outside this range. For example a Gaussian plume model may be satisfactory in rather uniform flow over level terrain, but may work poorly in strong shear flow over irregular terrain.

Many dispersion models have an element of parametrization, that is when some aspect of the physics or chemistry involved in the process being modelled is represented by some approximate algebraic or empirical expression which may include one or more adjustable constants. Sometimes these parametrizations yield solutions whose properties are in some respect physically unacceptable. For example, solutions of the classical eddy diffusion equation are known to give unrealistic plume widths close to the source, except when it is at ground level. Other parametrizations are in danger of being just too simplistic. For example it is clearly perilous to represent the often complex consequences of stability in the boundary layer by a single stability parameter as is usually done through necessity. Moreover many models contain parameters which cannot be measured directly in the atmosphere but have to be inferred indirectly or represented empirically with consequent loss of precision.

The actual solution of model equations can generate errors either by mathematical instabilities or by having too coarse a resolution. The former can nearly always be overcome with care but the latter are more insidious and can generate totally artificial dispersion of plumes and clouds for example unless very deliberate steps are taken to prevent this.

All models depend on input data, and these data may be insufficient, unrepresentative or erroneous. Data may be insufficient for assessing the possible consequences of future accidents and their associated probabilities This is particularly true for meteorological data in which so many scenarios are possible. In general at least 10 years reliable data are required to provide a reasonably representative data set. Rare conditions which lead to extreme concentrations or depositions are usually very difficult to quantify and may require even longer data sets.

The problem of representativeness is one of the biggest "headaches" facing anyone setting up meteorological instrumentation near a source of potentially hazardous material. The source is usually on or near large buildings, the local terrain is often heterogeneous and consequently all atmospheric properties, such as wind speed, are varying significantly from place to place. There is no simple solution to this problem since there is no single place which is representative of the changing conditions affecting the plume as it advects downwind. Faced with this dilemma, perhaps the best one can do is to site anemometers and windvanes in a nearby meadow with the maximum uniform fetch in every direction at a height of 10 metres, or, if the fetch is good at a height corresponding to half the stack height. Meteorological observing stations, whilst maintaining good quality instruments, may be too far away to give an adequate picture of local flow conditions, even in flat countryside. Vanderborght et al.(7) have given results which emphasise this from a study of wind speed and direction differences between sites at Beerse and Mol separated by some 22km on the very flat northern Belgium plain. The standard r.m.s. differences were about 15° and 2 ms^{-1}.

It must be stressed that all meteorological instruments should be properly maintained and recalibrated faithfully at intervals recommended by the manufacturers. More robust instruments require less maintenance of course but may lack responsiveness in certain critical conditions, for example in light winds.

Finally, careful consideration should be given to systems which will present meaningful and useful information to the appropriate person clearly and quickly in the case of an accidental release. Raw data from a windvane for example could be quite misleading and certainly confusing under such circumstances. Suitably averaged values of windspeed and direction, and turbulence levels, should be immediately available with the aid of a dedicated data-processor. The details of the data averaging and sampling should be linked to the safety issues involved and not to some arbitrary standard time like 1 hour.

Having said all this, it would be inappropriate not to remind ourselves that in spite of all these potential sources of error, the largest source of error, especially in accident situations at nuclear installations is likely to be associated with the amount, duration and type of release.

4. Sensitivity to errors and inherent variability

(i) It is often assumed that the mean values of the various parameters appearing in a model algorithm are the most appropriate, and these will give the best estimate of the mean concentration. This is clearly not so. For example if the Gaussian plume model were to be applied to dispersion from a source near ground level and it was known that the wind speed was fluctuating rather slowly between 3 ms^{-1} and 7 ms^{-1}, say, with a uniform probability distribution over the range, then $\bar{u} = 5$ ms^{-1} and, if no other parameter varied in this fictitious scenario, since $C \propto 1/u$, the value of C calculated using $\bar{u}$ would underestimate the true average C by some 6%. This is fairly small admittedly, but if the plume were also buoyant, the effect of the fluctuating wind on plume rise and hence on downwind ground-level concentrations would be much greater. Edwards and Misra (15) have studied this point. They consider the effect of mesoscale horizontal wind fluctuations on the effective entrainment rate of ambient air into a buoyant plume, and hence on its total rise. From their results it is easily shown that if $\bar{u} = 10$ ms^{-1} and $\sigma_u = 2.5$ ms^{-1}, and the average final virtual source height is 500m then the maximum ground level concentration is likely to vary tenfold in response to the fluctuations in u from about $6\bar{C}$ in strong "gusts" to about $0.77\bar{C}$ in relative lulls. ($\bar{C}$ is the maximum concentration assuming a single steady 10 ms^{-1} wind speed = $\bar{u}$).

To sum up it is important to remember that concentrations do not depend linearly on the basic input parameters, and that some of the parameters may be essentially inter-correlated.

(ii) As Sykes (16) points out, any measure of concentration within a plume is a time-averaged sample from a stochastic field. There is therefore a random component to the measurement which depends on the statistical properties of the instantaneous concentration field. He shows that the reduction in the variance s_T of the mean concentrations evaluated over a series of periods of duration T, as T increases, depends not only on T/T_e, where T_e is the Eulerian velocity timescale, but also on the variance s_o of the instantaneous concentrations, and on the mean concentration $\bar{C}$. On the centreline of a highly intermittent plume, Sykes shows that for large T:

$$\frac{s_T}{\bar{C}^2} = \frac{T_e}{T} \ln\left[1 + 2\frac{s_o}{\bar{C}^2}\right]$$

By "intermittent" Sykes means the plume is meandering about the long-term centreline, sometimes covering the receptor and sometimes not. Gifford (17) was the first to provide a useful theory of fluctuating plumes in his classic and still very important paper of 1959.

Draxler (18) has studied ^{85}Kr air concentrations measured at 13 sites located 30 to 150 km from the source at the Savannah River Plant in South Carolina, over a two year period. Comparing these with a long-term sector average Gaussian dispersion model which takes into account the varying meteorology, he showed that the root mean square error e of ratios of observed to calculated concentration (or calculated to observed, if the latter is smaller than the former) for different averaging times ranging from 1 week to the full 2 years, took the following values:

averaging time	1 wk	1 mth	3 mths	1 yr	2yrs.
e	5.2	3.6	2.6	1.8	1.7

Venkatram (19) and Hanna (20) have also studied the magnitude of concentration fluctuations within a plume. Hanna shows that the ratio of the standard deviation σ_c of these fluctuations to the mean concentration $\bar{C}$ varies across the plume and increases towards the edges since the entrainment of "clean" ambient air around the edges means σ_c decreases more slowly than $\bar{C}$ as one moves away from the centreline. Hanna also quotes measurements made near to 50m - 100m high stacks in which the concentration showed peak-to-mean values around 50-100 at the ground, and values 1-5 at the same level as the source. At much greater downwind distances $\sigma_c/\bar{C}$ appeared to vary like (L/σ_o) , where L is the Eulerian length-scale and σ_o the virtual size of the plume at the source.

Venkatram has shown that the standard deviation of the natural log of the concentration $\sigma(\ln C) \propto (u^{\frac{1}{2}} z_i^{1/6} h x^{-1})$, in a convective boundary layer, where z_i is the mixing depth and h is the height of the source (assumed greater than $0.1z_i$). Deardorff and Willis (10) confirm the quite rapid fall-off of concentration variance with downwind distance x in their measurements of buoyant and non-buoyant plumes in water tank experiments.

(iii) Smith and Readings (12) and Venkatram (19) have both considered the effect of a non-zero mean vertical velocity during the time of plume sampling on the ground level concentration. Smith and Readings do this very simply by looking at turbulence records on a single convective day and inferring the magnitudes of the vertical velocity associated with appropriate sampling and averaging times, and show that on this one occasion the 10-minute concentration 1200m downwind from a 100m stack would be expected to vary by some 15% (s.d.) about the mean due to this cause. Venkatram considered the problem from a theoretical standpoint and showed that the standard deviation would increase with stack height and decrease with downwind distance.

Smith and Readings also considered the sensitivity of the ground-level concentration on this day to variations in other parameters. As found by others, the variance increases with lateral distance from the centreline of the plume, with fluctuations in u (here $\sigma_c \sim \frac{1}{3}\bar{C}$), and with fluctuations in small-scale σ_w which only produced $\sigma_c \sim 0.03\,\bar{C}$.

(iv) Increasing the averaging time decreases the variance of any stochastic quantity such as concentration. NCRP Report No.76 (13) summarises the findings from many field measurements over flat terrain and in many stability conditions in terms of ranges in the ratios of observed to predicted concentrations, the latter based on Gaussian plume models. Figure 1 is a rough attempt to integrate these results. If R is defined as that number for which 90% of the observed concentrations within the plume lie within 1/R and R of the predicted concentrations, then the Figure expresses R as a function of distance downwind x and averaging time T.

(v) It was mentioned earlier that to gain adequate meteorological data measurements need to be made over at least 10 years. A study by Simpson and Jakeman (9) emphasises this point. They have attempted to link observed daily SO_2 concentrations at various sites in Newcastle, Australia, to wind speed. They show that they can explain about half the range of the annual maximum daily-SO_2 concentrations over a 10-year period (the maximum concentration varies by a factor of about 3-4) in terms of an inverse relation between SO_2 concentration and wind speed, and a log-normal probability distribution for the latter. The remaining variation they link to variations in emissions and other meteorological factors.

(vi) Finally in this section, the effect of rainfall must be mentioned. For any species that is removed by rain the inherently variable and patchy nature of rain can generate very irregular air concentration fields and deposition patterns. To predict the statistics of such patterns is made difficult for example by the complex air motions within clouds, and as a result of our lack of total knowledge of how radionuclides and other species are incorporated into raindrops. In a long-term statistical sense variations in single rain-event deposition at any point due to emissions from a constant source may be linked most strongly to variations in rainfall amount (which of course can vary by two or three orders of magnitude). In practise on any one occasion the problem is often to know where rainfall occurred and how the moving plume experienced the moving rain system. These are major problems.

5. Conclusions

We have attempted to review some aspects of modelling uncertainty. The review does not claim to be anything like exhaustive however. Emphasis has been placed on the following needs:

(i) to assess inherent variability and its consequences, whenever possible.

(ii) to take into account fluctuations in the meteorological input data (even when the time-mean value may be zero).

(iii) to make meteorological measurements close to the source over as uniform a site as can be found, and with full appreciation of appropriate sampling and average times, and to make this information available in a simple and clear way to Safety Officers in real time.

(iv) to improve models, reducing systematic errors.

(v) to assess input data errors and biases, whilst recognising that inherent errors may be unavoidable, especially in modelling future scenarios.

6. References

(1) J.F.Clarke,T.L.Clark,J.K.S.Ching,P.L.Haagenson,R.B.Husar & D.E.Patterson. Assessment of model simulation of long-distance transport. Atmos.Environ. 17,12. 1983. pp 2449-2462.

(2) D.J.Moore. A comparison of the trajectories of rising buoyant plumes with theoretical/empirical models. Atmos.Environ. 8,5,1974.pp441-458.

(3) G.E.Willis and J.W.Deardorff. On plume rise within a convective boundary layer. Atmos.Environ. 17,12,1983. pp2435-2448.

(4) G.A.Briggs. Plume rise and buoyancy effects. In"Atmospheric science and power production",Ed. D.Randerson (1983). Tech.Inform.Center.

U.S. Dept. of Energy, Oak Ridge, Tennessee.
(6) T.V.Crawford. Atmospheric transport of radionuclides. In "Proc. of Workshop on evaluation of models used for the environmental assessment of radionuclide releases",1978. Workshop held in Gatlinburg,Tenn. 1977. CONF-770901.
(7) B.Vanderborght, I.Mertens and J.Kretzschmar. Comparing the calculated and measured aerosol concentrations and depositions around a metallurgical plant. Atmos. Environ.17,9,1983. pp1687-1702.
(8) R.M.Harrison and H.A.McCartney. A comparison of the predictions of a simple Gaussian plume dispersion model with measurements of pollutant concentration at ground level and aloft. Atmos.Environ.14,5,1980. pp589-596.
(9) R.W.Simpson and A.J.Jakeman. A model for estimating the effects of fluctuations in long term meteorology on observed maximum acid gas levels. Atmos.Environ.18,8,1984. pp1633-1640.
(10)J.W.Deardorff and G.E.Willis. Groundlevel concentration fluctuations from a buoyant and a non-buoyant source within a laboratory convectively mixed layer. Atmos.Environ.18,7,1984. pp1297-1310.
(11) J.E.Fackrell and A.G.Robins. Concentration fields associated with emissions from point sources in turbulent boundary layers, Pt.III: Concentration fluctuations and fluxes. 1980. CEGB Memorandum MM/MECH/TF No.260.
(12) F.B.Smith and C.J.Readings. Inherent variability in turbulence and its consequences for pollutant-concentration estimates. In"proc. of the Workshop on the Parameterization of mixed layer diffusion". Ed. R.M. Cionco. 1982. New Mexico State Univ. Las Cruces.
(13)The National Council on Radiation Protection and Measurement NCRP Report No.76. Radiological Assessment.Predicting the transport, bioaccumulation and uptake by man of radionuclides released to the environment. 1984. NCRP 7910 Woodmont Ave, Bethesda,MD 20814.
(14) G.A.Briggs and McDonald. Prairie Grass revisited: Optimum indicators of vertical spread. Proc. of 9th Int.Tech.Meeting. NATO/CCMS Report103 Atlantic Council, Washington,D.C. 1978.
(15) G.C.Edwards and P.K.Misra. The effects of mesoscale turbulence on plume rise in a land/lake environment. Bound.Layer Met.29,2,1984. pp155-166.
(16) R.I.Sykes. The variance in time-averaged samples from an intermittent plume. Atmos.Environ.18,1,1984. pp121-124.
(17) F.Gifford. Statistical properties of a fluctuating plume dispersal model. Adv.Geophys. 6,1959. pp117-137.
(18) R.R.Draxler. An improved Gaussian model for long-term average air concentration estimates. Atmos.Environ.14,5,1980. pp597-602.
(19) A.Venkatram. The uncertainty in estimating dispersion in the convective boundary layer. Atmos.Environ.18,2,1984. pp307-310.
(20) S.R.Hanna. Concentration fluctuations in a smoke plume. Atmos.Environ. 18,6,1984. pp1091-1106.

TABLE I Uncertainty in Model- Predicted Concentrations

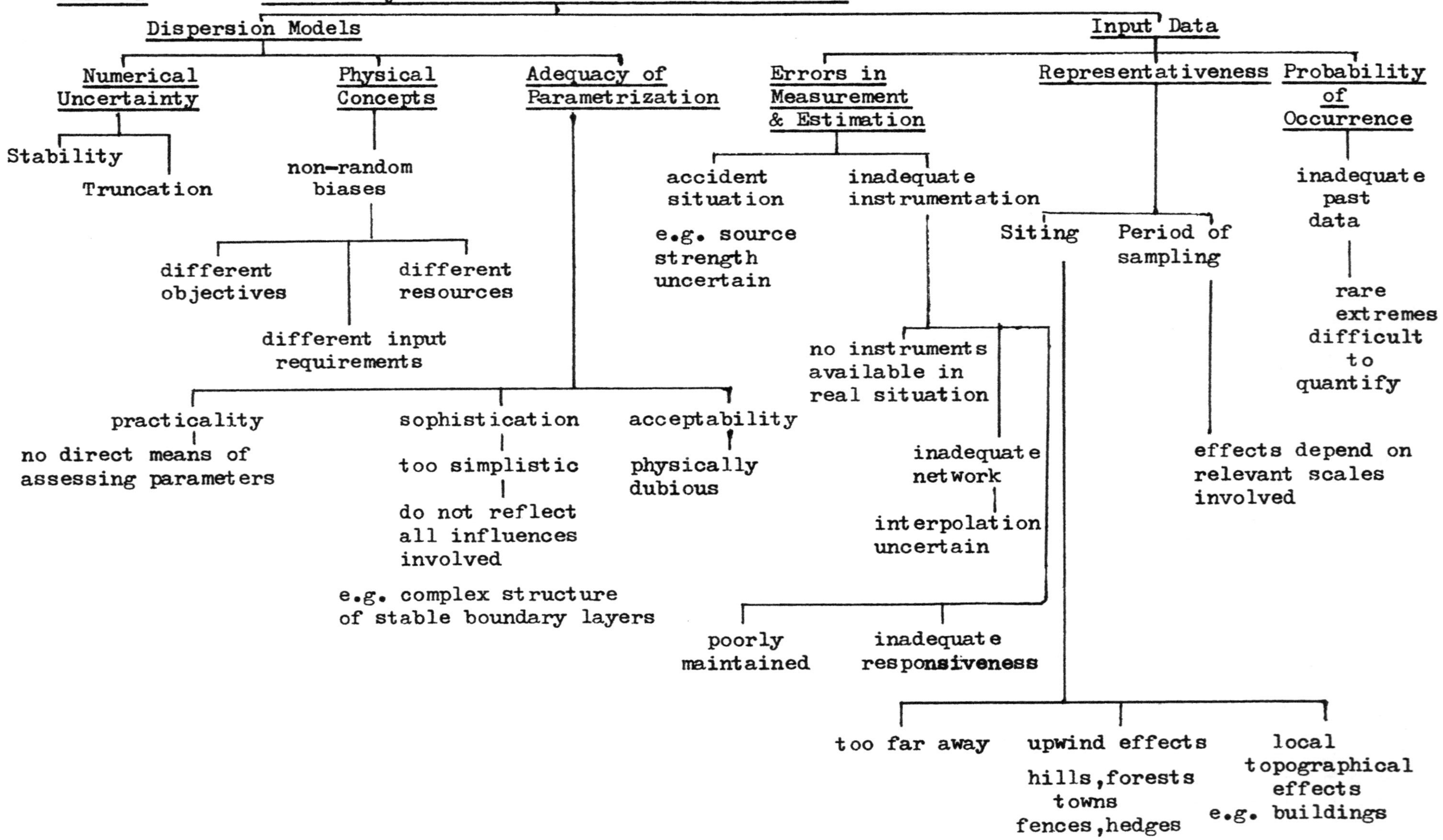

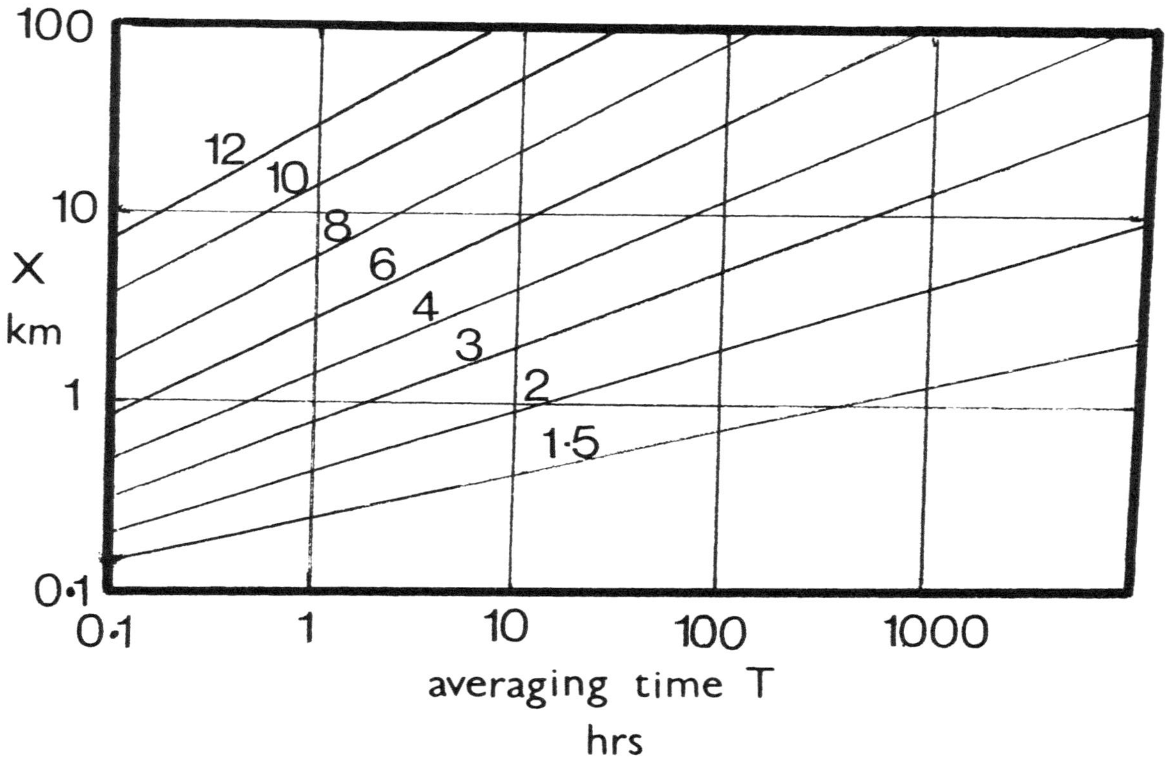

Figure 1. Contours of the scatter parameter R for a non-buoyant plume, under all stability conditions over a flat plain. Using Gaussian models, 90% of observed data will lie within 1/R and R of the predicted data.

Discussion

Question: B. Crabol - CEA/CEN-Far

You mentioned the importance of the Lagrangian time-scale. Do you think that this time-scale is enough to describe the whole turbulence, or do you associate it with a particular shape of the correlation function (or atmospheric spectrum)?

Answer: B. Smith - Brit. Met. Office

The mention of the Lagrangian timescale was to indicate by example a parameter which is conceptually important but which in practice is difficult to assess by direct measurement.
If, however, a good estimate of the timescale can be inferred it does give a good basis for predicting plume spread from a continuous source in homogeneous turbulence. Further knowledge concerning the correlation-function or spectrum provides only marginal improvements in our ability to predict spread, unless we are dealing with instantaneous releases where such knowledge would be vital.

Question: F. Desiato - ENEA/Roma

The question is about the method of evaluating the horizontal standard deviation of a plume.
Do you think that, providing the sigma-theta is computed over a time comparable with the sampling time, the use of a semi-empirical formula derived from the Taylor's theory (like the Draxler formula) is recommended for a sequential-puff model?

Answer: B. Smith - Brit. Met. Office

The value of σ_θ (sigma theta) sampled over the duration of release and averaged over a time appropriate to the time of travel (or distance from the source) would provide a good indicator of the spread of the plume. The component corresponding to the width of the sequential puffs would in principle, be included in this, provided the original puff size was small compared with the Eulerian lengthscale. If this is not the case the problem is rather more complex.

Question: P.M. Foster - CEGB/CERL

With reference to your statement that temperature and wind speed should be monitored at 2 heights at nuclear sites, can you make some suggestion as to what these heights should be?

Answer: B. Smith - Brit. Meteorol. Off.

In my paper I emphasize the impossibility of placing meteorological instruments at any place which is totally representative of conditions experienced by the plume. As very much a compromise I state that a site on a nearby open field with as good a fetch in every direction of wind should be chosen. I would recommend placing the instruments at standard Met. Station height (e.g. 10 metres for anemometers) and a second set of wind instruments at half the height of the source. Temperature sensors should be linked to give a direct measure of the temperature gradient and

I would suggest heights of 1.2 metre (normal screen height) and 10 metres. The upper wind instruments will measure values typical of the average experienced by pollution reaching the ground more or less directly from the source.
Longer range dispersion cannot be estimated reliably from measurements at a single rough site.

Question: H. ApSimon - Imp. Col./London

Do you think there is a message for legislators here? If for example legislators recognised the diffficulties of estimating the few worst situations in a year would they set standards based on concentrations not to be exceeded more than once or twice a year with the resulting complications for EIA, etc.?

Answer: B. Smith - Brit. Meteorol. Off.

In one sense it is reasonable for legislators to set a criterion of the kind you describe if they wish to maintain concentration levels well below some known threshold, above which health effects are known to occur. The lesson I believe has to be learnt by the planners who must assess how to meet this criterion most effectively. The lesson is that whilst simple models have very positive uses, they may be insufficiently sophisticated to take into account the sources of variability and uncertainty I have discussed in my paper, and which may lead to occasional very high groundlevel concentrations over some period of time.

Question: J.A. Hertig - E.P.F./Lausanne

Some influences only shortly mentioned are those of thermal structure of the atmosphere, the measurement of temperature and the associated complex flow structure that flows in opposite direction, cross flow, etc.

Answer: B. Smith, Brit. Meteorol. Off.

Yes, I agree that much could be said about the uncertainty likely to be introduced by such complex flow fields. I imagine these occur mainly in stable conditions; (except, of course, near bluff structures - and then only locally to the structure - when separated flows in general stability conditions can give rise to strong shears).
Over open terrain however, such marked shears are likely to be associated with poor vertical mixing. If this is the case, the the plume will remain largely unaffected by them.

Question: J.P. Schon - Ec. Centr./Lyon

Does the idea of gust effects on the plumes need some new descriptions of the flow field (like for example the concepts used in large eddy simulation).

Answer: B. Smith - Brit. Meteorol. Off.

I am postulating that it may be profitable in some circumstances to have an understanding of gusts, just as those concerned with the safety of aircraft need to understand gusts so that they can be sure the aircraft will react stably to them. How frequent are they, how intense are they,

what are the accelerations involved, what is the shape of the gust signature, and how are they formed? These may be relevant questions.
I would agree that large eddy model simulations are akin to this. A single large eddy is studied. The small-scale eddies are not perhaps totally linked to the large eddy as one would expect in practise, and so the simulation of gusts from such modelling would be only very approximate.

Concluding Plenary Session

Chairman: S. Finzi (CEC/BXL)

Session secretary: E. Skupinski (CEC/BXL)

OVERVIEW AND CONCLUSIONS ON THE RESULTS OBTAINED IN RESEARCH AREA A: THE LOSS-OF-COOLANT ACCIDENT (LOCA) AND THE FUNCTIONING AND PERFORMANCE OF THE CORE COOLING SYSTEM

K.F. Wolfert
Gesellschaft für Reaktorsicherheit (GRS) mbH

1. INTRODUCTION

Within the Research Area A of the first CEC Shared Cost Action Safety Research Program fundamental work on thermal-hydraulics and heat transfer for the refill and reflood phase of a loss-of-coolant accident has been carried out. The aims of the program have been to carry out useful fundamental research, to encourage collaboration between Member States and to avoid duplication of effort.

The work was concerned in particular with the quench front propagation process, the effect of blockages, two-dimensional effects in a reactor core during reflood, and with the counter current flow situation at the top nozzle of fuel assemblies. Most of the work was experimental, completed in some cases by subsequent modelling.

Area A involves eight distinct projects, implemented in different national laboratories of the Member States.

The paper gives an overview on the projects and on the main results obtained and summarises the conclusions of the general discussion which followed the presentations on research projects and invited papers in Sessions A1 to A4.

2. QUENCH FRONT PROPAGATION PROCESS

To learn about the rewetting of the Zircaloy cladding of PWR fuel rods during the reflooding process experimental and analytical work has been carried out within the project "Rewetting Propagation over Zircaloy under Bottom Flooding Conditions" by M. K. Denham and D. Blackburn at United Kingdom Atomic Energy Authority, AEE Winfrith.

The experiments were performed in the REFLEX Rig, in which the inside of a hot, electrically heated Zircaloy-2 tube (4.2 m long, 12.7 mm inside diameter) was reflooded from below.

The experimental program started with the test section in a clean, unoxidised condition and a series of 15 refloods was performed . This covered a range of pressure (1-4 bar), flooding rate (1-16 cm s^{-1}), and inlet subcooling (0-80 K), appropriate to a large break LOCA. Then, after examining the tube using an intrascope and a specially developed oxide thickness probe, the test section was oxidised in steam at 650°C until a layer of suitable thickness had grown. The series of 15 refloods was then repeated. In this way, layers of 1, 15, 25, 58 and 75 µm average thickness were studied in turn. Finally, the test section was cut up for detailed examination.

The systematic study of the effect of oxidation of the Zircaloy has shown that oxidation has an effect (over the range of oxide thickness likely during a loss of coolant accident) which is relatively small compared with the effect of flooding rate or pressure.

A straightforward comparison of the Zircaloy tube results with earlier REFLEX results for an Inconel tube of similar dimensions has confirmed that a Zircaloy tube rewets approximately twice as fast as a similar tube of Inconel. The quench front speeds were in inverse proportion to the thermal capacities of the two tubes, a ratio of 2.18 in this case. This agreed with previous workers' findings.

With respect to a detailed data evaluation the results have been compared with the Clement-Regnier K2 quench front speed correlation, taking into account heat conduction both parallel to and normal to the cladding surface. The results were generally consistent, but an effect of test section power was apparent in the Winfrith results, which is not accounted for by the K2 correlation. It was found that this difficulty could be overcome by including the wall temperature ahead of the quench front in the correlation.

Furthermore, calculations have been carried out to assess the effect of oxidation on the quenching of PWR fuel rods. The calculations have indicated that the effect of oxidation on the rewetting of a fuel rod would be equally small. Under conditions of practical importance, a small increase in quench front speed can be expected, and this should be beneficial from the point of view of reactor safety.

To study the influence of the composition of electrically heated fuel pins on the reflooding process, experimental and analytical work was carried out within the project "Study of Rewetting and Quench Phenomena by Single Pin Out-of-pile Experiments with Special Emphasis on the Effect of Pin Composition" by N. E. Kaiser and O. Rathmann at Risø National Laboratory, Denmark.

Within the experimental program a series of comparative experiments with electrically heated fuel pin simulators of different compositions were performed under identical boundary conditions. The following three types of pins were investigated: a hollow tube, directly heated by Joule heating, a conventional indirectly heated pin and a more advanced type of indirectly heated pins. This last type is based on a central heater surrounded by UO_2 ring pellets and with Inconel cladding, simulating very closely a real fuel pin. The pins are placed in a circular shroud, which may be unheated or heated by radiation from an outer shell. The annular space between pin and shroud has a hydraulic diameter corresponding to a real reactor core geometry. The tests were performed within a pressure range of 1 to 10 bar.

The difference between the 3 types of rods is only small. The directly heated rod quenches somewhat faster than the indirectly heated rods. But there is no significant difference between the two indirectly heated rods. The higher thermal capacity of the filler material in the UO_2 rod is apparently compensated by the poorer heat transfer from the inner area of the rod across the gas gap to the cladding. An increased quench time (decreased quench front velocity) was obtained by increasing initial temperature, increasing coolant inlet temperature and increasing power supply during the quenching, or by decreasing coolant flow rate and decreasing pressure.

The second task of the project consists of an analysis of the experimental results. This analysis includes a series of comparative calculations using the computer code NORCOOL-II, which is developed at Risø National Laboratory. Despite the discrepancies between the experimental and the calculated results due to inadequate modelling of flow regime effects and the gap heat transfer and due to the lack of a rod-to-shroud radiation model, the entire set of calculations shows that NORCOOL-II predicts correctly the effects of increased rod power, increased flooding rate and increased pressure qualitatively, and mostly also quantitatively.

To improve the understanding of the prequench heat transfer and quench front propagation, influenced by grids, cold walls, and radial non-uniform power and/or temperature profile, an experimental program has been carried out within the project "Heat Transfer to a Dispersed Two-Phase Flow and Detailed Quench Front Velocity Research" by T. C. de Boer and S. B. van der Molen at Netherlands Energy Research Foundation ECN. The project can be divided into two parts.

In the first part of the project experiments have been performed in three well instrumented test sections, a tube, an annulus, and a 4-rod bundle with a heated length of 1.5 m. The objectives of this part have been to investigate the influence of an unheated wall on the prequench heat transfer and to compare heat transfer coefficients derived from experimental data with values calculated using existing heat transfer correlations.

The experiments performed have shown an early wetting of the unheated shroud, which leads to a quenchfront velocity increase and an increase in liquid transport downstream of the quenchfront. Furthermore, an increased efficiency of the inverted annular flow regime in the cooling of the heater pin is found. For the inverted annular flow regime the extended Bromley correlation gives rather good agreement with the experimental data.

In the second part of the project boildown and reflood experiments using a test section with a 36-rod bundle within a rectangular shroud have been performed. The bundle consists of 32 axial uniform indirectly heated rods of 3 m heated length and 4 unheated corner rods. The tests have provided a data base for code development and have given in particular information on the influences of core inlet conditions, spacer grids, radial non-uniform power and/or temperature profiles on the reflood process and prequench heat transfer.

The experiments performed have shown that considerable wall-temperature differences in the radial direction, either existing initially or obtained during the quench process, deform the quench front. Under these conditions, a high increase of prequench heat transfer is observed on the hotter surfaces, suggesting an inverted annular or transient flow regime to be induced. This intensive heat transfer considerably levels the wall temperatures just before quenching and helps to smooth out the quench-front velocity differences. Heat transfer improvement induced by the presence of spacer grids was observed.

3. EFFECT OF BLOCKAGE

To demonstrate the variability of failure strain with temperature, heating method and heat transfer from the cladding surface, experimental and analytical work was carried out within the project "Out of Pile Ballooning Tests on Zircaloy Cladding - The Effect of Heat Transfer Conditions on Failure Strain" by L. Lombardini and

R. Vesprini at E.N.E.A. Casaccia and P. M. Jones and F. Casadei at J.R.C. Ispra, Italy.

Within the experimental program tests with a fuel rod simulator (0.4 m long, 9.5 mm outside diameter) consisting of Zircaloy clad and boron nitride pellets have been performed. The fuel rod simulators have been mounted within a directly heated stainless steel flow shroud and could be directly heated by Joule heating of the clad or indirectly by means of an inner Tantalum rod. Cooling has been adjusted by controlled flow of superheated steam. The prevailing heat transfer characteristics were determined from pre-operational steady state tests and analytical models accounting for the forced convective cooling and the radiant heat exchange between rod, shroud and outer vessel. Nominally constant temperature and temperature ramp tests have been carried out, the times to fail have been determined by the internal rod pressures.

The heat transfer conditions were chosen to cover the range expected in "real" accidents to enable prediction of those conditions leading to large balloons to be checked.

In spite of difficulties with the repeatability of the heat-transfer characterization tests, meaningful results demonstrating general trends have been achieved.

The test results at about 800°C confirmed the expected difference in clad failure strain between direct and indirect heating. They also confirmed current ideas and modelling predictions on the effect of clad heat flux on circumferential temperature gradients and on the failure strain and its distribution in indirectly heated cladding.

The measurements confirmed that the maximum failure strains obtained in a steam atmosphere are to be found in the high alpha-phase temperature regime. The existence of superplasticity was confirmed at low strain rates in the mixed phase temperature range in an inert atmosphere but not in steam at 900°C and higher. In the high temperature (beta-phase) region the results confirmed that high strains are not to be expected in steam and that oxidation can prevent failure completely.

The experimental program has been assisted intensively by analyses applying the RODSWELL code developed at J.R.C. Ispra.

The calculations, although showing qualitative agreement with the experiments indicated necessary model improvements.

The work of this project reinforces the prediction that balloon strains, sufficient to be a potential problem, should only be expected at temperatures near to 800°C during the quasi-adiabatic heat up phase of a large-break LOCA or during the quasi-steady-state phase late in a small-break LOCA. Such balloons are expected to be localized axially, reducing the potential for coolant channel blockage.

To assist in the understanding of the coolability of severe core channel blockages, experiments on the THETIS test rig have been carried out within the project "An Experimental Investigation of the Effect of Clad Ballooning on the Effectiveness of PWR Emergency Core Cooling" by C. A. Cooper and K. G. Pearson, United Kingdom Atomic Authority, AEE Winfrith. The test assembly consisted of a full length, electrically heated 7x7 rod fuel cluster containing a blockage simulating very severe clad ballooning. The blockage was formed by fitting thin-walled Inconel balloons with a length of 455 mm to a 4x4 group of rods. These reduced the subchannel flow area by 80% over a length of 200 mm.

Within the experimental program single-phase cooling, forced reflood, gravity reflood and level swell experiments have been performed.

The single-phase cooling tests with air have been performed to provide basic single phase heat transfer data including data concerning the enhancement of single-phase heat transfer at and downstream of spacer grids for inlet Reynolds numbers in the range 1000 to 9000. The experiments provided excellent data on the level of heat transfer within the cluster and the enhancement produced by turbulence created by the spacer grids and allowed checking, via repeated tests, of the conservation of the blockage shapes.

With the forced reflooding tests the effects on reflood heat transfer of varying pressure, flooding rate, inlet water temperature, initial rod temperature and cluster power were investigated. The experiments covered a pressure range of 1.3 to 4 bar and have led to a better understanding of the rewetting processes during bottom reflooding, of the very important influence of spacer grids in two-phase flow and of the complicated heat transfer processes within the blockage. The insight obtained has been used to develop a mechanistic model which gives quantitative support to the explanations which have been put forward.

The gravity reflooding tests investigated the effects of flow oscillations between the simulated downcomer and the test bundle on reflood heat transfer. The experiments covering a pressure range of 2 to 4 bar and indicated that essentially the same phenomena are important in both forced reflood and gravity reflood. The variation in inlet flow produced by the variations in system parameters was the dominant influence. The inlet flow oscillations which occurred in gravity reflood appeared not to influence significantly overall rewetting and heat removal performance.

The level swell tests have been carried out over a wide range of pressure (2 to 40 bar) to provide voidage and heat transfer data for comparison with correlations used in the analysis of the effects of core uncovery in a small break LOCA. The correlation of Gardner showed the best overall agreement with the data, followed by that of Cunningham and Yeh. Jowitt's correlation was best up to 10 bar but overpredicted the void fraction at higher pressures. The correlations of Wilson and of Zuber and Findlay seriously overpredicted the void fraction at all pressures.

To improve basic knowledge on heat transfer effects of localized coplanar ballooning in an electrically heated rod bundle under conditions of single- and two-phase cooling, the project "Studies on the Effects of Blockage Upon LWR Emergency Core Cooling Systems" has been carried out by S. A. Fairbairn and B. D. G. Piggott at Central Electricity Generating Board, Berkeley Nuclear Laboratories, United Kingdom.

The test section consisted of 44 electrically heated rod simulators of 1 m length. The rods were arranged on a square pitch enclosed in a cylindrical transparent silica shroud. Balloons were simulated by preformed sleeves attached to the central 4x4 group of rods towards the end of the heated length. Four different types of sleeves were used. The first two were machined and gave a 90% maximum subchannel blockage, and the final two were hydraulically formed at UKAEA Harwell having thin walls to give representative fuel pin balloon thermal capacities and produced maximum subchannel blockage of 90%

and 61%. Air, steam or water flows could be introduced into a lower plenum chamber and passed over an unheated entrance length of 28 equivalent diameters before the beginning of the heated section. Water could be injected, close to the start of the heated length, through capillary tubes located at the centre of each subchannel. Extensive high speed cine and video investigations were carried out during the steam and droplet tests which yielded information on drop size and velocity distributions at 5 axial positions in the bundle. A novel pulse heated thermocouple probe was developed to measure subchannel steam temperatures and was used with some success.

Each bundle has been subjected to a series of atmospheric pressure tests of increasing complexity including air flow tests, steam heat transfer tests, steam and droplet tests, and conventional reflood tests.

The tests have shown that single phase flow diversion can be up to 85% for a 90% local flow blockage. In two phase conditions the majority of the water is observed to pass through the blockage. For the severe blockage there is evidence of significant coalescence and entrainment in the blockage region. The reflood tests demonstrate enhanced coolability of the blockages such that quenching at the blockage elevations occurs earlier than in an unblocked bundle provided that the blockage shapes have a representative low thermal capacity.

Models for flow diversion and steam and droplet cooling have been developed which show good agreement with the experimental data.

In the concluding discussion regarding the effect of blockages it was pointed out in particular

- that the results of the blockage tests carried out within Area A are consistent with the results obtained in other experimental programs, (e.g. FLECHT-Westinghouse, FEBA-Kernforschungszentrum Karlsruhe),
- that no problem has been found regarding the coolability of the blockages tested, which have ranged up to 90% subchannel blockage,
- that there is no chance to predict with a system code local clad temperatures with an accuracy of 10-15 K, which would be required for a realistic determination of ballooning processes,
- that Best-Estimate calculations with realistic boundary conditions indicate that during a loss-of-coolant accident the clad temperatures will not rise to temperature levels, which have the potential to create fuel behaviour problems.

4. TWO-DIMENSIONAL EFFECTS

To study two-dimensional phenomena in the reactor core during the reflood process the project "2D Effects in the Core During the Reflooding Phase of a LOCA" has been conducted by R. Deruaz, P. Clement and J.M. Veteau at CEA-CEN Grenoble, France.

The experiments have been carried out on the PERICLES facility with a rectangular test section involving three full length 7x17 heater rod assemblies which are electrically supplied by two independent power sources. The upper plenum corresponding to each assembly is equipped with a steam-water separator. The steam flow rate is measured at each outlet while water is either collected and its flow rate measured, or is drained on the upper core plate where it can either accumulate or fall back into the core through the upper tie plates. The effect of various parameters like pressure, inlet flow rate and partition, subcooling, radial position of the hot assembly and radial power peaking factor (varied from 1 to 1.85) have been investigated.

Boil up and reflood tests (constant and oscillating flow rate) have been carried out to study 2D-effects in well defined conditions, and to provide the needed data base to help writing and adjusting models for safety codes.

In the boil up experiments the liquid inventory has been held constant during the test by equalizing inlet and outlet flow rates. Tests were performed at 3 and 6 bar. Main parameters were the radial peaking factor, varied from 1 to 1.85, the liquid inventory and the power level. The tests have shown that the two-phase level is identical in all the assemblies and depends on the average power and not on the radial power distribution. A simple model assuming perfect radial mixing in the wet region predicts heat transfer in the dry region well.

Reflood tests have been carried out to study the effect of main test parameters and to investigate physical effects like cross flows of steam and water, quench front propagation in hot and cold assemblies, or water fall back from the upper plenum.

Important results have been provided. The tests have confirmed e.g. that cross flows of steam and water are essential mechanisms in case of a 2D- reflood process. With respect to quench front propagation the tests have shown e.g. that the quench front velocity in the hot assemblies is reduced in the initial phase of the transient but enhanced in the major part of the transient, resulting finally in nearly the same turn-around temperatures in the hot and cold assemblies. Significant water fall back was not measured in the test conditions chosen for the PERICLES program.

In the general discussion it was pointed out in particular

- that cross flow effects must be taken into account in modelling reflood processes in large cores, either by multi-channel representation with horizontal flow connections in 1D-codes or by use of 3D-codes.

5. COUNTER CURRENT FLOW

To improve the knowledge on the fluiddynamic effects of counter-current flow , experimental and analytical work was carried out within the project "Fluiddynamic Effects in the Fuel Element Top Nozzle Area during Refilling and Reflooding" by R. Spatz, H. Kröning, A. Hawighorst, D. Mewes, F. Mayinger* at the University of Hannover, Federal Republic of Germany.

Within the experimental program tests with perforated plates, a 4x4 rod bundle with tie plate, and with a complete fuel element dummy were performed. The selected fluid systems were air-water and steam-water. Within the steam-water test series the effects of subcooled water injection above the fuel element dummy, simulating the hot leg injection of a PWR-system, were investigated. In order to identify the flooding mechanisms, photographs of the flow patterns were taken and the pressure drop along the test section was measured.

*F. Mayinger, Institut für Thermodynamik A, Technische Universität München

The results of the experiments indicate similar mechanisms for counter-current flow in tubes and square ducts with and without internals. In all cases a distinct gradient in pressure drop and a remarkable appearance of flow instabilities near the stagnation point were observed. A larger wetted perimeter of a given cross section results in a flooding phenomena that takes place at a lower gas superficial velocitiy. For subcooled water injection flooding occurs at higher steam superficial velocities in comparison with saturated water. However, the effect of water subcooling is only of importance if the subcooled water reaches the area where flooding takes place.

An analytical effort leads to a flooding model which is based on the stability of the wavy interfaces. The maximum differential velocity between the phases, where a disturbance does not change the wavy boundary layer, results from the Orr-Sommerfeld equation of a frictionless flow.

In the derived equations for the calculation of the volumetric mass fluxes the gas fraction is a parameter. The influence of friction is considered by an exponential equation for the liquid flooding velocity. The influence of geometry is considered by a correlation for the gas flooding velocity.

The experimental data obtained from this project and those from the literature are well represented by the derived model.

6. INVITED PAPERS

In addition to the presentations of the eight research projects of Area A three invited papers were given.

The first paper "The Role of Experiments in the Development and Qualification of Thermohydraulic Codes, as experienced in the Development of the CATHARE CODE" was given by J.L. Nigon from CEA/CEN Grenoble, France.

In this presentation a clear and convincing strategy for the development and qualification of thermalhydraulic codes has been shown. Important needs in model improvements with respect to the description of interfacial friction in rod bundle geometries, condensation of steam on subcooled water flow, heat transfer during reflooding, and phase separation in tees have been identified.

The second paper "Droplet Dynamics and Heat Transfer in Dispersed Two Phase Flow" was given by S.A. Fairbairn from CEGB, Berkeley Nuclear Laboratories, United Kingdom.

A mechanistic thermal and mechanical non-equilibrium model was presented which describes the fluiddynamic and heat transfer mechanisms in a partially blocked subchannel.

In the general discussion it has been pointed out that models describing in detail the physical phenomena at blockages or spacer grids require a high resolution in space. The implementation of these models in larger system codes would however increase significantly the overall computing time.

The third paper "Important Thermohydraulic Aspects During Refilling and Reflooding of an Uncoverd LWR-Core" was presented by

E. Hicken from Gesellschaft für Reaktorsicherheit (GRS), Federal Republic of Germany.

In the paper an overview on the reflood process has been given. The most important phenomena have been identified. It has been stated in particular that the main and relevant physical phenomena during refill and reflood are known. Multidimensional effects will be studied in 1:1 scale within the international 2D/3D project during the next few years.

7. FUTURE RESEARCH

Experts of group A, involving a number of European national laboratories and the Joint Research Centre, have specified in an ad-hoc meeting useful topics on thermalhydraulics appropriate for a future Shared Cost Action Research Program.

There has been an agreement that all future work should be justified by the need to develop, improve or validate models as input into advanced system codes and that highest priority should be given to small leak scenarios. The existing data base of experimental results, the existing models and codes, as well as the appropriateness of a topic for implementation in a future Shared Cost Action Safety Research Program have been taken into account.

Thirteen proposals have been considered to represent appropriate and useful work for a future program covering review studies, experiments, model development, model validation and the development and test of measurement techniques.

Within the field "Critical Review Studies, Experiments, Model Development and Model Validation" eleven topics have been specified.

- Heat transfer for the boiling curve which includes the minimum film boiling point
- Condensation effects related to the injection of ECCS-water
- Basic heat transfer above the mixture level in a partially uncovered core
- Two phase flow patterns (flow regime maps)
- Two phase flow behaviour in branches of horizontal pipes
- CCFL in horizontal and vertical pipes
- Hydraulic behaviour of a partially uncovered core
- Heat transfer characteristic of U-tube steam generators
- Two phase flow pump characteristica
- Single and two phase natural convection phenomena
- Flow oscillations during refill

All specified topics are in agreement with the most relevant phenomena listed in the phenomena list of the CSNI Validation Matrix for Small and Intermediate Leaks in Pressurized Water Reactors, which have been prepared by a task group of the CSNI Principal Working Group 2.

Within the field "Development and Test of Measurement Techniques" two topics have been specified

- In-pool level detection in test facilities
- Determination of superheated steam temperature under two-phase flow conditions.

In the concluding discussion of this meeting all main statements and conclusions on the above proposal were confirmed.

8. CONCLUSIONS

The work performed within research area A of the CEC Indirect Action Research Program has improved the understanding of the thermalhydraulic process during the refill and reflood phase of a loss-of-coolant accident. The work has extended remarkably the data base usable for the development, improvement and validation of refill and reflood models. A few particular projects have led to the development or improvement of analytical models.

The review group meetings of Area A, mostly upgraded by the invitation of one or two outstanding experts, have given the opportunity for profitable discussions and have enhanced the international cooperation and the establishment of a common understanding.

However, to derive the full benefit of the various projects and to contribute usefully to the improvement of analytical methods used for the analysis of reactor systems, it is absolutely necessary to transpose all findings relevant for thermalhydraulic system behaviour into the development, improvement or validation of models implemented in one of the large thermalhydraulic system codes.

Concerning a future Shared Cost Action Research Program in thermalhydraulics a list of useful topics covering critical review studies, experiments and model development was confirmed in the concluding discussion.

Discussion

Question: C.J. Wheatley - UKAEA/SRD

The ultimate objective of this work is to show that the core remains coolable during a LOCA. Has the work done in Area A brought you nearer to achieving this objective? How do you see this objective being achieved?

Answer: K. Wolfert - GRS

The work of Area A has contributed to a better knowledge and understanding of physical phenomena relevant for reflooding of an uncovered reactor core. The tests have extended remarkably the data base for the development, improvement and validation of refill and reflood models.

The required analytical work based on the provided data base for the improvement or validation of refill and reflood models to be incorporated in one of the large thermohydraulic system codes, has not been carried out within the projects. This work remains to be done. Then the general objective of the work - the improvement of the analytical methods used for the analysis of reactor systems - will be achieved.

OVERVIEW AND CONCLUSIONS ON THE RESULTS OBTAINED IN RESEARCH AREA B THE PROTECTION OF NUCLEAR POWER PLANTS AGAINST EXTERNAL GAS CLOUD EXPLOSIONS

A R Edwards
Chairman of Experts Subgroup Area B
Safety & Reliability Directorate/UKAEA

1. INTRODUCTION

In the transport and storage of flammable materials, such as liquified natural gas (LNG), composed mostly of methane, and liquified petroleum gas (LPG), composed mostly of propane or butane, it is remotely possible that an accident could occur resulting in the release of a large quantity of flammable vapour to the atmosphere. If this were to occur then the vapour cloud would be transported by the wind. If it were ignited before dilution to a concentration below the lower flammable limit (LFL), it could very well cause loss of life as well as significant damage to buildings and equipment near the point of ignition. This possibility is one of particular concern which has to be considered in the safety assessment of nuclear power plant (NPP) within the European Communities. It is a potentially serious hazard, partly because of the relatively close proximity of storage sites and transport routes of such flammable materials to NPP sites in general and partly because of the damage that could be caused by the large quantities handled. The concern may not necessarily be direct damage to the reactor building but possibly due to simultaneous damage to safety related equipment and power supplies on the site in general. In other words, it is a potential mechanism of common cause failure, which could initiate a sequence of events leading to loss of cooling of the reactor core and the eventual release of radioactive fission products to the atmosphere.

There are many uncertainties in assessing the significance of the potential hazards from flammable vapour clouds. A crucial question which has to be addressed is: given that a release of x tonnes of flammable vapour occurs, how far will the cloud travel before it is diluted below the LFL? In many cases it is sufficient to establish that this distance is sufficiently small that the consequences of ignition prior to dilution below the LFL can be ignored. Many NPP's within the European Communities are so situated that it cannot be established in every case that accidentally released flammable vapour clouds will be below the LFL before the NPP is engulfed. It is therefore necessary to consider the effects of possible ignition of the vapour cloud, the speed of burning and the consequential pressure rise inside the cloud. A rise in pressure inside a burning cloud of vapour will be felt as an external pressure field due to gas expansion effects. Locally pressure effects can be enhanced if wave reflections from walls of building occurs, these enhanced pressures might then cause structural failure.

Because of the uncertainties in the estimation of effects from flammable gas clouds, the Commission of the European Communities (CEC) initiated an indirect action programme on the safety of thermal water reactors with the aim of reducing these uncertainties. In the preliminary discussions, which took place prior to the formulation of the call for tenders, there was agreement that it was inappropriate, at that time, to study structural response. The dominant need was to obtain a better understanding of form of the pressure loading. Contracts were, therefore, only called for in three areas;-

1) dense gas dispersion (cloud formation)
2) cloud burning (flame acceleration effects)
3) external pressure fields (wave - obstacle interactions)

Work performed under contracts to the CEC in these three areas will be reviewed in subsequent sections of this paper.

2. GAS DISPERSION

Whilst the general study of gas dispersion in the atmosphere has to cover a density range from very light (relative to air) to heavy gases, the dominant concern for the safety of nuclear power plants lies with the dispersion of heavy gases. Contracts in the gas dispersion area of work were, therefore, restricted to denser than air vapours, because, whether the LNG or LPG is refrigerated or stored under pressure, any accidental release invariably leads to the formation of a denser than air vapour cloud. (For LNG this is because of the very low temperature of the vapour cloud formed).

As was mentioned in the introduction, a crucial question is: given that a release of x tonnes of flammable vapour occurs, how far will the cloud travel before it is diluted below the LFL? There is large uncertainty in estimating this distance. One reason for this was that field experiments on the dispersion of denser-than-air vapours which had been done to date had involved releases of, at most, only a few tonnes of vapour. However, the mass of flammable vapour of concern in NPP safety assessments extends to the order of 10,000 tonnes. This demands considerable extrapolation, which involves using models for the dispersion of such clouds. A comparison of these models shows a large variation of the predicted distance to LFL.

In this present indirect action programme, three contracts were initially placed. One with Safety and Reliability Directorate, UKAEA to review aspects of the dispersion of denser-than-air vapours relevant to gas cloud explosions; one with the Health and Safety Executive, UK, involving large scale field trials of dense gas releases; and a contract with Battelle Institute of Frankfurt, W. Germany, to carry out an initial comparison of existing computer code models with the data to be obtained from the HSE trials.

Because conditions of releases under accident conditions can vary very widely, experimental conditions must be carefully chosen and controlled to give good data for computer modelling and validation. The series of tests organised by the HSE were mainly instantaneous releases from a fixed volume container ($2000m^3$), although some continuous release tests were carried out as an additional test series. (This was a bonus to the community as they were not part of the original contract). The first phase of the test series involved unconfined releases over flat grass or concrete terrain, a second phase looked at the influence of partial confinement, walls of various porosity were placed around the release point. In general, the first phase of these tests involved conditions similar to earlier work, also organised by the HSE, which had involved releases from containers of 40 m^3 capacity, so information on the influences of scale could be expected to be obtained. Details of the large scale release work carried out under this contract have been presented in this conference.

To assimilate the large amount of data produced by this work involves a lot of time. Much useful information exists on photographic or video tape records, which are extremely expensive to manually reduce to a usable form. Late in this indirect action programme period, a further contract was placed, with the von Karman Institute, Brussels, to develop their methods of computer processing of visual records, and applying them to some of the HSE large scale test records. Typical results have been presented and discussed at this conference.

The work carried out by Battelle Frankfurt using a number of available models for gas dispersion has shown very few models give correct trends for all meteorological conditions. The work from SRD/UKAEA has identified many deficiencies in existing models regarding the treatment of turbulence and has led to recommendations on how modelling may be improved, without going to extreme elaboration which would be expensive in computing effort.

To summarise the present position, I would say that there now exists a better understanding of the limitations and shortcomings of the presently available models used for a sudden release of dense gases. A very important set of data now exists on one particular form of release of a dense gas, but the information has not yet been reduced to a form usable for analysis and development of improved models. This further work would be an ideal project for support by the community.

It should also be recognised that the test data only applies to isothermal conditions. Refrigerated releases are the main concern in NPP safety. Validation of heat transfer modelling as applied to dense gas dispersion is still required.

Whilst phase 2 of the HSE trials covers some aspects of the influence of obstacles on gas dispersion, adequate computer modelling does not exist. Here again, further work is required to complete our understanding of the phenomena involved and on whether we can sensibly extrapolate information obtained from small scale experimental work.

Because light water reactor systems tend to be enclosed in very strong containment buildings, it is argued that there is no need for a better understanding of gas cloud formation and dispersion, the current position is acceptable. However, risk assessments have highlighted the problems of power supplies and decay heat rejection where external events are concerned. Also it is not clear that fast breeder reactors necessarily have to be in such massive containment structures. If strong containment were required simply to deal with uncertainties in the outcome of external event, like the gas cloud explosion then there may, therefore, be a need for better information to avoid unnecessary over design.

3. CLOUD BURNING

The rate of reaction of a flammable gas mixture is very dependent on the local turbulence level. If laminar conditions are present, burning velocities are less than 1 m/s and the pressure rise generated by the energy release is very small, less than a millibar. In large scale systems, turbulence can easily be generated which can enhance the burning velocity by an order of magnitude and the pressure generated is correspondingly increased. If, in addition, obstacles are present and particularly if the gas cloud should be confined by the presence of walls or buildings, the acceleration effect may be very marked, and local pressure rises may be of the order 1/3 - 1 bar. In more extreme conditions, flame speeds increase to an unstable point when a sudden jump from deflagration to detonation can occur and the pressure rise becomes many bars. The need was, therefore, to arrive at a better understanding of the mechanisms and factors involved in flame acceleration and hopefully demonstrate extreme conditions would not apply to nuclear power plant layouts.

A total of five contracts were placed with Community organisations to study various aspects of the flame acceleration problem. As in the area of dense gas dispersion, contracts were placed to support both theoretical and experimental work and were hopefully directed at areas of large uncertainty. Insufficient funds were available to do all that we would have wished.

An important contract was with KWU (Erlangen) to review data available on gas cloud explosions with particular emphasis on the transition from slow deflagration to detonation. In general, we can say unless the possibility of a detonation is very small the safety of NPP's is doubtful. It is too expensive to design to resist detonation, so it was important to obtain confidence that the mechanisms involved in this transition were understood and that the conditions needed to cause transition could be avoided by suitable plant layout. As reported earlier to this conference, the conclusions of this study are affirmative and designers can continue to argue that only slow deflagration pressure rise effects need be catered for in design.

A contract was also placed with SRD/UKAEA Culcheth to study, via a one-dimensional computer code, the effects of various parameters on the pressure pulse produced in the burning of clouds of non-uniform mixtures. Real clouds are never uniform in composition, because air entrainment must always be from the outer edges. This contract also helped to support development of a two-dimensional cloud burning code (FLARE), which would allow the study of the influence of obstacles, or arrays of buildings on the development of the pressure field and allow for the influence of flow induced turbulence on rate of burning. Such calculations would also help to develop confidence that sudden acceleration to detonation conditions were very unlikely in reactor plant configurations.

Part of a contract placed with TNO/PML - Rijswijk also covered the validation and use of simple 1-d models of cloud burning to predict, from experimental flame front measurements, the pressure pulse and the influence of acceleration mechanisms. This involved an analysis of data obtained from small scale laboratory work at ENSMA/Poitiers and, later, data from large scale work by CEA/CESTA. A further part of this contract covered experimental work at TNO/PML on the effects of obstacle arrays on flame propagation initially on a small scale. The original contract was later extended so that further tests on a larger scale could be carried out, so giving increased confidence that the appropriate scale factors had been correctly identified.

Basic experimental data on the free field pressure wave, seen outside a gas cloud explosion, have been produced by a series of medium scale tests carried out under the contract with CEA/CESTA. Balloons of 12 m^3, filled with ethylene/air mixture, were ignited to give fairly fast deflagration conditions. Flame acceleration effects were studied for non-uniform situations by using two balloons, one inside the other with different mixture compositions. The final phase of test series used rectangular enclosures of 125 m^3 volume to give information on scale effects. Except for the rectangular enclosure data from all tests have been analysed under the contract with TNO/PML, but further analysis is possible and will be left unfinished with the end of the current Indirect Action Programme.

The fifth contract in this sub-area of work was placed with RWTH-Aachen for a fundamental study of the acceleration experienced by a deflagration wave in the wake flow region caused by the sudden area change in the flow field after an obstacle. In this work, the overall reaction rate was measured by optical means, so the flow field was undisturbed by the presence of instrumentation. A large amount of data now exists from this contract for ethylene-air and propane-air mixtures over a range of flow velocities up to 80 m/s. Tests have also been carried out using arrays of cylinders, similar in some ways to the TNO arrangement, but with through flow. Here again much data will exist at the end of the contract which will not have been adequately analysed.

In summary then, there is now a large body of data for small or medium scale cloud burning situations, much of which has been inadequately analysed. The basic mechanisms of flame acceleration are now better understood, but there is a lack of evidence to support extrapolation to the very large scale releases of concern to NPP safety. In general, data is for uniform cloud mixtures and not for the continuously varying composition of a real atmospheric release, one-dimensional code calculations imply this is not an important factor, but most calculations do not have a validated feed back mechanism between combustion rate, composition, and local turbulence. (This was the purpose behind the development of the SRD code FLARE).

Most experimental work is for spherical shaped clouds, in reality clouds of dense gas will be pancake shaped. The experimental data for free field pressure measurements will therefore be an over-estimate, currently by an unquantified amount.

To fully understand the influence of various parameters would need the use of a 3-d combustion code, which would be very expensive in computing time. Two-dimensional modelling is probably adequate for most situations. Further work is needed with codes which properly model the combustion process if the degree of over estimation in the free field pressure effects is to be assessed. Ideally large scale experimental data is required to confirm our current ideas on scale effects, but the practical problems are enormous and the economic value of such tests difficult to assess.

4. EXTERNAL PRESSURE FIELDS AND WAVE-OBSTACLE INTERACTIONS

As was mentioned in the previous section on 'Cloud Burning', obstacles and walls can cause flame acceleration when they are inside the bounds of the flammable mixture. Even when they are outside the cloud, they will still have an effect on the pressure waves and flow field produced by the general expansion of the hot gases pushing the cold air out of the way. Whilst simply pressure wave reflection is well understood, the complex interactions from multiple objects, both flat and curved, are more difficult to assess particularly when 3-d effects have to be accommodated. Three contracts were placed to develop a better understanding of this area of gas cloud explosion phenomena.

If the pressure rise is relatively small, a few millibars, linear response is observed with wave effects being additive (acoustic theory). For example, a simple plane wave reflected off a flat wall doubles in amplitude. When dealing with more intense pressure waves, non-linear effects become important and the pressure rise behind the plain reflected wave is more than twice the pressure rise of the incident wave. For conditions of concern in nuclear plant safety, non-linear effects may be expected. However, analytical methods for non-linear systems involve long computing times, it is therefore useful to know how reliable the simpler faster linear analysis methods would be when applied to nuclear plant safety arguments. Two contracts were placed, both involving the development of three-dimensional flow field codes. One contract, with EDF-Chatou, involved the development of a water channel for two-dimensional wave simulation, using the hydraulic jump analogy to the acoustic wave propagation equations. Measurements from the water channel experiments were compared with 2-d calculations of the linear analysis code, with very good results. The advantages of a water channel lie in its ability to give quick comparative indications of, for example, the influence of building spacing or orientation, on local pressure wave focussing effects so that undesirable layouts can be avoided. The water channel results can then be checked by two-dimensional calculations and the actual plant layout studied by three-dimensional code calculations to confirm that no excessive local pressure loads exist on the building walls.

In parallel to the EDF work, CEA/CEN Saclay developed their own version of a three-dimensional acoustic code and this was validated against tests using explosive charges. A further test, using a gas filled balloon to produce a more typical deflagration pressure wave shape on a model of a typical PWR layout was also carried out at CEA-CESTA. Free field measurements gave the input for the code calculations, good comparisons of measurements and code output were obtained, leading to confidence that the simple 3-d calculations can be used in many practical situations.

The third contract, placed with RISO, Denmark, was primarily concerned with an experimental investigation of the influence of atmospheric conditions on wave reflection from ground surfaces and obstacles such as earth mounds or buildings. Whilst some initial work was carried out at a small scale to develop the experimental techniques, the major tests series, however, used gas filled balloons of 1 to 1.5 m diameter. Weather conditions ranging from unstable through neutral, to stable were studied. Once again a large amount of data has now become available to the community members, with virtually no analysis having yet been carried out. The initial conclusions of this study indicate that atmospheric turbulence will give rise to scatter in test measurements which increases with distance from the driving source. Some departure from acoustic theory was observed when incident wave pressures were greater than 100 mb.

To summarise, it appears that computer code calculations based on simple acoustic theory are adequate for many practical situations, particularly as it is believed that the amplitude of the input driving wave is overstated. This overestimate arises from a simplified treatment of the gas cloud burning, where combustion rates have been artificially enhanced.

The external pressure wave produced by a real gas cloud situation would be very different from the idealised ones produced by exploding balloons filled with ethylene-air mixtures. In particular, the after-pulse (negative phase) is an important parameter in structural loading, response and eventual damage and this is poorly represented by the current experimental methods. Somewhat similarly, the case of an exploding engulfing cloud has not yet been considered.

As was mentioned in summarising the section on 'Gas Dispersion', there appears little need for further information to be able to argue the safety of the light water reactor system with a strong containment building. However, for other reactor systems, design savings might be possible if some of the current conservative assumptions could be quantified and removed.

5. FINAL REMARKS

In my opinion, the work in Area B sponsored in this Indirect Action programme has been very successful. In each of the sub areas collaboration between the various contractors has resulted in useful cross-fertilisation and development of analytical and experimental techniques. Because of the breadth of the subject area being covered it has been difficult for the group as a whole to indulge in detailed discussions, the time taken to just present all the progress reports at any one meeting being a significant factor, however, particularly on the experimental side useful discussions of the interested parties did take place.

A large body of experimental data is now available to the community countries, which still requires a lot of time and effort to properly analyse. It would be a great pity if this effort cannot be found or that the analysis only gets done in a fragmented way and not published.

In the overall context of the safety of nuclear power plants it is probably fair to say that we know enough to justify the present safety position on LWRs. It would appear to be biased in the conservative direction, but unless further analysis of the available data is undertaken the margin cannot be quantified. The data obtained and analytical methods developed in the programme of work have obvious applications to any industrial plant not just to nuclear system and needs to be widely publicised.

Despite the time taken up in meetings by the presentation of progress reports, I personally feel that the time was well spent. It allowed the various contractors a chance to emphasise what they thought was an important point and spurred comment and discussion on any controversial aspect.

Discussion

Question: W. Vinck - CEC/BXL

Your statements "we want to show that destructive conditions do not exist, we have reasonable confidence that destruction can be discounted", seems to be a similar situation as that with regard to possibility of steam explosions with respect to early containment failure. Rather than examining if there is an absolute (or deterministic) way did you try a probabilistic approach?

Answer: A.R. Edwards - UKAEA

Essentially, the work of the group has been deterministic, i.e. under what conditions can a detonation occur. In the application to nuclear plant safety a probablistic approach can be used, i.e., the available data on cloud formation and the data on the initial conditions that lead to detonation are considered and a judgment made of the chance of occurance of a detonation close to the plant buildings.

OVERVIEW AND CONCLUSIONS ON THE RESULTS OBTAINED IN RESEARCH AREA C : THE RELEASE AND DISTRIBUTION OF RADIOACTIVE FISSION PRODUCTS IN THE ATMOSPHERE FOLLOWING A REACTOR ACCIDENT

J.G. KRETZSCHMAR*
SCK/CEN
B-2400 Mol, Belgium

Summary

The paper presents the scope of work in all research projects of this area and gives a short description of selected results obtained, with a conclusion on the progress obtained in the area.
In this sub-programme, a total of 13 research projects was executed in different national laboratories of the EC.
It deals with the transport and diffusion in the atmosphere and with the possible ground deposition of active fission products released from the damaged reactor core during the course of a severe accident. Most of the projects are dealing with the study of a number of "partial effects" contributing to transport, diffusion and deposition. This area covers experimental and theoretical studies : modelling and code development and validations/assessments by field trials or by simulations in wind - and water tunnels.

1. INTRODUCTION

During 1980 the CEC initiated a "Cost Sharing Programme" of research work in the area of LWR safety. The aims of this five year programme were to improve communications between research teams, to encourage collaboration between member states and to avoid duplication of effort. As all cost sharing programmes, this one too was intended to be complementary to the current national programmes and to the existing programmes of the CEC Joint Research Centres.

The following three subject areas were covered by the programme :

- area A : the loss of coolant accident (LOCA) and the functioning and performance of the emergency core cooling system (ECCS);
- area B : protection of nuclear power plants against external gas cloud explosions;
- area C : release and distribution of radioactive fission products in the atmosphere following a reactor accident.

The total budget, including staff and administration, was 6.3 MUCE for 5 years. Taking into consideration that of this amount 5.5 MUCE was available for distribution to the contractors, and that the CEC contribution to contracts' budgets was between 20 % and 50 % of the actual costs, the total volume of the research programmes' budgets was somewhere between 15 and 20 MUCE.

---

* currently at Janssen Pharmaceutica, B-2340 Beerse

During the execution of the programme the Commission, as well as the ACMP (Advisory Committee on the Management of the Programme), have taken advice from three groups of national experts relevant to each area in order to set up priorities, to monitor work in progress and to give advice with respect to future work.

In what follows the findings of the experts group for area C are summarized and highlighted by some specific examples.

2. SCOPE OF AREA C

The thirteen projects carried out in area C of the CEC Indirect Action Programme on the Safety of Thermal Water Reactors are listed in Table I. In a first order approximation they can be subdivided into the following four categories :

- review work : project 1C;
- theoretical work aiming at the improvement or development of dispersion simulation codes : projects 3C, 4C, 8C, 10C (partly) and 12C;
- experimental work based on physical modelling in wind or water tunnels : projects 2C (partly), 9C, 11C and 13C;
- field trials to obtain input data for local models or to check their applicability under specific conditions by means of tracer releases : 2C, 5C, 6C, 10C (partly) and 14C.

More specifically project 1C had to review a set of some 13 specific effects such as plume rise, dry and wet deposition, coastal and/or topographical effects, influence of time varying meteorology, lateral spreading, variation of mixing height, urban heat islands, etc. in order to summarize the present state of knowledge, to establish the importance of each of the effects, to identify where validating experimental data are missing and to suggest where future experimental and theoretical efforts may most usefully be concentrated.

In the mathematical modelling group, project 3C aimed at the production of a general computer method for predicting the near-field dispersal of hot self-heating releases from a reactor building, given the local atmospheric conditions, the shape of the building and the nature of the release. With respect to the latter, project 12C focussed on the development of a (near-field) model for internally heated jets, requiring a review of existing models, an extension of an already available integral model for steady buoyant jets and a parametric study to determine the influence on the plume behaviour of radioactivity, humidity, atmospheric stability and some initial conditions.

Table I : Research projects in area C : The release and distribution of radioactive fission products in the atmosphere following a reactor accident

Index	Contract	Contractor	Title of the research project	Project Team
1C	011 SRUK	United Kingdom Atomic Energy Authority, S.R.D. Culcheth	Review of specific effects in atmospheric dispersion calculations	B.Y. Underwood, P.J. Cooper, N.J. Holloway, G.D. Kaiser and N. Nixon
2C	012 SRUK	Central Electricity Generating Board, Berkeley Nuclear Laboratories	Effects of buildings on low-level atmospheric discharges	P.M. Foster and A.G. Robins
3C	016 SRUK	Imperial College of Science and Technology, London	Predicting lift-off of major hot self-heating releases under the influence of a building	A.J.H. Goddard, R.W. Benodekar and A.D. Gosman
4C	014 SRUK	Imperial College of Science and Technology, London	Development of a prototype mesoscale computer model, incorporating treatment of topography	H.H. ApSimon, K. Kitson, M. Fawcett and A.J.H. Goddard
5C	010 SRF	Commissariat à l'Energie Atomique, C.E.N. Fontenay-aux-Roses	Evaluation of the atmospheric dispersion of fission products released from a nuclear reactor following an accident, under meteorological conditions characterized by low wind speeds, with or without high temporal and spatial variability in wind direction	B. Crabol and G. Deville-Cavelin
6C	027 SRI	E.N.E.A./Casaccia	Local-scale atmospheric diffusion at a coastal site, in the presence of breeze effect	P. Cagnetti, V. Ferrara and A. Pellegrini

Table I : continued

Index	Contract	Contractor	Title of the research project	Project Team
8C	015 SRF	Electricité de France, Direction des Etudes et Recherches	Development of a numerical model for the transport of pollutants at meso-scale with the aid of a calculation of the wind field by a method of objective analysis	J.Y. Caneill, J.P. Granier, P. Racher and R. Rosset
9C	017 SRF	Electricité de France Direction des Etudes et Recherches (with Ecole Centrale de Lyon)	Wind tunnel modelling of turbulent diffusion of pollutant puffs	J.P. Schon, O. Debbarh, P. Mejean, R. Morel and J.P. Granier
10C	028 SRB	Nuclear Energy Research Centre SCK/CEN - Mol	Sensitivity, applicability and validation of bi-gaussian off- and on-line models for the evaluation of the consequences of accidental releases in nuclear facilities	J.G. Kretzschmar, I. Mertens and B. Vanderborght
11C	018 SRF	Commissariat à l'Energie Atomique, C.E.N. Grenoble	Physical model of the dispersion of a radioactive contaminant in the atmosphere above a "heat island"	J.A. Toly and D. Tenchine
12C	013 SRF	Commissariat à l'Energie Atomique, C.E.N. Grenoble	Development of an integral model of a radioactive jet released accidentally from a nuclear reactor	D. Grand and A. Badr

Table I : continued

Index	Contract	Contractor	Title of the research project	Project Team
13C	031 SRF	Ecole Centrale de Lyon (with Ecole Polytechnique Fédérale, Lausanne)	An application of the method of concentration determination by digital image processing : wind tunnel simulation of atmospheric dispersion in stable conditions at a real site	J.A. Hertig, P. Liska, M. Ayrault, F. Ladhari and J.P. Schon
14C	032 SRB	Nuclear Energy Research Centre SCK/CEN - Mol	Interlaboratory comparison of SF_6 determinations used within the framework of tracer releases	B. Vanderborght

Taking into account that impact assessment studies, especially for the intermediate 0 to 10 km region, are commonly based on Gaussian dispersion models or "fix-ups" of the former, task 1 of project 10C aimed at a systematic sensitivity study taking into account actually measured hourly meteorological conditions combined with the diffusion typing schemes and sets of dispersion parameters that are commonly used for such studies in the official or recommended procedures in the EEC and OECD member states.

In contrast with the previous projects, dealing with the dispersion in the vicinity of the sources, projects 4C and 8C were dealing with the far-field as they had to develop prototype mesoscale computer models incorporating treatment of (complex) topography. While project 8C was focussing on the development of a finite element method to provide optimal adjustment of a given interpolated windfield to generate non-divergent flow over topography, project 4C was aiming at the development of a full-scale mesoscale transport and dispersion model, including a three-dimensional wind field code and a particle-in-cell or Monte-Carlo type code for dispersion, as well as at its validation on the basis of available data sets and its intercomparison with the results of other models e.g. the one of project 8C.

The study of the effects of an urban or industrial heat island upon the characteristics of the atmospheric boundary layer was the main objective of project 11C, tackling this problem by physically modelling the heat island by means of a heated flat plate in a water flume with appropriate stratification. Within the context of wind tunnel modelling of the turbulent diffusion of pollutant puffs, project 9C proposed the design and construction of a laser based flow visualisation system for the recording, analysis and interpretation of instantaneous concentration fields of pollutant puffs released from a point source.

The goal of project 13C was to use project 9C's instrumentation and methodology on a wind tunnel simulation of the atmospheric dispersion in stable conditions at a real site, namely the Cadarache site for which actual field data were obtained within the framework of project 5C.

The latter project's principle objectives concerned the further development and verification of the ICAIR code by means of in-situ SF_6-tracer experiments at the Cadarache site, under low wind speeds and high temporal and spatial variability of the wind direction. As well SF_6 as smoke were used as tracers in project 2C investigating the effects of buildings on low-level atmospheric discharges and their dispersion in the near field. Specific topics of this project were the development of a plume-mapping technique, measurement of building entrainment effects and comparison of the modelling results with wind tunnel data and field data from the tracer release experiments at the Oldbury power plant site. The development and evaluation of the possibilities of on-line dispersion models for real-time use in an emergency plan of a nuclear installation formed the second task of project 10C. Here too SF_6 tracer releases, simulating accidental releases, were the appropriate means to check the operationality of the emergency procedures and to evaluate the usefulness of on-line models within the framework of an emergency plan, as well as the support that they can give to the intervention teams and the emergency plan's headquarter staff. With already three projects in this programme using SF_6 tracer releases, and taking into account the widespread use of this tracer in many other projects all over the world, it was felt appropriate to start an interlaboratory comparison of SF_6 determinations used within this context. Project 14C tackled therefore the problems of making transferable standards, harmonizing calibration and analysis procedures as well as testing the comparability and reproducibility of analyses within the 0,1 to 100 µg SF_6/m^3 range. The observation, the analysis and the modelling of local scale atmospheric diffusion at a coastal site in the presence of breeze effects, the research topic of project 6C, was the only one focussed on specific and detailed meteorological measurements.

3. SELECTED RESULTS

Within the framework of project 1C the Safety and Reliability Directorate (SRD) produced a comprehensive and very extensive review covering eleven, less or more mutually coupled, specific effects that have an influence upon the "end-points" of consequence modelling or probabilistic risk assessment studies. In general the different authors started from a systematic evaluation of how the specific effects are physically treated and mathematically translated into the first generation computer codes actually in use. They noted that in general the models do account for the phenomena under consideration at only a very rudimentary level, despite the fact that the subject areas have undergone considerable development in the last few years.

Although conclusions and recommendations are to a certain (variable) degree specific for each topic, the following rather common themes emerge :

- simple models for consequence calculations are still "a must";
- in this group "fix-ups" of the Gaussian model do hold a strategic position;
- pragmatic "fix-ups" must go hand-in-hand with more complex, and hopefully more realistic models;
- increased realism nevertheless leads to a requirement for substantial amounts of extra data, much of which is not readily available or very expensive to obtain or not obtainable at all.

Future theoretical as well as experimental work has to focus on code validation by means of field experiments and/or physical modelling as well as on improving our understanding of the basic physics and the translation of this understanding into computer codes.

If a ranking has to be made on the basis of the conventional consequence modelling end-points (number of acute and delayed fatalities, areas of land contaminated and number of people moved) the following priorities do appear :

- high : improvements in modelling wet deposition and time varying meteorology;
- moderately high : dry deposition and plume rise;
- fairly low : cross-wind shear, mixing height, building wake.
- undecided (due to lack of sufficient evidence) : urban, coastal and topographical effects.

As already mentioned two projects dealt with the development or improvement of numerical 1D-, 2D- or 3D-codes able to simulate transport and dispersion phenomena in the immediate vicinity of the release point or surface. In project 3C a general computer method has been developed to predict the flow and dispersion around a building which, when fully developed, may be applied in risk assessment. The prediction procedure has been tested successfully for the complex fields around cuboid buildings submerged in normally incident neutral boundary layers. Flow fields and dispersion of passive point source releases have been modelled. In addition, a selected number of cases from the wind tunnel experiments at the Warren Spring Laboratory, where buoyant gas was released from various faces of a simulated reactor building have been simulated. Near-passive releases and releases with a full scale equivalent of 80 MW were simulated. Predictions are generally in good agreement with the wind tunnel experiments and represent a marked advance in 3-D computing of bluff body flows. Resolving remaining discrepancies is likely to require improvement of the turbulence model. Recent work during the contract extension has led to atmospheric stratification being incorporated. Future development of the procedure will permit the treatment of non-orthogonal approach flows, non rectilinear building shapes, source momentum, etc... In project 12C the already existing one-dimensional integral model MICAR for a jet released into a horizontal cross flow or a stagnant stratified medium has been extended to take into acount internal heat sources, internal and external humidity, vertical wind gradients and inclination of the released jet. A parametric study has indicated that initial conditions, such as jet orientation, temperature, velocity, ... as well as ambient conditions, such as stability class, wind velocity,... are very important, while humidity can be neglected in unsaturated conditions. For a (highly) hypothetical energy efflux of 16250 W/s the following results were noted :

- in stable stratification a pronounced height increase in the calculated trajectory with corresponding decrease in the ground- level concentration values;
- in neutral atmosphere the lifting of the jet's trajectory is less pronounced;
- while in unstable atmosphere rather chaotic trajectories could be produced under the influence of the high atmospheric turbulence;
- the observed effects decrease with decreasing energy content of the jet.

The results of task 1 in project 10C, evaluating the influence of the choice of a specific "turbulence typing scheme - dispersion parameter set" combination upon the cumulative frequency distributions of the calculated relative concentrations in receptors up to ten kilometers downwind a ground-level or elevated release, can be summarized as follows.

The intercomparison of hourly stability classes, determined simultaneously in ten different, but widely used, turbulent diffusion typing schemes, taking into account the same 1977-1979 meteorological data base (hourly measurements on the 120 m tower at the SCK/CEN, Mol and the corresponding synoptic observations at a nearby military airport) illustrated very clearly that systematic correlations between different systems are indeed rare. The majority of the two by two combinations showed less than 50 % agreement with respect to the simultaneous occurrence of the same stability category (A, B, ...). Deviations from this synchronism were sometimes systematic, sometimes stochastic.

The practical consequences of the uncertainties in specifying the stability class at a given moment, as well as the influence of the chosen dispersion parameter set σ_y, σ_z (Pasquill-Gifford, Vogt, SCK/CEN, Briggs, Smith-Hosker, Doury) were evaluated in twelve different combinations "turbulence typing scheme - dispersion parameter set". For each of the combinations yearly cumulative frequency distributions were calculated for respectively 1h, 8h, 16h, 3d and 26d-average ground-level concentrations. For all the statistical parameters (yearly averages, percentiles and extremes), normally used in PRA's, major differences (factor 4 to 10 to 100) in the results were noted. These differences were almost entirely due to the differences in the numerical values of the various sets of dispersion parameters $\sigma_y(x)$ and $\sigma_z(x)$, while for the overall statistics, the specific procedure to determine the stability class at a given moment was of almost no importance. As a conclusion it can be stated that the results of different impact assessment studies are not directly comparable even if they are based on the same simple bi-Gaussian dispersion approach.

The modelling of transport, dispersion and deposition phenomena over large distances (≈100 km) in complex topography basically requires two different but mutually coupled codes. First there is the prediction of 3D-windfields making optimum use of the limited wind data likely to be available in the region and secondly the dispersal and deposition of the pollutant in the inhomogeneous windfield must be simulated. This was successfully realized in project 4C by means of the WAFT model (Windfields Adjusted For Topography) in combination with the TOMCATS model (Topography Oriented Monte-Carlo Atmospheric Transport Simulation). The windfield model WAFT (similar to ARAC's MATHEW, but with a better resolution) refines an initial windfield obtained by straightforward interpolation from the (measured) wind data to obtain a final windfield which is consistent with the conservation of mass in the flow around hills and orographic contours. In TOMCATS the release is followed as an assembly of pseudo-particles, each with a parameter representing the probability that the particle has not yet been deposited or decayed.

Although substantial improvements have been made in comparison with previous models, it had to be concluded from several realistic case studies that considerable uncertainties do persist in the predictions of such a model(s). On the question "Do more complex methods produce significantly better predictions than simple Gaussians?" the investigators answered that this may not be so. They therefore conclude that at present simple procedures, based on Gaussian puffs or plumes, are still the best option in emergency situations, while the complex WAFT/TOMCATS type of approach could be useful in post factum analysis or in training or rehearsal scenarios.

The calculation of wind fields in complex topography and on a regional scale was also the aim of project 8C. In contrast with the finite difference approach, used in the previously discussed WAFT model, the french teams based their model on a finite element variational procedure. Comparison with the WAFT model on the basis of testcases in the hilly Bristol Channel region in the U.K. on one hand, and in the mountainous Vosges region in France on the other hand, showed qualitative agreement between both models. The already mentioned caveats do of course hold for both codes.

With respect to physical modelling in wind and water tunnels a new laser-based visualisation and data treatment and analysis system for plumes and flows has been developed (project 9C) and successfully tested as well on the study in a water tunnel of the "urban heat island effects" (project 11C) as in a large wind tunnel simulating complex drainage flows under different atmospheric stratifications (project 13C). The operational system visualizes instantaneous as well as average plume and concentration profiles, derives different statistical parameters for the former and visualizes differences due to changes in source characteristics, atmospheric conditions and terrain features.

In the already mentioned water tunnel simulation of the two dimensional flow above an urban heat island (11C) no significant influence of the heat island upon the average velocity and concentration fields was noted, although the internal random fluctuations in the flows seemed to increase.

Although no large scale field trials could be included in the first phase of the programme, experimental field work has successfully been initiated by means of a number of projects. At an Italian coastal site (project 6C) a methodology, based on off-shore and on-shore meteorological measurements and soundings, has been worked out to follow the initiation, formation and elimination of internal boundary layers (IBL) as a function of time. Relevant meteorological data for the study were obtained by means of 4 different measuring campaigns at the coastal site and 2 campaigns on sea.

The analysis of the data showed the development of a well-defined IBL during on-shore flow only in spring and summer, while for anticyclonic situations an inversion layer was detectable aloft independently of the season.

A simple model - actually a fix-up of the bi-Gaussian fumigation approach - to evaluate in a statistal way the probability of exceedance of a given short or longer term concentration level, at distances up to 10 km from the source, was proposed.

At an english reactor site (Oldbury Power Station) smoke as well as SF_6 were used simultaneously as tracers to study the complex flows in the immediate vicinity ($<$ 1 km) of a large reactor building (project 2C). By using the SF_6 ground-level measurements to calibrate the corresponding smoke measurements obtained by means of a Lidar, it became possible to translate the qualitative Lidar information on the vertical (smoke) concentration profiles into quantitative information. Approximately 20 tracer release experiments were made of which about one half were released at pile cap height (25 m) from an outside balcony, the remainder being split between ground-level and roof level releases. Complementary wind tunnel investigations were conducted in the CEGB wind tunnel at its Marchwood Laboratory. Comparisons between full scale (field) and model scale (wind tunnel) plume section contours were good. Tests on some popular simple analytical models showed that they are no good close to the source but give acceptable results further downwind with predictions of peak concentrations within a factor of three.

While the data have not yet been completely analysed the modelling efforts and SF_6 tracer experiments carried out at the Cadarache site (project 5C) have highlighted the difficulties and uncertainties when dealing with dispersion phenomena under low wind speeds and the typical local topography. A slow temporal evolution of the concentration fields has been noticed during the experiments. High concentration values were recorded in the experimental 0 to 2 km downwind range a long time (up to 6 hours) after the end of the release. This phenomenon has not yet found a plausible explanation. The tri-Gaussian puff model calculations showed reasonable agreement for the maximum concentrations but the observed plumes were narrower than the calculated ones, using Doury's $\sigma_y(t)$ and $\sigma_z(t)$ values. This is probably linked to the local topography. Further on-site experiments, as well as simulation in a wind tunnel (project 13C) will be needed to clarify these problems together with the question if punctual measurements of wind speed and direction are sufficient to characterize the mean transport of the pollutants under variable and low windspeed conditions at a site of the Cadarache type.

The possibilities of using in real time (simple) mathematical models as an operational support to the Emergency Plan of a nuclear site - or in general whatever industrial installation having the possibility for accidental releases with an environmental impact - were investigated at the belgian nuclear energy centre in Mol (project 10C - task 2). An on-line trajectory model, continuously fed by on-site short-term (10 sec.) wind direction and windspeed measurements at various levels, in combination with a tri-Gaussian puff subroutine was developed, implemented and tested by means of simulated accidental releases using once again SF_6 as tracer.

By means of this so-called CAERS (Computer Aided Emergency Response System) system it was shown that real time modelling is a support for the emergency plan headquarter's staff, for the intervention teams in the field, for the coordination between both and for filing and interrelating events as they occur during routine or accidental release. Last but not least a CAERS system, in combination with tracer releases, gives a very high degree of realism to the exercises that must be organized to keep whatever emergency plan operational. People are more motivated for doing real measurements in the field than for simulating them.

As all field trials in this programme were, at least partly, using SF_6 as tracer to verify models or to study dispersion phenomena, an intercalibration at laboratory level was undertaken (project 14C) in order to evaluate the feasibility of making reliable exchangeable standard (calibration) samples and to check the reproducibility and comparability of the techniques and the data of the participating laboratories.

Of the nine participants four showed excellent agreement, two were systematically too low, two had problems with the lowest concentrations (ppt range) being outside their normal working range, while the last lab had problems with contaminated samples. More work is surely to be done in this area, not only under lab conditions but also field intercomparisons.

4. CONCLUSIONS AND RECOMMENDATIONS

What preceeds is only a very incomplete summary of what can be read in the respective final reports already published or in press. What will not be in those thousands of pages is the continuous exchange of ideas and experiences that took place during the successive "expert group" meetings or during bilateral or trilateral formal and informal contacts.

The spirit of international collaboration, as well as the perception of the real need for it, will not easily disappear within group C.

With respect to the future it would also be highly inefficient not to continue and round-up the research that has been initiated during the now terminated first phase of the CEC indirect action research programme on safety in thermal water reactors. Budgettary constraints, being quite realistic at the present time, must not be overemphasized and a pragmatic way-out does certainly exist. Our Europe of to-morrow will be based anyway on cooperative efforts, combining and coordinating the available means and expertise.

Therefore the group strongly supports the idea of putting together the tools and the knowledge, that we have now, to carry out a limited number of larger scale projects tackling actual real situations that are of interest to the member states. The already available complex codes, the commonly used more simple models, the wind tunnel expertise, meteorological instrumentation and knowledge and the tracer release techniques can be brought together to carry out experimental work on real, actually existing configurations and problems in the zero to 80 km range.

Finally the group feels that, up to now, the transfer of knowledge from the non-nuclear field to the nuclear field has been almost exclusively left to the initiative of each individual (organisation or lab), although for atmospheric transport and dispersion problems there is no wall dividing the atmosphere in a nuclear and a non-nuclear compartment.

A trivial possibility to optimize the needed cross-fertilisation is to bring the many different CEC and national programmes and actions as close together as possible, as well in the conceptual as in the executive phase.

Acknowledgements

The group acknowledges the continued efforts of J. Vilain and B. Tolly, both of DG XII, CEC-Brussels, to stimulate over the past four years the work and the spirit in group C.

Discussion

Question: W. Vinck - CEC/BXL

The most practical application of the work in consequence modelling relates to actual emergency preparedness. Do you feel that time has come to reflect the findings in sophisticated codes or code validation, into practicalities such as requirements for (modèle) measurements techniques. Of importance in the same context is computer-aided emergency response. Attention must be paid to the education in the use of codes by those confronted with an actual accident situation.

Answer: J.G. Kretzschmar, SCK/CEN

Within the framework of an emergency plan, on-line modelling in real time of the transport, dispersion and deposition of (accidentally or routinely) released fission products can be a support for the staff of the headquarters and the field workers of the intervention team only if,

a) there is a continuous link and feedback between the model and the mobile as well as the stationary field measurements and observations;

b) concerned staff has been trained by means of as realistic as possible excercises (e.g. making use of on-site tracer releases simulating accidental releases);

c) the models are as simple as possible, but validated (see b) and the outputs are user friendly and obtainable in realtime under hardcopy format for the staff of the headquarters. Continuous display on a large screen would be very welcome. Measured as well as calculated data must be comparable in real time.

THE CEC RESEARCH ACTION PROGRAMME ON REACTOR SAFETY 1984-1987

H.F. HOLTBECKER
Commission of the European Communities D.G.XII

Summary

This report gives a brief outline of the Commission Reactor Safety programme. Following the Commission decision on a "European Scientific and Technical Strategy" the activities to be performed either by J.R.C. or by national organisations form a coherent programme which should cope with the requirement to make this contribution as valuable as possible.
In the LWR programme particular emphasis is given to the development of methodologies for reliability and risk assessment which encompasses the problems of operator behaviour and man/machine interaction studies.
The experimental effort is centered on the study of loss coolant accidents, the study of materials and of structures properties. Modelling work and code development are conducted in parallel.
Source term studies have been adequately considered.
The Commission LMFBR Safety programme still reflects the emphasis with which power excursion accidents have been analysed in the past. Expensive out- of-pile tests are conducted at J.R.C. and in-pile tests are proposed to be supported in S.C.A.'s. However a number of projects are adressing the question of accident prevention and efforts are being made to better understand potential accident initiating events such as the partial loss coolant flow.
The need of a medium size shaking table for seismic studies is being examined.

1. STRUCTURE OF THE COMMISSION RESEARCH PROGRAMMES

In COM (82)865 final (1) the Commission has made proposals for a European Scientific and Technical Strategy. This paper is intitled "On the Framework Programme (FP) for Community Scientific and Technical Activities 1984/1987.

The FP sets out Community goals and related scientific and technical objectives which serve for the preparation of so called Action Programmes (AP) (2).

The Reactor Safety programme belongs to the AP "nuclear fission" and corresponding goals are:

- Promoting industrial Competitiveness
- Improving the management of energy resources
- Improving living and working conditions

All Commission activities are executed either by the Joint Research Center (direct action) or by national organisations which provide part of the necessary budget (Shared Cost Activities).

2. OBJECTIVES OF THE COMMISSION REACTOR SAFETY PROGRAMME

The Commission programme covers both the Light Water Reactor Safety and Liquid Metal Fast Breeder Reactors. The action taken by the Commission for both reactor types pursue the following main objectives (see also (1)):

- exploit to the full the operating data from the many reactors currently in service in the Member States which provide an irreplaceable data base for improving our knowledge of the reliability of the plant, system and components, and of the effect of human factors on safety. Using these data, probabilistic methods - whose development should be continued - will enable risk assessment to be improved;
- develop different accident prevention techniques and methodologies, focusing in particular on monitoring systems, instrumentation and inspection.
- improve our knowledge of the mechanisms governing accidents, developing the calculation codes used for safety analyses. The comparison and verification of the different codes, the dissemination of information and the presentation of the codes to their potential users are all factors that should help a gradual harmonization of nuclear safety analyses at European level;
- improve, by means of in-pile and out-of-pile experiments, the knowledge of accident sequences, with a view to establish rules for the behaviour of operators in the event of an emergency.

All these objectives can only be attained through a balanced combination of theoretical and experimental research. The latter aspect somemtimes requires the use of large facilities and multidisciplinary teams of a high standard, and should be carried out with the widest possible international cooperation. Such cooperation should be reflected at Community level in a pooling of efforts around major experiments, conducted either in Community research centres or in national centres with Community participation. This would also facilitate international collaboration, the results of which would be available to all the Member States.

In the now following description of research areas, no distinction will be made between Direct (JRC) and Shared Cost Actions (SCA), however the activities will be identified separately.

3. LIGHT WATER REACTOR SAFETY

3.1 Accident prevention

In the area of accident prevention the major Commission effort is devoted to the following research items:

3.1.1 Risk and Reliability Evaluation

Methodologies are being developed with the aim to model and analyse accident sequences. This action includes the collection of data and information from operating power plants.

More in particular the JRC will develop further and put into operation the data banks which constitute the european reliability data system. (Component Event Data Bank, Abnormal Occurrences Reporting System, Operational Unit Status Report, Generic Reliability Data Bank).

In order to reduce the uncertainties of probabilistic risk assessment,new techniques are developed which describe each physically possible accident sequence as a result of random and dynamic interactions between the physics of the transient and the time evolving aspects of the component states. In addition the feasibility of an accident simulator will be explored.

In order to support the programme mentioned above and to get more and more organisations involved in assessing advantages and disadvantages of the probabilistic risk assessment, a series of benchmark exercises will be organised by J.R.C. with the support of SCARP on subjects to be agreed e.g. procedures to construct event tree, common cause/mode failure, etc.

3.1.2 Human Factors and Man-Machine Interaction

This programme shall be executed as SCA primarily. It is designed for improving the understanding of operator behaviour in normal and abnormal conditions and of the effect of environmental factors.

3.1.3 Project for the Inspection of Steel Components (PISC)

International cooperation in the field of inspection of LWR components has existed for a long time and is contralized in a working party under the joint responsability of CSNI (OECD-NEA) and the Commission. The PISC I project successfully identified existing gaps and uncertainties when applying Non Destructive Examination (NDE) techniques. The PISC II programme is a logic continuation of the PISC I exercise to test and compare new techniques and procedures which should better cope with identified requirements.

The JRC is the operating agent also for PISC II and actively participates in the execution of parametic studies which aim to better characterize the measuring equipment, the defects, and to study the influencee of vessel cladding on NDE measurements.

The SCA are designed to support national organisations in their active participation in PISC II activities and to help in assessing the NDE techniques applied by them.

The first part of PISC II will end in 1984 by the destructive examination of four plates. A second phase should spread over 1985 and will be dedicated to integrate the results of parametric studies and Round Robin Test.

To round off the PISC exercise the application and verification of NDE techniques on real defects has been decided in principle and is being actively prepared by the experts.

Already in may 1983 it was agreed to instrument 3 existing ATFI cells in the ESSOR nuclear plant:

- 1 cell for dismantling of steel components up to 4 tons weight;
- 1 cell for X-ray examination of 25 cm thick plates;
- 1 cell for Ultrasonic examinations.

3.1.4 LWR Primary Circuit Components Life Prediction

NDE examination techniques constitute an important tool to localize and monitor the size of defects periodically.

However the long term objective is a systematic approach which allows, by relying on different types of data (including NDE), to determine the residual life time of a structure.

The J.R.C. will contribute to solve this problem by making efficient use of a laboratory in which a series of already available 1 to 5 scale models of PWR pressure vessels will be subjected to fatigue tests. This activity is being carried out in collaboration with national laboratories which are invited to use the J.R.C. experimental facilities for testing their NDE procedures including acoustic emission and for verifying their residual life prediction codes. The code COVASTOL developed jointly by J.R.C., C.E.A. and FRAMATOME will be improved to check the validity of combined modelling and non destructive inspection techniques when estimating the life expectancy of structures.

The SCA are programmed to support the parametric studies mentioned above. In addition the thermal stresses problem will be investigated further by e.g. measuring their distribution, analysing their relaxation during post weld heat treatment and evaluating their influence on stable crack growth.

3.2 Accident Analysis, Control and Mitigation

In this area the Commission intends to improve the knowledge of mechanisms governing the accident evolution. Merging experimental and analytical work is essential to assure that the results of expensive experiments are used to verify models and codes.

For each of the below described particular investigations, an effort will have to be made to define the acceptable error band for the calculated predictions of a postulated accident scenario.

3.2.1 Study of Abnormal Behaviour of LWR Cooling Systems

Work concentrated on large break studies of the reactor cooling loops before the TMI accident and attaches considerable importance to small leak scenarios and special transients since.

Because of the many implications of this activity on e.g. reactor instrumentation and operation, operator training, development of simulators, the Commission considers it necessary to continue the fundamental work undertaken in the field of thermohydraulics and heat transfer under two phase flow conditions as well as the execution of large scale well instrumented experiments. Both activities are designed to define the range of applicability and to increase the precision of results of available system codes.

The list of fundamental studies to be executed primarily in the frame of SCA has been drawn up after analysis of system code weaknesses. High priority items have been identified to be:

- interfacial transfer mechanisms i.e. heat, mass and momentum transfer;
- counter current flow in horizontal and vertical pipes;
- two phase natural convection flow including reflux condenser operation of steam generators;
- emergency core cooling water mixing and condensation.

The direct action programme is centered on the integral system LOBI facility built and exploited in cooperation with the BMFT.

The plant has been upgraded recently to better cope with the requirements of small leak and special transients accident scenarios. Experimental results are used for improving large system computer codes and the associated modelling work.

3.2.2 Degraded Core and Severe Fuel Damage

Following the discontinuation of Super Sara experiments the Commission, convinced of the necessity to solve the problems connected with damage of fuel under severe accident conditions, whishes to continue its efforts on three levels which are: execution of separate effect experiments, participation in in-pile tests and support to the development of best estimate codes. Studies shall concentrate mainly on so called recoverable situations for which the conditions for the in-situ coolability of damaged fuel elements must be defined.

The Commission sponsored activities will be executed primarily using the SCA approach. The intervention of J.R.C. is limited to technical expertise and to the publication of research work executed in this area previously at J.R.C.. The separate effect studies, included in the programme, propose a participation in the analysis of chemical and physical processes during material interactions, i.e. high temperature and pressure oxidation kinetics of Zircaloy, clad/fuel liquefaction, candling and relocation. They also foresee the study of particulate formation under various cooling regimes, the effects of quenching, the cooling by two phase mixtures.

The participation to in-pile test programmes is intended to broaden the spectrum of information and to provide experimental data and results for comparative bench mark calculations.

3.2.3 Source Term

Evaluation of the source term requires knowledge of the inventory of fission products that exists potentially in a nuclear power station at the moment of, and during the various phases of, an accident as well as knowledge of the fission product release and trasport process in the primary circuit, then in the containment and finally to the environment. Knowledge of the source term plays a central role in realistic risk assessment and makes it possible to define emergency countermeasures. Because of the conservatism of assumptions made in safety reports, several OECD countries started in-pile and out-of-pile research programmes in recent years which have produced important results already. The Commission has a strong interest to disseminate results of ongoing activities and to conduct confirmatory research wherever necessary.

In order to establish an appropriate computation capability, some computer codes already available at J.R.C. or availbale in the open literature are being implemented, used and improved.

SCA activities have been proposed in the areas:

- Release of fission products from the fuel. Such data should be collected from tests conducted at temperatures between 1500 and 1800°C either in-pile or out-of-pile.
- FP/aerosol transport in the primary cooling system.
- FP/aerosol transport in the containment, diffusiophoretic deposition and resuspension effects.
- Comparison of predictive calculations, which should lead to a critical review of models and identification of additional research needs.

4. LIQUID METAL FAST BREEDER REACTOR SAFETY

The Commission LMFBR Safety programme has been devoted to the analysis of hypothetical severe accidents in the past. The results so far obtained have demonstrated both, that design features of this reactor produce a large safety margin against uncontroled power excursions and that the conversion of thermal into mechanical energy during a power excursion is much less than initially expected.

Profiting from this situation, reactor designers are proposing to eliminate large accidents from design specifications. If this proposal is accepted the degree of necessary detail and precision in the knowledge of LOF and TOP type accident scenarios will have to be reconsidered and emphasis on some of the proposed research items may well change. In particular the identification of accident initiators and the prevention of growth of small accidents will become research items of major concern.

4.1 Accident prevention

Under this heading, it should first of all be mentioned that the effort of development of probabilistic risk analysis, human factors and man/machine interaction covered under & 3 are generally relevant also to LMFBR's and are therefore not repeated here. In addition the Commission efforts concentrate on the following research topics.

4.1.1 LMFBR Material Properties and Behaviour of Structures

LMFBR materials are normally operating at low pressure and high temperature. Structures are thin walled and made of stainless steel. As a consequence material properties and inspection techniques are often different from those used in LWR's.

Activities of J.R.C. deal mainly with creep phenomena, creep fatigue interaction, fracture mechanics, dynamic material behaviour and structural analysis. These studies are performed using materials which have been aged, welded and irradiated. Experiments are evaluated for the development of constitutive laws. The analytical capability of the Eurdyn series of codes are extended by including non linear dynamic analysis (large strain formulation) and by developing a 3 D-ALE code for dynamic fluid structure analysis.

The Large Dynamic Test Facility is being used to investigate the effects of strain rate on elasto-plastic fracture process and on dynamic mechanical behaviour of structures..

Biaxial dynamic tests are conducted on virgin and damaged materials using small specimens.

The SCA in this research area represents an important extension of the J.R.C. programme and can be summarized as follows:

- validation of constitutive laws on structures using loading combinations which are representing real operating conditions;
- life prediction exercises for selected austenitic steel components e.g. a pipe elbow, a tubing section;
- structural defect propagation tests usign large specimens and model structures (including weldings) subjected to typical mechanical and thermal loading conditions;
- behaviour of structures under dynamic loading.

4.2 LMFBR Accident Analysis Control and Mitigation

The objective is to participate in solving specific thermohydraulic problems which are important during the initiating phase of an accident. A significant effort is being devoted like in the past, to the study of the most severe but very unlikely accidents leading to core damage.

Increased emphasis will be given to a more realistic description of accident scenarios.

At J.R.C., the European accident code will be developed further. Its modular structure has already enabled to include various validated stand-alone programmes. The future work includes modelling of fuel dynamics, fission gas release, clad failure, in-pin fuel motion, in-channel dispersive fuel and clad motion, improvement of structural feed back, improvement of neutronic modules and possibly the introduction of a primary loop module.

In the existing well equipped sodium loop test rigs will be installed to measure in-subassembly temperature and flow pattern at low inlet flow and low power conditions. Specific phenomena to be studied are heat transport by natural convection including boiling.

Transient experiments will be conducted in high powered bundles to obtain data on transient flow pattern and possibly clad melting.

To reduce the probability of energetic U02-Na interactions further the stability of sodium film boiling will be investigated.

At J.R.C. a multi purpose 100 kg U02 melting facility has been built and tested (FARO). It is now being prepared to perform experiments in the areas: U02 plugging and freezing, steel plate ablation and perforation by fuel jets, fuel fragmentation and settlement in a Na pool, thermomechanical loading of structures.

In addition to the above mentioned out-of-pile test programme, the J.R.C. is conducting in-pile experiments for post accident heat removal (PAHR) studies at SANDIA, Grenoble and Mol. The final objective is the demonstration that fuel debris settled on vessel internal structures can be cooled without putting in danger the integrity of the reactor vessel. Computer codes are developed in parallel to better understand all phenomena involved and to predict the behaviour of fuel debris in real reactor conditions.

The SCA have been designed to substantially increase the value of the Commission activities in accident analysis. It foresees the Commission participation in several european in-pile test programmes which are relevant for:

- the investigation of fuel and cladding behaviour prior to pin failure in transient overpower accidents;
- the investigation of a loss of heat sink accident;
- the study of thermal subassembly to subassembly propagation at nominal power and in the presence of local blockage;
- the investigation of thermal and mechanical loading of a subassembly in contact with a pool of molten fuel at decay power.

5. EVALUATION OF A VIBRATING TABLE

Following the request of the Council, J.R.C. is executing a series of studies and actions which have the following objectives:

- Assess the feasibility and the interest of a medium size shaking table.
- Identify research needs in seismic analysis.
- Execute eventually a general design for a vibrating table.

This task is being carried out by analysing the needs of industrial plants in general and of civil engineering.

6. CONCLUSIONS

The Commission is providing continuous support to Reactor Safety studies. Safety research presents an indispensable back up to a Community approach as regards safety criteria and standards for construction and operation of nuclear power plants. The results are made available to reactor designers, constructors, utilities and licensing authorities to ensure operational reliability of the plants, to prevent or limit the consequences of accidents and to develop calculation tools for the analysis of very low probability events.

The combination of direct and shared cost programmes in one coherent set of activities is a novelty introduced via the Commission framework programme and is expected to have a positive impact on the results.

An indicative percentage break down of resources already allocated to J.R.C. and requested for SCA is given in tables 1 and 2.

7. LITERATURE

(1) COM(82)865 final. On the Framework Programme for Community Scientific and Techynical Activities 1984/1987.

(2) COM (83)300 final. Research Action Programme on the Development of Nuclear Fission Energy 1984/1987.

CEC REACTOR SAFETY PROGRAMME 1984/1987

Table 1

Indicative Break-down of J.R.C. Resources

Research area	Budget %	Manpower %
1. Reliability and Risk Evaluation	5,3	8,5
2. PISC	4,7	7,0
3. LWR Primary Circuit Components Life Prediction	3,7	4,3
4. Study of Abnormal Behaviour of LWR Cooling Systems	22,6	25,7
5. Source Term Studies	2,1	2,3
6. LMFBR Accident Modelling	13,7	28,8
7. PAHR in-pile	31,6	7,4
8. LMFBR Material Properties and Structural Behaviour	7,9	15,2
9. Evaluation of a Vibrating Table	8,4	0,8
TOTAL	100	100

CEC REACTOR SAFETY PROGRAMME 1984/1987

Table 2

Indicative Break-down of the proposed SCARP Resources

Research Area	Proposed SCARP MioECU	Proposed SCARP %
1. Reliability and Risk Evaluation	1	4,2
2. Human factors and Man/Machine Interface	1	4,2
3. PISC	0,6	2,5
4. LWR Primary circuit components Life Prediction	2,4	10,1
5. Study of Abnormal behaviour of LWR Cooling System	2,2	9,2
6. Degraded Core and Severe Fuel Damage	4,8	20,2
7. Source Term	3,5	14,7
8. LMFBR Accident Modelling	4,8	20,2
9. PAHR in-pile	-	-
10. LMFBR Material Properties and Structural Behaviour	3,5	14,7
TOTAL	23,8	100

DEVELOPMENTS IN REACTOR SAFETY RESEARCH

A. Birkhofer
Gesellschaft für Reaktorsicherheit (GRS) mbH
(translated from German)

Summary

Status and trends of reactor safety research and technology in the following fields are addressed:

- thermohydraulics and fuel behaviour
- severe accidents
- fission product behaviour
- probabilisitic methods
- optimization of instrumentation and control systems, including man/machine interaction
- operational experience and its influence on reactor safety research
- interaction between licensing and research.

Aspects of system behaviour under accident conditions are discussed mainly with respect to pertinent computer codes and test facilities. In the context of severe accidents and fission product behaviour in-vessel and ex-vessel phenomena, issues of source term reduction, and accident management to prevent core damage are highlighted. The probablistic approach is presented as an important supplement to the approved deterministic methods of safety assessment. The basis for sensible applications of probabilistic methods has to be improved by evaluation of operational experience. The developments of different countries in the field of instrumentation and control are discussed. To detect disturbances as early as possible by advanced I$C-systems is one of the future aims.

"Safety has been an important consideration from the very beginning of the development of nuclear reactors." This is the first sentence of the well-known book on "The Technology of Nuclear Reactor Safety", published by T.J. Thomson and J.G. Beckerly in 1964.

A number of fundamentals did remain unchanged over four decades of reactor development. So, reactor safety is still based to a high degree on the theoretical analysis of postulated accidents, in order to anticipate and thus to avoid bad experiences. In spite of these and other conformities, a lot has changed from the safety design of early small scale reactors until that of modern commercial nuclear power plants.

Before dealing with technical questions, I would like to say a few words concerning the financial cost of safety research. Figure 1 shows the reactor safety research budget of the BMFT (Federal Ministry of Research and Technology) and the USNRC (US Nuclear Regulatory Commission) in recent years.

An appreciable increase in resources between 1970 and 1980 can be observed, together with a further rise resulting from the TMI-2 accident. Expenditure has been declining since 1982.

According to estimates made by a DOE (Department of Energy) advisory committee, the free world spends approximately US $800 million/year on R&D activities relating to LWR safety. This figure includes expenditure by both governmental agencies and industry. This expenditure can be set against the value of the electric power produced by nuclear power stations, which amounted to approximately 800 TWh in 1983. Figure 2 shows the present position and the expected development. If one assumes a price of 5 cent/KWh, it is clear that only a very small percentage of the revenue from nuclear electricity production is spent on reactor-safety research. This figure corresponds roughly to the relative expenditure on safety research in the automobile industry. It must, therefore, be asked whether, in view of the importance of reactor safety, an assessment of the present trend is justified.

After this brief digression on financial matters, I shall now turn to technical issues. Reactor safety research and reactor safety technology cover a wide field. I will, therefore, concentrate on the following points:

- thermohydraulics and fuel behaviour with reference to transients, including loss of coolant accidents (LOCA)
- physical phenomena and system behaviour during severe accidents
- probabilistic methods in reactor safety
- optimization of instrumentation and control systems, including man-machine interaction
- importance of operational experience for safety research and safety technology
- interaction between research and licensing.

I will not address material problems since these will be dealt with in a series of special meetings. At present, I only wish to point out that a great deal of work has been done on the integrity of the pressurized containment, so that in future the emphasis should lie more on the long-term effectiveness of components. For reasons of time, I must also renounce to discuss the behaviour of fission products in the atmosphere.

1. Thermohydraulics and fuel behaviour with reference to transients, including LOCAs

The aim of the research on thermohydraulics and fuel behaviour is to assess the effectiveness of the reactor-control and emergency cooling systems during an accident. This requires a detailed description and simulation of the physical phenomena and of system behaviour.

The physical phenomena occurring during operating transients including LOCAs have been adequately investigated and simulated with the help of expensive computer models, with the exception of certain issues having three-dimensional aspects.

System behaviour during operating transients and LOCAs have been also successfully simulated in test installations. In addition to the results of 400 integral experiments which are already available, some further 150-200 experiments will have been conducted by the end of this decade.

The construction and operation of thermohydraulic-test facilities is very expensive. Nevertheless, from the long-term standpoint the attempt should be made to keep at least one existing test facility in operation, possibly as a joint venture by a group of interested countries. This would enable a group of trained experts to respond rapidly to unforeseen events.

Concerning fuel behaviour a CSNI (Committee on the Safety of Nuclear Installations) has prepared a statement, concluding that the essential research in this field has been completed and that suitable computer codes were available.

Computer codes in thermohydraulics are a major factor in reactor safety assessment. Codes for detailed assessments, e.g. for design purposes or prediction of complex accident conditions, are needed in addition to fast running codes for parametric studies and for use in simulators or plant analysers. To decrease the verification effort the same code should be used for both purposes with - perhaps - coarse nodalisation and large time steps for faster running applications.

To avoid misleading results codes should - as far as possible - calculate best-estimate behaviour. Conservative assumptions - if any - should be included in the initial conditions (e.g. power level), boundary conditions (e.g. availability of systems) or design limits. Best-estimate assumptions are first of all necessary for codes to be used in simulators or plant analysers.

The following are regarded as the best ways of further improving the reliability of computer results:

1. Use of a verification matrix prepared by the CSNI which enables various codes to be compared and clarifies their potentialities and limits.

2. Setting up of user groups in an effort to optimize code use, since experience shows that the programme user frequently represents the principal uncertainty factor. In my view, this would give the European Community the opportunity to promote and preserve European codes which could be kept available for all Member States in the JRC at Ispra. This means that the JRC should also participate in the verification of these models. Such a procedure would certainly lead to even greater understanding on the part of national experts of the significance of the physical processess that occur during transients.

In the past, safety requirements were frequently laid down before the results of relevant research projects were available. I believe that optimisation of safety design of nuclear power plants can be achieved on the basis of the existing know-how. This can also mean a reduction in the number of active systems if the flexibility of other systems is correspondingly increased.

To sum up, I should like to say that the research objectives set in respect of thermohydraulics and in the field of fuel behaviour have been attained as far as DBA scenarios are concerned. Some questions, however, call for more detailed research. These problems were dealt with in other contributions during the last two days.

2. Physical phenomena and systems behaviour during accidents ("Severe accidents")

In order to improve plant safety further, analyses and experiments based on the assumption of the partial, temporary or complete failure of safety systems are conducted.

In this way it is hoped to identify and, where possible, further reduce potential risks.

Like the TMI-2 accident, risk studies provided the impetus for a more intensive investigation of conceivable accident sequences. Thorough analyses of actual systems behaviour during accidents must be carried out before it can be decided whether and, if so, which additional measures are appropriate for the purpose of further reducing technical risks.

In analysing the processes that occur inside the reactor pressure vessel, it is useful to define three types of core damage:

Category I: although core temperatures and core damage remain within design limits, licensing requirements are violated;

Category II: the core is partly degraded; cooling can, however, be restored by flooding;

Category III: the coolability of the core cannot be restored.

Adoption of these categories makes it possible to assess the effectiveness of measures against core melt-down (accident management) more adequately in contrast to a rough "melt/non-melt" distinction.

Important experimental and analytical studies on reactor-core behaviour during accidents have already been conducted or are currently underway. The processes that occur in the case of major core damage, up to core melt-down, were examined in out-of-pile and in-pile experiments.

Although there are no experiments capable of revealing the level of damage at which the core can no longer be cooled (point-of-no-return), comparatively simple computer programmes can be used to investigate these phenomena.

Steam-explosion phenomena have been intensively studied. Although the individual mechanisms are still incompletely understood, it can be assumed on the basis of existing knowledge that steam explosions inside the reactor pressure vessel, which could result in containment damage, are extremely unlikely.

Figure 3 shows the times at which the various categories of damage could be expected to occur, assuming the failure of all safety systems after a malfunction. In the case of transients and small leaks, which are assumed to occur with greater frequency than other incidents, there is a relatively lengthy period before substantial core damage must be expected. During this time, suitable countermeasures could be taken, as shown by an example in Figure 4. If, for instance, the active emergency-cooling systems fail under high system pressure, the pressurizer valves could be opened. This would allow the system pressure to be efficiently reduced so that water from the accumulators could be injected. As a result, the period until core uncovery would be extended by more than one and a half hours, thus enabling the operating staff to take further action.

Analytical studies have shown that even in extreme cases such as occurring during the TMI-2 accident, the flexible use of available systems can be extremely effective in restoring core cooling and preventing core melting.

To sum up, it can be said that measures taken in the context of incident and accident management are fundamentally suitable for a substantial reduction in core-melt frequency.

The processes which would occur inside the containment after melt-through of the reactor pressure vessel are influenced by a variety of phenomena. These include rapid interaction between the molten material and water, interaction between molten material and concrete including crust formation, hydrogen distribution and explosion, and the overall and local temperature and pressure conditions. The understanding of these phenomena will be significantly improved by the important research activities currently in progress.

Because of the safety margins which must be provided in the design of the containment, these structures would not fail immediately, even if design values were exceeded. In general, therefore, provided there are no leaks, early containment failure is not to be expected during core-melt accidents. Moreover, the load on the containment can be further reduced by the use of the available systems. Of particular interest in this concern are measures which can be taken after the restauration of the power supply following core damage after a loss of power.

Since fission-product behaviour during an accident can now be assessed with far greater accuracy than was the case a few years ago, less pessimistic assumptions can be made. The extent to which the release rates can be reduced relative to earlier assumptions depends on the reactor type and the particular accident sequence. An international agreement on the analysis of fission-product behaviour during an accident could be reached in the next few years. First tentative steps in this direction were apparent at the last ANS/ENS meeting in Karlsruhe.

The main tasks for the future are as follows:

- containment failure modes and leak behaviour need to be analysed in greater detail;
- international groups of experts should seek to achieve a "common understanding" as regards the significance of the physical phenomena that occur and the ways of dealing with them;
- incident and accident management should be further improved to allow the appropriate use of available components and systems.

3. Probabilistic methods in reactor safety

The aim of probabilistic methods is to make possible the in-depth study of reactor-safety questions so as to develop a quantitative measure of safety and to improve reliability.

Very high functional reliability of safety related systems was always a fundamental requirement of reactor technology. Deterministic criteria, such as that of individual failure, and the principles of redundancy, diversity and fail-safe design have been and are still being applied in order to comply with reliability requirements.

Even at an early stage efforts were made to arrive at a quantitative assessment of reliability. To this end, probabilistic methods were developed and on an increasing scale. There are various incentives for the use of these methods:

- in attempting to develop highly reliable systems, it is helpful to compare different system designs. The most appropriate way of doing this is to quantify and compare the individual system-reliabilities;
- although a high safety level can certainly be attained through the use of deterministic criteria, they do not really allow the balance of the safety concept to be examined. A comprehensive probabilistic analysis which takes account of interactions between various safety and operating systems can reveal weak points in sensitive areas and also unnecessary efforts for less important areas;
- as well as in the determination of system reliability, probabilistic methods are used to a considerable extent in the context of risk analyses. Such studies have frequently been conducted in order to evaluate the risk to which the population would be exposed during accidents. Up to now, more than 30 probabilistic risk analyses (PRA) have been performed in various countries.

The probabilistic approach is now mainly regarded as an alternative, and in many respects as an extremely powerful, method of technical safety assessment. For a variety of reasons, which will not be discussed here, risk studies play only a subordinate supporting role in political decision-making.

At the technical level, however, probabilistic safety analyses have proved useful in many respects in the case of plant-specific and fundamental decisions. They provide data on the ways in which core damage occurs and, as a result, on how it can be prevented. They also allow a relative ranking of safety matters. In many cases, qualitative data established in the course of the analysis are more important than quantitative bottom line results.

Probabilistic and, especially, risk analyses represent an attempt to prognosticate the behaviour of complex systems. Like all forecasts, they are inevitably affected with uncertainties, the causes of which range from imprecise knowledge of plant accident behaviour to a lack of data on component or system reliability and on human behaviour. Some of these problems can be overcome by further research, but essentially by an intensive evaluation of operating experience. Other problems are inherent to the method. Hence, it will always be necessary to incorporate subjective judgment in probabilistic analyses, particularly in considering extremely improbable accident sequences.

To sum up it can be said that effective probabilistic methods have been developed and used in numerous cases in an effort to gain a better insight into safety questions. They are widely accepted as a valuable instrument used in addition to the approved deterministic methods.

Nevertheless, it is necessary to enlarge the reliability data base. For this purpose, operating experience, including repair and maintenance activities, must be evaluated. Confidence in the probabilistic methods should be improved by comparing operating experience and analytical results at systems level. In order to gain a more realistic understanding of human influence and possible common-mode failures, these methods must be further developed and, in particular, empirical data must be collected.

4. Optimization of instrumentation and control systems

The optimization of instrumentation and control systems is intended to improve plant availability while maintaining a high safety standard. In this connection, the most important factors are the use of advanced instrumentation and control systems (I&C systems), the availability of computer systems to assist plant management and the early detection of incipient failures.

The availability and safety of nuclear power plants are influenced to a large extent by automation, by the efficiency of I&C systems, by the quality of maintenance and, above all, by the competence of the operators.

As a consequence of the TMI accident, extensive studies of human influences on plant operation and safety have been conducted in the US in recent years. These studies led to improvements in the conception of control rooms, to the optimization of operating procedures and to improved operator training. Further research is necessary, particularly with a view to the more efficient organization of alarm systems, to an increased transparency of complex systems and to improved operator aids.

The development of different Safety Parameter Display System (SPDS) concepts, modification of control rooms, more intensive research on human factors and the development of modified I&C concepts characterize the present situation in the USA.

Work conducted in France was particularly concentrated on the development of system-oriented response procedures for unforeseen events. The introduction of the so-called U- and H-procedures, together with their associated systems, and investigations on organizational aspects indicate the leading role in this field.

Great efforts have been made in Japan to develop advanced control room designs and computerized operator aids.

The situation in Germany is not directly comparable, because of the early introduction of a high degree of automation within the operational systems and also within the engineered safeguards. However, I should like to make particular mention of the limitation systems. These systems have been shown to be extremely effective in limiting malfunctions and thereby preventing unnecessary scrams, component stress and unnecessary containment isolation.

An additional important aspect is the use of redundant control systems which enable abnormal functioning due to control-system failure to be largely avoided. The efficiency and reliability of the limitation and control systems will be further improved by an increased use of microprocessors. A task which I should like to mention only briefly in this context is the development of advanced control rooms with computerized operator aids.

In order to detect component failures as early as possible, diagnostic methods like loose parts monitoring and on-line vibration analysis are being developed. Loose or detached parts can be detected by monitoring operating noise. Also damage to inaccessible components can be identified and localized in good time by means of this technique.

To sum up, it can be said that:

- progress in plant management, particularly as regards instrumentation and control, has improved plant availability and safety;
- systems should be designed to enable the earliest possible detection of malfunctions and the avoidance of difficult countermeasures. Design should be such as to permit a flexible response to malfunctions.

5. Significance of operating experience for safety research and safety techniques.

As I have tried to point out at several points in this report, operating experience has a great influence on safety research and safety techniques. Expenditure on safety facilities, which is often regarded as disproportionate by the utilities has at the same time increased the availability of nuclear power plants. The availability figures reveal that the increased expenditure has been rapidly recovered. In particular, major damages have had a lasting influence on systems technology and safety research. It should be remembered, however, that operating experience plays an important role in the continuing development and improvement of systems. I should like to show this on just two examples: direct measures taken in response to the accident at the Würgassen nuclear power station included the installation of nozzles in the relief pipes, he strengthening of the pressure suppression pool and the consideration of the suppression-chamber water temperature as a reactor-safety indicator. The accident also provided the incitation for theoretical and physical studies of condensation processes and the resultant loads on the surrounding structure.

For a number of years problems with steam-generator tubes of PWRs were encountered. This experience led to continuous improvements in the choice of materials, mode of operation, as well as the procedures for maintenance and inspection. In addition, provision was made for improved accident management on the basis of modern processes in the event of steam generator tube leaks.

6. Interaction between research and licensing

In most countries, responsibility for reactor safety research is, to a large extent, separate from responsibility for licensing and operation surveillance. Whereas it is the task of research to provide the know-how required, licensing is concerned with the definition and application of necessary damage prevention measures in accordance with the state of the art in science and technology.

Account must be taken of the results of reactor safety research in order to reach balanced decisions in licensing procedures. Nevertheless, reactor-safety research must not be exclusively orientated towards the requirements of the licensing procedure, but must be broadly based. The conservative assumptions adopted in licensing procedures may produce meaningless results from the standpoint of the physical phenomena involved. In this connection, it is the task of research to analyse conservative assumptions, in order to create the conditions for reduction of excessive safety provisions.

The licensing of nuclear power plants is largely based on laws, rules and directives. These must be both sufficiently rigid to allow of adequate plant and cost calculations within the framework of the authorization procedure and, on the other hand, sufficiently flexible to enable further reasonable technical developments to take place.

International cooperation in the field of safety research has proved as fruitful. Further efforts should be made to achieve a "common understanding" regarding fundamental licensing questions.

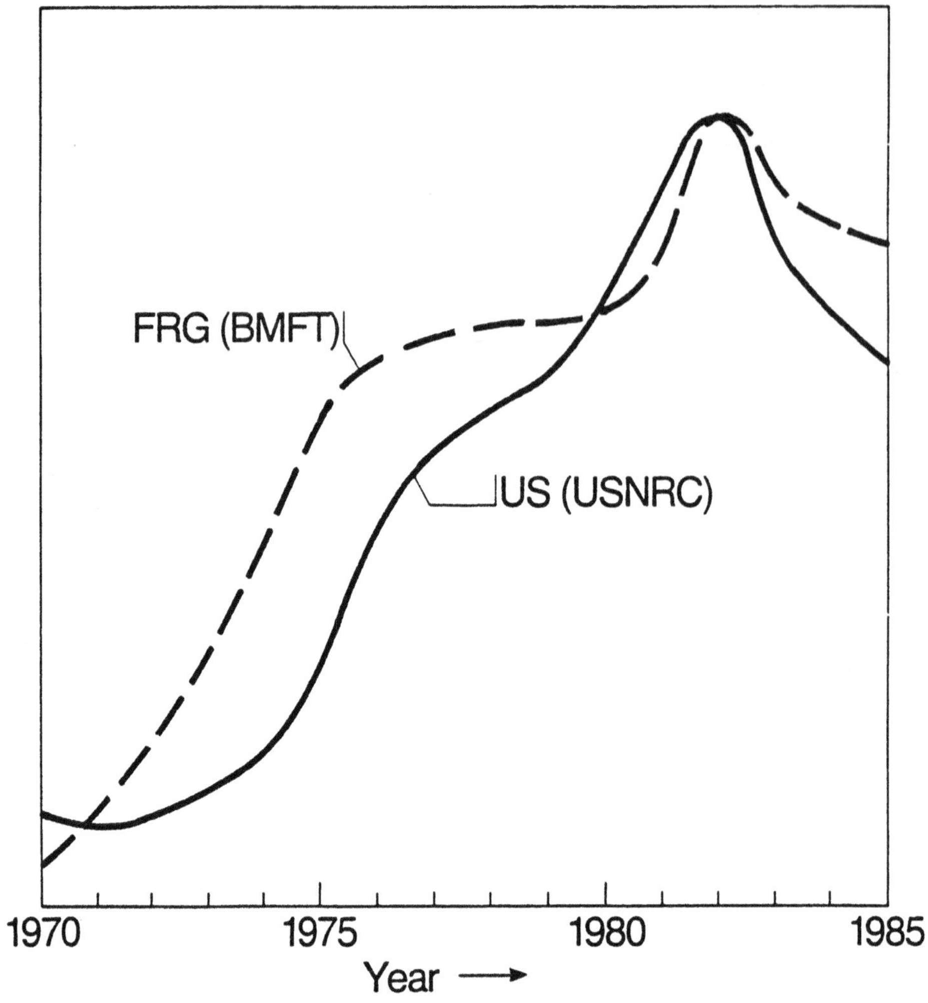

Fig. 1: Relative Budget Development in Selected Organisations

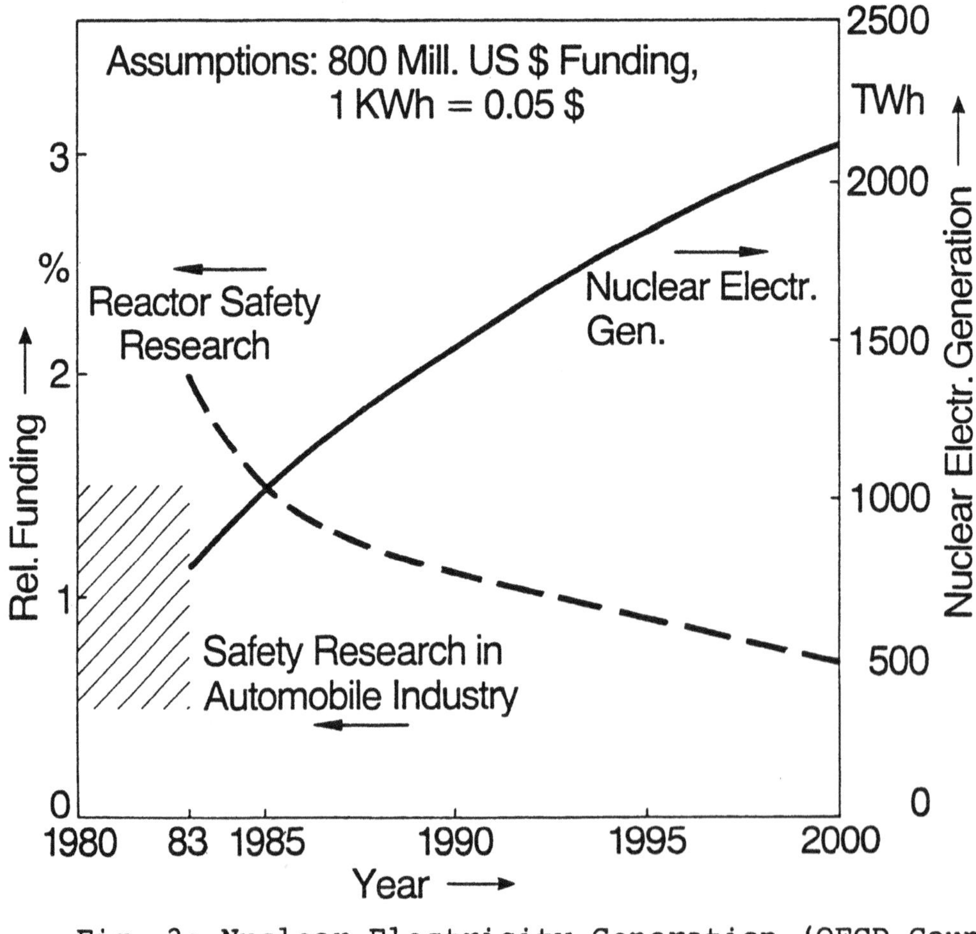

Fig. 2: Nuclear Electricity Generation (OECD Countries) and Relative Funding for Reactor Safety Researc

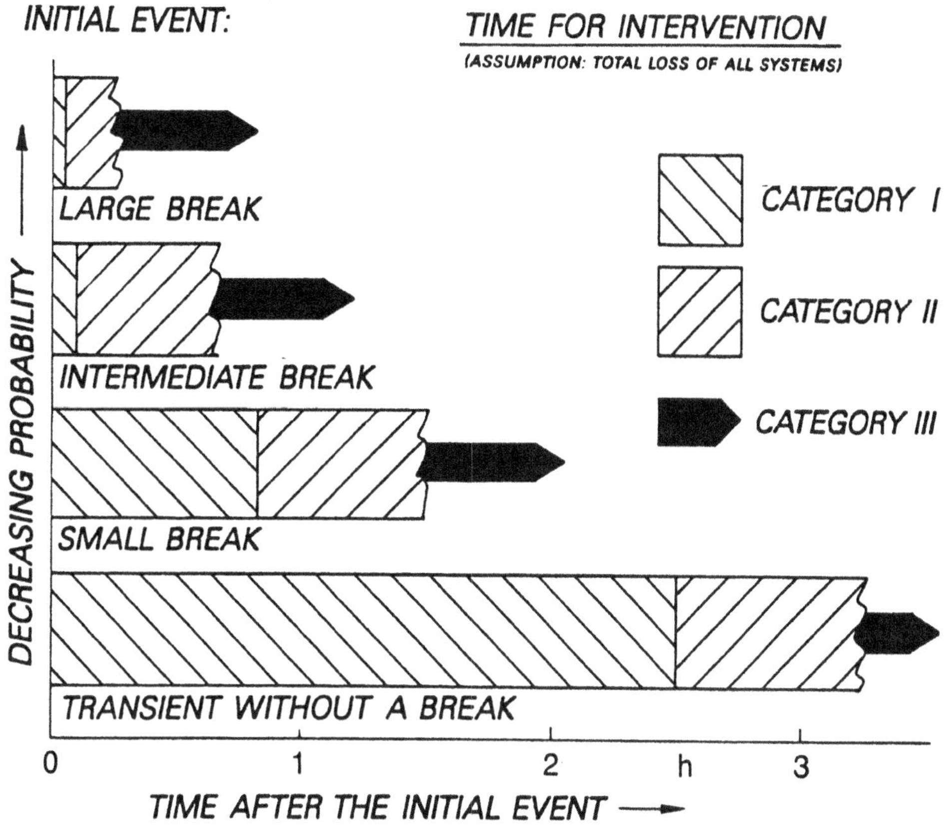

Fig. 3: Categories of Accidents

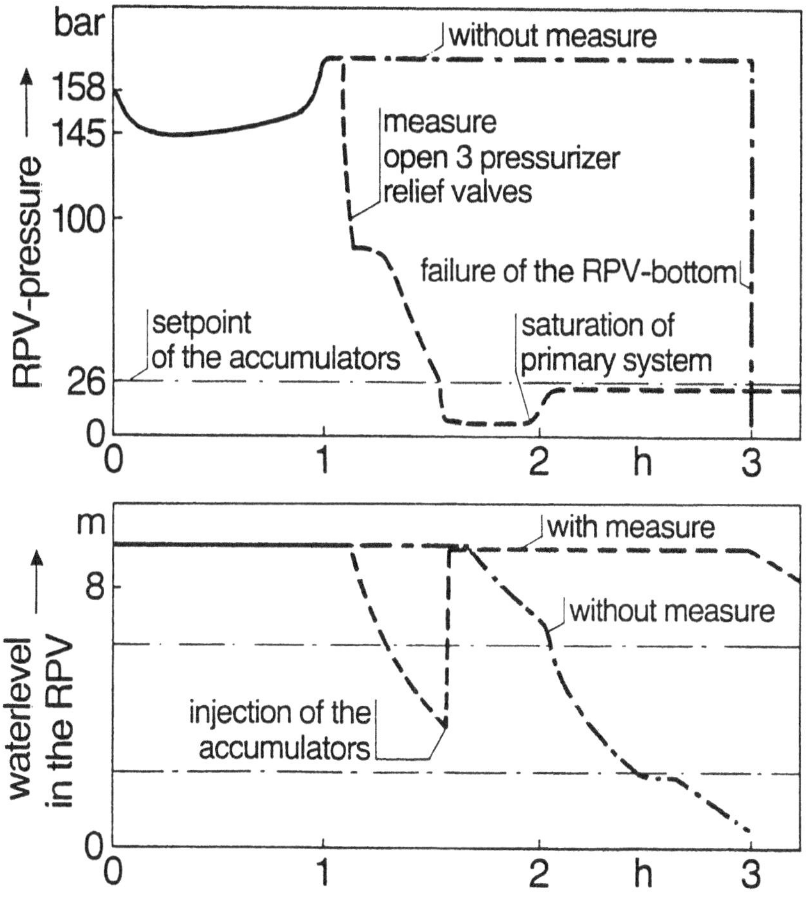

Fig. 4: Example for an Unspecified Action
Event: Station black out
Measure: Open 3 pressurizer relief valves

CONCLUDING PANEL

Evaluation of the results of the present Indirect Action Research Programme Safety of LWR. State-of-the-art in LWR Safety Research. Identification of useful topics for future research, with special consideration of Community programmes.

Panel moderator: S. Finzi (CEC/Bxl)

Members:

A. Birkhofer (GRS/München)

D.H. Davies (CEC/Bxl)

A.R. Edwards (UKAEA/SRD-C)

H. Holtbecker (CEC/JRC)

J.G. Kretzschmar (Janssen Pharmac./Beerse)

R.B. Minogue (USNRC (Wash)

H. Teague (UKAEA/SRD-C)

J. Vilain (CEC/Bxl)

W. Vinck (CEC/Bxl)

A. Wolfert (GRS/München)

SUMMARY OF CONTRIBUTIONS AND COMMENTS BY PANEL MEMBERS AND OF DISCUSSION FROM THE FLOOR

S. Finzi

I welcome the participants to this panel, and I suppose you know all of them. Starting from my right, then going to my left, I will recall the names and the organizations of the participants to this panel.

Mr. Davies, Deputy Director General of DG XII, Science, Research and Development of the Commission.
Dr. Minogue, Director of the Office of Nuclear Regulatory Research in the Nuclear Regulatory Commission of the United States.
Dr. Petrangeli, responsible for research in the Inspectorate for Protection and Safety of the ENEA, Italy.
Dr. Vilain, who is responsible for the programme we have discussed these two and a half days, and the three chairmen of the relevant experts groups that you have met this morning.
Dr. Wolfert from GRS.
Mr. Edwards from SRD of the UKAEA, and
Dr. Kretzschmar from Janssen Pharmaceutica.
Prof. Birkhofer, Director of the Gesellschaft für Reaktorsicherheit.
Mr. Pelcé, responsible for the research in the IPSN-Institut pour la Protection de la Sécurité Nucléaire of the CEA.
Mr. Teague, Deputy Director of the Safety and Reliability Directorate of the UKAEA.
Mr. Holtbecker, Director of the Department B in the JRC in Ispra who spoke already to you this morning and
Mr. Vinck, Head of the Division Safety of Nuclear Installation in DG XII.

I will just remind you that we intend to have a discussion on the results of this meeting, and on the perspective for future research. Before we start a real discussion on this point, I want to ask Mr. Davies to explain what is the philosophy and the approach that the Commission is giving to research based on shared cost action. Then I will ask Dr. Minogue to give his impression on the European cooperation on Reactor Safety and on plans for research in the Nuclear Regulatory Commission.

H. Davies

Firstly, I would like to make a few general comments on the mission of the Community Research. Broadly speaking, the aim is to carry out research which satisfies certain conditions, serving the interest of all Community states which make up the European Communities and of a nature which is better carried out in common or in a coordinated way, rather than separately and with the possibility of duplication. Those conditions apply to all of the fields of research which we support and engage in.

In the case of atomic energy, these general considerations are added to by the specific terms of the European Atomic Energy Community Treaty which includes specific obligations to provide advisory functions relating to the exploitation of nuclear power and to promote the development of nuclear power in the Community.

Secondly, let me make just a comment on the context in which this mission is pursued. At the outset, Euratom was conceived as an instrument for European collaboration, at that time thought to be indispensable because of the size, cost and difficulty of the task of promoting nuclear power. However, in practice, national nuclear agencies and industries in many states have achieved a high level of autonomous capability and this now makes international collaboration a matter of choice, rather than of necessity. I believe that it is important that meetings such as this should explore the question of whether that choice should be made in favour of cooperation. Another consequence of this trend, of course, is that the actors on the Community research scene are mainly large state or semi-state laboratories or institutions and this has consequences for the type of relations, that we have.

Lastly, in this set of general comments, let me mention the current political context for the pursuit of research in the field of nuclear energy, and most particularly in the field of reactor safety. In spite of its excellent safety record and the exceptional quality of the design and engineering, which this audiance is particularly well qualified to appreciate, the nuclear industry now finds itself the object of vociferous criticisms by minority groups, both in respect to its safety and its environmental impact; more widely, these criticisms have engendered a certain public aprehension. Paradoxically, these criticisms and conerns become an obstacle to the whole pursuit of research and reactor safety; it is our own private version of Catch 22: the argument being that if reactor safety research is needed, this must mean that the present reactors are unsafe. We have to be very careful in the treatment of this point. This is certainly not a fanciful argument. We have actually experienced its effect very directly in the on-going discussions of the Commission's proposals for the new reactor safety programme. Those, then, are the few comments I wanted to make about the broad background against which the Commission seeks to define and execute appropriate programmes of research in the field of reactor safety. Now I would like to speak briefly about the research on reactor safety as you have already been reminded by Mr. Holtbecker, we believe that it is most efficient to pursue the Community research by a combination of intramural work in the JRC and shared-cost contract research in the laboratories in Member States. As regards the structure of the overall programme, the Commission strongly believes that Community research should satisfy a number of conditions. It should certainly seek to involve major centres of excellence working in the various fields, which set the general standards of the subject. It should try to be constituted in such a way as to provide a programme framework to serve the needs and interests of different member countries and also to recognize the varied interests and concerns of the different parties involved: system designers, vendors, utilities, licensing and regulatory authorities.

And, lastly, it should have a management framework, which assures the full coordination of work under the programme with the related work which takes place in the Member States, and which also effectively integrates the external work with that done intermurally in the Joint Research Centre.

Within that overall pattern of Community research, the Commission regards the shared-cost research element as very important in bringing together the vast expertise which has been accumulated within the Member States, both in the introduction and exploitation of thermal reactors and in the development and demonstration of advanced reactors, notably Fast Breeders. The major aim is to establish and consolidate a common and coherent technical basis for the benefit of the whole Community. This will facilitate a harmonious approach to nuclear safety issues in the Community as well as well-balanced and steady progress in the introduction of nuclear energy, including again Fast Breeders. Within the shared cost contract research, we seek a number of what I think are important aims. We seek the secure Community access to the wide based scientific and technical knowledge, which already exists in national laboratories and institutions within the Community. We seek to encourage the pooling of efforts needed to pursue research actions which are particularly costly, either in terms of facilities, or in the use of human expertise. We seek to encourage the development and the strengthening of the network of contacts and cooperative information exchange between those of you responsible for work on those complex and extremely important matters in the Member States. We seek, as I have said, to provide a framework for the harmonious and effective integration of JRC research activities into the wider field of Community safety research, and we seek to make available, should it be needed, a mechanism for concertation of views for unifying action in the case of cooperation with countries outside the European Communties. These, I think, are all very valid and very important aims, which I think have been already to some degree fulfilled in the programme being discussed.

The evidence of this seminar clearly supports this view of the benefits that can come from shared cost research. It is for this reason that the overall pattern of Commission programme proposals, presented by Mr. Holtbecker this morning, displays the slight shifts of direction of work from the Joint Research Centre to strengthen the external research actions. We believe that by doing so, we should achieve an even more effective and rewarding Community research programme in the coming years. But, that is a view on which we would certainly welcome your comments during the remainder of this afternoon.

R.B. Minogue

I appreciate the invitation to attend and participate in this seminar. It is particularly appropriate in view of the recent signature of an extension of the arrangement between Euratom and NRC for research coordination and cooperation. This extension reflects an increasing trend towards joint projects and greater cooperation in planning of programs. Although this trend is largely driven by budget constraints, it is, I believe, also improving the quality of the research. Nuclear safety issues are complex and they require multi-disciplinary, highly interactive, efforts in order to resolve them.

They need the cross fertilization that comes by having a number of participants. Further, we all have a parallel interest in the prompt resolution of nuclear safety research safety problems as they may arise. A problem for one is a problem for all.

What I propose to do in the few moments that I have in this crowded program is to go through five major areas, which I think are the principal reactor safety areas requiring research, and briefly identify for each what I believe to be the principal problems to be solved, and the topics where research is required.

The first area relates to the mechanisms and the significance of aging and degradation in service. In our country we have a large number of plants that are becoming mature, and as these plants mature we have encountered a number of phenomena of corrosion, radiation embrittlement, fatigue and other effects which have raised serious questions about the continued safety and viability of the plants and, in particular, of the primary coolant pressure boundary. These have led to a perception of a need to improve techniques of non-destructive examination, both in terms of detecting flaws and properly characterizing flaws. An important development of some time ago was the role of the PISC program in flagging the deficiencies in ultrasonic examination techniques. Fortunately, because these were identified early there was time for a major effort to resolve some of these questions by improvements in non-destructive examination techniques, which has come to be very, very helpful to us in recent years as we have had to make judgements, when flaws are detected, as to what kind of inspection is required, when to require it, whether plants could be allowed to operate prior to those inspections and also as we have to make assessments of the significance of flaws and possible mechanisms of flaw propagation. An item touched on earlier as being considered in a PISC-III program would be an assessment of the applicability of data taken from artificially flawed specimens to real flaws, an effort which I think would be very constructive.

Steam generator problems are so widespread and so serious, they hardly need to be discussed.

I do think there has been one major success. The pressurized thermal-shock issue, I believe, is largely resolved for the present plants. However, we face the fact that there may be degradation of other components or currently unrecognized degradation effects as plants age, and the United States NRC has recently made arrangements, with the cooperation of the Naval Reactors Staff of the Department of Energy, to examine components from the Shippingport reactor to determine whether insights can be obtained from that plant regarding degradation mechanisms not previously recognized. We have also given a great deal of attention to environmental qualification programs, for both electrical and mechanical equipment, to develop testing techniques by which one can better predict performance throughout life.

The second area is that of system response to complex operational and thermal-hydraulic transients. It has been discussed at great length in this programme, so I will be brief.

In the first years of commercial nuclear power, work in this area was largely oriented towards the large break LOCA issues and reflected an underlying perception that the whole range of transients could be adequately characterized by analysis of an extreme limiting case. The TMI accident showed us all that this is a questionable assumption, and in recent years the emphasis has shifted much more to small break LOCAs and other transients towards best estimate calculations.

As the range of transients considered has been broadened, it has also become clear, at least to many of the industry people in our country, with whom I interface, that there are implications of this research, as you orient more towards best estimate codes and wider treatment of transients, that affect system reliability and on-line availability and thus can offer a significant potential of savings for the industry beyond that of simply resolving safety questions. Certainly in our country the utility industry and the manufacturers have participated heavily in these programs in the last few years.

The proper approach to the thermal-hydraulic research program, as I see it, is first to understand the basic phenomena, which means that the smaller scale and separate effects programs are extremely important and, second to demonstrate and validate the codes with large scale integral tests and, third to have the ability to assess the applicability of those codes to new problems as they may arise.

Particular applications of the results are in the evaluation of plant transients, training of operators, simplification of operating procedures, and the use of these sophisticated codes in the development of a basis for plant analyzers which can be used widely in assessing operations and in developing operating procedures. A particular end use in this area that, I think, is most important in the United States, relates to the wide variation in designs among U.S. reactors. Here the lack of standardization creates an additional problem that, unexpected transients may occur on a plant and we then must evaluate these as they may apply to many other reactors of different configurations. It is because of the likely continuing need to assess code applicability in such analyses that we are seriously considering the desirability of a long term highly flexible integral facility that could be used to continue to do testing of the type now done on the Semiscale facility.

Moving on to the third area, that of severe accident research, in which I would include the source term questions, I would begin by saying that the proper first objective of severe accident research is always to prevent the accident. That is often forgotten. The work that one does looking at thermal-hydraulic transients in that sense is part of severe accident research.

With that qualification, we see the rest of the severe accident work as being in two categories. (This will closely parallel the breakdown that Prof. Birkhofer made earlier).

The first category is accidents which involves degradation of the core in an arrested sequence, where the problems really are those of fission product release, plate-out, and things of that type.

The phenomenology of that category is near solution, at least in a regulatory sense, that is, in regard to what is pertinent to questions of siting and emergency planning. There is, of course, a longer term problem in such accidents, which was touched on earlier today, and also by Prof. Hicken yesterday, relating to the activity distribution within the plant and its systems in such arrested sequences. That is a longer term problem requiring much more work that, in my perspective as staff member of a regulatory agency, is really more of an industry problem, although it is also of some concern to the regulators.

The second category under severe accident research is that of severe core degradation. Here, I think the issues are: coolability of degraded cores; loads on containment that occur when you get core meltdown, for example, from gas production or effects of core concrete interaction; and containment failure modes. The last of these, we think, is particularly critical. Too often people think that the containment is a simple pressure vessel and think of failure modes in margins in that context. In reality containments are full of penetrations and irregularities, and it is there where failure mechanisms can be important. Early containment failure is a major element in assessing risk.

This whole area of severe accident research is one where we do see a need to improve the quantification of residual risks and foresee potential problems of ultimate public acceptance. We think it is important that in-pile capability for fuel damage work be retained. Examples would be the capabilities of PHEBUS, PBF or the ACRR in the U.S., and NRU reactor in Canada, all of which have significant capability to do in-pile work.

Let me now move to the fourth area. I apologize for going through these so quickly, but time is limited. The fourth area is that of research related to reactor operations and probabilistic risk assessment. Here the main issues relate to the application of the insights obtained from the two areas which I have just discussed to plant operations, to assessments of operational capabilities, instrumentation, monitoring, control room design, operator procedures and operator training, technical specifications, and generally to incorporation of these insights into the human factors areas.

As to the probabilistic risk assessments (PRA's) themselves, there clearly is a need for significant improvement in PRA's if they are ever to be fully useful. I see these improvements as breaking into two categories. First, an improvement in the data base. That means better data on component reliability, better data on maintenance experience, and better data feedback from operating plants. These are all points that Prof. Birkhofer made earlier. I also see an important source of information for the data base for PRA's to be an incorporation into PRA's of the insights and understanding that have developed from the experimental and thermal-hydraulic programs. Another important aspect of PRA's, which needs improvement, is the methodology itself, by incorporation of human factors and common cause failure modes such as might arise from earthquakes, internal and external flooding, or sabotage.

The uses of PRA are first, as is now being done in the United States, its application to specific licensing decisions, to decisions related to a particular design or to prioritizing unresolved safety issues and research needs. Ultimately with significant improvement in the methodology and data base, some application in terms of safety goals or establishment of quantified levels and acceptable risks will follow.

Going on to the fifth and last area, which is seismic safety margins research, here I think our problem in the United States may be somewhat different from the problem in Europe. The U.S. by comparison with much of Europe is a very seismic area.

With the continuing improvement in understanding of geology and seismology in recent years, we have come to realize that a number of the older plants may have been designed for seismic hazard levels that are not high enough. Thus, although we believe that there is considerable seismic safety margin in present designs, we see a need for well-validated improved seismic analysis methods, which could be used to reassess these currently operating plants, to assure that margins do exist and ideally to quantify these margins to some degree. Also, the PRA's which have been done have shown that seismic risk is a significant contributor. Last we have the perception that the design approach that has been used in the past is not a good approach; it is not cost-effective, and in fact in some respects may be counter productive.

The following research is needed to deal with these points. First, a better approach to defining seismic hazards on a regional basis. The technique used in NRC's regulations (which I believe is used in other countries as well) is unduly conservative when applied in areas where the seismicity cannot be related to specific geological structures.
The second thing that is needed is hard data, that can be used to validate the non-linear seismic response models which have been developed. This is the sort of research that requires extensive and expensive shake testing. This topic also was touched on by earlier speeches. We have recently concluded an agreement with Kernforschungs zentrum Karlsruhe to participate in the HDR program in which they will shake that facility; we are also negotiating with the Japanese to participate in the Tadotsu facility. I have a meeting scheduled Friday with some people from Ispra to learn about Ispra's planning in the same context.

The last thing that is needed in seismic research is a simplified design approach that could be compared against, or benchmarked against, these validated non-linear methods I have just mentioned. I think such simplified techniques are necessary in order to take advantage of possible simplifications as modifications may be made in present plants, and hopefully for future plants, to lead to a much simpler and more straightforward design approach.

I close with a general comment. I have been struck in listening to the previous speakers, particularly to Prof. Birkhofer, that there is broad international consensus as to the areas that require research. The differences that I hear are differences of relative emphasis only.

I think this consensus is important, because if we all agree on what the problems are, it will be a lot easier to try to develop a consolidated prògrams or complementary programs, where duplication is avoided and a general improvement in quality and reduction in cost for all concerned will be achieved.

S. Finzi

It was very interesting to hear directly from you about the position of USNRC. We had this morning a report on the state of the art on three research areas, which were the object of the programme just finished. We had an overview on future plans and on the present work of the Commission in this field. We had the benefit of a review of the open problems and of the achievements reached already in reactor safety, and this afternoon, a statement on the policy of the Commission. What I intend to do now is to ask the four participants which have not yet had the opportunity to present their statements, namely, Mr. Pelcé, Mr. Teague, Mr. Vinck and Mr. Petrangeli to make their comments, taking into account the different communications we have had already.

We should have essentially in mind the tentative of answering to these three main questions: the first question is whether or not this programme now finished was a contribution to a certain progress in the knowledge in nuclear safety in Europe? The second question: is this kind of programmes appropriate to improve collaboration in the Community? Third question: are the problems which remain to be solved, or the research which remains to be done, also an appropriate motivation for a collaboration within the Member Countries stimulated by the Commission's programme?

J. Pelcé

I will treat two points. One is related to work executed in the Member States, and the other will be a very brief outline of what was said before.

Maybe I should start off by saying that we really must congratulate all the participants in this work. The contractors, because they have shown results of excellent quality, and we should also congratulate the staff from the Commission, as they have led and managed things so well.

Now, there is a point I would like to highlight. I do particularly agree with the conclusion of one of the speakers, Mr. Wolfert, and it is a conclusion in question form. When this morning he asked to integrate all the information obtained in the course of this programme into the major evaluation process, which is necessary to analyze the safety of reactors. We agree that this sort of effort should continue between the Member States. However, if this sort of effort were to be increased then surely that means that we need to multiply the effort, and I will explain what I mean.

If you want to make a step towards improving the integration of results into the instruments of safety evaluation by improving the links between the partners to increase everyone's benefits from this exercice, is it not desirable to concentrate such efforts on projects of a sufficient scale?

This would enable a common assessment of obtained results due to evaluation work by using directly all available tools for evaluation needed for the analyse of reactor safety and currently under development in some Member States. This would encourage a direct dialogue between experts of the JRC.

As Dr. Minogue mentioned, everyone agrees on areas of priority, including France. However, perhaps can I stress in which points we are particularly interested in France, among these, which were touched upon. It is true that many results have been obtained today and are virtually finalized and that logically justify a slowing down or even stopping of some R&D programmes.

Nevertheless, the experience obtained with already constructed nuclear reactors is rather limited. So people may pose the question as to what will happen in future years or even decades. So, it will be necessary to maintain a minimum level of competence to cope with the possible occurence of serious problems with nuclear power plants.

So, in France, to illustrate this aspect, thermalhydraulics in the primary circuit is considered as a key area, and there is general agreement about this at EDF, FRAMATOME and at the CEA/IPSN. This area is very specific for PWR's, conversely to such topics as technology, human factors, etc., which have concern also to other installations.

In this respect, we provide in France that the teams working in the area of thermalhydraulics at the Research Centre in Grenoble will continue to work at a minimum level to be defined, on the development of the CATHARE code and the system loop BETHSY. This work could be open for collaboration with countries being interested in this area.

<u>H. Teague</u>

I was, of course, aware of the existence of the programme which is just ending and was indeed familiar with some isolated parts of it. But, I must say that the experience in the last two days have been quite a revelation in having seen it all put together. It strikes me as a remarkably effective programme which has made very worthwhile contributions and, which I think, fits very well the standards and criteria which Mr. Davies outlined towards the joint action under the Community. I feel, that I welcome the work particularly because it illustrates a general point which causes some concern to us who have been in the safety business a long time and that is that the further one goes into an accident, the greater the possibilities exist for more complication and more complexicity, and there is practically no limit to the amount of detail one may reach.

And yet, the further one goes into an accident, the more unlikely reaching that stage becomes. There is a certain inconsistency surely in allowing a more and more improbable part of the accident sequence to consume ever increasing amount of experimental and theoretical effort. If we are talking about the design basis accident, then one is relying on the successful operation of engineered safety features, and the probability that they will be required is quite high and, therefore, it is only sensible to devote a lot of effort to ensuring that they can meet the duty that is imposed upon them. This explains the sort of thermohydraulic programme we have had in the past. I agree with Dr. Birkhofer's conclusion that effectively the state of the art in that particular area is pretty good, and I see now the interest arising in that rather hazy region where one asks the question "yes, we have a good safety system and we think it is going to work, but what happens if it doesn't quite work?". Do we get rapidly into a completely untameable situation where the price of being slightly wrong in your predictions is an enormous increase in the consequences of an accident? This is sometimes called the cliff edge approach in Britain, if there is a cliff edge awaiting us just beyond the bounds of the Design Basis. Now, I think that there are a couple of important areas here. Having designed a reactor to cope with a number of foreseen accidents, the design basis arrangement, one makes quite conservative assumptions about the performance of safety devices. I think we all know that, in reality, if we did push things a little further, then the resiliance of the plant, its capacity to cope with a partial failure is much higher than the conservative sort of safety case put up for the licensing authority. This brings me to the idea that even if a severe accident started, then there is a lot that the operator could do within the balance of the present system to intercept it and terminate it. This has been referred to by previous speakers.

Now, where I am coming to is that, if we are trying to evaluate this sort of capability, we do not need a great deal of assurance in the correctness of our research for severe accidents in the same way that we do for the design basis. The programme which is just ended contains a number of elements where we have, in fact, gone beyond the design basis. In the case of atmospheric dispersion, clearly we have gone well beyond that. (In case of the gas cloud explosion area it seems to me, considering the general balance of detail and the amount of effort which has been done, and the chairman has not been afraid to state this, we cannot rule this particular effect out, we cannot say that a detonation of a gas cloud is physically impossible, but, in the light of the work done, we could make a balanced engineering judgment, that we can confidently proceed. I think I can recognize this sort of characteristic throughout the programme which has been presented to us in the last few days. In general I would agree with the points that have been made by Prof. Birkhofer and by Mr. Minogue and their ranking of the sort of safety issues which are currently facing us. I would only wish to reinforce the point that I have made that, where they go beyond the design basis accident, let us beware that we do not get trapped into a morass of ever increasing details.

By our own actions we may have implied that it is necessary to have this high degree of assurance and that we understand every detail. Let us rather look to this sort of concerted programme where the balance seems to be just about right and, I think it emerges from the papers and from the chairmen's summaries that there has been a high degree of that cross fertilization which has been referred to earlier as an objective of this type of work.

So, the only one further point I would wish to refer to is that in several of the reports of the chairmen, reference was made to the existence of a large volume of useful data which has been generated in the programme and which really ought to be analyzed. We have rightly heard some emphasis on the need to improve our data base, and I think it would be very unfortunate, indeed, if the unanalyzed data from this programme were allowed to moulder unseen, and I do hope that nothing will, in fact, prevent that analysis from being done.

S. Finzi

Thank you Mr. Teague for reminding that the work is not finished in some areas and, in fact, that the analysis of a large amount of available data should deserve a big attention.

W. Vinck

I plan to focus essentially on the third question you have raised Mr. Chairman. First, I would like to speak in general terms and I would like to highlight, as Prof. Birkhofer and Mr. Minogue already did, that not only regulatory requirements should give a partial orientation for priorities in research programes, but also operational feedback, including abnormal occurences. This should perhaps be done more than in the past, also bearing in mind, after all, accident prevention should remain the main objective.

First let me take a recent example chosen from Germany, mainly the report on special events in NPP in Germany during 1983. Among others, the following has been noted from the analysis of these events: More than 31% of the events relate to auxiliary and emergency systems, 11% to reactor protection and shutdown systems, and one third of the events originates in defect components, mainly due to wear-out, one fourth originates from errors in repairs, inspection, assembly and operation.
Similar indications are obtained from the paper presented by Mr. Gauvenet at the recent Karlsruhe meeting on Thermal Reactor Safety, analyzing e.g. the reasons of unavailability of NPP's all over the world.
Twenty to twenty-five percent of the unavailabilities are due to component failures, i.e. from pumps, valves, circuits. Amongst system failures, up to seventy percent of the cases are related to the secondary circuit and the non-nuclear part. This provides already an indication, where prime attention should go in safety research and in technology for practical reasons, namely: the system interaction problems, which can be solved by an appropriate redundancy definition and with the aid of reliability analysis.

Secondly, generally to mechanical and material problems and to the improvement of inspection techniques and procedures.

Thirdly, to man-machine interaction problems.

However, we know that the tendency has been since TMI 2 to focus also on the examination of severe accident situations, with a varying degree and mode of connections with licensing and regulatory requirements in the various countries. This has led to an increased attention to less practical but more public concern oriented subjects like, core thermal-hydraulics and associated phenomena in SFD conditions, F.P. behaviour (and transport in primary circuit and containment), containment isolation and performance (in SFD challenges e.g. influence of time effects on releases) and connections between source term and consequence modelling.

By the way, part of these problems also are supported by reliability and risk analysis techniques. This broadly speaking.

Let me turn to the specifics, related more directly to today's presentations and discussions. I will be very brief on the thermalhydraulic side, because most has been said, with emphasis on small size LOCA's and transients. It is clear that this must be oriented to code assessment and validation to real plant geometry and conditions, incorporating effects such as presurizer behaviour, steam generator modelling, two-phase behaviour, effect of non-condensibles, etc. I may add that counterpart testing seems to be a valuable tool for assessing the scaling effects.

In the area of containment loading and failure under SFD conditions the connected source term impacts on consequence modelling. Broadly speaking the low pressure case has been fairly well studied experimentally and theoretically. What still needs further investigation are other core degradation and melt down sequences at high system pressure and the herewith associated phenomena such as, aerosol plate-out and possible resuspension. Further improvement of F.P. behaviour generally focussed on both involuntary leakage or controlled venting. Further the effects of steam spikes in particular on aerosols. The effect of hydrogen distribution and explosion phenomena, but not so much from the standpoint of containment failure as from the standpoint of safety equipment survivavility and again of the effects on aerosols.

Long term melt concrete behaviour, especially from the point of view of crust formation, inhibiting cooling and pressure build-up.
At last, but not least, leakage behaviour of containment, especially its penetrations, beyond design basis conditions, since this is the more likely type of loss of containment performance under SFD conditions.
In the area of consequence modelling, related to source terms derived from SFD conditions, examples of items remaining for investigations are in my view: Define to what extent more or less sophisticated calculations can practically be applied for emergency response. As specific problem here is the training of the users of such models.

Further the quantification of uncertainties in consequence models, especially in the case of PRA's, sensitivity studies indicate indeed that main factors contributing to the uncertainties would be: the importance of the source term and the mode or location of containment failure, the emergency response, wet versus dry deposition, dose response relationships, short term and latent effects.

Further the large uncertainties in modelling urban contamination. Theoretical work and possibly experiments related to wet versus dry deposition. And, finally, the impact of susbsisting uncertainties in this area, even more so than in the technological part of the SFD are on having a sound basis for decisions with social and economic impact that have to be made in the case of an emergency.

In Conclusion.

In the present circumstances of knowledge and incentives, a fair distribution in research work seems to be in my opinion to emphasize:

One, human engineering and plant emergency procedures and control room concepts.

Two, PRA techniques and improvements in their capabilities, for example benchmarks which have already been mentioned by Mr. Holtbecker.
Third, mechanical engineering for the plant, (e.g. seismic effects), primary systems integrity, electrical equipment integrity, process control (systems interactions).

Four, a point which has not been mentioned in these discussions, occupational radiation protection.

Further to de-emphasize:

One LOCA termalhydraulics and reorient them towards special phenomena and increased international cooperation in code assessment and use.

Two, man-made external impact consequences (e.g. aircraft crash), but perhaps reorient this towards vibrational impacts on reactor structures and systems.

I will recommend a provisional status quo on SFD experimental and analytical work in fuel behaviour and coolant interaction, F.P. behaviour, containment performance, and fourthly the associated consequence models.

Mr. Chairman, it is not surprising that most of the issues are addressed in national research programmes with varying emphasis and availability in experimental equipment and qualified manpower for each of the subjects. For several of these teams, it seems clear that there must be interest, not only to simply exchange views or information about the results, but also to cooperate in effect more closely by pooling the efforts.

That is the true sense of the CEC/JRC programme and the proposed SCA programme on reactor safety, which are both aimed at stimulating and facilitating this pooling of efforts on both subjects, where it is warranted or easier to accomplish.

G. Petrangeli

I agree to what has been said before on how useful this programme has been, as described in the last couple of days. I want to be very brief, as we do not have much time and I want to say right away that I do wish to express my appreciations on behalf of my country on the work accomplished and that we hope that a new programme will be approved very soon.

There have been a number of arguments put forward concerning discontinuation of work on atmospheric dispersion on gas cloud explosions. I think that there will certainly be a possibility available for going on promoting studies like these in the Community. Indeed studies in this field which are going on in the different countries should be done in some form of collaboration in the Community context.

As concerns the results obtained in atmospheric dispersion, I just want to underline their relevance for treating in particular emergency management problems. In this respect, I would like to welcome an initiative by the Commission of the European Communities in order to form a joint pool for the use of pertinent installations which we have in the countries of the Community. That is something that we would absolutely support by contributing with the availability of our ARIES (Accident Release Impact Evaluation System), which includes both hardware and software for emergency management.

Now, I would like to come to the issues which have been touched on in particular by Dr. Birkhofer this morning. I would to say, on behalf of my organization, that we entirely agree with what has been said. I would just like to emphasize some of the points made and make a few additional comments.

First of all, we wonder if technologically speaking it would not be time to organize some activity aimed at a simplification of the present emergency systems that we have in application at the moment, in particular, this should be considered in reference to future reactors. This sort of simplification is very important and necessary in the light of the long term future. There are some indications, which come from research that has already been carried out and which seems to show us that simplification could be possible.

For emergency core cooling for example, excess capability is usually provided and we feel it would be better to try to move towards improved systems, passive systems for example, which of course would continue to give enough protection to the core, without exceedingly high flow rates but at the same time would also result in a simplification of hardware and in increased basic reliability. That is a long term future goal, which we would really like to put forward for consideration.

Now as concerns severe accidents, apart from what has already been said, we would like to underline the importance of these issues to be studied, in a Community context. Here we feel that the Commission should do its utmost to study the phenomena involved, as well as questions of accident management, of any possible emergency situation and use of risk studies. I think it is most important to consider also more simpler design and management measures.

For example, I mention the optimized choice of concrete for the containment floor and the environmental specifications for the electric penetrations and for other weak points of the containment.

Looking at the weak points of containments in particular, it seems to us that research has perhaps not dealt enough with this matter and that it has not considered enough the various types of interactions between environment and for instance, leaking electric penetrations

The heat released by deposited fission products and also the chemical effects as well should be considered besides pressure and temperature.

As concerns instrumentation I just want to underline how important it is for the Community to support the developments which are going on in many countries at the moment in this field. In particular I refer to instrumentation which is directly devoted to core water level measurement. We are finally particularly interested in the study of the integrity of the pressure bearing components and in non-destructive testing.

I think this is all I have to say, thank you very much.

S. Finzi

I was impressed by the statement of Dr. Petrangeli on the need of simplification of reactor systems. This is particularly important coming from an organization which is dealing with licensing.

I think finally that a large consensus was expressed on the identification of useful research to be done, as was expressed by Prof. Birkhofer, and a consensus was also expressed on the programme, which was presented by Mr. Holtbecker today. Nevertheless I want to give to these gentlemen the opportunity to make further comments.

A. Birkhofer

I think that one could reflect again about the objectives of a future cost shared programme. I see two diverging arguments, one that has been put forward by Mr. Vinck and I think somewhat seconded by Dr. Petrangeli: We have to pool national efforts. I do not believe that this is true in most cases. I feel we should give, in selecting topics for the shared cost programme, highest priority to the objective that we maintain and or even reach a common understanding on relevant important phenomena. I believe this common understanding can only be reached if national organizations under the auspices of the Commission perform common tasks. I will try to give an example: Messrs Holtbecker and Vinck mentioned the benchmark calculations. I do not believe that the calculation itself needs to be sophisticated, but it brings the people together and they select systems where they have new items like common mode failures and they try to discuss and see how they can model them.

Based on my old experience and on my present experience, this type of studies, is much more beneficial to the experts than going to meetings and presenting their own task. So I would like to see also this element as an objective in the cost shared programme, because it must be our goal that in all countries we have the same amount of experience in order to license reactors at the same level.

I would hesitate somewhat on doing important research to simplify systems. I think this is a question for the industry. We could participate usefully along the lines I have tried to explain, to some research where we optimize our present situation, but not to work on newly designed reactors or safety systems.

I listened very carefully to Mr. Pelcé's argument on the responsability of national and international organizations like the Commission to preserve competence of experts in fields where actual research may diminish. We also think in our country how we have to overcome this problem in the medium term range, and I think it is worthwhile to consider this also on the Community level.

H. Holtbecker

I would like to raise two questions which I have already mentioned this morning.

The first is what are we going to do with the research programmes B and C in the future. Both items in my opinion are not only concerned with nuclear safety but they are directly addressing questions also relevant for industrial safety, and if you like, more generally for the environment protection. Therefore what we can do when we evaluate the data, so far obtained, is to see, if we cannot make profit those other users from our work and, from there, get the necessary support.

The second point I would like to underline is related to the qualification of themalhydraulic codes. Several times it was mentioned that with best estimate calculations we are able to satisfy the requirement of accuracy of results.

Now, if you ask two researchers to make best estimate calculation based on the same boundary and initial conditions, you will get different results. I therefore think that this is not the correct way to solve this problem. On the contrary, the proposal which has been made by the CSNI working group to run a whole series of tests for which experimental data exist reproducing conditions similar to those occuring in a reactor should be actively pursued.

The other suggestion is to perform sensitivity calculations to check the validity of our models in the variety of conditions which we find in various accident scenarios.

R.B. Minogue

I would like to make two brief comments, if I may. First, Mr. Pelcé made a point, which I think is very important. In each of the areas that require research which we are discussing, I think it is important to realize that there may be surprises in the future, and there needs to be some continuing minimum level of competance maintained. As regards the comment that Dr. Petrangeli made, I am also a regulator as well as a researcher and I agree with him regarding the desirability of applying the lessons learned from the thermalhydraulic work to the simplification of ECCS systems and ECCS operations. I think simpler systems are more reliable systems and are safer systems. There is an aspect of this where industry should take the lead. That has, in fact, be the case in the U.S. In our recent experimental programmes, the industry participation has been quite substantial, in terms both of planning the programmes and conducting the experiments, as well as funding the programmes. In fact, the two most recent thermalhydraulic loop programmes in the United States were both funded approximately sixty-percent by industry. I think that reflects a commitment of industry and an intent to apply these results to simplify operations. Thank you.

S. Finzi

Now I want to open the discussion to the floor. But I will first give to Mr. Vilain and to the three chairmen of the relevant experts groups the opportunity to comment.

Dr. K. Wolfert

I will give a comment in relation to a statement of Mr. Holtbecker concerning the confidence level of thermohydraulic analyses. I cannot agree completely with Mr. Holtbecker's statement, saying that presently with best-estimate calculations we have not yet reached a high level of confidence. When we evaluate analysis results obtained with advanced best estimate codes, done by experienced users, in general a high level of confidence can be stated. For example, comparing calculations for a double-ended cold leg break, performed at GRS, LANL and by industry, we have found a difference in the peak cladding temperature lower than 100K. Also experimental results can be predicted within this margin.

Before going in a detailed discussion of the confidence level already obtained, the question has to be answered, what confidence level we need exactly.

A.R. Edwards

I would like to make just one point which, is not in my role of chairman of area B, but a more general one.

It is very obvious that not only do we have to get experts of a particular discipline to talk together but we have to get experts from different disciplines to talk to each other to understand the interaction between their problem areas. This has been touched on today as an obvious problem of fission products distribution. Where they go depends on the thermalhydraulics, and to some extent the thermalhydraulics will depend on where the fission products are, because they are a heat source. Similarly, the response of the containment will depend on where the fission products have gone. In the past, there has been far too much separation, compartments drawn around particular areas, and I think the time is now right for trying to break down some of these, and this is another area where I feel a Community sponsored grouping, or even just sponsored discussion groups could have some very good benefits.
This again can lead to a simplification of systems, because people may realize that they designed a plant for the wrong objective or maybe you can handle two objectives without actually providing extra equipment.

S. Finzi

In fact, I certainly share this opinion. What is also to be said is that a research programme in reactor safety must have a scope such that it gives the possibility of looking across a number of problems in order to obtain a good cross-fertilization.

May I have comments from the floor?

J.M. Quinauld

I would certainly agree with the previous speaker's comments. In the area of atmospheric dispersion an important issue, which has not been tackled at all, is the status of the source term, because it influences the parameters of transport and diffusion in the environment. It is important to know the evolution of the physical and chemical form of the releases, i.e. of the iodine. Dry or wet deposition is strongly influenced by that.

G.L. Shires

I feel that the panel has not quite answered one very fundamental question that seems to have arisen in all the areas and that is the question of how do we get the most out of the work that is done in this Community funded programme. The question, as you are all aware, that has been asked by all the chairmen, is how do we convert the data, the information, or the understanding and apply it to methods of assessment? Particularly how do we incorporate it in codes? I can see that there is a fundamental problem here in that each country itself is going along its own route, with its own particular codes, whether they are best estimate or evaluations, and it will, of course, select the data as it wants from the Community programme, apply that in its own way.

But my question is perhaps, could I ask the panel if they have any ideas in what form of management of the future programme we could ensure the best uses made of the information and data that is obtained from the programme?

S. Finzi

As a matter of fact, we must recognize that the question of a good use of information and data obtained in the programme is not only something specific for a Community programme but also a question which concerns the relation between the researcher and the users. But certainly one of the major concerns for the management of the future programme will be the problem mentioned by Mr. Shires.

J. Vilain

I think that you are touching a fundamental difficulty of any and every Community programme. They are born abruptly and they die abruptly. Beyond that, it is not only the programme, but all people participating who lose legal existence and this is one difficulty which we have to face in the use of the data. This is referring for example to the follow-on of Thorney Island experimental interpretation, where we have no funds to follow on with the work. There is a need for continuity in that sense, when a programme finishes, every activity should not stop so abruptly, and there is also a need for continuity forward in the preparation of follow on work.

For example, if you want to imagine large field trials, and it has been mentioned that we were thinking of it in Group C, these things take probably two years of convincing to be accepted. These things take a considerable time. The Thorney Island programme has taken three years for that.

In a new programme, the call for tenders may bring together completely different people, which makes you lose completely continuity in work and teams.

J. Pelcé

I will try to say something about the future use of the information we have, on the insertion into code development. I think it is important to find some way in the future to make use of this technical debate and dialogue and bring it to the level of the actual codes, being used for the purposes of interpretating experimental programmes. Otherwise, people will have their own personal sets of results and they will use some of this for their purposes and will end up with the unsatisfactory situation you have just referred to.

W. Vinck

Just on the same question, Mr. chairman, I would like to bring up a comment, less pessimistic than what I have heard so far on this problem. Although, I would admit that probably in the future we would have to be more systematic in the transfer of results, I think the mere fact that experts of the Member States have been working together can also incite them to work together in the future, for example on benchmark exercises. The nature of the common activity already brings people together.

S. Finzi

It seems to me that it is very important that the activity performed by the Commission staff be efficient. In fact, one of the difficulties for this kind of programmes is that the procedures are sometimes rather heavy.

The competence of the staff in charge of the management of the programme should play a very important role for the efficient transfer of results.

It is important that the results of the research be presented in a well documented manner, so that they can be used in codes used both by licensing bodies, designers and operators.

T.C. de Boer

I have two points: The first, I would like to ask is, how to make more fruitful the use of experimental data for code verification work. Various experimental facilities exist in Europe, such as PHEBUS, LOBI, etc. I propose the setting up in JRC Ispra of a data base of experimental data from such European facilities. Also to form a specialized group centred on Ispra of European code users to verify such codes using this data base. This information could then be available for other countries.

The second point concerns containments. The major problem perhaps is the current, normal containment itself, constructed with various penetrations. Is the major concern the existing penetrations or a flaw due to an accident? Also, is safety philosophy moving in the direction of vented containments or to look in more detail at current containment design?

R.B. Minogue

I will try to answer the second question. Perhaps, because I was pressed for time in my remarks I did not make clear the sequence of activity in the containment programme to assess the margins in presently constructed containments which might be exposed to pressures beyond their design basis. This effort was initially in terms of treating the containment as a simple pressure vessel. As that work proceeded it became clear that the more likely failure modes, exceeding the design basis, were related to penetrations or irregularities in the construction. The flaw in a penetration may consist of the fact that someone has left a valve open or may be due to some other cause not really related to the accident. Another source might be some sort of degradation mechanism over the life of the plant, in some sealing compound, for example, which leads to a failure under accident conditions, or failure at a penetration because of deformation under the pressure load. As we have worked on this problem of containment failure modes, we have come to realize that it is a larger part of the overall severe accident issue than we had perceived at first because in the risk assessments which have been done it appears that if you do not have early containment failure in most sequences you have no major problem in terms of public exposure. I hope I have answered the question.

A. Birkhofer

I have also a short comment to the two questions. Concerning implementation and availability of data. To my knowledge discussions based on standard problem exercises have been very beneficial and efficient. Specific test data are made available to all participants calculating with different models, and I believe that discussions after the calculation are always very fruitful.

On the second question concerning containment behaviour and filtering. I believe the evaluation of containment behaviour in the high pressure range is a typical task that you have to perform with the detailed design of the relevant containment. I do not think that it is very apt for international cooperation. The tools to analyze that are conventional tools. No additional model is necessary.

In our country we certainly are not in favour of filtered containments because we want to keep the fission products in place. You may remember that we always calculate a fairly low increase in pressure and that it may take a few days before we get overpressure failure. Recently we found out that this is the case because there is a lot of concrete in the containment as our reactors are designed for accessability during full power operation. Therefore we have an enormous amount of concrete and it might be due to this difference to other containments that we have a low pressure increase. That means that fission products stay a fairly long time in the containment atmosphere. Therefore deposition is prevailing at that time, so we certainly do not think of filtering and having our fission products outside and so having the problem of handling filters.

G. Petrangeli

I am not completely convinced that the containment integrity in accident condition is a matter of conventional analytical tools, but this may be going too much into the details.

S. Finzi

Apparently the floor is well satisfied about the things which were said. Perhaps I can restate that there was in general a good appreciation of the work which was done in this programme. Technical results which were achieved are significant and that the stimulation of the cooperation between laboratories was efficient. This is outstanding in view of the very limited budget which was involved in this work for the contracts and for the management by the Community staff. I think I can state that general agreement was expressed that such Community shared cost action programmes enhance efficiently technical collaboration. An objective which is particularly important and which was outlined several times and particularly by Prof. Birkhofer, is the one of reaching of common understanding through collaborative work. Common understanding can be achieved by putting together people working on subjects of the same nature for the execution of contracts in specific areas. Another remarkable suggestion was made by Mr. Pelcé concerning the importance of maintaining an efficient standby of knowledge and competence associated to suitable installations in order to be able to answer in short time relevant safety questions which can arise in the future. A collaborative effort in the Community can help in this direction.

Concerning the cooperation with the United states, I speak about that in particular, because we have the honour to have Mr. Minogue here, I feel that the cooperation from Europe across the Atlantic is rather efficient. In this collaboration the Commission has expressed its goodwill on its part. Mr. Minogue mentioned the recently confirmed agreement of cooperation between the CEC and the USNRC, and this proves the intention of taking profit of possible future collaboration as an efficient way of solving the problems, saving money and reaching results faster.

There were expressed certain regrets on the lack of decision for the programme proposals which are on the table of the Council of Ministers. I do not want to insist too much here on this, nevertheless, I should perhaps should emphasize again what Mr. Holtbecker said in closing his exposé this morning: This programme, which we conceive as an integrated programme between the JRC and shared cost action, is now walking on only one leg. We hope that this situation will be changed soon. Now I would like to ask Mr. Davies to close the meeting.

Closing Address

D.H. Davies (CEC/Bxl)

Before I say the last few words of thanks and acknowledgement for the efforts which have gone into making this, a most succesful seminar, I should just add one or two words more on the subject just mentioned by Dr. Finzi - the interruption which action in this field suffers at the present time. There is nothing that can be said about that situation which is good; and I will not pretend that there is. It has had unfortunately consequences which Mr. Vilain and others have referred to in respect of the interruption in the continuation of useful actions, but I hope that these will remedied in the not too distant future. The point I should make is that the difficulty which you see at the present time is certainly not confined only to the field of reactor safety. I suppose you are very well aware there is a wide range of Community areas of research in which the Commission has made proposals for either new or renewed programmes where the same generic budget problem has prevented a prompt decision by Ministers and by the Parliament. I think it is worth pointing out that the particular problem which has given rise to this situation is not something which is likely to happen every year, year in year out. The problem is that we have got a naturally expanding level of activity in the Community. The Community is still quite young after all. It is still growing. It is still learning the things which it is useful to do and the things which is not useful to do, and most of the lessons show that in any joint operation you discover many new possibilities from working with people. So the tendency for the range of the activity of the Community is to grow. There is also of course a tendency towards inflation and Parkinsonian growth, but there is a genuine and well founded justification for a tendency to growth. There is however a fixed threshold built into the legal financial structure and as the cost rises when you come to the threshold: this we have just done with the result that these is a discontinuity in our affairs at this point. We would be foolish to ignore the fact there is a serious hiatus and there may still be trouble before we come out of it: but we are making progress I think it is also worth recognizing that it is the consequence of a healthy tendency to growth (apart from the over vigorous growth of plant crops of course) and I hope that the problem is something which we will pass through and out into the sunshine on the other side, before too long.

So much then for financial philosophy. Let me now turn to the process of winding up the proceedings. I should thank the various people who have helped to make this a very successful and useful event. All the people have spoken and contributed. The programme committee would like to thank particularly the raporteurs Messrs Wolfert, Kretzschmar and Edwards for the quick reaction work that is needed to develop useful summaries in very limited time. I would like to thank our keynote speakers of today, Prof. Birkhofer and Dr. Minogue for their contribution which have given very valuable senior insights and overviews on the situation. I think that you would all want me to thank the people responsible for the management and running of the programme which we have been discussing here, Mr. Vilain and particularly I think you would want me to thank Mr. Skupinski for what I can tell you has been a prodigious effort of organization, administration, to keep the show on the road. I should like to thank the interpreters for the splendid efforts with which we take very much for granted. So, I think with those general words of thanks to the many people who have helped to make this a successfully event I will declare the meeting closed.

Seminar on The results of the indirect action research programme, safety of thermal water reactors (1979-1983)
Brussels, 1-3 October 1984
List of participants

A		
AKSAN, S. NURSET	S. FED. INST. R. RES.	WUERENLINGEN SCH-5303
ANDERSEN, SVEND IB, PH.D.	RISØ N.L.	ROSKILDE DK-4000
APELQVIST, GOERAN A	SWED. STATE POWER B.	VAELLINGBY S-16287
APSIMON, H., DR	IMP.COL./LONDON	LONDON UK-
B		
BAEVERSTAM, ULF	N.I. RAD. PROTECTION	STOCKHOLM S-10401
BALZ, W.	CEC	BRUSSELS B-1049
BENODEKAR, R., DR	IMP.COL./LONDON	LONDON UK-
BERG, VAN DEN, BERT	T.N.O./P.M.L.	RIJSWIJK NL-2280
BERRIAUD, CHARLES	CEA/CEN	SACLAY F-91191
BIRKHOFER, A, PROF.DR.	GRS	GARCHING, FRG-8046
BOECK, DE, BENOIT	ASS. VINCOTTE	BRUSSELS B-1180
BOER, DE, TEUN	ECN	PETTEN NL-1755
BONNAURE, PIERRE	CEC/JRC	ISPRA I-21020
BOORN, VAN DEN, H.	CEC	BRUSSELS B-1049
BOSSTRAETEN, VAN, CYRIEL	SCK/CEN	MOL B-2400
BRADFORD, PHILIP	NII	LONDON UK-SW1P 4QJ
BRAUNS, J.	KWU	OFFENBACH FRG
BROSSARD, JACQUES, PROF	U/ORLEANS	BOURGES F-18028
C		
CALABRESI, FRANCO	ENEL-DCO	ROMA I-00198
CAGNETTI, P., DR.	ENEA	ROMA I-00144
CANEILL, J.Y.	EDF/CHATOU	CHATOU F-
CARRARO, GIUSEPPE	CEC/JRC	ISPRA I-21020
CHAINEAUX, JACQUES	CERCHAR	VERNEUIL EN H. F-60550
CLEMENT, PATRICE	CEA/CEN	GRENOBLE F-38041
COOPER, COLIN A., DR.	UKAEA	WINFRITH UK-DT2 8DH
CRABOL, B.	CEA/CEN	FONTENAY A-R. F-92260
D		
DAUWE, RUDI	DOW CHEMICALS	TERNEUZEN NL-4530
DELLA LOGGIA, E., DR.	CEC	BRUSSELS B-1049
DEMENY, RAYMOND	SOLVAY LAB. CENTRAL	BRUSSELS B-1120
DENHAM, MICHAEL, DR.	UKAEA	WINFRITH UK-DT2 8DH
DERUAZ, ROGER	CEA/CEN	GRENOBLE F-38041
DESIATO, FRANCO,	ENEA	ROMA I-00144
DEVILLE-CAVELIN	CEA/CEN	CADARACHE F-13115
DI FABIO, NOE,	EDF	VILLEURBANNE F-69628
DOURY, ANDRE	CEA/CEN	FONTENAY A-R. F-92260
DRENCKHAN, W., DR.	KWU	ERLANGEN FRG-8520
DUCO, J.	CEA/CEN	FONTENAY A-R. F-92260
DUPONT, MICHEL	MINIST. AFF. ECON.	BRUSSELS
E		
EDWARDS, ANTHONY R.	UKAEA	WARRINGTON UK-WA3 4NE
EECKEN, VAN DER, DIRK	SCK/CEN	MOL B-2400
EMERSON, M.	TECHNICA	LONDON UK_WC1 H 9LT
ERLENWEIN, PETER, DR.	GRS	KOELN 1 FRG-5000
ESSERS, J.A., PROF.	U/LIEGE	LIEGE B-4000

EVERETT, MAURICE R.	UKAEA	HARNWELL UK-OX11 ORA
F		
FAIRBAIRN, STEPHEN, DR.	CEGB/BNL	BERKELEY UK-GL13 9PB
FASOLI-STELLA, P.	CEC/JRC	ISPRA I-21020
FAYL, GILBERT	CEC	BRUSSELS B-1049
FINZI, S., DR.	CEC	BRUSSELS B-1049
FISCHER, MILO,	TUV BADEN	MANNHEIM I FRG-6800
FOSTER, P.M.	CEGB	LEATHERHEAD UK-KT22 7SE
FOX, ALAN,	NAT. NUCL. CORP.	WHETSTONE UK-LE8 3LH
FRANK, M.	CEC	BRUSSELS B-1049
FRASER, G.	CEC/DG V	LUXEMBOURG
FRIEDMANN, PETER	KRAFTWERK UNION	MARLOFFSTEIN FRG-8525
G		
GAERTNER, MICHAEL,	KRAFTWERK UNION	ERLANGEN FRG-8520
GARNIER, J.L.	CEA	FONTENAY A-R. F-92260
GEIGER, W., DR.	BATT. I.E.V.	FRANKFURT FRG-6000
GIOT, PROF.	U.C./LOUVAIN	LOUVAIN LA N. B-1348
GIRI, ALAIN,	FRAMATOME	COURBEVOIE F-92084
GODDARD, A.J.H., PROF., DR.	IMP. COL./LONDON	LONDON UK-
GOVAERTS, PAUL	SCK/CEN	MOL B-2400
GRAND, D.	CEA/CEN	GRENOBLE F
GREEF, DE, ROBERT	SOLVAY	BRUSSELS B-1120
GRYNING, SVEN-ERIK	RISØ N.L.	ROSKILDE DK-4000
GROLLIER-BARON N.	IFP	SAINT-DENIS F-69390
H		
HALL, STEVEN F.	UKAEA	WARRINGTON UK-WA3 4NE
HAUGUEL, ALAIN	EDF	CHATOU F-78400
HERKENRATH, H., DR.	CEC/JRC	ISPRA I-21020
HICKEN, E., PROF. DR.	GRS	GARCHING FRG-8057
HIRANO, MASASHI	GRS	GARCHING FRG-8057
HOFBAUER, MICHAEL, DR.	RUHRGAS	ESSEN 1 FRG-4300
HOLTBECKER, HELMUT	CEC/JRC	ISPRA I-21020
HOUSIADAS, M.	CEA/CEN	GRENOBLE F-38041
HERTIG, J.	E.P.F.	LAUSANNE CH-1015
J		
JEGU, JEAN	FRAMATOME	COURBEVOIE F-92084
JOHANNSEN, KLAUS, PROF.DR.	TU/BERLIN	BERLIN 10 FRG-1000
JONES, JAMES ARTHUR	NAT.RAD. PROT. BOARD	DIDCOT UK-OX11 ORQ
JONES, PHILIP	CEC/JRC	ISPRA I-21020
JOUKÓFF, ALEXANDRE	INST. ROYAL MET.	BRUSSELS B-1180
JEULINK, J.	MIN. HOME AFFAIRS	'S GRAVENHAGE NL-2500
K		
KAISER, NIELS, ERIK	RISØ N.L.	ROSKILDE DK-4000
KALRA, S. PAL	EPRI	PALO ALTO USA-95136
KARWAT, H., PROF. DR.	T.U./MUENCHEN	MUENCHEN FRG
KRETZSCHMAR, DR.	SCK/CEN	MOL B-2400
L		
LAAKE, HANS-JOERG	U/HANNOVER	HANNOVER 1 FRG-3000
LAMY, J.L.	CEC	BRUSSELS B-1049
LANNOY, ANDRE	EDF	SAINT-DENIS F-93206
LARSEN, GUNNER	RISØ N.L.	ROSKILDE DK-4000

LEVEN, DIETRICH, DR.	GRS	KOELN 1 FRG-5000
LEVY, MICHEL E.	MIN. DE L'INDUSTRIE	PARIS F-75700
LINDER, CHRISTIAN	BEICIP-RUEIL	R. MALMAISON F
LOMBARDI, L.	ENEA	ROMA I-00144
LYCK, ERIK	RISØ N.L.	ROSKILDE DK-4000
M		
MAERE, DE, GUY	TRACTIONEL	BRUSSELS B-1040
MALET, LOUIS	ROYAL MET. INSTITUTE	BRUSSELS B-1180
MARKGRAF, JOACHIM F.W.	CEC/JRC	PETTEN NL-1755
MARKOVINA, A.	CEC/JRC	ISPRA I-21020
MAURER, H.	CEC	BRUSSELS B-1049
MCQUAID, JAMES, DR.	HSE	SHEFFIELD UK-S3 7HQ
MCMILLAN, ROBERT N.H.	UKAEA	WARRINGTON UK-WA3 4NE
MEWES, DIETER, PROF. DR.	U/HANNOVER	HANNOVER 1 -FRG-3000
MINOGUE, R.B., DR.	USNRC	WHASINGTON D.C. U.S.A.
MOLEN, VAN DER, S. DR.	ECN	PETTEN NL-1755
N		
NIEUWLAND, E.R.	PROV. ZEEUWSE ENERG.	MIDDELBURG NL-4330
NIGON, JEAN-LOUIS	CEA/CEN	GRENOBLE F-38041
O		
ODGAARD, ERIK	NORSK HYDRO A.S.	HØVIK N-1322
P		
PANITZ, HANS-JUERGEN	KERNFORSCHUNGSZENTRUM	KARLSRUHE 1 FRG-7500
PAPPAS, JAN, A.	NORSK HYDRO A.S.	HØVIK N-1322
PAESLER, JUERGEN, DR.	KERNFORSCHUNGSZENTRUM	KARLSRUHE 1 FRG-7500
PELCE, J.	CEA/CEN	FONTENAY A-R. F-92260
PELE, J.P.	CEC	BRUSSELS B-1049
PELLEGRINI, A., DR.	ENEA	ROMA I-00144
PERROT, JACQUES	CEA/CESTA	BELIN-BELIET F-33880
PETRANGELI, G., DR.	ENEA/DISP	ROMA I-00144
PFOERTNER, H., DR.	FRAUNH.I.	PFINZTAL FRG-7507
PHAM, V. THINH	G.R.S.	GARCHING FRG-8057
PICCIOLO, G.	ANIC	MILANO I
PIGGOTT, B.D.G.	C.E.G.B.	BERKELEY UK-GL13 9PB
POCHARD, RAYMOND	CEA/CEN	FONTENAY A-R. F-92260
POLET, FERNAND	TRACTIONEL	BRUSSELS B-1180
PORRACCHIA, ALAIN	CEA/CEN	CADARACHE F-13115
PRATO, CLAUDIO	NIRA	GENOVA I-16100
Q		
QUINAULT, J.M.	CEA/CEN	CADARACHE F-13115
R		
RANDLES, J.	JRC	ISPRA I-21020
RENARD, ALFRED	BELGONUCLEAIRE	BRUSSELS B-1050
RIETHMULLER, MICHEL, PROF.	VON KARMAN INST.	RHODE ST GENESE B-1640
ROCH, M.	ELECTROBEL	BRUSSELS
ROEBUCK, BRIAN	HSE	SHEFFIELD UK-S3 7HQ
RORIZ, ISABEL MARIA	MINISTERIO DA INDUSTRIA	LISBOA P-1000
RUEDIGER, BODO	BATTELLE INST.E.V.	FRANKFURT A/M FRG-6000
ROESEMS, L.	CEC/DG XIII	BRUSSELS B-1049

S		
SAINT-CLOUD, JEAN-PIERRE	ENSMA	POITIERS F-86034
SARTO, SERGIO	ENEA/DISP	ROMA I-00144
SCHLENKER, HANS-VOLKER, DR.	RWE	ESSEN FRG-4300
SCHNATZ, G.	BATT. I.E.V.	FRANKFURT FRG-6000
SCHON, J.P., PROF.	EC.CENTR/LYON	ECULLY F-69130
SCHMIDT, KARL-E, DR.	PREAG	HANNOVER FRG-3000
SCILLY, NORMAN	HEALTH SAFETY EXECUTIVE	BOTTLE UK-L20 3QZ
SHIRES, GEORGE LISTER	UKAEA	WINFRITH UK-DT2 8DH
SIEBERTZ, ANDRE	CEN/SCK	MOL B-2400
SKUPINSKI, E., DR.	CEC	BRUSSELS B-1049
SMITH, F.B., DR.	METEOROLOGICAL OFFICE	BRACKNELL UK-RG12 2SZ
SPATZ, ROLF	U/HANNOVER	HANNOVER FRG-3000
STALPAERT, J.	CEC/DG XIII	BRUSSELS B-1049
STEINKAMP, DIETER, DR.	BROWN-BOVERI REAKTOR	MANNHEIM FRG-6800
STEPHENSON, WILLIAM, DR.	NAT. RESEARCH CORP.	WHETSTONE UK-LE8 3LH
T		
TAMER, A., DR.	KWU	OFFENBACH FRG
TEAGUE, H.	UKAEA	CULCHETH UK-WA3 4NE
THENHAUS, G., DR.	PREAG	HANNOVER FRG-3000
THOMAS, JAMES, DR.	BRITISH GAS	SOLIHULL UK-B91 2JW
TOLLEY, B.	CEC	BRUSSELS B-1049
TOLY, J.A.	CEA/CEN	GRENOBLE F-
TONON, PIERRE	SEC. STATE FOR ENERGY	BRUSSELS B-1040
U		
UNDERWOOD, BRIAN, DR.	UKAEA	CULCHETH UK-WA3 4NE
UTTON, DONALD BRIAN, DR.	NAT. NUCLEAR CORP.	WHETSTONE UK-LE8 3LH
ULDEN, VAN, A.	KNMI	DE BILT NL-3730
V		
VANDERBORGHT, BRUNO	SCK/CEN	MOL B-2400
VANHOENACKER, LUC	TRACTIONEL	BRUSSELS B-1040
VESTIBULE, DE, L.	CEC	BRUSSELS B-1049
VESPRINI, R., DR.	ENEA	ROME I-00144
VETEAU, JEAN-MICHEL, DR.	CEA/CEN	GRENOBLE F-38041
VILAIN, J., DR.	CEC	BRUSSELS B-1049
VINCK, W.	CEC	BRUSSELS B-1049
VOLCHER, J.	CEC	BRUSSELS B-1049
W		
WEBBER, D.M., DR.	UKAEA	CULCHETH UK-WA3 4NE
WHEATLEY, C., J., DR.	UKAEA	CULCHETH UK-WA3 4NE
WILLMS, HERBERT	RWTH	AACHEN FRG-5100
WINGERDEN, VAN, KEES, DR.	TNO	RIJSWIJK NL-2280
WITTEK, PETER	KERNFORSCHUNGSZENTRUM	KARLSRUHE FRG-7500
WOLFERT, K., DR.	GRS	GARCHING FRG-8057
Z		
ZEEUWEN, JAN PIETER, DR.	TNO	RIJSWIJK NL-2288

Author Index

GPSR Compliance

The European Union's (EU) General Product Safety Regulation (GPSR) is a set of rules that requires consumer products to be safe and our obligations to ensure this.

If you have any concerns about our products, you can contact us on ProductSafety@springernature.com

In case Publisher is established outside the EU, the EU authorized representative is:

Springer Nature Customer Service Center GmbH
Europaplatz 3
69115 Heidelberg, Germany

Batch number: 10825074

Printed by Printforce, the Netherlands